Fundamental Sciences
of Lithium Battery

锂电池
基础科学

李 泓 主编

化学工业出版社

·北京·

内容简介

本书重点介绍了化学储能电池理论能量密度的估算，电池材料缺陷化学，相、相变与相图，电池界面问题，离子在固体中的输运，锂离子电池正极材料、负极材料，非水液体电解质材料，全固态锂离子电池，锂空气电池与锂硫电池，表征方法、电化学测量方法，锂二次电池材料的计算研究。同时对锂离子电池基础研究的科学问题、存在的难点、发展趋势等进行了详尽分析。本书内容详实丰富，涵盖了锂离子电池基础科学的关键问题，兼顾实际工程技术问题，努力为我国锂离子电池产学研从业者提供一本从事该领域工作的工具书。本书适合从事锂电池研发的相关人员参考，也适合作为高等院校相关专业师生的教学参考书。

图书在版编目（CIP）数据

锂电池基础科学/李泓主编. —北京：化学工业出版社，2021.9（2025.3重印）
ISBN 978-7-122-39582-5

Ⅰ.①锂… Ⅱ.①李… Ⅲ.①锂电池-研究 Ⅳ.①TM911

中国版本图书馆CIP数据核字（2021）第144934号

责任编辑：郝向丽　　　　　　　　　　文字编辑：向　东
责任校对：宋　玮　　　　　　　　　　装帧设计：韩　飞

出版发行：化学工业出版社（北京市东城区青年湖南街13号　邮政编码100011）
印　　装：三河市航远印刷有限公司
787mm×1092mm　1/16　印张27　字数619千字　2025年3月北京第1版第7次印刷

购书咨询：010-64518888　　　　　　　售后服务：010-64518899
网　　址：http://www.cip.com.cn

凡购买本书，如有缺损质量问题，本社销售中心负责调换。

定　　价：198.00元　　　　　　　　　　　　　　　　版权所有　违者必究

《锂电池基础科学》
编写人员

主编： 李　泓

其他参与编写人员（按姓氏拼音为序）：

曹文卓	陈立泉	褚　赓	丁黎明
高　健	胡勇胜	黄何俊宁	黄　杰
蒋礼威	李文俊	李西阳	凌仕刚
刘亚利	卢　侠	罗　飞	吕迎春
马　璨	牛慧丹	彭佳悦	孙　洋
王丽平	王少飞	王雪龙	吴娇杨
肖睿娟	杨　伟	张杰男	张　舒
赵清清	郑　浩	郑杰允	祖晨曦

前　言

锂电池是能在化学能与电能之间相互转换的一种电化学储能器件。从单位质量或体积储存能量的角度考虑，化学能储存的能量密度仅次于核能，高于其他能量储存形式。电化学储能器件，既有主要完成一次转换的，如一次锂电池（Li/MnO_2、Li/FeS_2、$Li/(CF_x)_n$、$Li/SOCl_2$、Li/SO_2、Li/H_2O 等电池），也包括能可逆充放电的二次电池和电化学超级电容器，如各类锂离子电池、锂硫电池、锂空气电池、电化学双电层电容、混合电池电容、锂电容等。

目前，电化学储能已经成为能源、信息、交通、医疗、航空航天、智能制造、先进装备、智能建筑、资源环境、国家安全等领域的关键支撑技术。目前一次和二次锂电池在各类电化学储能器件中市场占比最高、研究最活跃。但无论是一次锂电池还是二次锂电池，在关键技术指标、成本、安全可靠性方面还不能满足很多应用不断增长的要求，需要显著提高锂电池的电化学性能，发展新型的高能量密度、高功率密度、长寿命、高能量效率、快响应速度、本质安全，具有较强的环境适应性、较低成本的电池技术，同时实现智能化、高效率、模块化的系统集成、全寿命周期状态可预测可监测、梯次利用、资源回收和再利用的技术体系。

锂电池的开发既包括创新链：从提出原始概念，逐步经过实验室测试、验证、小试、中试、量产和商业化，技术成熟度从 1 到 9，不断提高；也包括产业链：从矿产、原材料、前驱体材料、电池材料、电芯、模组、系统、应用、检测、回收再利用。无论是创新链还是产业链，新材料、新体系、新设计、新工艺、新装备、新产品、新应用的实现都依赖于对其中的科学和技术的准确、系统、前瞻的理解。锂电池涉及的电化学反应和化学反应跨越多个时间尺度和空间尺度。由于存在着上述复杂性，研究和理解这些问题需要综合多种科学知识。锂电池的研究开发目前正在从根据已有的知识、经验经过大量试验、优选材料的试错研究模式过渡到通过需求导向、高通量计算、正向设计的理性研究模式，这些变化都需要对锂电池中涉及的主要基础科学问题深入了解。

从 2013 年开始，中国科学院物理研究所清洁能源实验室纳米离子学与纳米能源材料研究组在《储能科学与技术》杂志上撰写了锂电池基础科学问题系列文章。希望通过对热力学、动力学、界面反应、尺寸效应、材料体系、电池体系、计算方法、表征技术等方面的讨论，促进研发人员对锂电池基础科学问题的关注和思考。我们也在中国科学院物理研究所、中国科学院大学开设了针对研究生的课程。为了便于后续学生和读者能更系统地了解相关的讨论，我们在已发表的系列文章的基础上，结合最近的研发进展，特编写此书。

锂离子电池涉及的知识面非常广泛，涉及的基础科学问题很多还没有形成定论，本书中涉及的内容，并没有覆盖锂离子电池基础研究的所有重要方面，书中观点有很多方面不够深入、细致，远未到利用已有知识，融汇贯通地理解锂电池问题的程度。抛砖引玉，期望能启发读者深入思考，促进科学研究。为此，在每一章节之后，我们还留下了10道思考题。欢迎专家、学者、学生以及对锂电池感兴趣的朋友来函交流，批评指正。后续我们将通过天目湖储能学堂，创办"锂电池基础科学问题"网上课程，更加广泛地和世界范围的研究人员和学生、学者以及工程师进行充分交流。本书的出版，也将成为储能专业学科建设的一本参考书。希望本书再版的时候，能够更深入全面，并能抓住本领域的重要的基础科学问题给予进一步的阐述。

在此特别感谢陈立泉、黄学杰两位老师的教诲和E01组以及中国科学院物理研究所各位老师、同学十分有益的讨论及长期合作带来的知识和启发。特别感谢《储能科学与技术》杂志编辑部，从2013年以来，一直持续不断的鼓励促成了本书的出版。

2021.5

目 录

第1章 化学储能电池理论能量密度的估算　1

1.1 能量密度的计算公式 …………………………………… 3
1.2 不同电池能量密度的比较 ……………………………… 6
1.3 采用不同负极的锂离子电池能量密度 ………………… 8
1.4 电池的实际能量密度 …………………………………… 10
1.5 电池与电极材料的电压 ………………………………… 12
1.6 电极材料的理论容量 …………………………………… 13
1.7 本章结语 ………………………………………………… 15
参考文献 ……………………………………………………… 16

第2章 电池材料缺陷化学　17

2.1 锂离子电池材料中常见的缺陷形态 …………………… 17
2.2 缺陷产生的基础 ………………………………………… 19
2.3 电极材料中的缺陷 ……………………………………… 21
　　2.3.1 TiO_2 中嵌锂过程 ………………………………… 21
　　2.3.2 $FePO_4$/$LiFePO_4$ 电极 …………………………… 22
2.4 本章结语 ………………………………………………… 27
参考文献 ……………………………………………………… 28

第3章 相、相变与相图　31

3.1 相 ………………………………………………………… 31
3.2 相变 ……………………………………………………… 32
　　3.2.1 相变的热力学描述 ……………………………… 32
　　3.2.2 合成制备中的相变研究 ………………………… 32
　　3.2.3 电解质中的相变 ………………………………… 38
　　3.2.4 电极材料脱嵌锂过程中的相变 ………………… 46
3.3 相图 ……………………………………………………… 61

　　　　3.3.1　相图与相律 …………………………………… 61
　　　　3.3.2　典型材料的相图 ………………………………… 62
　　3.4　相图的计算 …………………………………………… 66
　　　　3.4.1　CALPHAD ……………………………………… 66
　　　　3.4.2　第一性原理计算 ………………………………… 66
　　　　3.4.3　相图的高通量计算 ……………………………… 68
　　3.5　相、相变与相图的实验研究方法 …………………… 70
　　3.6　本章结语 ……………………………………………… 72
　　参考文献 …………………………………………………… 73

第4章　电池界面问题　84

　　4.1　锂离子电池界面问题 ………………………………… 84
　　4.2　SEI膜结构及生长机理 ……………………………… 86
　　4.3　SEI膜表征手段 ……………………………………… 89
　　　　4.3.1　SEI膜形貌 ……………………………………… 89
　　　　4.3.2　SEI膜的组成分析 ……………………………… 92
　　　　4.3.3　SEI膜热电化学稳定性 ………………………… 94
　　　　4.3.4　SEI膜力学特性及覆盖度分析 ………………… 95
　　　　4.3.5　锂离子在SEI膜中的输运 ……………………… 98
　　　　4.3.6　SEI膜的动态生长过程 ………………………… 99
　　4.4　界面改性 ……………………………………………… 101
　　4.5　本章结语 ……………………………………………… 104
　　参考文献 …………………………………………………… 105

第5章　离子在固体中的输运　109

　　5.1　离子输运的相关参数 ………………………………… 110
　　5.2　离子在晶格内的输运机制 …………………………… 115
　　5.3　离子在晶界处的输运机制 …………………………… 118
　　5.4　无序态与电导率 ……………………………………… 122
　　5.5　锂离子在电极材料中的输运 ………………………… 124
　　　　5.5.1　锂离子在正极材料中的输运 …………………… 124
　　　　5.5.2　锂离子在负极材料中的输运 …………………… 127
　　5.6　锂离子在固体电解质中的输运 ……………………… 128
　　5.7　离子在电极/固体电解质界面 ………………………… 131
　　　　5.7.1　固体电解质/负极与混合输运 …………………… 132
　　　　5.7.2　固体电解质/正极与空间电荷层 ………………… 133

5.8 影响离子输运的因素 ………………………………… 134
  5.9 实验表征方法 ……………………………………… 135
    5.9.1 晶体结构与锂原子占位 ………………………… 136
    5.9.2 锂扩散通道 …………………………………… 138
    5.9.3 电导率和扩散系数 …………………………… 138
    5.9.4 全频电导分析 ………………………………… 141
  5.10 本章结语 ………………………………………… 142
  参考文献 ……………………………………………… 144

第6章 锂离子电池正极材料 154

  6.1 正极材料概述 ……………………………………… 154
  6.2 典型的锂离子电池正极材料 ……………………… 155
    6.2.1 六方层状结构 $LiCoO_2$ 正极材料 …………… 156
    6.2.2 立方尖晶石结构 $LiMn_2O_4$ 正极材料 ………… 158
    6.2.3 正交橄榄石结构 $LiFePO_4$ 材料 ……………… 159
  6.3 其他正极材料 ……………………………………… 159
    6.3.1 层状结构正极材料 …………………………… 160
    6.3.2 高电压尖晶石结构正极材料 ………………… 162
    6.3.3 聚阴离子类正极材料 ………………………… 163
    6.3.4 基于相转变反应的正极材料 ………………… 164
    6.3.5 有机正极材料 ………………………………… 165
  6.4 本章结语 …………………………………………… 165
  参考文献 ……………………………………………… 166

第7章 负极材料 173

  7.1 典型的锂离子电池负极材料 ……………………… 174
    7.1.1 层状石墨类负极材料 ………………………… 175
    7.1.2 立方尖晶石结构 $Li_4Ti_5O_{12}$ 负极材料 …… 177
  7.2 小批量应用的负极材料 …………………………… 178
    7.2.1 硬碳负极材料 ………………………………… 178
    7.2.2 软碳负极材料 ………………………………… 179
    7.2.3 高容量硅负极材料 …………………………… 180
    7.2.4 SnMC 合金负极材料 ………………………… 183
  7.3 其他负极材料 ……………………………………… 184
    7.3.1 其他合金类负极材料 ………………………… 184

		7.3.2　LiVO$_2$层状负极材料 ················ 185
		7.3.3　过渡金属氧化物负极材料 ············· 185
	7.4　负极材料的基础科学问题小结 ················· 187
	7.5　本章结语 ······························ 188
	参考文献 ································· 189

第 8 章　非水液体电解质材料　　199

	8.1　液态电解质的性质 ························· 199
		8.1.1　离子电导率 ······················ 199
		8.1.2　离子迁移数 ······················ 200
		8.1.3　电化学窗口 ······················ 201
		8.1.4　黏度 ························· 201
	8.2　液态电解质在锂离子电池中的反应 ················ 202
		8.2.1　负极表面形成 SEI 膜的反应 ············· 202
		8.2.2　与正极之间的反应 ·················· 203
		8.2.3　过充反应 ······················· 203
		8.2.4　受热反应 ······················· 204
	8.3　溶剂、锂盐和添加剂 ······················· 204
		8.3.1　溶剂 ························· 205
		8.3.2　锂盐 ························· 207
		8.3.3　添加剂 ························ 212
	8.4　离子液体电解质 ·························· 213
	8.5　凝胶聚合物电解质 ························· 214
		8.5.1　聚丙烯腈 ······················· 215
		8.5.2　聚氧化乙烯 ······················ 215
		8.5.3　聚甲基丙烯酸甲酯 ·················· 215
		8.5.4　聚偏氟乙烯 ······················ 216
	8.6　本章结语 ······························ 216
	参考文献 ································· 217

第 9 章　全固态锂离子电池　　227

	9.1　全固态锂离子电池概述 ······················ 228
	9.2　固体电解质材料 ·························· 230
		9.2.1　无机固体电解质 ··················· 231
		9.2.2　聚合物固体电解质 ·················· 243
		9.2.3　聚合物复合电解质 ·················· 244

9.3 高通量计算在固体电解质材料筛选中的应用 ··· 245
9.4 全固态锂电池的界面问题 ·················· 248
　　9.4.1 固态电解质/正极界面 ············· 248
　　9.4.2 固态电解质/金属锂负极界面 ······· 249
9.5 全固态锂电池性能参数 ·················· 251
9.6 本章结语 ····························· 252
参考文献 ································· 253

第 10 章　锂空气电池与锂硫电池　　261

10.1 锂空气电池 ························· 261
　　10.1.1 锂空气电池基本工作原理 ········ 261
　　10.1.2 锂空气电池组成 ················ 265
　　10.1.3 锂空气电池中的科学问题 ········ 273
　　10.1.4 其他锂空气电池体系 ············ 274
10.2 锂硫电池 ··························· 275
　　10.2.1 锂硫电池基本工作原理 ·········· 275
　　10.2.2 锂硫电池存在的基本问题 ········ 276
10.3 本章结语 ··························· 277
参考文献 ································· 278

第 11 章　表征方法　　286

11.1 元素成分及价态 ····················· 288
　　11.1.1 电感耦合等离子体 ·············· 289
　　11.1.2 二次离子质谱 ·················· 290
　　11.1.3 X 射线光电子能谱 ·············· 295
　　11.1.4 电子能量损失谱 ················ 297
　　11.1.5 扫描透射 X 射线显微术 ·········· 298
　　11.1.6 X 射线近边结构谱 ·············· 299
　　11.1.7 杂质测量 ······················ 300
　　11.1.8 俄歇电子能谱仪 ················ 301
11.2 形貌表征 ··························· 302
11.3 材料晶体结构表征 ··················· 308
　　11.3.1 X 射线衍射 ···················· 309
　　11.3.2 扩展 X 射线吸收精细谱 ········· 313
　　11.3.3 中子衍射 ······················ 313
　　11.3.4 核磁共振 ······················ 315

11.3.5 球差校正扫描透射电镜 …………… 316
11.3.6 Raman 光谱研究晶体结构 ………… 317
11.4 物质官能团的表征 …………………………… 318
11.4.1 拉曼散射光谱 ……………………… 318
11.4.2 红外光谱 …………………………… 319
11.4.3 色谱技术 …………………………… 320
11.5 材料离子输运的观察 ………………………… 322
11.6 材料微观力学性质 …………………………… 324
11.7 材料表面功函数 ……………………………… 325
11.8 绝热加速量热仪在锂电领域中的应用 ……… 325
11.9 互联互通惰性气氛电池综合分析平台 ……… 330
11.10 其他实验技术 ……………………………… 333
11.11 本章结语 …………………………………… 334
参考文献 ……………………………………………… 334

第 12 章 电化学测量方法　345

12.1 电化学测量概述 ……………………………… 346
12.1.1 测量的基本内容 …………………… 346
12.1.2 测量电池的分类及特点 …………… 346
12.1.3 参比电极的特性及分类 …………… 346
12.1.4 研究电极的分类及特性 …………… 347
12.1.5 电极过程 …………………………… 347
12.1.6 极化的类型及影响因素 …………… 348
12.2 测量方法 ……………………………………… 349
12.2.1 稳态测量 …………………………… 349
12.2.2 暂态测量 …………………………… 349
12.3 典型的测量方法及其在锂电池中的应用 …… 352
12.3.1 锂离子电池电极过程动力学及其
测量方法 …………………………… 352
12.3.2 稳态测量技术——线性电势
扫描伏安法 ………………………… 353
12.3.3 准稳态测量技术——交流阻抗谱 … 355
12.3.4 暂态测量方法（Ⅰ）——电流
阶跃测量 …………………………… 357
12.3.5 暂态测量方法（Ⅱ）——电势
阶跃测量 …………………………… 361

12.3.6 暂态测量方法（Ⅲ）——电位弛豫技术 …… 363
12.3.7 不同电化学测量法的适用范围与精准性 …… 364
12.3.8 影响电极过程动力学信息测量准确性的基本因素 …… 370
12.4 本章结语 …… 371
参考文献 …… 372

第13章 锂二次电池材料的计算研究 376

13.1 原子尺度的模拟 …… 377
 13.1.1 基于密度泛函理论的第一性原理计算 …… 377
 13.1.2 分子动力学 …… 387
 13.1.3 蒙特卡罗方法 …… 390
13.2 介观尺度的模拟 …… 393
 13.2.1 相场模型 …… 393
 13.2.2 分子力学 …… 394
13.3 宏观尺度的模拟 …… 395
 13.3.1 有限元方法介绍 …… 396
 13.3.2 有限元方法在锂电池研究中的应用 …… 396
13.4 本章结语 …… 397
参考文献 …… 398

第14章 总结和展望 401

14.1 锂离子电池中涉及的学科领域 …… 403
14.2 锂离子电池中基础科学问题讨论 …… 403
 14.2.1 固态电化学 …… 403
 14.2.2 复杂的构效关系 …… 404
14.3 锂离子电池共性基础科学问题研究难点 …… 405
 14.3.1 SEI膜 …… 405
 14.3.2 结构演化 …… 406
 14.3.3 多尺度复杂体系输运 …… 406
 14.3.4 材料表面反应 …… 407
 14.3.5 高倍率问题 …… 407
 14.3.6 正负极材料的电压调控 …… 408

14.3.7　电荷有序 …………………………………… 408
　　14.3.8　离子在固体输运中的驱动力 ………… 409
　　14.3.9　寿命预测与失效分析 ………………… 409
　　14.3.10　材料的可控制备 ……………………… 410
14.4　锂离子电池基础研究发展趋势讨论 ……… 410
　　14.4.1　创新驱动 …………………………………… 410
　　14.4.2　指标驱动 …………………………………… 412
　　14.4.3　方法驱动 …………………………………… 414
　　14.4.4　需求驱动 …………………………………… 416
14.5　本章结语 ……………………………………… 417
参考文献 ……………………………………………… 418

第 1 章
化学储能电池理论能量密度的估算

随着消费电子、电动交通工具、基于太阳能与风能的分散式电源供给系统、电网调峰、储备电源、绿色建筑、便携式医疗电子设备、工业控制、航空航天、机器人、国家安全等领域的飞速发展,迫切需要具有更高能量密度、更高功率密度、更长寿命的可充放储能器件。未来还将出现透明电池、柔性电池、微小型植入电池、耐受宽温度范围、各类环境的各类电池。无线充电技术、自充电技术或许将成为标配,图 1-1 显示了电池的典型应用以及电池应用需要考虑的性能。各类不同的应用对电池的各方面性能要求不尽相同,需要有针对性地开发适合的电池体系。三种常见的不同应用场合锂离子电池的性能参数见表 1-1[1]。其中电池的能量密度,是最受重视的性能参数。

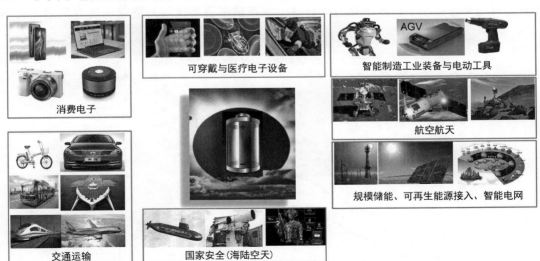

图 1-1 电池的应用及需要综合考虑的主要性能

从伏打电池、铅酸电池、镉镍电池、镍氢电池,再到锂离子电池,化学电源技术在过去 200 年取得了长足发展,能量密度也显著提高。与其他商业化的可充放电池比较,锂离子电池具有能量密度高、能量效率高、循环寿命长、无记忆效应、快速放电、自放电率低、

表 1-1 三种常见的不同应用场合锂离子电池的性能参数[1]

项目	3C 设备	动力电池	储能
质量能量密度/(W·h/kg)	260~295	240~300	140~200
体积能量密度/(W·h/L)	650~730	500~600	320~450
循环寿命/周	1000	1500~3000	5000~15000
倍率	3C	1~3C	0.2~0.5C
工作温度/℃	-20~55	-30~55	-30~55
成本/[元/(W·h 电芯)]	1.2~2.0	0.5~1.2	0.5~0.8
电池容量/A·h	3~20	3.2~200	50~400
工程能力指数(Cpk)		1.33~1.66	
安全性(欧洲汽车危险等级)		4	

工作温度范围宽和安全可靠等优点,因而成为世界各国科学家努力研究的重要方向[2-4]。如今的小型商品锂离子电池的能量密度可达到 295W·h/kg,但还不能满足日益增长的不同产品的要求。锂离子电池能量密度发展史如图 1-2 所示,其中《中国制造 2025》计划提出的能量密度发展目标用五角星标示[1]。其他国家对于能量密度的发展也制定了相应的规划,例如,为了提高纯电动车以及混合动力汽车电力驱动部分的续航里程,日本新能源和工业技术发展组织(NEDO)在 2008 年制订了目标:希望在 2030 年将电池的能量密度提高到 500W·h/kg,继而实现 700W·h/kg 的目标[5],以便达到或接近汽、柴油车一次加油的行驶里程。这些目标能否实现?电化学储能技术的能量密度是否存在极限?锂离子电池、锂电池是电池研究开发的终极方向吗?对于热点的化学电源,其理论与实际能量密度大致能够达到什么水平?基于热力学计算,本章试图回答这些与能量密度有关的热力学问题。

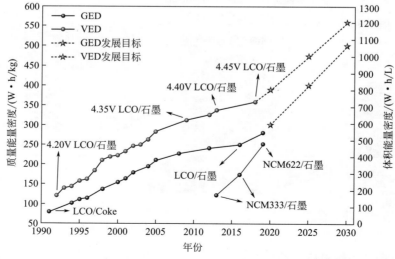

图 1-2 锂离子电池能量密度发展史及发展目标(其中 LCO 指钴酸锂,NCM 指 Li[$Ni_x Co_y Mn_z$]O_2;后面的 333 表示 $x=y=z=1/3$;622 表示 $x=0.6$,$y=0.2$,$z=0.2$)[1]

1.1 能量密度的计算公式

电池是能够实现化学能与电能相互转换的装置。对于一个化学反应体系反应前后的化学能变化情况，可通过该反应的 Gibbs 自由能进行描述：一个化学反应在标准状态下所释放或吸收的能量，是产物的吉布斯生成能（$\Delta_f G^s$）减去反应物的自由能，即

$$\Delta_r G^s = \sum \gamma_i \Delta_f G_i^s \tag{1-1}$$

如果 $\Delta_r G^s$ 为负值，且反应存在氧化还原（电子转移），则该反应可以自发地发生电化学反应，可以作为电化学储能系统考虑。

如一个一般形式的化学反应

$$\alpha A + \beta B \longrightarrow \gamma C + \delta D \tag{1-2}$$

在标准条件下该反应的 Gibbs 生成能可表示为

$$\Delta_r G^s = \gamma \Delta_f G_C^s + \delta \Delta_f G_D^s - \alpha \Delta_f G_A^s - \beta \Delta_f G_B^s \tag{1-3}$$

在等温等压条件下，当体系发生可逆变化时，体系 Gibbs 自由能的减小等于对外所做的最大非体积功，如果只有电功，则

$$\Delta_r G^s = -nFE^s \tag{1-4}$$

式中，n 为每摩尔电极材料在氧化或还原反应中转移电子的量；F 为法拉第常数（$F=96485 C/mol$）；nF 为转移总电荷量；E^s 是标准条件下的热力学平衡电位，也称为电化学驱动势（electromotive force，EMF），该方程式为 Nernst 方程式。

对不同体系电池的能量密度进行理论计算，可以为选择电极材料和电池体系提供理论依据，同时有助于阐明电池能量密度的极限。电池的能量密度可以用两种方式表示：质量能量密度（W·h/kg）和体积能量密度（W·h/L）。

质量能量密度定义为

$$\varepsilon_M = \Delta_r G^s / \sum M \tag{1-5}$$

体积能量密度定义为

$$\varepsilon_V = \Delta_r G^s / \sum V_M \tag{1-6}$$

式中，$\sum M$ 是反应物摩尔质量之和；$\sum V_M$ 是反应物摩尔体积之和。

对于给定电极材料，其充放电比容量可通过式(1-7)计算

$$\text{Capacity} = nF/(3.6M) \tag{1-7}$$

式中，M 是反应物的摩尔质量，g/mol。

从式(1-2)~式(1-6) 可以看出，当反应物具有较低吉布斯生成能而生成物具有较高吉布斯生成能时，电化学体系将具有较高的能量密度。对于标准状态下物质的吉布斯自由能数据可通过热力学手册查找。对于吉布斯自由能尚不清楚的物质，如果已知所有参与反应物质的晶体结构，可以通过基于第一性原理的密度泛函方法，计算出材料的吉布斯自由能；如果不知道晶体结构，也可以通过第一性原理计算先获得弛豫后的晶体结构，然后计算获得。如果已知所有材料的吉布斯生成能，当反应体系为封闭体系时，则可以计算由该

反应物组成的电池按照预计反应式工作时的理论能量密度（图1-3）。理论电压可以通过式(1-4)计算，电极材料的理论储锂容量可以通过式(1-7)计算。

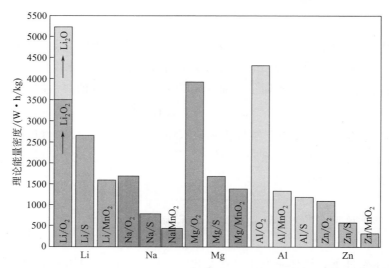

图1-3 不同金属负极的 M/O_2、M/S、M/MnO_2 等电池的理论质量能量密度比较

需要说明的是，上述计算中，如果反应物为气体，为了方便计算，按照反应式的摩尔计量比计算。在实际应用中，气体来自于外界，如 O_2，有些作者在计算理论能量密度时不考虑气体的质量，计算出的理论能量密度会显著高于考虑气体质量的计算方法。此外，对于固体反应物和产物来说，由于计算采用的吉布斯生成能一般为不含缺陷的体材料（perfect bulk material）的测量数据，实际材料由于存在缺陷和尺寸效应，导致生成能会偏离理想材料的生成能，因此需要考虑各类缺陷能的贡献，如

$$\Delta_f G^s(\text{real material}) = \Delta_f G^s(\text{perfect material}) - \sum \Delta_f G_i^s(\text{defect}) \tag{1-8}$$

对于偏离理想情况的问题将在后续讨论中涉及。

通过式(1-1)～式(1-7)可以对1683种较为典型的化学反应体系的理论质量能量密度、体积能量密度、电化学反应的理论电压、电极材料的理论容量进行计算[1]。

能量密度计算的结果表明，在所有计算的封闭体系的化学储能系统中，Li/F_2 体系具有6294W·h/kg 的最高能量密度。Li/O_2 体系按产物为 Li_2O 计算，能量密度为5217W·h/kg，排名第二，如果按照产物 Li_2O_2 计算，理论能量密度为3457W·h/kg。这两类电池的理论能量密度较高，是由于反应物的生成能较低，产物的生成能较高。由于氟不便于利用，因此产物为 Li_2O 的 Li/O_2 电池是理论能量密度最大的电池，从质量能量密度考虑，Li/O_2 电池是化学储能器件的终极目标体系。

为了从1683种电池体系中筛选出能量密度较高、有实用化前景的体系，我们使用了如下标准：①理论质量能量密度（TGED）＞1000W·h/kg；②理论体积能量密度（TVED）＞800W·h/kg（若正极是气体，或反应物密度未知，则不使用该标准筛选）；③电化学驱动势（EMF）＞1.50V；④廉价，不含贵金属元素；⑤低毒害，低腐蚀性，低燃烧性。采用这五项标准最终筛选出51种电池体系，结果见表1-2[1]。

表 1-2　51 种低毒害、TGED>1000W·h/kg 和 TVED>800W·h/L、EMF>1.50V 的电池体系[1]

正极/负极	TGED/(W·h/kg)	TVED/(W·h/L)	EMF/V
Li/O_2	5217	—	2.91
Al/O_2	4311	—	2.73
Mg/O_2	3924	—	2.95
$Li/O_2$①	3457	—	2.96
Li/H_2O	2824	2044	1.68
Li/CO_2	2818	—	1.89
Li/S	2567	2843	2.20
Al/CO_2	2294	—	1.71
Al/H_2O	2241	3262	1.50
Mg/CO_2	2232	—	1.93
Mg/H_2O	2181	2881	1.72
Mg/S	1684	3223	1.77
Na/O_2	1683	—	1.95
Li/CuF_2	1646	3799	3.54
Li/FeF_3	1643	3223	2.73
$Na/O_2$①	1595	—	2.32
Li/MnO_2	1592	2642	1.70
Li/MoO_3	1520	2598	1.75
Li/CrO_2	1435	2319	1.50
Mg/MnO_2	1380	4153	1.74
Li/Co_3O_4	1379	2849	1.91
Al/MnO_2	1333	5382	1.53
Mg/MoO_3	1332	3983	1.80
Li/Fe_2O_3	1299	2412	1.63
Li/FeS_2	1295	2519	1.79
Li/CuO	1287	3115	2.24
Al/MoO_3	1283	5018	1.58
Mg/CuF_2	1279	4239	3.00
Li/GeO_2	1261	2180	1.56
Na/S	1245	1542	1.81
Mg/CrO_2	1244	3655	1.54
Mg/Co_3O_4	1235	4382	1.95
Li/Fe_3O_4	1190	2296	1.59
Al/Co_3O_4	1186	5617	1.73

续表

正极/负极	TGED/(W·h/kg)	TVED/(W·h/L)	EMF/V
Mg/FeF$_3$	1180	3516	2.19
Mg/CuO	1176	4595	2.28
Li/MoS$_3$	1163	1943	1.69
Mg/Fe$_2$O$_3$	1153	3709	1.67
Al/CuO	1137	5754	2.07
Mg/GeO$_2$	1119	3263	1.60
Li/NiO	1097	2617	1.81
Li/CoO	1085	2562	1.80
Li/NiCO$_3$	1077	2693	2.66
Al/CuF$_2$	1075	4189	2.40
Mg/Fe$_3$O$_4$	1066	3483	1.63
Li/CuSO$_4$	1056	2605	3.42
Li/MoO$_2$	1051	2281	1.53
Li/CoCO$_3$	1036	2533	2.57
Na/CuF$_2$	1132	2334	3.11
Zn/O$_2$	1094	—	1.66
K/O$_2$	1080	—	2.22
Na/FeF$_3$	1018	1844	2.30
Mg/NiO	1004	3962	1.85

① 反应产物是 Li$_2$O$_2$ 或 Na$_2$O$_2$。

1.2 不同电池能量密度的比较

考虑到锂电池大规模应用的发展趋势，锂资源的局限性和矿藏分布的不均衡性以及锂电池回收工业经济回报较差的特点，一些学者对锂电池工业可持续发展的前景存在着一定的顾虑，因此其他类电池，如 Na、Al、Mg、Zn 等电池的研究日趋活跃。图 1-3 所示是以金属 Li、Na、Mg、Al、Zn 做负极的电池体系计算出的质量能量密度图。计算结果可以看出，对于相同正极的体系，金属锂电池相比其他金属电池具有更高的理论能量密度。如果 Li/O$_2$ 电池的产物是 Li$_2$O$_2$，则 Al/O$_2$ 电池成为质量能量密度最高的化学储能体系，其计算值为 4311W·h/kg，其次是 Mg/O$_2$ 电池，能量密度计算值为 3924W·h/kg。Na/O$_2$ 的理论能量密度为 1683W·h/kg，Zn/O$_2$ 的能量密度为 1094W·h/kg，远高于锂离子电池的理论能量密度 360W·h/kg（按 graphite/LiCoO$_2$ 电池脱出 0.5Li 计算）。从体积能量密度考虑，Al 电池的理论体积能量密度最高，为 5382W·h/L，高于 Mg/MnO$_2$（4153W·h/L）、

Li/MnO_2 (2642W·h/L)、Na/MnO_2 (709W·h/L)、Zn/MnO_2 (1738W·h/L)。

早在20世纪70年代,就有大量的工作致力于理论能量密度极高的可充放金属锂电池。后来由于金属锂负极存在的锂枝晶刺穿等安全问题,研究者将目光转向了锂离子电池。1972~2019年间由金属锂电池过渡到锂离子电池正负极材料的研究历程可以参见图1-4[6]。对锂电池而言,从能量密度逐年增长的角度考虑,可充放锂电池今后的发展趋势可能是:①采用高容量正极、高电压正极、高容量负极的新一代锂离子电池,如以$LiNi_{1/2}Mn_{3/2}O_4$、xLi_2MnO_3、$(1-x)LiNi_{1/3}Co_{1/3}Mn_{1/3}O_2$为正极,高容量Si基材料为负极的锂离子电池。②以金属锂为负极的可充放锂电池。氟化石墨$(CF)_n$的工作电压在2.9V,储锂容量为800mA·h/g,Li/$(CF)_n$电池具有较高的质量能量密度,但是循环性能较差。其他锂电池,如Li/FeF_3、Li/CuF_2、Li/MnO_2、Li/FeS_2电池,有计算结果表明这几种电池体系有望在电芯层级提供1000~1600W·h/kg和1500~2200W·h/L的能量密度,但是循环性、安全性等综合性能还不能全面满足应用的要求[6]。预计首先实现的有可能是以金属锂为负极,采用现有锂离子电池正极材料的可充放锂电池。这方面还有待于进一步研究。③最终发展的高能量密度电池应该是以金属锂为负极,O_2、H_2O、CO_2、S为正极的可充放锂电池,如图1-5所示。这些电池目前的研究无论从科学还是技术方面看都很不成熟,是研究者追求的终极目标。从目前的进展看,在中短期内,Li-S电池获得较高的质量能量密度最有竞争力。关于上述电池的科学与技术问题,将在后面章节中详细讨论。

从资源、环境保护、价格因素考虑,Al、Na、Mg、Zn都是值得发展的电池体系,但是这些电池的材料体系、综合性能、技术成熟度还不能和锂电池竞争,还需要经过长期的努力。

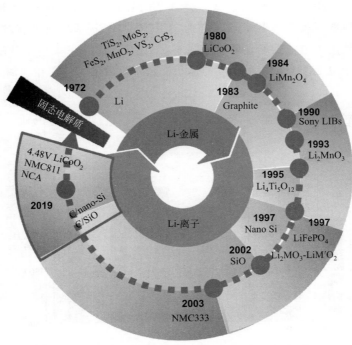

图1-4　1972~2019年间由金属锂电池过渡到锂离子电池正负极材料的研究历程[6]

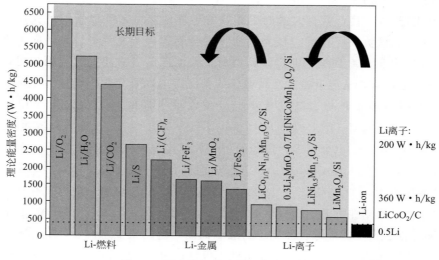

图 1-5 可充放锂电池的可能发展体系

1.3 采用不同负极的锂离子电池能量密度

目前的锂离子电池主要采用石墨类碳负极,其形成 LiC_6 的理论容量为 $372mA \cdot h/g$。近年来,充放电过程中体积变化较小(0.2%)的 $Li_4Ti_5O_{12}$ 逐渐受到关注,有望在储能电池、汽车启动电源、高功率电池方面得到应用。Si 与 Sn 基材料具有较高的储锂容量,有望进一步提高锂离子电池的能量密度。图 1-6~图 1-9 比较了采用 4 种不同负极与 7 种正极材料的锂离子电池的质量能量密度与体积能量密度的估算值。在锂离子电池中,由于

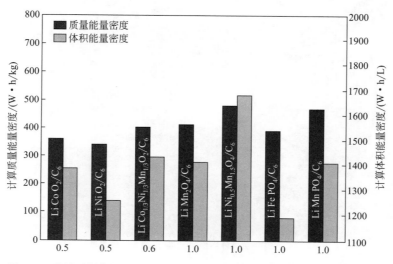

图 1-6 采用石墨负极的锂离子电池的计算质量能量密度与体积能量密度

多数材料脱锂或嵌锂的热力学数据欠缺,计算能量密度主要根据平均开路电压和理论容量获得,存在一定的误差。特别是含锂层状化合物,最大脱锂量与多种因素有关,并不确定。在本计算中,脱锂量在横轴标注,计算结果供参考。可以看出,采用了高容量的Sn/Si负极材料,质量能量密度与体积能量密度相对于石墨负极,最大增长接近1倍。电池实际上能够实现的能量密度显著低于基于活性物质计算的理论能量密度。采用Sn/Si负极的实际能量密度是否能达到石墨负极锂离子电池能量密度的2倍,特别是体积能量密度,还需要进一步研究确认。

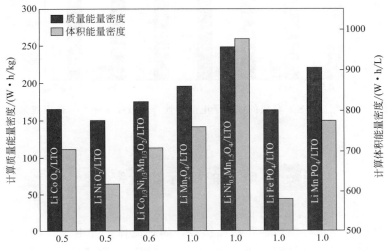

图1-7 采用钛酸锂($Li_4Ti_5O_{12}$)负极的锂离子电池的计算质量能量密度与体积能量密度

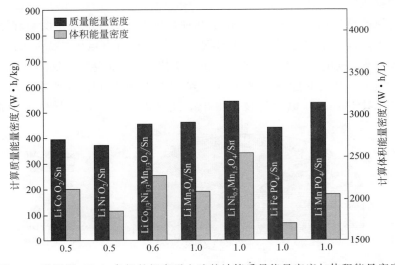

图1-8 采用锡(Sn)负极的锂离子电池的计算质量能量密度与体积能量密度

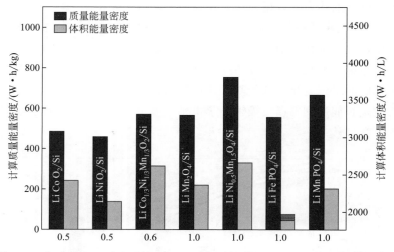

图 1-9 采用硅（Si）负极的锂离子电池的计算质量能量密度与体积能量密度

1.4 电池的实际能量密度

在实际电池电芯中，存在多种非活性物质，如集流体、导电添加剂、黏结剂、隔膜、电解质溶液、引线、封装材料等。图 1-10 是一个典型动力锂离子电池电芯的材料质量分布图。在不计入引线、封装材料的情况下，正负极活性物质的质量分数为 61%。从 1990 年到现在，电池实际能量密度的提高主要是提高了正负极活性物质在电池中的质量比例，降低了非活性物质的质量比例。但是经过 30 年的努力，通过技术的进步已经很难再提高正负极活性物质的质量比例。表 1-3 列出了几种常见电池实际能量密度与理论能量密度的比值。可以看出，锂离子电池的这一比值 R（61%）是所有电池中最高的。按照这一比值，其他电池在不更换电极材料体系的情况下，从技术角度仍然有发展的空间。如果是按照 61% 的比例，Li/O_2 电池的能量密度可以达到 3182W·h/kg（Li_2O 产物）或 2109W·h/kg（Li_2O_2 产物）。对于 Li/O_2 电池来说，由于空气电极需要大量的导电添加剂和催化剂，

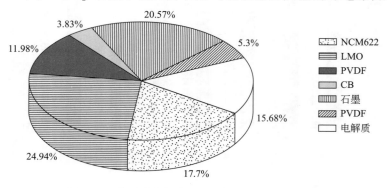

图 1-10 典型动力锂离子电池中材料的质量比例

显然 R 值不可能达到 61%。因此，化学储能电池的能量密度不可能超过 $3182W \cdot h/kg$。对于容量较大的电池来说，还需要包括电池管理系统、线缆、冷却系统、传感器、固定框架或保护罩等，R 值还会显著降低。目前，对于 Li/O_2 电池的电芯而言，预计达到的质量能量密度估值约为 $500 \sim 700W \cdot h/kg$。因此，实现 NEDO 提出的 $700W \cdot h/kg$ 技术指标的化学储能体系，实际上屈指可数。

表 1-3 典型电池的计算能量密度与实际能量密度的比值

电池	电化学反应	计算质量能量密度/(W·h/kg)	实际能量密度/(W·h/kg)	实际/计算(R)/%
Pb-acid	$Pb+PbO_2+2H_2SO_4 \rightleftharpoons 2PbSO_4+2H_2O$	171	25~55	15~32
Na-S	$2Na+3S \rightleftharpoons Na_2S_3$	792	80~150	10~19
Ni-MxH	$1/5LaNi_5(1/2H_2)+NiOOH \rightleftharpoons Ni(OH)_2+1/5LaNi_5$	240	50~70	20~29
Li-ion	$2Li_{0.5}CoO_2+LiC_6 \rightleftharpoons 2LiCoO_2+C_6$	360	150~220	42~61
Li-S	$2Li+S \rightleftharpoons Li_2S$	2674	350	9~11
Li-MnO$_2$	$2Li+2MnO_2 \rightleftharpoons Li_2O+Mn_2O_3$	970	100~220	10~23
Zn-O$_2$	$Zn+1/2O_2 \longrightarrow ZnO$	1094	150~200	14~18
Li-(CF)$_n$	$nLi+(CF)_n \longrightarrow nLiF+nC$	2189	200~300	9~13

以上所说的是使用传统有机电解液的商业化锂离子电池的情况。近几年，固态电解质逐渐兴起，然而距离商业化还有一段路要走。有文章估计了固态电池的实际能量密度，如图 1-11 所示，以使用聚环氧乙烷（PEO）基的 Li/FeS_2 电池为例，使用 $100\mu m$ 厚的 FeS_2 正极，电芯层级估算能量密度为 $995W \cdot h/kg(1550W \cdot h/L)$。固态电池有望使得锂负极得以应用，故在能量密度方面可能会有进一步的提升[6]。

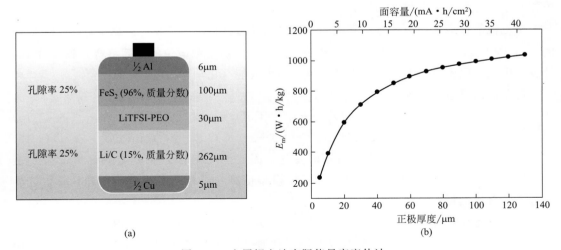

图 1-11 金属锂电池实际能量密度估计

(a) FeS_2 正极固态电池电芯模型；(b) 随 FeS_2 正极厚度和面容量的变化而变化的估算电芯层级能量密度[6]

1.5 电池与电极材料的电压

根据 Nernst 方程 [式(1-4)]，一个电池中电化学反应的理论电压可以通过反应的吉布斯自由能计算。对于典型的基于相转变反应的电池，如 Li/MnO 电池，其反应式如下

$$MnO + 2xLi \longrightarrow xLi_2O + xMn + (1-x)MnO \tag{1-9}$$

其电池的理论电压 E 通过如下公式计算

$$-2xEF = \Delta_r G = x\Delta_F G(Li_2O) + x\Delta_F G(Mn) - x\Delta_F G(MnO) - 2x\Delta_F G(Li) \tag{1-10}$$

可以看出，该电池的电压与 x 值无关，为 1.028V 定值。这一电压 E 的意义是由体相的 MnO 和锂组成的电池生成体相的 Li_2O 与 Mn 的热力学平衡电位。在实际电池中，由于反应物和产物的状态显著偏离理想材料，导致 E 值不是定值，这一问题将在后续的尺寸效应部分讨论。

如果单看电极电位，按照如下考虑：

正极：
$$MnO + 2Li^+ + 2e^- \longrightarrow Li_2O + Mn \tag{1-11}$$

$$-2\varphi^+ F = \Delta_r G = \Delta_F G(Li_2O) + \Delta_F G(Mn) - \Delta_F G(MnO) - 2\Delta_F G(溶液中Li^+) - 2\Delta_F G(MnO电极内e^-) \tag{1-12}$$

φ^+ 为由体相的 MnO 电极与体相的 Li_2O 与 Mn 作为一对氧化还原电对的热力学平衡电极电位。

负极：
$$2Li \longrightarrow 2Li^+ + 2e^- \tag{1-13}$$

$$-2\varphi^- F = \Delta_r G = 2\Delta_F G(溶液中Li^+) + 2\Delta_F G(Li电极内e^-) - 2\Delta_F G(Li) \tag{1-14}$$

φ^- 为由 Li 与 Li^+ 组成的氧化还原电对的热力学平衡锂电极电位，标准状态下，该电位相对于标准氢还原电位（SHE）为 $-3.04V$。

全电池的反应式(1-9)为反应式(1-11)与反应式(1-13)之和，假设 $2\Delta_F G$（Li 电极内 e^-）与 $2\Delta_F G$（MnO 电极内 e^-）相等，对于全反应电池电位的计算将合并到式(1-10)计算，因此可以不需要考虑电子以及锂离子的生成能（化学势）。

对于嵌入反应，例如

$$LiCoO_2 \longrightarrow Li_{1-x}CoO_2 + xLi \tag{1-15}$$

$$-xEF = \Delta_r G = \Delta_F G(Li_{1-x}CoO_2) + x\Delta_F G(Li) - \Delta_F G(LiCoO_2) \tag{1-16}$$

由于 $\Delta_F G(Li_{1-x}CoO_2)$ 随 x 值不断变化，因此该反应的 E 值随着脱锂量 x 发生变化。$Li_{1-x}CoO_2$ 的生成能可以通过点阵气体模型估算，或者通过第一性原理计算，或者通过实验直接测量。

通过上述计算方法，对于类似于反应式(1-9)的相转变反应，假设由不同的二元过渡金属化合物 NX 与金属 M 形成电池 M/NX，则通过计算 M/NX 电池的电压，可以比较由相同金属、不同材料 NX 组成的电池的电压高低。事实上，这些电池的电压存在着一般性规律[7]：对于以相转变反应储能的同系列 NX 材料，X 相同，M 相同，过渡金属 N 不同，且 N 具有相同的化学价，则基于相转变反应储能电池电压高低的顺序是按照元素周期表

的逆序，Cu＞Ni＞Co＞Fe＞Mn＞Cr＞V＞Ti。对于同系列的 NX 材料，过渡金属 N 相同，M 相同，X 不同，则相转变反应电压高低的顺序是氟化物体系＞氧化物体系＞硫化物体系＞氮化物体系＞磷化物体系。对于同样的 NX，不同的 M，相转变反应电压高低的顺序是 Li 体系＞Na 体系＞Mg 体系＞Al 体系。

相转变反应的理论工作电压不随充放电过程发生变化，但是对于嵌入化学反应，由于生成能不断变化，电压在反应过程中存在一定的电压范围。从电器使用的角度，电压变化便于判断电池荷电量，或称之为充放电深度（state of charge，SOC）。但电压范围太宽不利于电器的使用，一般电器对放电截止电压有要求。从比较不同电池的角度，比较平均工作电压或中点电压有一定的参考价值。对于基于嵌入反应储能的体系，这方面的系统研究在文献中还不多，主要的原因是对产物生成能的准确估算需要第一性原理的计算或者精确的热力学测量，而这方面的工作目前开展得还较少，不过根据相转变反应计算的电极材料电压的高低顺序，也基本适用于嵌入化学反应电位高低的定性判断。

1.6 电极材料的理论容量

通过式(1-7)，可以计算材料的理论电化学容量（常用单位为 mA·h/g）。

对于锂离子电池的负极而言，需要知道在金属锂析出电位之上，该材料最大能储存的锂的量。例如，对于锂离子电池的 Si 负极，锂最多可以形成 $Li_{22}Si_5$，相当于每摩尔 Si 原子储存了 4.4 个电子与 4.4 个锂离子，按照 Si 的摩尔质量计算，其理论容量计算为 4200mA·h/g。对于合金类反应，由于能够与锂形成合金的材料的相图都已经测量了，可以方便地根据合金相图和无机晶体学数据库来计算理论容量。对于相转变反应，如 MnO 中 Mn 的化合价为二价，其储锂反应 MnO 最多可以还原到 Li_2O 与 Mn，因此该反应可以转移两个电子，按照相转变反应储能的电极容量可以方便地计算出来理论容量。图 1-12 显示了典型的锂离子电池的负极材料的计算容量和电压（实际电压范围）。在含有液体电解质的锂电池中，由于在低电位还存在形成固体电解质中间相（solid electrolyte interphase，SEI）的反应以及界面储锂（interface charging）反应，实际储锂容量有时高于单纯按照主要的电化学反应计算的理论容量。

对于正极材料，以锂离子电池为例，仍以相转变反应为例，类似于负极，可以通过其最大还原反应消耗的电子来估算。例如，FeF_3 中 Fe 为三价，储锂的反应最多可以还原到 3LiF/Fe，因此该反应的 n 值为 3。这类反应计算出的容量很高，如氟化石墨 $(CF)_n$ 还原到 LiF/C 的理论容量为 864.6mA·h/g。因此，基于相转变反应的正极材料引起了广泛的注意，但是室温下相转变反应存在能量效率较低、循环性较差的缺点。此外，该类正极材料不含锂，必须发展提供锂源的负极与之匹配，这方面的研究还在进行中。

对于含锂的正极材料，电极材料的容量取决于最大脱锂量和最多可转移电子的量。以相转变反应 $LiFePO_4$ 正极材料为例，Fe^{2+} 可以氧化为 Fe^{3+}，对应一个电子，同时允许一个锂离子脱出，反应产物是 $FePO_4$，因此理论容量可以按照式(1-7)准确计算。类似的

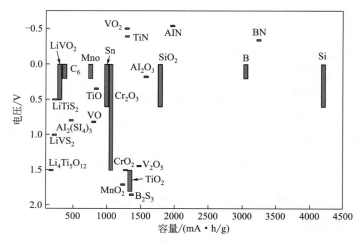

图 1-12 锂离子电池负极材料储锂容量和电压范围

$LiNi_{0.5}Mn_{1.5}O_4$、$LiMn_2O_4$，其可转移电子数与可脱出锂离子数相等，理论容量可以同样计算。

对于层状化合物，如 $LiCoO_2$，Co^{3+} 可以氧化为 Co^{4+}，对应一个电子，但脱出一个锂离子，会导致结构的不可逆相变，因此该类材料的容量取决于在不影响结构可逆变化的前提下实际可脱出锂的最大的量，这一点往往不能准确地通过理论计算估计。多数含过渡金属正极材料的理论容量是基于过渡金属最大可以转移的电子数以及最大可以脱锂或嵌锂的量来决定的。而最近对含 Li_2MnO_3 的材料研究表明[8~10]，O 可以参与电子转移。如果阴离子也可以参与电子转移，则材料的最大理论容量对于锂源正极材料来说，完全取决于可以脱锂的量。图 1-13 的部分计算，如 Li_5FeO_4、Li_6CoO_4，每摩尔材料最大可以脱锂离子的量为 5 个和 6 个，因此按此计算的理论容量较大，该类反应的本质实际上已与 Li/O_2 电池相近，为 Li_2O 的可逆断键与成键。

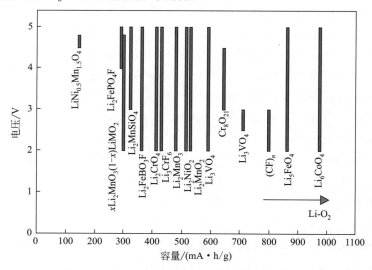

图 1-13 锂离子电池的正极材料的理论容量和估计电压范围

根据图 1-13 和表 1-3 可以看出，对于锂离子电池而言，单纯提高正极材料 1 倍的储锂容量，在平均电位不下降的前提下，提高锂离子电池的质量能量密度最大约为 40%；提高负极材料 1 倍的储锂容量，提高电池的质量能量密度最大约为 20%。由于电极储锂容量提高一般伴随着体积变化，单纯通过提高电极材料的储锂容量来提高电池的体积能量密度，应该很难超过 40%。目前的水平是 750W·h/L，也就是未来锂离子电池很难达到 1050W·h/L 以上的水平。实际上，在过去 20 年里，锂离子电池的能量密度每年稳步增长 3%[7]，主要依赖于增加活性物质比例技术方面的进步，但这种方法也逐渐接近极限，这两年能量密度的提高有减缓的趋势。由于电池应用的重要性，不断出现电池能量密度显著提高的新闻，采用本书的分析方法有助于去伪存真。纵观电池发展的历史[7]，电池能量密度的提高往往是突变性的。采用新的电池材料体系和新的结构设计，是提高电池能量密度的关键，但由于需要兼顾图 1-1 中指出的其他综合性能，实际上电池的改性是非常具有挑战性且十分漫长的任务。

1.7 本章结语

热力学理论计算有助于了解化学储能的理论极限，为估算实际电池的能量密度，开发新的电极材料、电池体系，了解化学储能器件能量密度的极限提供一定的理论参考依据。计算结果表明，Al/O_2、Li/O_2 和 Mg/O_2 电池的理论能量密度在化学储能器件中最高。从电池能量密度提高以及技术成熟度的角度，预计今后电池发展的顺序依次是采用高容量电极材料的下一代锂离子电池；采用金属锂负极，嵌入化合物作为正极的可充放锂电池；采用金属锂、铝、镁、钠为负极，S、H_2O、O_2 为正极的金属燃料电池。由于使用电池的电子设备对电池的性能要求存在显著差异，能量密度不是各类电池研发追求的唯一目标，针对性地开发电池体系与充电方式，也将会为未来电池的研发和创造新的市场开拓出更广泛的空间。

本章需要进一步思考的基础科学问题：

1. 半电池装配后测量到的开路电压数值为何与正负极的电化学势相差较大，代表着什么电化学反应？
2. 电能转化为化学能的原子尺度的微观机制是什么？
3. 共有 8 种储锂机制，其开路电压曲线有何特点？理论能量密度如何计算？不同储锂机制可以共存吗，哪些可以共存？
4. 导致小尺寸材料的生成能与体材料的生成能存在差异的原因主要有哪些？如何计算？
5. 在已知理论能量密度的前提下，如何准确估算电芯层级的实际能量密度？
6. 用固态电解质替代传统的有机电解液后，电池能量密度会不会提升？
7. 不同储锂机制（例如嵌入反应，相变反应），其实际电压曲线有什么特点？与热力学平衡电位有什么关系？

8. 为何有时电极材料的实际测得容量会超过理论计算容量值？

9. 如何提高商业化电池能量密度？

10. 是否存在理论能量密度超过锂电池的电池体系？

参考文献

[1] Cao Wenzhuo, Zhang Jienan, Li Hong. Batteries with high theoretical energy densities [J]. Energy Storage Materials, 2020, 26: 46-55.

[2] Armand M, Tarascon J M. Building better batteries [J]. Nature, 2008, 451: 652-657.

[3] Goodenough J B, Kim Y. Challenges for rechargeable Li batteries [J]. Chem Mater, 2010, 22: 587-603.

[4] Li Hong, Wang Zhaoxiang, Chen Liquan, et al. Research on advanced materials for Li-ion batteries [J]. Adv Mater, 2009, 21: 4593-4607.

[5] NEDO. Secondary Battery Technology 2008 [EB/OL]. [2009-05-29]. http//app3. infoc. nedo. go. jp/informations/koubo/other/FA/nedothernews. 2009-05-29. 2374124845/30ed30fc30de30c389e38aacP_516c958b7248518 d65398a02.

[6] Wang Liping, Wu Zhenrui, Zou Jian, et al. Li-free Cathode Materials for High Energy Density Lithium Batteries [J]. Joule, 2019 (3): 2086-2102.

[7] Zu Chenxi, Li Hong. Thermodynamic analysis on energy densities of batteries [J]. Energy Environ Sci, 2011, 4: 2614-2624.

[8] Thackeray M M, Johnson C S, Vaughey J T, et al. Advances in manganese-oxide "composite" electrodes for lithium-ion batteries [J]. J Mater Chem, 2005, 15: 2257-2267.

[9] Koyama Y, Tanaka I, Nagao M, Kanno R. First-principles study on lithium removal from Li_2MnO_3 [J]. J Power Sources, 2009, 189: 798-801.

[10] Xiao Ruijuan, Li Hong, Chen Liquan. Density functional investigation on Li_2MnO_3 [J]. Chem Mater, 2012, 24: 4242-4251.

第 2 章

电池材料缺陷化学

2.1 锂离子电池材料中常见的缺陷形态

理想的晶体材料具有无限大的有序周期性点阵结构。真实存在的材料包含多种缺陷，在一定程度上偏离了理想材料的对称性和周期性。空间结构存在着包括点缺陷（如空位、填隙、取代、色心、互占位等）、线缺陷（如刃位错与螺位错）、面缺陷（如晶界与表面）、体缺陷（如孔、洞、缺陷簇、畴结构）等各类缺陷[1~3]。和具备长程有序的晶体材料相比，准晶与无定形材料为短程有序或无序材料。电子结构方面，电荷与自旋的分布可以形成电荷有序或者电荷无序的畴结构。缺陷的存在对生成焓、构型熵、稳定性、熔点、硬度、介电、空间电荷层等热力学性质以及输运、储存、相变、反应、激发等动力学过程均有显著影响。

缺陷是固体物理与固体化学中的重要研究内容。Kröger 和 Vink[4,5]在 20 世纪 60 年代建立了比较完整的缺陷化学理论，主要用于研究晶体内的点缺陷。其基本假设为：用连续溶液模型来处理晶体结构，将缺陷以溶质化处理，用经典物理学，尤其是热力学和统计物理学的方法，来研究缺陷在不同条件下的产生、湮灭和平衡。这种近似的处理有一定的适用范围。缺陷浓度超过某一临界值，缺陷之间的库仑相互作用更加显著，导致缺陷出现配对、缔合，进一步产生缺陷簇、超结构和中间相等。虽然早期的点缺陷理论不能很好地处理存在复杂相互作用的缺陷簇，但基于点缺陷的缺陷化学理论仍然是认识材料微观结构以及与电池材料相关的离子输运等动力学过程的重要基础。

图 2-1 给出了固体中典型缺陷的原子结构示意图，包括 Frenkel 缺陷，即格点正常位置的一个原子扩散至间隙位置，留下一个空位，空位与间隙原子形成一个缺陷对；Schottky 缺陷，即格点原子由于热振动等因素扩散到晶格表面，内部留下空位，一般是阴离子与阳离子空位成对出现；取代缺陷，即外来原子取代了其中的一个格点上原来的原子，如果被取代原子与取代原子电荷不等，需要产生空位或间隙原子，或者被取代原子得失电子，以达到电荷平衡。图 2-1 中还包括由于缺失一列原子引起的边位错，空位串以及杂质原子局部凝聚等缺陷。

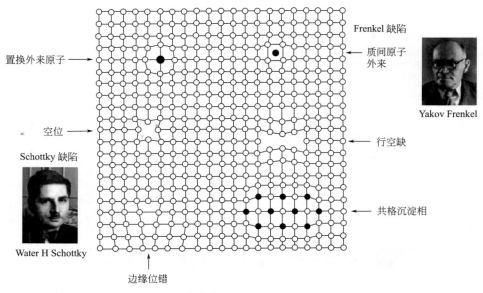

图 2-1 固体中典型缺陷的原子结构示意图

点缺陷化学是电化学活性粒子的化学，即离子激发的化学[6]。在只有少数活性的游离粒子存在的溶剂里亦或是在离子晶体里，绝大多数的天然离子间都是强库仑力结合的，并且在电化学上是不活跃的。这种情况类似于半导体中的电子载流子，其中绝大多数处于更深层次的电子是不活跃的。人们将其称为传导电子和空穴。点缺陷化学为活性粒子之间的相互作用及它们的浓度与控制参数化学计量学、掺杂量和温度的关系提供了一个广义的热力学图像[6]。与化学计量相比，在开放系统中使用组分分压或在电池中使用电池电势更为方便。

在锂电池研究的材料中，除图 2-1 中所示的缺陷类型外，较为常见的缺陷还包括同一晶体结构中不同原子之间的互占位（antisite），非整比（nonstoichiometry），至少两种以上原子无序占据相同位置的固溶体（solid solution），晶界（grain boundaries），有序与无序共存的纳米结构（nanostructure），包括微孔、介孔、大孔的孔结构（porous structure），异质结（heterogeneous junction），表面修饰层（surface coating layer），梯度结构（gradient materials），核壳结构（core-shell）等。如果同一个晶体结构同一个位置同一种原子存在两种以上的价态，还可能存在电荷无序或电荷有序的结构。由于锂电池一般在室温附近工作，目前还没有讨论材料中的自旋有序或无序结构，但确实有些含过渡金属的电极材料由于多种价态原子的共存，可能存在着分布不均的磁畴，这方面的研究在锂电池领域还较少涉及。

锂是一种非天然成分，因此常常"自动"视为一种附加成分（对应于间隙缺陷）和活性粒子出现[7]。如果缺少了缺陷化学，我们将很难理解单相存储的 S 形容量-电压曲线特性，更不用说对于 $FePO_4/LiFePO_4$ 中接触的电荷转移动力学的理解了。因为在上述情况中，必须考虑到 Li 空位的存在。精确地处理方式必须要考虑到缺陷化学涉及的物质。即使在富锂情况下，忽略化学缺陷的处理方式都会导致得出错误的结论。

电池材料中存在着哪些缺陷，以哪一类缺陷为主及一定温度下缺陷的浓度属于热力学问题，与材料的组成、晶体结构、制备过程有关。除了纳米效应等特殊问题外，锂电极处理的困难主要源于低温。在低温状态下，平衡缺陷热力学只适用于电子和锂离子交换的过程。而所有其他缺陷浓度则被视为"冻结态"。J. Maier[8]详细讨论了缺陷热力学从"平衡态"到"冻结态"的转变。这增加了研究的复杂性，但同样增加了材料研究的自由度。材料中缺陷的存在同样会显著影响电子与离子在固体内部、界面、表面的输运、储存与反应性质。考虑界面电荷转移动力学时，空间电荷效应是决定性的因素[9]。在这种情况下，详细了解边界区的缺陷化学（作为掺杂、温度和电荷状态的函数）是必不可少的[10]。

离子的电导率与可参与输运的自由离子的浓度和离子的迁移率有关。离子在固体中的输运一般需要通过空位、间隙位、晶界、表面等介质和通道进行传输。占满格点位、没有空位和间隙位的离子晶体，在较低的温度下，离子电导一般较低。金属中的缺陷会引起电子的散射，半导体中的缺陷既有可能通过引入杂质能级降低电子跃迁的激发能，提高电子和空穴的浓度，提高电子的迁移率，也有可能增大电子的散射等。电极材料中同一种缺陷的存在，对于离子传导和电子传导的推动作用既有可能相同，也有可能相反。缺陷结构的存在还有可能提供额外的锂的存储位置。锂在缺陷位置的存储有一定的可逆性，例如锂在空腔中的欠电位沉积、在界面处的界面储锂、在表面的表面电容，均为可逆的储锂机制[11]。材料表面的某些悬挂键、表面的新相，有可能引起不可逆的储锂。

缺陷结构的实际材料与理想结构的体材料相比，由于缺陷的存在，材料的生成能发生了变化，会影响材料的稳定性和反应性。理解了材料的缺陷化学，通过可控增加或减少材料中的缺陷，可以有目的地调控材料地输运、储存、稳定性和反应性。对材料中缺陷结构的控制、确认、分析以及建立起与之相应的准确、全面的物理、化学、电化学储锂特性之间的构效关系，是锂电池材料研究中的核心基础科学问题，也是实现材料理性设计（rational design）的关键。

2.2 缺陷产生的基础

与外界无粒子交换的孤立的固体材料一般满足：①电荷守恒原则（conservation of charge），即体系中正负电荷相等，不能凭空增减；②质量守恒原则（conservation of mass），即参加反应的原子数在反应前后一致；③晶格位守恒原则（conservation of structure）。缺陷的引入和产生也必须服从上述三原则。基于以上基本原则产生的缺陷，在实际过程中，当达到一定浓度时，存在着相互作用，其主要特点为：带有相反电荷的缺陷容易缔合成对；缔合以后会导致局域晶体场的极性变化；这种极性变化进一步诱导相邻的缺陷对构成缺陷簇，更进一步形成超结构等[12]。那么，从能量的变化考虑，缺陷产生过程可以简单地按如下热力学表述。

定义：ΔH 表示一个空位生成的能量。则：

$$\Delta H = E_{vac} + \mu_{defect} - E_{bulk} \tag{2-1}$$

式中，E_{vac} 表示产生空位后的体系总能量；μ_{defect} 为缺陷物质的化学势；E_{bulk} 为完美晶体的总能量。其中，ΔH 越大，则缺陷越难生成。

含有 N 个原子的晶格中产生 n 个空位时，体系可以看成含有 $(N+n)$ 个格点，则空位总的生成能可以近似为 $n\Delta H$。由空位引起的体系熵变可以表述为单个空位的振动熵 ΔS_{vib} 与构型熵 ΔS_{conf} 之和：

$$\Delta S = \Delta S_{vib} + \Delta S_{conf} \tag{2-2}$$

则体系的总能量变化为：

$$\Delta G = n\Delta H - T\Delta S = n\Delta H - T(\Delta S_{vib} + \Delta S_{conf}) \tag{2-3}$$

令缺陷产生反应的反应常数 $K = n/(N+n)$，对于构形熵有 $\Delta S_{conf} = K\ln[(N+n)!/(N!\cdot n!)]$，这里利用关于阶乘的 Stirling's Formula 公式 $\ln(x!)x\ln x - x$，则：

$$\Delta S_{conf} = Kn\ln[(N+n)/n] \tag{2-4}$$

那么 n 个空位体系在温度 T 下的稳态要求总能量与 n 的一阶导数为 0，即 $d\Delta G/dn = 0$ 即 $\Delta H - T\{\Delta S_{vib} + k_B\ln[(N+n)/n]\} = 0$，则有

$$K = n/(N+n) = \exp[\Delta H/(k_B T) - \Delta S_{vib}/k_B] = \exp[(\Delta H - T\Delta S_{vib})/(k_B T)]$$

令 $E_{act} = \Delta H - T\Delta S_{vib}$，则

$$K = \exp[E_{act}/(k_B T)] \tag{2-5}$$

图 2-2 为缺陷产生的热力学原因分析。从图 2-2 的示意图可以看出，缺陷产生的热力学驱动力来自于材料自身的熵变。在一定的缺陷浓度下，由于构型熵和振动熵的存在，缺陷的产生可能是热力学有利的。是否会出现这种情况，与具体材料焓的增长与熵降低的相对数值有关，材料中不一定必须出现缺陷结构。实际材料的生成能可以按照式(2-6)考虑

$$\Delta_f G_{real\ material} = \Delta_f G_{perfect\ material} - \Delta_f G_{defect} \tag{2-6}$$

合并式(2-3) 与式(2-6)，得：

$$\Delta_f G_{real\ material} = (\Delta_f H_{perfect\ material} - n\Delta_f H_{defect})$$
$$- T[\Delta S_{vib(defect)} + \Delta S_{conf(defect)} - \Delta S_{perfect\ material}] \tag{2-7}$$

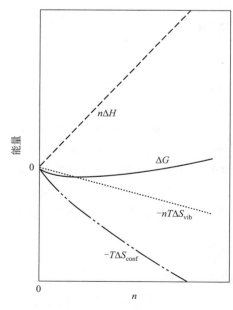

图 2-2 缺陷产生的热力学原因

在上述的讨论中，假定缺陷结构的引入不引起材料焓的变化，这在多数情况下是没有问题的，即少量缺陷的存在不会引起本体材料多数原子之间，特别是远离缺陷的原子结合能的变化。但当材料逐渐从单晶结构过渡到微晶结构，再过渡到纳晶结构，最后到无定形结构或者团簇，本体材料中的大量原子逐渐偏离了原来的理想晶体结构的周期势，材料的生成焓因此偏离具备理想结构的材料的生成能，即 $\Delta_f H_{perfect\ material}$ 这一项可能也会发生变化。

在第 1 章，提到了 Nernst 方程，一个有关电化学反应的电位、能量密度与材料和产物生成能的关系式[13]。此前计算的结果均是基于理想材料或者热力学数据库中根据体材料测试的热力学数据[11]。从式(2-6)、式(2-7)中可以看出，对于含有大量缺陷的材料，生成能会发生变化，因此能量密度与理论电压相对于理想的体材料会出现一定的

差别，其大小取决于缺陷能。材料缺陷能的实验测量与理论计算均不易进行。以表面能为例，材料表面能 σ 可以通过 $(2\gamma/r)V_M$ 估算，其中 γ 为表面张力；r 为粒子半径；V_M 为摩尔体积。以纳米颗粒材料为例，如果材料的表面张力能为 2J/mol（金红石类氧化物材料），晶粒尺寸的大小从 5nm 减小到 1nm，由于表面能的贡献引起的电化学反应电位的变化从 50mV 提高到 700mV，可见对尺寸非常敏感[14]。从实验上看，以相转变反应为例，在嵌脱锂过程中，材料由初始的多晶材料转到纳晶材料及无定形材料，前后开路电压的差别可以达到 200～600mV[15,16]。有实验曾经通过开路电压法直接测量了 25nm 和 2μm 尺寸金红石 TiO_2 表面能的差别，其引起的电位差为 62mV[17]。

锂电池的实际性能与理论性能间的差异通常是由过电压所反映的动力学问题引起的。虽然这种电化学电阻具有一定实用价值，但由于极化现象而产生的容量效应相对于化学储能容量而言是很小的，因此通常可以被忽略，更值得我们关注的是固体电活性粒子的传输特性。为了实现储能过程，Li^+ 和 e^- 的输运需要同时进行，而输运过程又是由缺陷而不是由离子或电子来实现的。例如，对于一个"过剩储存"的粒子：$\sigma_{ion}=\sigma_i=Fu_ic_i$，$\sigma_{eon}=\sigma_n=Fu_nc_n$，其中 u 是迁移率；F 是法拉第常数。当这两个值相差很大时，粒子的填充过程是扩散控制的。化学扩散系数 D^δ 完全由缺陷化学公式决定[18]：

$$D^\delta \propto \frac{\sigma_i\sigma_n}{\sigma_i+\sigma_n}\left(\frac{X_n}{c_n}+\frac{X_i}{c_i}\right) \tag{2-8}$$

值得注意的是，粒子内部的化学计量极化效应的瞬态同样由 D^δ 确定。在固态电解质中，任何有关电子的贡献通常都会导致内部短路；而对于液态电解质，电子浓度（通过氧化还原对）通常很小，但它们的反离子贡献很大，导致浓度极化[19]。在液态电解质中，所有离子理论上都是可移动的，但缔合效应起着重要作用，系统地降低了总电导率。J. Maier 的研究表明，离子缔合作用包括高阶聚集效应、导电性和极化行为。另一个重要的极化是离子从电解液转移到电极上，特别是对于重溶剂化的锂离子。这种电荷转移与非对称自由能分布的传导不同，通常是由于大场效应引起的[20]。此过程可应用 Butler-Volmer 方程：

$$i_0=[\vec{k}\Pi(electrode)\overleftarrow{k}\Pi(electrolyte)]^{1/2} \tag{2-9}$$

在 Butler-Volmer 方程中，$\Pi(electrode)$ 是电极相平衡缺陷浓度的乘积（以化学计量指数加权），$\Pi(electrolyte)$ 指电解液中与电荷转移反应有关的载流子浓度的产物。考虑锂离子从 $LiFePO_4$ 转移到电解液中的过程，如果在局部保持 $LiFePO_4/FePO_4$ 两相平衡，$LiFePO_4$ 会表现出一定的平衡缺陷，即 $Li_{1-\varepsilon}FePO_4$。因此，离子交换率与 $[V'_{Li}]$（即 ε）成正比，而不是总的 Li^+ 浓度。

2.3 电极材料中的缺陷

2.3.1 TiO_2 中嵌锂过程

J. Y. Shin 等[21,22]探讨了 TiO_2 中 Li 嵌入过程中涉及的缺陷化学。可以通过在较高的

温度下利用 H_2 气氛退火来增加 TiO_2 中的氧空位，从而提高 n 型电导率。基于 $Li_xTiO_{2-\delta}$ 氧空位和活锂量的函数可以半定量地给出离子和电子的电荷载体浓度间的关系。这些关系均遵循缺陷化学定律。图 2-3(a)、(b)、(c) 展示了 δ 与 x 之间的关系。而在室温时的一个重要特征是离子和电子载流子的结合，例如氧空位（$V_O^{\cdot\cdot}$）与过剩电子（e'）及间隙锂离子（$Li_i^{\cdot}$）与过剩电子之间的结合。平衡关系式 $Li_i^{\cdot} + e' \rightleftharpoons Li_i^x$ 可以解释 H_2 气氛处理对储能的影响，容量的变化如图 2-3(d) 所示，随着 δ 的增加，不仅 σ_{eon} 会随之增加，σ_{ion} 同样会随之增加。

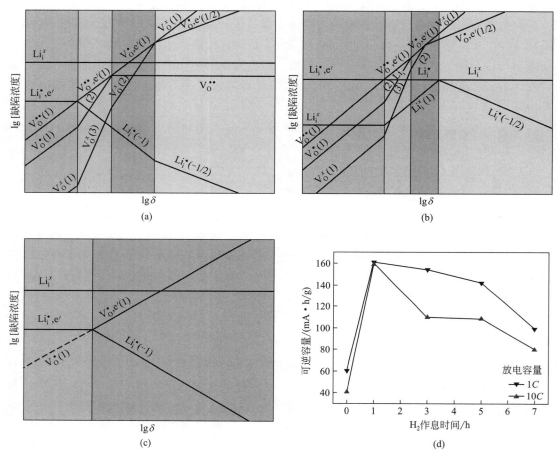

图 2-3 不同初始条件下，缺陷浓度随氧化学计量学 δ 的变化

（a）强和（b）弱锂离子/e^- 缔合；(c) 考虑了强 Li^+/e^- 和强 $V_O^{\cdot\cdot}/e^-$ 缔合的特殊情况，导致 Li_i^x 和 $V_O^{\cdot\cdot}$ 的显著形成；(d) 氢还原二氧化钛电极在 1C 和 10C 放电过程中的可逆容量与缺氧的关系[21,22]

2.3.2 $FePO_4$/$LiFePO_4$ 电极

当电极材料体系确定后，电极材料改性的方法如下。

① 包覆：可以选择导电介质，如碳、导电聚合物、无机导电层（如 ITO、FTO、

AZO、TiN 等）、金属层包覆来改善颗粒之间的电接触；惰性介质，如 Al_2O_3、AlF_3、$AlPO_4$、MgO、ZrO_2、ZnO 等提高材料的界面稳定性；离子导体，如 $LiNbO_3$、$LiLaTiO_3$、$LiNiPO_4$、LiPON、Li_2CO_3、Li_3PO_4 等提高材料的界面离子传输特性及界面稳定性；混合离子导体，如包覆 $LiNi_{0.5}Mn_{1.5}O_4$、$LiCoO_2$、$Li_4Ti_5O_{12}$ 于内核正极材料的表面，提高电子、离子输运特性及界面稳定性。

② 降低颗粒与晶粒尺寸：在界面副反应不严重的情况下，假设电解液能充分浸润每一个颗粒，降低颗粒尺寸可以缩短离子输运的路径，增大电化学反应面积，通过增加与导电添加剂接触的面积提高电子输运特性。晶粒与颗粒尺寸的下降还可能有利于释放材料中的应力。

③ 掺杂：有时文献中所指的掺杂（doping），既包括简单的物理混合与复合，也包括外来原子替代或占据主体材料中的格点位置。

在实际应用中，简单的物理混合由于通过电解液的复杂离子交换反应，也可能带来意想不到的积极的效果，如 Li（$Ni_{1/3}Co_{1/3}Mn_{1/3}$）O_2 与 $LiMn_2O_4$ 的混合可以提高电池的循环性和高温特性。物理混合还可能有助于提高电接触，缓解应力。文献中更普遍的掺杂是指第二种情况。如果掺杂或替代元素的量较多，且不产生第二相，主体材料的结构不发生变化，这时候也叫固溶体（solid solution）。例如少量 Na 在 $LiFePO_4$ 中的掺杂，Li（$Fe_{0.99}Na_{0.01}$）PO_4，摩尔分数为 50% 的 Mn 与 50% Fe 形成的 Li（$Fe_{0.5}Mn_{0.5}$）PO_4 固溶体。一个材料可以实现的最大掺杂量或形成固溶体的范围，与主体材料晶体结构、晶粒尺寸、掺杂原子的价态、离子半径等因素有关。小尺寸材料易于形成更大的掺杂量。判断材料是否形成晶格掺杂或固溶体以及外来原子在晶格中的占位、占有率，需要通过严格的结构分析来验证。

以 $LiFePO_4$ 为例，简单介绍该材料研究中涉及的缺陷化学。$LiFePO_4$ 材料最早在 1997 年由 Goodenough 等[23]将其作为锂离子电池的电极材料进行了研究，发现该材料具有可逆的储锂性能，但其可逆容量只达到了理论容量 170mA·h/g 的 60%。1999 年 Armand 等[24]提出了碳包覆 $LiFePO_4$ 的设想，显著提高了其电化学活性，可逆容量接近理论容量，碳包覆的作用似乎主要是改善了颗粒之间的电接触。半导体掺杂可以显著提高材料的电子电导。因此通过掺杂来提高 $LiFePO_4$ 的电子与离子输运特性，获得了广泛的研究。2002 年，Chiang 等[25]发表了 $LiFePO_4$ 掺杂少量金属离子（Mg^{2+}、Al^{3+}、Ti^{4+}、Zr^{4+}、Nb^{5+} 和 W^{6+}）的研究工作，使掺杂后 $LiFePO_4$ 的电子电导率在室温下提高 8 个数量级至 4.1×10^{-2}S/cm，所合成的材料在低倍率充放电时的放电比容量可接近理论比容量（170mA·h/g）；即使在高达 6000mA/g 的电流密度下进行充放电时，仍可保持着可观的放电比容量，极化作用很小。他们认为掺杂少量金属离子后 $LiFePO_4$ 电子电导大大提高的主要原因是由于 $LiFePO_4$ 中 Li 或 Fe 的缺陷出现了 Fe^{3+}/Fe^{2+} 的混合价态，从而导致形成了 p 型半导体。该结果发表后引起了广泛争议，特别是高价的 Zr^{4+}、W^{5+} 是否能在 Li^+ 位置掺杂，是否掺杂进入晶格，掺杂是否能显著提高输运和电化学特性。2003 年 Armand 等[26]重复 Chiang 等的试验后认为是残余碳的作用提高了表观电子电导。2004 年 Nazar 等[27]认为电子电导的提高不是由于混合价态 Fe 的出现，而是因为表面生成了金属磷化物的导电网络。施思齐等[28]在2003年通过第一性原理GGA方法研究了$LiFePO_4$掺杂Cr元素,从理论上预测了对$LiFePO_4$的Li位进行Cr掺杂可以减小带隙,有利于提高电子电导。之后欧阳楚英等[29]在2004年的第一性原理分子动力学计算表明,$LiFePO_4$为一维离子导体,锂位掺杂将有可能阻塞锂离子输运[30],如图2-4所示。

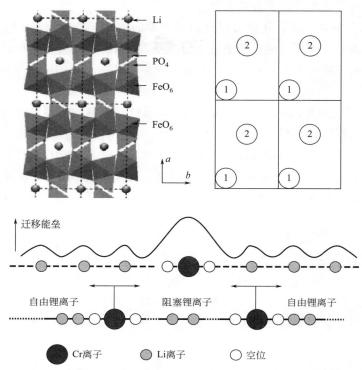

图2-4 $LiFePO_4$中Li位掺杂Cr后离子输运阻塞效果示意图[31]

从$LiFePO_4$为一维离子导体考虑,掺杂有可能在提高电子电导的同时降低了离子电导,而离子电导一般认为比电子电导低3~5个数量级,因此掺杂需要兼顾电子与离子的输运特性。根据Ceder等的理论分析,$LiFePO_4$中电子的输运是通过Fe^{2+}/Fe^{3+}小极化子传输[32],因此在材料中适当增加Fe^{3+}将有利于电子的输运。通过引入杂质能级,而不是增加Fe^{3+}提高电子电导的思路在理论上似乎存在一定的问题。考虑到体相$LiFePO_4$材料以两相反应为主,则电子的输运一定需要借助Fe^{2+}/Fe^{3+}小极化子,如果材料没有掺杂,电子输运只能发生在两相界面,因此电子的输运应该也可能是一维或二维。

2004年,中科院物理所提出了氧空位$LiFePO_4$材料的设想,其做法有两种。①在Fe位掺杂一价碱金属元素,在Fe^{2+}位置掺杂一价元素的效果,或者是形成氧空位,或者是同等数量的Fe^{2+}氧化为Fe^{3+};②N取代O,有关方案参见文献[33]。通过实验,以5%(摩尔分数)Na占据Fe位为例,通过中子衍射实验确认了采用高纯试剂合成的固相目标化合物中实际组成为$Li_{0.952}Fe_{0.952}Na_{0.047}PO_{4-\delta}$。通过穆斯堡尔谱证明了4.8% Fe^{3+}的存在。根据Julien等[34]在其分析文章中的说法,Na^+在Fe位的掺杂是唯一在物理上合理的掺杂方案。

关于高价元素是否能在 Li^+ 位掺杂一直存在争议。Islam 根据缺陷平衡方程，利用第一性原理计算了不同元素在 $LiFePO_4$ 材料中的溶解能，其计算表明，从热力学方面考虑高价元素溶解于 $LiFePO_4$ 材料中能量不利[35,36]。但 2009 年 Chiang 等[37]证明 12%（摩尔分数）的 Zr^{4+} 可以掺杂进 $LiFePO_4$，其晶胞参数随 Zr^{4+} 掺杂量的变化符合 Vergerd 规律，如果结果属实，这是高价元素掺杂可以进入 $LiFePO_4$ 晶格的最有利的实验证据。而 2008 年之前提出表面 FeP 电导增强机理的 Nazar 等[38]通过中子衍射等手段也证明 3%（摩尔分数）的 Zr 可以掺杂进去，并且宣称由于晶胞参数只膨胀 0.3%，不影响锂离子的输运。之后 Nazar 还和 Chiang 出现了关于最大掺杂量的争论。从实验结果看，早期 Islam 的计算存在一定的问题，也许因为没有考虑尺寸效应，而 Chiang 等的样品为纳米尺寸，相对于大晶粒尺寸材料，一般说来对外来原子掺杂具有增强溶解度的效果。如何从理论上考虑尺寸效应对掺杂极限的影响目前还是一个很有挑战的问题。

对于一维离子导体而言，锂位掺杂可能会影响离子的一维输运特性。Maier 团队[39]对 $LiFePO_4$ 单晶的输运实验表明无论是电子电导还是离子电导均为二维输运特性，与之前的分子动力学理论计算结果矛盾。而中子衍射谱的数据证明了 $LiFePO_4$ 材料离子的一维输运特性[40]，软 X 射线吸收谱证明了电子的一维输运特性[41]。之前的单晶光学照片显示为棕色，说明其应该存在杂质。之后 Maier 团队仔细研究了 Al、Si 掺杂样品的输运特性，也显示了二维输运特性。因此可能是杂质元素在 Li^+ 位的存在影响了一维输运通道上 Li^+ 的输运，导致了快离子通道的输运受到阻挡，表观上出现了二维输运行为，这还有待于进一步验证。$LiFePO_4$ 中容易出现 Li-Fe 互占位，特别是对于低温水热合成的样品。研究结构中的互占位在实验上有一定难度。中子衍射数据的拟合可以提供一定的占有率信息。近些年发展的球差校正扫描透射电镜技术同样能够实现对轻原子的直接观测，为研究互占位提供了有利的工具[42]。图 2-5 为 $LiFePO_4$ 颗粒球差校正电镜的环形暗场照片[43]，暗场照片的衬度大约跟原子序数的 1.7 次方成正比，也就是说原子序数越大的原子越容易被观测到。沿（010）方向看，在暗场下，首先可以清晰地看到 FeO_6-PO_4 组成的一个近似六元环的结构，而六元环的中心就是 Li 的位置，如果在完美晶格中，Li 是没有衬度的，但是在图中却零零星星地看到一些 Li 的栏上出现了衬度。与之前关于 Li-Fe 互占位的各种讨论对比之后，在右下角模拟的照片中，通过人为设定一定量的 Li-Fe 互占位，可以得到与实验类似的图像。也就是说，在该 $LiFePO_4$ 颗粒中确实存在着互占位，并且通过最直观的手段看到了这种现象。

在早期的研究中，由于充放电曲线显示了一个清楚的电位平台，因此提出 $LiFePO_4$ 在充放电过程中经历了一个两相反应的 $LiFePO_4 \rightarrow (FePO_4 + Li)$ 过程。对于该两相反应的微观机制，提出了多个反应模型，其中的核心问题是 $LiFePO_4$ 和 $FePO_4$ 的界面结构的细节。谷林等[44]利用先进的球差校正扫描透射环形明场成像技术（ABF-STEM），首次在部分充电的 $LiFePO_4$ 单晶纳米线（$d=65nm$）中观测到了锂离子的隔行脱出，类比于石墨中存在的"2 阶结构"的现象。这一发现与之前提出的相边界推移、核壳结构等各类两相反应模型均不一致，为一单相结构。$LiFePO_4$ 中存在单相结构也被软 X 射线吸收

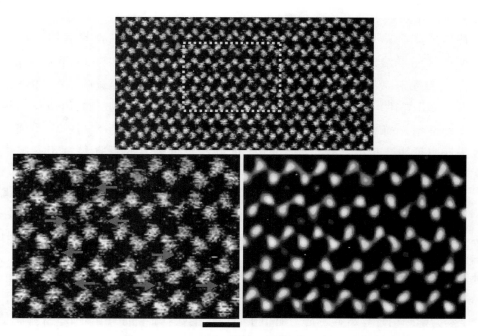

图 2-5　$LiFePO_4$ 放大的球差校正透射电镜环形暗场照片以及模拟图[40]

(图中箭头所指的地方为 Li 的位置，在暗场下本不应出现衬度，但是却明显有一定的衬度变化，而模拟的照片上用部分 Li 和 Fe 互占位的晶格做出来的)

谱数据证实。索鎏敏等[45]在 200nm 的 Nb 掺杂的 $LiFePO_4$ 脱锂样品中，清楚地看到界面处的阶结构，并且证明阶的方向垂直于 b 轴。这进一步支持了沿着 b 轴的一维离子输运模型。电镜照片还显示了 Nb 在点阵中的不均匀团聚。M. Rahul 等[46]的理论计算支持单相的存在。孙洋等[47]的计算表明两相分离是热力学有利的，阶结构在界面的出现是 Fe^{3+} 与 Li^+ 相互作用的结果，动力学上有利。阶结构是动力学的非稳态结构，其普遍性及其对动力学性能的影响目前还不清楚。

此外，R. Amin 等对氧化还原偶 $LiFePO_4/FePO_4$ 进行了广泛的缺陷化学分析[48~52]。$LiFePO_4$ 在转变为具有相同结构的 $FePO_4$ 前，Li 损失约为 2%。虽然我们指的是相同结构（室温下的宏观晶体）通道中的 Li，但是 $LiFePO_4$ 中的 Li 缺陷是由锂空位引起的，而 $FePO_4$ 中的 Li 过剩是由 Li 填隙物引起的。在第一种情况下，锂空位（V'_{Li}）由空穴补偿；在第二种情况下，间隙（$Li^·$）由过剩电子补偿。$LiFePO_4$ 中微空的价带成为 $FePO_4$ 中微填充的导带。图 2-6 所示的缺陷图的高度对称性反映了这种紧密的对应关系，其中 $FePO_4$ 和 $LiFePO_4$ 的离子和电子缺陷浓度是作为施主和受主浓度的函数绘制的。由此可以得知：①离子和电子载体间具有拮抗作用；这两种载体对于储能来说都很重要，因此掺杂不总是有益的；通过缩小尺寸来减少扩散长度，并通过涂层包覆来增加电子传输是更可取的做法；②相界可能与 p-n 和 V-i 跃迁有关；③缺陷图对任何晶体取向都是有效的；$Fe^·_{Li}$ 浓度同样也会影响 1D 传导通道，破坏通过模拟预测验证的 b-c 平面中的传导各向异性[53]。

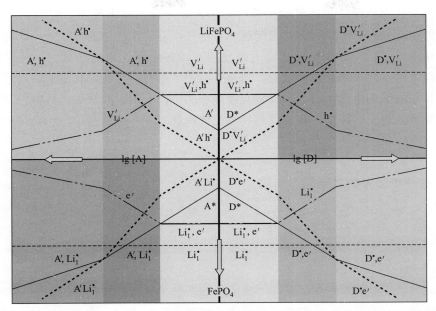

图 2-6　$LiFePO_4$ 和 $FePO_4$ 中缺陷浓度的对数随受体 [A] 或供体 [D] 含量的变化而变化，表明材料具有广泛的改性性质，说明电池材料中缺陷平衡的复杂性[53]

从上文的讨论可以看出，在材料研究中，清楚地证明掺杂、理解掺杂对结构的影响以及掺杂引起的物理化学性质的变化，需要大量的实验与理论工作的结合。除了广泛采用的掺杂，氧化物的制备与气氛的控制密切相关，含有氧空位的材料通过相应的电荷补偿机制，材料的性质发生了显著的变化，这方面典型的例子是 $LiNi_{0.5}Mn_{1.5}O_4$。氧空位的存在影响了材料的晶体结构，从 Ni、Mn 有序占位的 $P4_332$ 转变为无序占位的 $Fd\bar{3}m$ 结构材料，其离子、电子、反应、稳定性、充放电曲线形状都有显著的差别[54]。对于固体电解质而言，快离子导体的发展也是建立在丰富的对缺陷化学系统理解的基础上。

2.4　本章结语

缺陷化学是处理固体电极和电解质的决定性基础科学，因为电活性成分（例如 Li）通常是通过产生或消除离子和电子点缺陷来调节的。电池的理论容量和电池电压都可以精确地表达在点缺陷的水平上，它们的浓度反映了热力学活动。动力学性质则在很大程度上取决于输运和转移动力学，其描述直接涉及载流子的化学，即缺陷化学。本章简单总结了锂电池材料研究中涉及的缺陷化学。主要以常见的 $LiFePO_4$ 电极材料为例，讨论了缺陷态在锂离子电池中的重要作用以及对电化学特性的可能调制作用。相对而言，由于 LiFePO_4 材料的重要性，$LiFePO_4$ 的缺陷化学问题得到了深入研究。但一些基本的科学问题，如掺杂是否能在碳包覆、较小尺寸、降低游离过渡金属含量、提高结晶度等策略后能进一步提升材料的动力学特性、稳定性，电子与离子的混合输运特性及其与材料尺寸的关系，相界面结构的普遍规律，相边界的推移动力学，多颗粒之间嵌脱锂后锂在颗粒间的再平衡，嵌脱锂引起的材料的应力演化等问题一直都还没有明确的结论。在基础研究还存在很多疑问的情况下，$LiFePO_4$ 材料已广泛地使用在动力电池、储能电池中，这也是锂离子电池材料研究的普遍现状。管中窥豹，锂离子电池各类材料中缺陷化学问题还远没有研究清楚，在单个缺陷的电化学行为、缺陷与锂离子的协同作用以及缺陷在材料稳定性的作用等方面亟须深入研究，相信这些问题的日渐澄清能够为今后材料的进一步优化和改性提供新的机会。

本章需要进一步思考的基础科学问题：

1. 从热力学的角度考虑，缺陷会改变材料的哪些性质和状态？
2. 离子的输运性质与点缺陷、线缺陷、面缺陷、体缺陷有什么关系？如何设计缺陷结构来调控离子的输运性质？
3. 如何通过缺陷工程调控电极材料中的电子结构及电子输运特性？
4. 电极材料体相、表面的缺陷结构有哪些各自的特点？
5. 电极材料的颜色与缺陷结构有关系吗？
6. 如何判断一个电极或电解质材料能否掺杂引入另一种异价元素？如何引入？如何判断已经引入？一般应关注掺杂对材料的哪些物理、化学性质的影响？
7. 亚稳态材料是否一定包含缺陷？缺陷材料是否一定是亚稳态材料？
8. 充放电过程中缺陷结构是否能稳定存在？是否会产生新的缺陷结构？具有超长循环寿命的材料是否应选择高结晶度材料？
9. 在锂离子电池电极材料中，使用纳米材料会带来哪些影响？这些影响是如何产生的？

参考文献

[1] Crawford J H, Slifkin L M. Point Defects in Solids [M]. New York：Plenan Press，1975.
[2] 黄昆，韩汝琦. 固体物理学 [M]. 北京：高等教育出版社，1988：529-552.
[3] Kittel C. Introduction to Solid State Physics [M]. 8th ed. New York：John Wiley & Sons Inc，2005：597-618.
[4] Kröger F A. Defect chemistry in crystalline solids [J]. Annu Rev Mater Sci，1977，7：449-475.
[5] Kröger F A, Vink H J. Solid State Physics [M]. New York：Academic，1956，3：307.
[6] Frenkel J. Über die Wärmebewegung in festen und flüssigen Körpern [J]. Z Phys，1926，35：652.
[7] A V ander V en, Ceder G. Lithium diffusion in layered Li_xCoO_2 [J]. Electrochem，Solid-State Lett，2000，3：301.
[8] Maier J. Complex oxides：high temperature defect chemistry vs. low temperature defect chemistry [J]. Phys Chem Chem Phys，2003，5 (11)：2164.
[9] Maier J. Kröger-vink diagrams for boundary regions [J]. Solid State Ionics，1989，32：727.
[10] Fu L J, Chen C C, Samuelis D, Maier J. Thermodynamics of lithium storage at abrupt junctions：modeling and experimental evidence [J]. Phys Rev Lett，2014，112：208301.
[11] Zu C X, Li H. Thermodynamic analysis on energy densities of batteries [J]. Energy Environ Sci，2011，4：2614-2624.
[12] Smyth D M. The Defect Chemistry of Metal Oxides [M]. USA：Oxford University Press，2000.
[13] Peng Jiayue, Zu Chenxi, Li Hong. Fundamental scientific aspects of lithium batteries（Ⅰ）—Thermodynamic

calculations of theoretical energy densities of chemical energy storage systems [J]. Energy Storage Science and Technology, 2013, 2 (1): 55-62.

[14] (刘兆君). 锂离子电池电极材料中热力学和动力学问题的第一性原理研究 [D]. 北京: 中国科学院物理研究所, 2010.

[15] Delmer O, Balaya P, Kienle L, Maier J. Enhanced potential of amorphous electrode materials: Case study of RuO_2 [J]. Adv Mater, 2008, 20: 501.

[16] Zhong K F, Zhang B, Luo S H, et al. Investigation on porous MnO microsphere anode for lithium ion batteries [J]. J Power Sources, 2011, 196: 6802-6808.

[17] Balaya P, Maier J. Thermodynamics of nano-and macrocrystalline anatase using cell voltage measurements [J]. Phys Chem Chem Phys, 2010, 12: 215-219.

[18] Maier J. Mass Transport in the Presence of internal defect reactions-concept of conservative ensembles: Ⅰ, Chemical diffusion in pure compounds [J]. J Am Ceram Soc, 1993, 76: 1212.

[19] Maier J. Salt concentration polarization of liquid electrolytes and determination of transport properties of cations, anions, ion pairs and ion triples [J]. Electrochim Acta, 2014, 129: 21.

[20] Bockris O'M, Reddy A K V. Modern electrochemistry: an introduction to an interdisciplinary area [J]. Modern Electrochemistry, Plenum Press, 1970.

[21] Shin J Y, Samuelis D, Maier J. Defect chemistry of lithium storage in TiO_2 as a function of oxygen stoichiometry [J]. Solid State Ionics, 2012, 225: 590.

[22] Shin J Y, Joo J H, Samuelis D, Maier J. Oxygen-Deficient $TiO_{2-\delta}$ Nanoparticles via Hydrogen Reduction for High Rate Capability Lithium Batteries [J]. Chem Mat, 2012, 24: 543.

[23] Padhi A K, Nanjundawamy K S, Goodenough J B. Phospho-olivines as positive-electrode materials for rechargeable lithium batteries [J]. J Electrochem Soc, 1997, 144: 1188.

[24] Ravet N, Goodenough J B, Besner S, et al. Improved iron based cathode material [C] //196th Meeting of the Electrochemical Society, Honolulu, Hawaï: 1999.

[25] Chung S Y, Bloking J T, Chiang Y M. Electronically conductive phospho-olivines as lithium storage electrodes [J]. Nat Mater, 2002, 1: 123-128.

[26] Ravet N, Abouimrane A, Armand M. On the electronic conductivity of phospho-olivines as lithium storage electrodes [J]. Nat Mater, 2003, 2 (11): 702-704.

[27] Herle P S, Ellis B, Coombs N, et al. Nano-network electronic conduction in iron and nickel olivine phosphates [J]. Nat Mater, 2004, 3 (3): 147-152.

[28] Shi S Q, Liu L J, Ouyang C Y, et al. Enhancement of electronic conductivity of $LiFePO_4$ by Cr doping and its identification by first-principles calculations [J]. Phys Rev B, 2003, 68: 195108.

[29] Ouyang C Y, Shi S Q, Wang Z X, et al. First-principles study of Li ion diffusion in $LiFePO_4$ [J]. Phys Rev B, 2004, 69: 104303.

[30] Ouyang C Y, Shi S Q, Wang Z X, et al. The effect of Cr doping on Li ion diffusion in $LiFePO_4$ from first principles investigations and Monte Carlo simulations [J]. J Phys Cond Mat, 2004, 16 (13): 2265-2672.

[31] 欧阳楚英. 锂离子电池正极材料离子动力学性能研究 [D]. 北京: 中国科学院物理研究所, 2005.

[32] Maxisch T, Zhou F, Ceder G. Ab initio study of the migration of small polarons in olivine Li_xFePO_4 and their association with lithium ions and vacancies [J]. Phys Rev B, 2006, 73: 104301.

[33] Li H, Wang Z X, Chen L Q, et al. Research on advanced materials for Li-ion batteries [J]. Adv Mater, 2009, 21: 4593-4607.

[34] Axmann P, Stinner C, Wohlfahrt-Mehrens M, et al. Nonstoichiometric $LiFePO_4$: Defects and related properties [J]. Chem Mater, 2009, 21: 1636-1644.

[35] Islam M S, Driscoll D J, Fisher C A J, et al. Atomic-scale investigation of defects, dopants, and lithium transport in the $LiFePO_4$ olivine-type battery material [J]. Chem Mat, 2005, 17: 5085-5092.

[36] Fisher C A J, Prieto V M H, Islam M S. Lithium battery materials $LiMPO_4$ (M=Mn, Fe, Co, and Ni): Insights into defect association, transport mechanisms, and doping behavior [J]. Chem Mat, 2008, 20: 5907-5915.

[37] Meethong N, Kao Y H, Speakman S A, et al. Aliovalent substitutions in olivine lithium iron phosphate and impact on structure and properties [J]. Adv Func Mater, 2009, 19: 1060-1070.

[38] Wagemaker M, Ellis B L, Luetzenkirchen-Hecht D, et al. Proof of supervalent doping in olivine $LiFePO_4$ [J].

Chem Mater，2008，20：6313.
[39] Maier J，Amin R. Defect chemistry of LiFePO$_4$ [J]. J Electrochem Soc，2008，155：A339-A344.
[40] Nishimura S，Kobayashi G，Ohoyama K，et al. Experimental visualization of lithium diffusion in Li$_x$FePO$_4$ [J]. Nature Mater，2008，7：707-711.
[41] Liu X S，et al. Phase transformation and lithiation effect on electronic structure of Li$_x$FePO$_4$：An in-depth study by soft X-ray and simulations [J]. J Am Chem Soc，2012，134：13708-13715.
[42] He X Q，Gu L，Zhu C B，et al. Direct imaging of lithium ions using aberration-corrected annular-bright-field scanning transmission electron microscopy and associated contrast mechanisms [J]. Mater Express，2011，1 (1)：43-50.
[43] Chung S Y，Choi S Y，Yamamoto T，et al. Atomic-scale visualization of antisite defects in LiFePO$_4$ [J]. Phys Rev Lett，2008，100：125502.
[44] Gu L，Zhu C B，Li H，et al. Direct observation of lithium staging in partially delithiated LiFePO$_4$ at atomic resolution [J]. J Am Chem Soc，2011，133：4661-4663.
[45] Suo L，Han W，Lu X，et al. Highly ordered staging structural interface between LiFePO$_4$ and FePO$_4$ [J]. Phys Chem Chem Phys，2012，14：5363.
[46] Malik R，Zhou F，Ceder G. Kinetics of non-equilibrium lithium incorporation in LiFePO$_4$ [J]. Nat Mater，2011，10：587.
[47] Sun Y，Lu X，Xiao R J，et al. Kinetically controlled lithium-staging in delithiated LiFePO$_4$ driven by the Fe center mediated interlayer Li-Li interactions [J]. Chem Mater，2012，24：4693-4703.
[48] Amin. R，Balaya P，Maier J. Anisotropy of electronic and ionic transport in LiFePO$_4$ single crystals [J]. Electrochem Solid-State Lett，2006，10：13 .
[49] Maier J，Amin R. Defect chemistry of LiFePO$_4$ [J]. J Electrochem Soc，2008，155：339.
[50] Amin R，Maier J. Effect of annealing on transport properties of LiFePO$_4$：Towards a defect chemical model [J]. Solid State Ionics，2008，178：1831.
[51] Amin R，Lin C T，Maier J. Aluminium-doped LiFePO$_4$ single crystals Part Ⅱ. Ionic conductivity，diffusivity and defect model [J]. Phys Chem Chem Phys，2008，20：3524.
[52] Amin R，Lin C，Peng J，et al. Silicon-doped LiFePO$_4$ single crystals：Growth，conductivity behavior，and diffusivity [J]. Adv Funct Mater，2009，19：1697.
[53] Samuelis D，Maier J. in Chemical Energy Storage [M]. Berlin：De Gruyter，2013：225.
[54] Wang L P，Li H，Huang X J，et al. A comparative study of Fd-$3m$ and $P4_332$ "LiNi$_{0.5}$Mn$_{1.5}$O$_4$" [J]. Solid State Ionics，2011，193：32-38.

第 3 章

相、相变与相图

相变是电池材料基础研究中的重要问题。合成制备过程中对材料相变的准确认识有利于获得晶体结构与组成符合设计要求的目标材料;充放电过程中电极材料的相变与材料的储锂机制、储锂容量范围、电压曲线、储锂动力学、材料的体积变化以及吸放热等密切相关。通过高通量计算、制备、表征方法,获得材料在制备与充放电过程中的相组成、相结构演化及相图,对于全面理解材料体系及其充放电行为,开发新的电池材料体系具有十分重要的意义。本章小结了与锂离子电池相关的相变与相图研究。

3.1 相

热力学系统中,物理性质均匀的一个空间区域称为一个相,具体可见文献 [1~3]。锂离子电池的研究中,主要关注材料的物相,包括晶体结构、化学组成、结晶度、固液态等。电极、导电添加剂、集流体等材料一般为固态(电极也可以为液态);电解质材料可以为液态,也可以为纯固态或凝胶态;黏结剂、隔膜一般为聚合物(固态软物质)。从电子自旋态来看,由于 3d 电子常常具有自旋长程关联作用,含有这些元素的正极材料可能表现出不同的磁性状态,对第一性原理计算吉布斯自由能和输出电压具有较大影响。多数电池正极材料包含 Fe、Co、Ni、Mn 的一种或多种。充放电过程中 3d 层通常有未成对电子,从而表现为顺磁态。例如,尽管在完全放电情况下 Co^{3+} 的 d 电子配对,理论上表现为抗磁态,当材料充电时存在 Co^{4+} 的 d^5 离子,表现为顺磁态[4]。对于另一种常见的正极材料 $LiMn_2O_4$,材料包含混合价态的 Mn^{3+}(d^4)和 Mn^{4+}(d^3),同样表现为顺磁态[4]。未成对电子长程关联的时候,在低温下会表现为铁磁态或反铁磁态。例如,对于富锂材料 Li_2MnO_3[其中 Mn^{4+}(d^3)]表现为反铁磁态[5]。磁性状态与材料制备条件以及除 Fe、Co、Ni、Mn 以外的其他元素组成也有关系。例如,Yao 等报道,850℃烧结的 $LiFePO_4$ 和 $Li_{0.95}Mg_{0.05}FePO_4$ 均表现为反铁磁态,700℃烧结的表现为铁磁态 $Li_{0.95}Mg_{0.05}FePO_4$[6]。对于含 F 的聚阴离子正极,$LiNaFePO_4F$ 表现为铁磁态[7],$LiFeSO_4F$ 表现为反铁磁态[8]。研究电极材料的磁性可以帮助理解原材料以及充放电过程中过渡金属的电子结构和局域结构的有序与无序变化[4]，也可以用来检测电池材料中痕量的磁性杂质。

3.2 相　　变

3.2.1 相变的热力学描述

相变往往经历不同有序结构之间，以及有序与无序结构的相互转化。一般而言，相互作用导致有序和组织，热运动引起无序和混乱。保罗·埃伦费斯特（Paul Ehrenfest）首先对相变进行了分类[9,10]，其分类标志是热力学势以及其导数的连续性。热力学势定义为 $\varGamma = F + PV$，其中 F 为自由能，P 为压强，V 为体积。热力学势连续、一阶导数不连续的状态突变，称为一级相变，这类相变伴随着明显的体积变化和热量的吸放（潜热）。对于热力学势和它的第一阶导数都连续变化而二阶导数不连续的情形，称为二级相变，此时没有体积变化和潜热，但是比热、压缩率、磁化率等物理量随温度变化的曲线出现了跃变，如图 3-1 所示。大部分固态相变属于一级相变；只有少部分属于二级相变，如某些合金的有序-无序转变、磁性转变、超导态转变等[11]。以锂离子电池电化学脱嵌锂为例，两相反应电位曲线一般出现显著的平台区域，反映的相变为一级相变；充放电过程中结构保持固溶体形式的相变反应充放电曲线为 S 形，为二级相变。

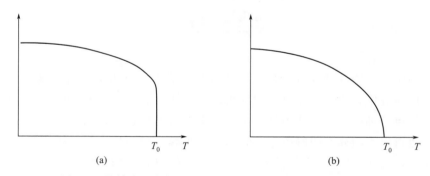

图 3-1　纵轴为一个物理量的一级相变（a）和连续相变（b）

3.2.2 合成制备中的相变研究

在锂离子电池中，无论是电极还是电解质材料，材料的纯度、均匀性、结晶度往往是影响其性能的重要因素。在合成制备过程中，从前驱体材料逐渐转变为目标产物，涉及一系列相变，此过程通常伴随着杂相的消失或者产生。确定材料制备过程中的相结构与相变，是实现材料可控制备的基础。可以采用衍射、散射、成像、磁性测量等方法，分辨目标相与杂相；通过测量变温过程中的吸放热、失重、晶体结构、电导率、磁性、织构等物性变化，确定相变温度或温度范围。温度和相组成是最重要的两个参量。一些材料的制备过程中，还需要施加高压、磁场、电场、机械搅拌、不同气氛等条件，因此还需要考虑其他物理和化学参数。以下通过几个具体例子，简述锂电池材料制备中的相变特点。

3.2.2.1 高压法以及固相合成法制备 $LiAlO_2$

偏铝酸锂（$LiAlO_2$）具有较高的化学稳定性和热稳定性，可应用于核聚变反应作为氚增殖材料[12-16]，以及熔融碳酸盐燃料电池（molten carbonate fuel cells，MCFC）的基体（matrix）[17-20]。此外，细颗粒状 $LiAlO_2$ 也可以用于锂离子电池中，例如，作为惰性陶瓷添加入聚合物电解质[21]，或者作为复合固体电解质的组分[22]，来提高离子电导率，增加与锂金属电极的相容性以及改善形貌和力学性质[23,24]。这些应用所需要的性质，包括热稳定性、化学稳定性以及形貌可控性，同样适用于锂离子电池，包括应用于固体电解质本身或者固体电解质与电极材料界面相的优化。其中，α-$LiAlO_2$ 具有 α-$NaFeO_2$ 型六方层状结构（其空间群为 R-$3m$），与常见的层状结构正极 $LiMO_2$（M=Co，Ni，Mn，等）相同，有希望提供更好的物理和化学相容性。例如通过掺杂 Al 提高 Li（M，Al）O_2（M=Co，Ni）正极的充放电稳定性[25]。$LiAlO_2$ 也被用作正极、负极的包覆材料[26-29]。正极表面包覆，可根据包覆的形貌分为粗包覆（岛状包覆）、核壳包覆以及超薄薄膜包覆[30]。然而实际应用中很难严格鉴别区分包覆的类型。而且，仅从文献所提出的合成手段和表征结果来看，甚至很难精准地确认包覆层 $LiAlO_2$ 的相。

$LiAlO_2$ 已知有 α、β、$(\beta)^b$、γ、δ、ε 六个相，见表 3-1[31]；研究表明，α、γ、$(\beta)^b$、δ 为高压稳定相[32~35]。在常压下，加温可以使 α 相不可逆转变为 γ 相[36~40]。然而，这种材料目前尚无明确完整的一元组分温-压相图。1979 年，Byker 等[41]计算了 $LiO_{0.5}$-$AlO_{1.5}$ 在 1000K 以上的二元组分相图，并对 $LiAlO_2$ 的一元温压相图进行了估计，如图 3-2(a) 所示。2008 年，雷力等在温压范围为 300~1873K、0.5~5.0GPa 的条件下，实验得到高压相图，如图 3-2(b) 所示[42]。注意到，其中三种相比较常见，包括 α-$LiAlO_2$（六方相），β-$LiAlO_2$（单斜相）和 γ-$LiAlO_2$（四方相）[43]。其中，β-$LiAlO_2$ 在通常的温压条件下比较少见。而 γ-$LiAlO_2$ 为高温相，意味着其在高温条件下最稳定，且该相的锂离子电导率已经被研究过[44~46]。和常见层状正极材料结构相似的 α-$LiAlO_2$，其合成制备和电导率、离子输运机制也已经被研究[47]。

表 3-1 $LiAlO_2$ 的相和晶体结构[31]

相	晶体结构	空间群	配位数	晶格参数	密度/(g/cm³)
α	六方	R-$3m$	6	$a=2.799Å, c=14.180Å$	3.401
β	正交	$Pna2_1$	4	$a=5.280Å, b=6.300Å, c=4.900Å$	2.685
$(\beta)^b$	单斜	$P2/m$	4,6	$a=8.147Å, b=7.941Å, c=6.303Å, \beta=93.18°$	0.269
γ	四方	$P4_12_12$	4	$a=5.169Å, c=6.268Å$	2.615
δ	四方	$I4_1/amd$	6	$a=3.887Å, c=8.300Å$	3.51
ε	立方	$I4_132$	4	$a=12.650Å$	约 2.615

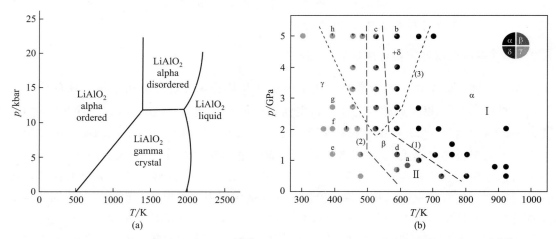

图 3-2 LiAlO$_2$ 温-压相图的预估（a）[41] 和 LiAlO$_2$ 在温-压空间中的实验结果总结（b）[42]

为了展示多维度下的合成相变与相图，以高压下用 γ-LiAlO$_2$ 合成 α-LiAlO$_2$ 为例，如图 3-2(b) 所示，在不同温度-压力条件下，可以得到不同的亚稳相，包括 α、β、γ、δ 相。Ⅰ 区为 α-LiAlO$_2$ 的稳定相区，在 0.5～2GPa 的条件下提高温度，LiAlO$_2$ 逐渐从 γ 相（杂相）转变为 β 相（杂相）再转变为 α 相（目标相）；Ⅰ 区和 Ⅱ 区分界线的斜率为负，提高压强可以使 α-LiAlO$_2$ 生成温度降低；压强提高到 2GPa 以上，在升温过程中，生成四方结构的 δ 相。

另外，以 α-LiAlO$_2$ 层状相的合成制备为例，表 3-2 小结了其制备方法，包括制备工艺复杂的高压法、水热法、溶胶沉淀法和溶胶-凝胶法。注意到，虽然在固相反应法[48-52]中可以得到该相，但是其合成时间通常较长，且产物通常伴随着杂相（更普遍的情况，甚至 α-LiAlO$_2$ 才是作为杂相出现）。而通过研究反应过程中的相变-温度对应关系，并控制中间相及其形貌，可以高效制备纯相[47]。从图 3-3（a）和（b），可以推断，反应的机制如下：

$$227.5℃:Al(OH)_3 \longrightarrow \alpha\text{-}AlO(OH)+H_2O$$

$$296.9℃:2\alpha\text{-}AlO(OH) \longrightarrow \gamma\text{-}Al_2O_3(纳米颗粒)+CO_2+H_2O$$

$$528.6℃:Li_2CO_3+\gamma\text{-}Al_2O_3(纳米颗粒) \longrightarrow 2\alpha\text{-}LiAlO_2(纳米颗粒)+CO_2$$

$$720.1℃:Li_2CO_3+\gamma\text{-}Al_2O_3(纳米颗粒) \longrightarrow 2\gamma\text{-}LiAlO_2(纳米颗粒)+CO_2$$

其中，反应温度来自 TG-DSC，反应产物推断于 X 射线衍射（X-ray diffraction, XRD）峰；纳米颗粒推断于该峰的展宽。之前研究认为，α-LiAlO$_2$ → γ-LiAlO$_2$ 的相变温度约为 900℃[53]，然而通过控制中间体的相和纳米形貌，可以在更低的反应温度得到更纯的相。

通过以上的研究可以看出 LiAlO$_2$ 结构的复杂性。明确不同相合成制备的条件，便于获得具有目标相结构的产物。作为包覆材料，电极表面包覆层出现的是哪一种结构，哪一种结构具备较高的锂离子电导率和稳定性，无定形相与晶态在输运性质上的区别，与被包覆物质的界面稳定性等问题，还需要进一步的研究。

表 3-2 α-LiAlO$_2$ 的制备方法

方法	特征	文献
Li$_2$CO$_3$+α-Al$_2$O$_3$ 600℃	超过 600℃，不可逆相变为 γ 相；产物有 γ 相杂相	[48]
Li$_2$CO$_3$+γ-Al$_2$O$_3$ 高压 35kbar；850℃；30min	前驱体需要研磨单晶 γ-Al$_2$O$_3$	[32]
600~700℃ Li$_2$CO$_3$+K$_2$CO$_3$+γ-Al$_2$O$_3$	α-LiAlO$_2$ 为主相	[54]
Li$_2$CO$_3$+γ-Al$_2$O$_3$ 600℃ Li$_2$CO$_3$+γ-AlO(OH) 600℃ Li$_2$CO$_3$+Al(OH)$_3$ 700℃	α-LiAlO$_2$（主），γ-LiAlO$_2$（杂） α-LiAlO$_2$ α-LiAlO$_2$（主），γ-LiAlO$_2$（杂）	[55]
Li$_2$CO$_3$+γ-AlO(OH) 600℃	倾向于生成 α-LiAlO$_2$；片状；1~25μm；<5m^2/g	[56,57]
[LiAl$_2$(OH)$_6$]$^+$(OH)$^-$·2H$_2$O+LiOH·H$_2$O 室温	氮气+水蒸气气氛下 <5m^2/g	[58]
溶胶-凝胶法 550℃	初始反应物残留（硝酸锂，异丙醇铝）， 直至烧至 1150℃	[39]
碳酸 Li/K 置换	需要清洗步骤，否则有杂相	[59]
利用有机物合成、利用熔盐合成 廉价法（固相反应）	原料便宜； 604nm~11.85μm；3.22~11.4m^2/g	[49]
高压，19GPa；1700℃	改变温压条件，可控合成 α、β、γ、δ 相；结晶度良好	[42]

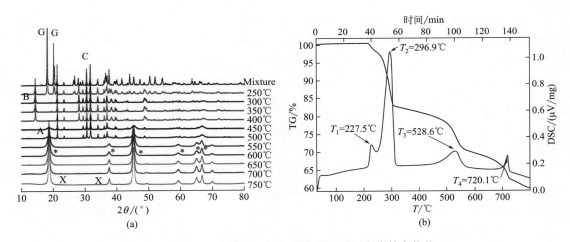

图 3-3 （a）2Al(OH)$_3$+Li$_2$CO$_3$ 混合物不同温度烧结产物的 XRD[G=Al(OH)$_3$，B=AlO(OH)，C=Li$_2$CO$_3$，A=Al$_2$O$_3$，*=α-LiAlO$_2$，X=γ-LiAlO$_2$]；（b）2Al(OH)$_3$+Li$_2$CO$_3$ 的 TG-DSC[47]

3.2.2.2 LiMn$_2$O$_4$、Li$_2$MnO$_3$ 的合成与相稳定性

尖晶石结构 LiMn$_2$O$_4$ 正极材料的电化学性能与制备条件密切相关，需要了解不同条件下制备的材料的相组成。Luo C. 等研究了 Li$_{1+x}$Mn$_{2-x}$O$_{4-\delta}$ 合成物随温度、锂嵌入组分、锂/锰比例和氧分压的变化[60]。

图 3-4 确定了尖晶石结构稳定存在的区域。可以看到，Li$_{1+x}$Mn$_{2-x}$O$_{4-\delta}$ 在其合成制

备中，当温度 T 在临界温度 T_{c1} 和 T_{cL} 之间（阴影区），只有单相的立方尖晶石结构生成；当温度 $T>T_{c1}$ 时，Li_2MnO_3 作为次相出现；当 $T<T_{cL}$ 时，Mn_2O_3 作为次相生成；T_{c1} 和 T_{cL} 均随着锂/锰比（r）的增加而连续下降。图 3-4 中的阴影部分（单相区）包括锂缺陷和锂过量尖晶石结构。当 $n_{Li}/n_{Mn}=0.5$，即合成物恰好是尖晶石 $LiMn_2O_4$ 时，没有不连续的变化。

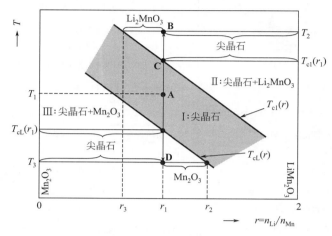

图 3-4　Li-Mn-O 尖晶石结构的稳定域相图以及升降温情况下的分解[60]

图 3-5 利用降温条件下原位 XRD 结构分析的方法，确定了杂相 Li_2MnO_3、Mn_2O_3 的消失和出现的温度，粗略地验证了图 3-4 的一些性质。图 3-5（a）为富锂条件下（$n_{Li}/n_{Mn}=0.564$）的 XRD 图，从图中可以看出，随着温度的降低，Li_2MnO_3 的峰逐渐消失；图 3-5（b）是在缺锂条件下（$n_{Li}/n_{Mn}=0.455$）测得，Li_2MnO_3 的峰同样会随着温度的降低而消失，而且在温度更低的情况下 Mn_2O_3 相逐渐生成。比较图 3-5（a）和（b），可以看到，当 n_{Li}/n_{Mn} 比较小的时候，Li_2MnO_3 在比较高的温度就开始消失，与图 3-4 中 T_{c1} 相界线斜率为负相符合。

利用 XRD 方法只能得到粗略的相变温度，当 Li_2MnO_3 相、Mn_2O_3 相含量少的时候很难探测。用热重分析（thermal gravimetric analysis，TGA）可以确定其相变温度，如图 3-6 所示，为 $p(O_2)=0.2bar$ 条件下，富锂（$n_{Li}/n_{Mn}=0.564$）和缺锂（$n_{Li}/n_{Mn}=0.455$）时，从室温加热到 1000℃ 相对质量的变化；内嵌图展示了 T_{c1} 和 T_{cL} 的确定方法（两侧曲线连续变化段切线的交点）。由于 Li_2O 在 1200℃ 以下的挥发可以忽略[61,62]，T_{c2} 以下相对质量的变化主要源于化学反应中结构的失氧以及从环境中得氧；温度达到 T_{c2} 的时候 $LiMnO_2$ 开始形成，Li_2O 挥发或者 Li_2O 与氧的共同损失是造成失重的主要原因，这一过程可以参照下面关于 Li_2MnO_3 的合成以及稳定性的讨论。

由上文及图 3-4 知，层状 Li_2MnO_3 的稳定域位于更高的温度，以下讨论其合成制备过程中的相变和热稳定性[63]。层状结构 Li_2MnO_3 与尖晶石结构 $LiMn_2O_4$，两相在制备过程中是共生的。图 3-7 是固相反应法制备的 Li_2MnO_3 样品 XRD 谱图（上）与理论计算谱图（下）。实验 XRD 谱比理论计算多 3 个峰，如箭头所示。其中两个分别是尖晶石结构的 $LiMn_2O_4$（111）峰（$2\theta=18.4°$）和（311）峰（$2\theta=35.3°$）；而 $2\theta=17.9°$ 峰，可以

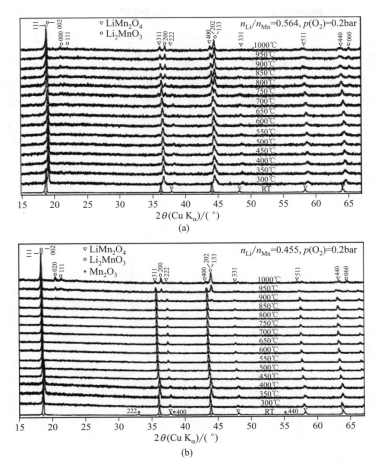

图 3-5 富锂（$n_{Li}/n_{Mn}=0.564$）(a) 和缺锂（$n_{Li}/n_{Mn}=0.455$）(b) 的尖晶石结构在降温条件下的原位 XRD 谱图[60]

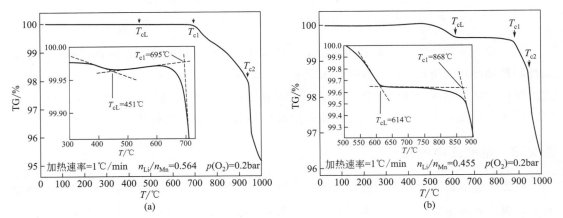

图 3-6 富锂（$n_{Li}/n_{Mn}=0.564$）(a) 和缺锂（$n_{Li}/n_{Mn}=0.455$）(b) 时，从室温加热到 1000℃ 相对质量的减少；嵌入图阐述了临界温度 T_{c1} 和 T_{cL} 的确定方法；T_{c2} 的时候 $LiMnO_2$ 开始形成[60]

从 Diffax 程序模拟得知，来自于 Li_2MnO_3 与 $LiMn_2O_4$ 的共生（intergrowth）。二者共生的温度条件以及二者相转变的机制，可以从图 3-8 推断：左图 0～400℃质量的减少源于样品中水的挥发；400～650℃有一段平台，可以认为水分除尽。下面解释 650℃以上质量变化原因，并阐述其相变过程。右图（b）的 XRD 谱图有 $LiMnO_2$ 的峰，说明在氩气气氛下，$Li_2MnO_3 \longrightarrow LiMnO_2 + 0.5Li_2O + 0.25O_2$；而右图（a）中 $LiMnO_2$ 消失，取而代之的是 $LiMn_2O_4$，说明在氧气中，经过足够长的时间，有：$3LiMnO_2 + 0.5O_2 \longrightarrow LiMn_2O_4 + Li_2MnO_3$，环境中氧气参与反应，对应着左图（a）降温至 990℃质量的增加，也说明了这一相转变温度起始于 990℃。左图（c）将升温温度控制在 900℃，没有经过 Li_2MnO_3 被氧化得到 $LiMn_2O_4$ 的过程，而右图（c）中没有 $LiMnO_2$ 峰，说明在这一温度下不会分解得到 $LiMnO_2$；因而左图（c）中 650～900℃质量的减少可能源于 Li_2O 挥发生成非化学计量的 $Li_{1-2\delta}MnO_{2-\delta}$。

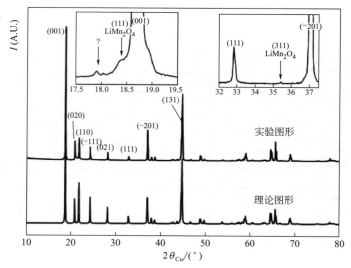

图 3-7　利用固相反应法，在 900℃条件下 4h 退火，制备得到的 Li_2MnO_3 样品的 XRD 谱图，与理论计算 XRD 谱图的比较

（三个箭头标识了相对于理论图谱额外的反射峰。80～120°（2θ）未在图中展示[60]）

从上面的举例介绍可以看出，理解在材料制备过程中的相的转变和演化过程，对于精确控制目标产物的组成，防止杂相生成具有重要的意义。在此基础上，可以进一步通过改变前驱体、烧结制度、烧结气氛、烧结压强等方法，控制材料的形貌、致密度、结晶度、结晶取向、粒度、表面结构等指标，制备出符合要求的目标产物的纯相。

3.2.3　电解质中的相变

锂离子电池目前主要采用液体电解质和聚合物胶体电解质。从安全性以及未来高能量密度负极等方面考虑，固体电解质是当前以及未来的重要发展方向。从液体电解质应用的角度考虑，希望电解质在宽温度范围内具有较高的电导率，一般不希望出现从液体转变为固体或气体的相变。从固态电解质的应用的角度，希望在高温超离子导体相可以稳定在电

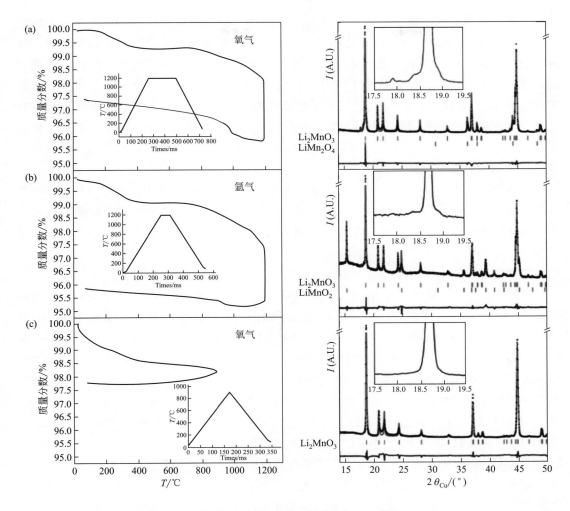

图 3-8　左侧是 TGA 曲线；右侧是 X 射线衍射谱，分别是 650℃ 合成的
粉末样品在氧气中 1200℃ 烧 4h（a）；在氩气气氛中 1200℃ 烧 1h（b）；
在氧气中加热到 900℃（c）[63]

池工作的实际温度。对电解质的相变加以利用，也可以发展新型电解质。

3.2.3.1　相转变电解质[64]

室温熔融盐具有很多优点，如较宽的液体温度范围、较高的热稳定性、很低的蒸气压、不易燃、无腐蚀性等。室温熔盐一般由较大的阴阳离子组成：其阴离子一般选择全氟化合物如 BF_4^-、PF_6^-、CF_3COO^-、$CF_3SO_3^-$、$N(CF_3SO_2)_2^-$、$N(C_2F_5SO_2)_2^-$、$C(CF_3SO_2)_3^-$，尤其 $N(CF_3SO_2)_2^-$ 阴离子对降低熔点效果显著；而阳离子一般含氮或磷，如季铵盐类、季鏻盐类、烷基吡啶类和烷基咪唑类等。但是上述室温熔盐由于具有较复杂的有机阳离子，使得它们的合成制备和提纯相对比较困难，而且价格比较昂贵。自 2000 年，物理所尝试开发低成本的室温熔盐电解质，梁宏莹博士和胡勇胜博士做了一系列开创性的研究工作。

梁宏莹首次发现二（三氟甲基磺酰）亚胺锂（LiTFSI）与尿素可以在室温下自发形成液体[65]，并对此体系的物理、热学及电化学性质进行了系统研究。随后，胡勇胜选择了氢键较弱的乙酰胺作为锂离子的配体，研究了由三种具有不同阴离子半径、结构类似的锂盐：二（三氟甲基磺酰）亚胺锂［LiTFSI，$LiN(CF_3SO_2)_2$］、二（全氟乙基磺酰）亚胺锂［LiBETI，$LiN(C_2F_5SO_2)_2$］、三氟甲基磺酸锂（$LiCF_3SO_3$）和乙酰胺组成的熔点在室温以下的电解质的物理化学性能，并对它们的物理化学性能作了对比分析，获得了更低熔点的体系，LiTFSI/乙酰胺（1∶4）低共熔点低至－67℃。其中 LiBETI 与乙酰胺的体系低共熔点可以低至－57℃。

在研究室温熔盐电解质的过程中，谢斌等发现将一定量的双草酸硼酸锂（LiBOB）与乙酰胺混合后加热，并没有像二（三氟甲基磺酰）亚胺锂（LiTFSI）与乙酰胺那样成为室温熔盐，而是得到了熔点在50℃左右的电解质[66]。此电解质体系在熔点以下为白色蜡状固体，而在熔点以上则成为透明的液体，熔点上下电导率相差很大，根据这一特性，可以将这种电解质称为相转变型电解质。

由图 3-9 可以看到，当 $LiBOB/CH_3CONH_2$ 的摩尔比为 1∶16 或 1∶4 的时候，阿伦尼乌斯（Arrhenius）曲线分为三段，每段有不同的斜率，这意味着每段温度范围内的活化能都不相同；而当 $LiBOB/CH_3CONH_2$ 的摩尔比为 1∶8 或 1∶6 的时候，Arrhenius 曲线则分为两段。Arrhenius 曲线的分段情况也从一个方面证实 $LiBOB/CH_3CONH_2$ 的二元相图情况，当摩尔比为 1∶16 或 1∶4 的时候，体系从低温到高温经历从固态到固液共存再到液态的过程，所以 Arrhenius 曲线体现了在三个相下的三种活化能；而摩尔比为 1∶8 或 1∶6 的时候，在相图上可以看到，体系只经历了从固态到液态的转变，所以 Arrhenius 曲线在这一温度范围内只有两个不同的活化能。在低温固相区，对于 $LiBOB/CH_3CONH_2$ 体系来说，我们可以认为离子在其中的传输与固态聚合物电解质中的离子传输类似。其离子迁移主要在固体中离子导电相区域内通过空位扩散或间隙位扩散进行。在高温液相区，这是一种液体电解质。在液体电解质中，普遍被人们接受的有两种离子导电

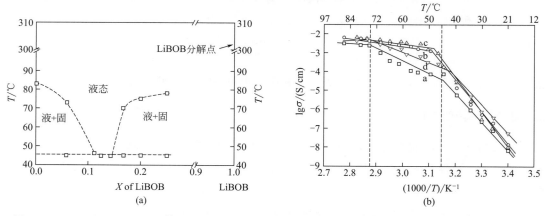

图 3-9　$LiBOB/CH_3CONH_2$ 体系相图（a），LiBOB 与 CH_3CONH_2 不同摩尔比 Arrhenius 曲线（b）
a—$LiBOB/CH_3CONH_2$（1∶4）；b—$LiBOB/CH_3CONH_2$（1∶6）；
c—$LiBOB/CH_3CONH_2$（1∶8）；d—$LiBOB/CH_3CONH_2$（1∶16）[66]

行为，一种是遵从 Arrhenius 方程，即离子从一个配位中心直接跳跃到另一个配位中心，离子电导率由离子跳跃的活化能所决定。从图 3-10 看，LiBOB/CH_3CONH_2 体系遵从 Arrhenius 方程，由于液体状态流动性好，所以离子电导率高，活化能低。对于 LiBOB/CH_3CONH_2 摩尔比为 1∶16 和 1∶4 的体系，我们注意到还有一个固液共存的状态。在这一状态下，固相分散于液相中，离子在其中既可从液相中传输，也可以从固相中传输，还可以沿着固相的晶界传输，最后体现的是一种平均的效果。因此由于固相的存在，离子电导率要比液相低；但又比固相的离子电导率高。

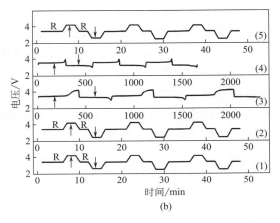

(a)   (b)

图 3-10　(a) 采用 LiBOB/CH_3CONH_2 摩尔比为 1∶8 的 Li/$LiFePO_4$ 电池在不同温度下的状态变化的照片；(b) 采用 LiBOB/CH_3CONH_2 摩尔比为 1∶8 的 Li/$LiFePO_4$ 电池在不同温度下的充放电曲线[66]

图中向上的箭头代表充电过程，向下的箭头代表放电过程，字母 R 代表弛豫过程；
(1) 25℃（升温）；(2) 40℃（升温）；(3) 60℃；(4) 40℃（降温）(5) 25℃（降温）

从图 3-10 可以看出，在室温 25℃时，电解质还是白色固体，此时电解质电导率很低，电池内阻很大，电池在这个状态下完全不能工作；而当环境温度升高到 40℃并恒温 12h 后，电解质电导率有所提高，但还没达到能使电池工作的足够高的电导率；而当环境温度升高到 60℃并恒温 12h 后，此时的电解质已经完全变成无色透明的液体，电导率也有较大的提高，能够满足电池正常工作的要求，此时的电池可以正常工作；然后把环境温度又降低到 40℃，并恒温 12h，此时由于电解质的过冷性质，我们可以看到在此温度下的电解质还保持有点透明的黏稠态，应处于过冷态，而此时的电解质仍然具有较高的电导率，电池仍然可以工作，但是相比于 60℃时，此时的电池具有较大的极化；最后我们把环境温度降低到 25℃恒温 12h，电解质又变成了白色的固体，此时的电池不能正常工作。根据此电解质的这一特性，我们设想可以将它应用于锂离子储能电池中，在低温下储存能量，自放电率低；稍微升温使电池达到电解质相变点以上，此时电池就能正常工作。

3.2.3.2　塑晶电解质[67]

在不同的温度和压强条件下晶态物体可以处于气相、液相和固相 3 种不同的状态。在低温下，物质形成位置有序和排列取向有序的固相。当温度升高时，分子晶体中的位置有序或取向有序就开始被破坏，可能出现下列情形[68]：①原子同时失去位置有序和取向有序，形成各向同性液体；②物质保持着固态，但是分子取向有序先遭到破坏，形成塑性晶体，到更高温度才破坏位置有序而形成各向同性的液体；③物质先失去位置有序形成液体，但是保留着取向有序，形成液晶，到更高温度才破坏取向有序而形成各向同性的液体。在第二种情形中，材料从固态晶体转变为塑性晶体的温度称为固固相变温度（temperature of transition，T_t），而从塑性晶体转变为液态的温度称为熔点（melting temperature，T_m）。

塑性晶体是 Timmermanns 在 1935 年首先发现的，其结构要素是晶体中分子规则地占据三维立体晶格中特定的位置点，塑性晶体依然有 X 射线衍射图谱，但只有少量的衍射峰[69]。同时塑晶还可以进行相对自由的热旋转运动，存在取向无序。此外塑晶从固态变为液态的熵变化均较小[<5cal/(deg·mol)]，这是由于塑性晶体取向无序的特性已经比较接近液体性质。如图 3-11 显示了塑晶材料在固相与塑晶相固相变时，其差示扫描量热曲线的特点，升温时首先出现一个吸热峰。然后随着温度进一步升高，在发生固液相变时，会出现另一个吸热峰[70]。

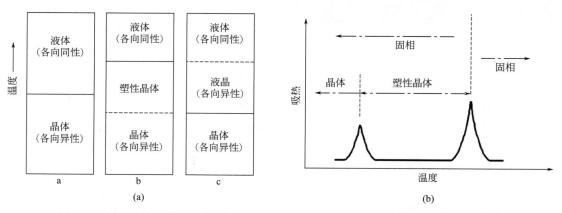

图 3-11　晶态物质的转变示意图（a）和塑晶材料的差示扫描量热（DSC）图（b）[70]
（当温度升高，塑晶材料从固相进入塑晶相时会出现一个吸热峰；从固态变成液态时又出现另一个吸热峰）

Angell 在 1986 年首先报道了在常温下锂离子电导率约为 10^{-5} S/cm 的塑晶材料[71]，之后塑晶材料的离子导电行为得到了进一步研究[72-75]。1999 年澳大利亚 Monash 大学 Forsyth 教授等报道了一种由甲基烷基吡咯烷（N-methyl-N-alkylpyrrolidinium）与三氟甲基磺酸亚胺［bis-(trifluoromethanesulfonyl) amide］形成的盐类物质中加入三氟甲基磺酸亚胺锂制备的塑晶材料，室温电导率可达 2×10^{-3} S/cm[76]。目前研究的塑晶电解质包括非离子型化合物塑晶和离子型化合物塑晶材料。非离子型化合物塑晶材料是在常温有机塑晶材料中加入锂盐制备得到，例如在丁二腈（succinonitrile，其塑晶相温度范围是 $-35\sim62$℃）中加入 5%（质量分数）的三氟甲基磺酸亚胺锂［lithium bis (trifluoromethanesulfonyl) amide, LiTFSA］或 2%（质量分数）的双腈胺锂（lithium dicyanamide, LiDCA），其室温电导率分别达到 3.4×10^{-4} S/cm 和 5.7×10^{-6} S/cm[77-79]。离子形塑晶材料的阳离子为含氮杂环的有机化合物，阴离子为聚阴离子，例如三氟甲基磺酸亚胺、六氟磷酸根和四氟硼酸根。虽然以塑晶材料为基质的锂离子固体电解质的离子电导率

达到 10^{-3} S/cm，目前还需要解决界面电阻、电化学稳定性问题。

3.2.3.3 Li-B-H 型硼酸盐固体电解质的相与相变

锂离子二次电池具有良好的综合性能，其在大型储能、电动汽车、消费电子等行业中的广泛应用，对安全性能、能量密度以及功率密度提出更高的挑战。与传统的有机电解质相比，无机固体电解质具有更优良的安全性能，且其中一些体系能够匹配高电压正极或金属锂负极，并提高电池的能量密度。然而，功率密度仍然是限制全固态电池应用的瓶颈之一。一种有效的解决方案，可以选用快离子导体，该类材料的离子电导率比普通离子导体高多个数量级。按照缺陷类型和相变机制，可以将固体电解质分为三类：①点缺陷。载流子缺陷符合阿伦尼乌斯关系的简单热激活行为，例如 Li_3N、LISICON 和 $\beta\text{-}Al_2O_3$。②一级相变。初始晶格载流子非迁移部分的重排，与骨架离子共同形成一个新的无序相，常见于传统快离子导体，包括银离子导体和氧离子导体；锂离子无机固体电解质中的 NASI-CON 结构 [例如 $LiZr_2(PO_4)_3$][81]，反钙钛矿结构锂离子导体[82,83] 和 $LiBH_4$[84]（以及与卤化锂复合[85]中）出现类似行为。③二级相变。玻璃转变，表现为非阿伦尼乌斯线性关系，源于随温度升高宽载流子从有序渐变至无序，在锂离子无机固体电解质中常见于硅酸盐、磷酸盐、硫化物玻璃等，$Li_{0.33}La_{0.56}TiO_3$ 和石榴石结构在足够宽的温度范围也有此趋势。不同材料体系电导率-温度倒数如图 3-12(a) 所示[86]；虽然从图上看，大部分晶态锂离子无机固体电解质符合阿伦尼乌斯方程，然而，在足够宽的范围内是否可以发生相变却并不十分明确。与离子传导态有关的热力学效应和熔化的热力学效应有一定关联[87]。固体电解质材料中的点缺陷、一级相变和二级相变，显示了材料的差异。文献[88]中介绍了朗道提出的现代相变理论，然而其中序参量的本质并不清楚；为了解释临界行为而构造的准化学模型或晶格气体模型均有其局限性。理论上虽然存在困难，然而可以从实验结

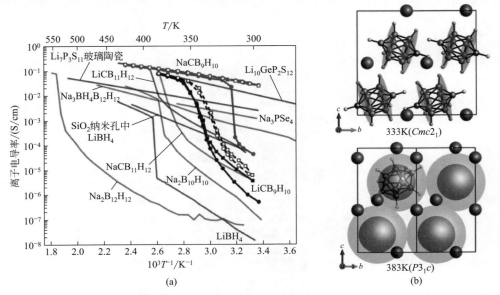

图 3-12 硼酸盐固体电解质的阿伦尼乌斯曲线（a）和低温（上）、高温（下）时 $LiCB_9H_{10}$ 的结构（b）[86]

果来研究临界点的物理性质行为,包括:①离子电导率和活化能的突变。固体电解质中最关心的性质,其表征手段见第5章中,电导率和扩散系数的测量。②比热容[82,83,89-91]。比热容反常通常为相变的关键标识。③与声子模相关的行为。声学性质体现了快离子导体相变的一些本质特征,包括存在一个与应变分量具有相同对称性的序参量时,导致此应变的弹性常数软化[92]和超声衰减反常[92-94]。研究临界现象,有利于确定序参量的特征,并研究离子-离子相互作用。

近期发现的新型高离子电导率硼酸盐固体电解质,具有大的、有取向性的 $B_{12}H_{12}^{2-}$,$B_{10}H_{10}^{2-}$,$CB_{11}H_{12}^{-}$ 和 $CB_9H_{10}^{-}$ 笼子[86,95-97]。这类电解质随温度的升高,会发生有序-无序相变,如图3-12(a)所示,阿伦尼乌斯曲线可以看到电导率随温度的陡增。此时,高温无序相为阳离子的扩散提供了宽敞的通道,具有高的离子电导率[98]。这里以 $LiCB_9H_{10}$ 为例,因为其具有最高的锂离子电导率(383K时为0.1S/cm,活化能为0.29eV)[86]。其中,用高价C取代B所引入的多余的空位,以及单价的阴离子笼子带来的弱的相互作用,均有利于电导率的提高。此外,温度升高的时候,C掺杂的聚硼烷笼子受温度驱动转动起来,并导致笼子的取向性消失[86,98],如图3-12(b)所示,这种现象被证明有助于促进电导率[99,100]。

3.2.3.4 NASICON型固体电解质的相与相变

目前较为常见的锂离子快离子导体中,NASICON(Na super ionic CONductor)型固体电解质具有高的离子电导率、可接受的晶界阻抗以及低廉的原料成本,受到广泛关注。1976年,Goodenough 和 Hong 等提出 NASICON 结构钠离子化合物,其具有三维离子通道和较高的钠离子电导率[101-103];1977年发现该结构的锂离子化合物同样具有较高锂离子电导率[104],其通式写作 $LiM^1M^2P_3O_{12}$,由 MO_6 八面体和 PO_4 四面体组成共价的 $[M^1M^2P_3O_{12}]^-$ 骨架(如图3-13所示)、导电锂离子在骨架中热力学统计分布并存在多种占位类型。

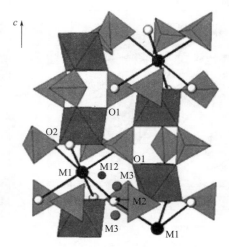

图3-13 NASICON结构中,锂离子可能占据的
M1,M2,M3以及M12位[105]

NASICON 结构中锂离子可占据位置远多于锂离子实际数目，Li 离子优先占据 M1 位。NASICON 在不同温度下会有不同的结构相，如图 3-14 所示。多数情况下，NASICON 结构是菱方结构，如图 3-14(c) 所示，称为"理想的 NASICON"结构；事实上，此时该结构为高温对称相，通常仅在掺杂条件下可以获得。由于该无序相，锂离子在不同位置都会有较高概率的占据，锂的占据位可以连成锂离子通道，因此该相为快离子导电相。经验可得，大多现有的快离子导电材料均有锂离子无序占据的现象，且高温无序相电导率更高。不同的掺杂取代、合成条件、环境温度条件的共同作用可能会导致晶格扭曲，产生具有不同的晶体结构的相。下面以 $LiZr_2(PO_4)_3$ 为例，因为这种材料常常表现为复杂的多晶结构。$LiZr_2(PO_4)_3$ 在室温时为三斜结构；而除去锂离子外其他离子为单斜结构；这是由于锂离子并非在 M1 的中心位置，其位置如图 3-14(a) 所示[106]，这个相称为 α′ 相。由于此时锂离子在一半的 M1 位中心对称分布，大量减少了可供跳跃的位置[107]，因而具有极低的电导率（10^{-9}S/cm）[108]。30～60℃的时候会发生第一次相变（此相变温度与制备样品的烧结温度有关[109,110]），成为菱方结构（$R\bar{3}m$），称为 α 相，如图 3-14(b) 所示[107]。在相变温度下，α↔α′ 相变为可逆的锂离子占据方式的变化，其中 α′→α 方向相变为有序-无序相变，无序相时锂离子可以在 M1 多面体内超过 6 个位置随机分布。锂离子这种高度无序的四配位结构，辅以大量可以供锂离子跃迁的空位，导致此相中电导率提高。高于 300℃时，伴随着骨架结构的变化，普通导电相 α 相转变为快离子导电相 β 相；β 相为正交晶系，如图 3-14(c) 所示[111]。经过 α→β 相变，电导率可以提高一个数量级，活化能从 0.76eV 突变到 0.44eV，如图 3-14(d) 所示[108]。另外，在略低于 300℃ 时另有一个单斜晶系[81]，空间群为 $P2_1/n$，称为 β′ 晶系[106]。

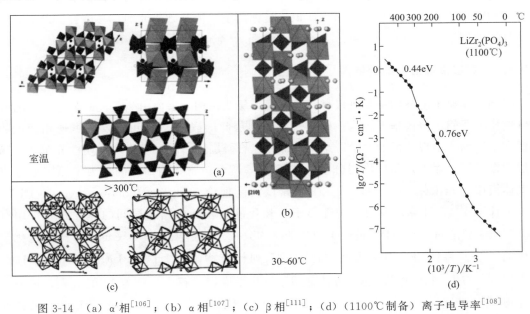

图 3-14 (a) α′ 相[106]；(b) α 相[107]；(c) β 相[111]；(d) (1100℃制备) 离子电导率[108]

注意到，NASICON 材料的相变不仅来源于锂离子在不同温度下的概率占据，也来源于其他离子多面体骨架的扭曲。这是因为 NASICON 型固体电解质具有热力学（thermodynamic) 柔韧的骨架结构[112]，如图 3-15 所示，对于 $AB^1B^2(PO_4)_3$ 而言，$B^{1(2)}O_6$ 八面体和 PO_4 四面体通过共享顶点的方式连接，因此具有一定的转动能力，且转动伴随着多面体扭曲。该特性提供了更高的掺杂容忍度和固溶体范围[113]，并使通过掺杂优化得到高电导率锂离子导体成为可能，其中 $Li_{1.3}Al_{0.3}Ti_{1.7}(PO_4)_3$ 具有最高的电导率，室温可达 $3\times10^{-3}\,S/cm$[114]。注意到，这种结构灵活性可以导致更低的热膨胀系数[115]，有希望在锂离子全电池应用中获得更良好的机械稳定性，防止由于工作产热引起形变而导致的断路和失效。

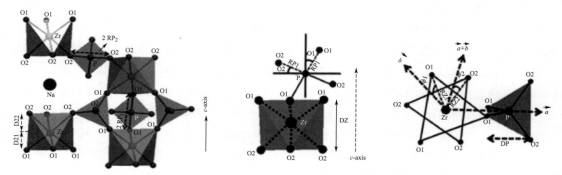

图 3-15 ZrO_6 八面体和 PO_4 四面体中键长与键角的变化关系[112]

3.2.4 电极材料脱嵌锂过程中的相变

锂离子电池的充放电过程伴随着电极材料中锂离子的嵌入和脱出，往往引起电极材料晶体结构的显著变化，对相变的深入认识，是锂离子电池基础研究的核心内容。以下通过几个具体的例子来说明。

3.2.4.1 石墨嵌锂形成的阶结构

石墨是锂离子电池的主要负极材料。石墨为层状结构，层内由 sp^2 轨道杂化形成的共价键连接，层间由范德华力结合。锂离子在石墨层间的嵌入-脱出可以发生可逆相变，随着锂含量增加，出现特征的"阶"结构，即锂在石墨层间沿着 c 轴出现周期性分布，每隔几层占据一层，而不是随机均匀分散，如图 3-16 所示。阶结构形成的热力学原因通过点阵气体模型（lattice gas model）[116,117]、密度泛函理论（density function theory，DFT）进行了计算[118]，计算认为与层间锂离子的长程静电相互作用及层间的弹性相互作用有关。实验主要通过原位 XRD 进行了初步的结构分析[119]。Woo 和 D. Guerard 等通过化学合成的方法，将锂插入石墨层间，得到一系列化合物，如 LiC_{24}、LiC_{18}、LiC_{12}、LiC_6 等[120,121]。Dahn 利用原位 XRD 方法证明了电化学嵌锂过程中，随着 Li 的嵌入，锂嵌入化合物（graphite intercalation compound，GIC）按照 1 阶、4 阶、3 阶、2L 阶（稀释的 2 阶）、2 阶和 1 阶的顺序发生相变，如图 3-16(a) 所示[122]。图 3-16(b) 标明不同阶的嵌锂结构示意图，以及阶与嵌锂电位的关系[123]。图 3-16(c) 表示随着 Li 嵌入引起的层间距的不连续变化[122]。电化学嵌锂导致"阶结构"的不断演化是一个非常有意思的现象，迄今为止，表面包覆固体电解质膜的嵌锂石墨不同阶结构之间的转换机制、阶结构与石墨尺寸、充放电倍率的关系、阶结构形成的动力学/热力学成因等基础科学问题仍有待深入研究。将计算与实验结合，可以用来研究相变机理。例如可以用蒙特卡洛方法，研究高温时 2 阶 LiC_{12} 相变为无序相的临界温度约为 354.6K[124]。再例如从 2 阶 LiC_{12} 到 1 阶 LiC_6 这一过程虽然从热力学可以发生，但是从动力学来说，锂离子倾向于团聚为"阶化合物"而非形成固溶体，在形成 1 阶 LiC_6 之前还要经历多个中间过程[125]。

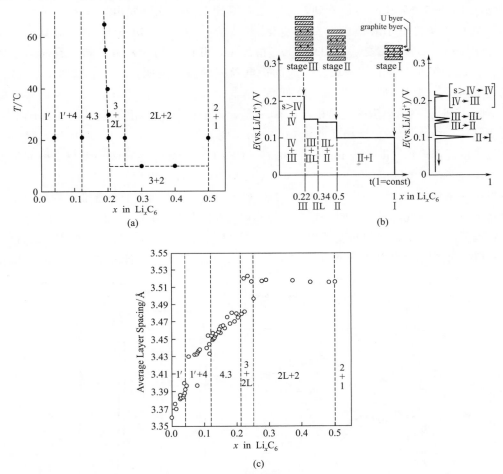

图 3-16 石墨阶结构嵌锂的电化学反应相图（a）[122] 与嵌锂电位（b）[123] 及 Li_xC_6 平均碳层间距随 x 变化（c）[122]

3.2.4.2 层状电极充放电过程的相变

层状结构 $LiCoO_2$ 是消费电子类用高能量密度锂离子电池的主要正极材料。采用传统方法高温合成的 $LiCoO_2$ 为 O3 结构，氧离子为立方密堆积（ABCABC…）。在 Li_xCoO_2 中，锂脱出后，相邻层间氧原子之间的排斥作用导致 c 轴膨胀。$LiCoO_2$ 具有工作电压高、充放电电压平稳、循环性能好的特点，不仅是最早商品化的正极材料，且与其他商品化正极材料磷酸铁锂相比具有更高的能量密度，与尖晶石材料相比不易产生高温的容量衰减，广泛应用于传统 3C 电子产品领域。$LiCoO_2$ 的理论容量为 274mA·h/g，实际容量为 140mA·h/g，这是因为 Li_xCoO_2（$0.5 \leqslant x \leqslant 1$）是当前主要的容量利用范围。为了得到更高的容量，希望可以脱出更多的锂离子。理解层状材料中锂离子脱出时相的变化，有利于从理论上预测新的掺杂方式，提高理论容量。当 $0.4 \leqslant x \leqslant 1$ 时，随着锂的脱出，$LiCoO_2$ 是以固溶体的形式连续相变，另有三个微弱的一级相变[126-128]。其中两个相变发生在 $x=0.5$ 左右，为有序-无序相转变[126,129]，伴随着晶格从六方结构扭曲至单斜结构，如图 3-17 所示。图 3-17(a) 中还可以看到，另一个一级相变发生在锂脱出量 $x=0.75 \sim 0.93$ 范围内，此相变来自电子效应而非结构变化[130,131]。1999 年，Menetrier 等对 $0.75 \leqslant x \leqslant 0.93$ 范围的 Li_xCoO_2 重复了 XRD 测试，指出 I，II 两相均为六方结构，且晶格参数相差不大，此一级相变并非结构相变引起的；电导率的测量证明，$x<0.75$ 时 Li_xCoO_2 表现为金属性质（温度升高，电导率降低）；热电功率（thermoelectronic power）测量表明，$x \leqslant 0.7$ 时材料为典型的金属相，$x \geqslant 0.9$ 是金属相和半导体相的两相区；^7Li MAS NMR 验证了这一绝缘体-金属一级相变源于 Co^{3+}-Co^{4+} 变化的电子效应[130]。2004 年，Marianetti 等利用密度泛函理论（density function theory，DFT），计算了 Li 脱嵌导致空位引起的能带变化，验证了这一 Mott 转变[131]。Van der Ven 利用第一性原理计算了低锂浓度 $0 \leqslant x \leqslant 0.5$ 范围发生一系列相变[132]。图 3-18(a) 为 $LiCoO_2$ 中锂离子与钴离子在密堆积的氧原子中的堆垛排列方式。以此为模型，图 3-18(b) 为 O1、O3 以及 O1 和 O3 的混合结构 H1-3 的自由能随锂离子浓度的变化。相转变依次为：①O1+H1-3 混合相；②H1-3 混合相；③H1-3 混合相+O3。嵌锂过程的放电曲线计算结果如图 3-17(c) 所示，此计算结果与实验结果相一

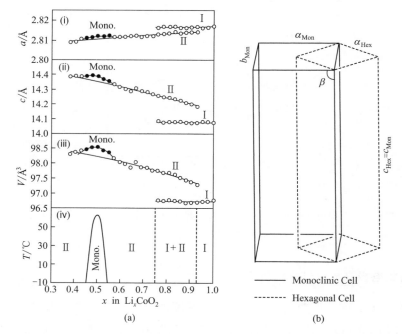

图 3-17　(a) Li_xCoO_2 的 i) a 轴、ii) c 轴的晶格常数以及 iii) 晶胞体积
随锂离子嵌入量 x 的变化；iv) 为综合 i)、ii)、iii) 和电化学结果得到的整体相图；
(b) Li_xCoO_2 单斜（实线）与六方（虚线）晶胞轮廓图[126]

致[126,128,133-135]。可以看到，当锂离子脱出至 $x=0.5$，均为 O3 三方层状结构，继续脱锂至 $x=0.15$，转变为 H1-3 混合结构，此过程预计不可逆[132]。

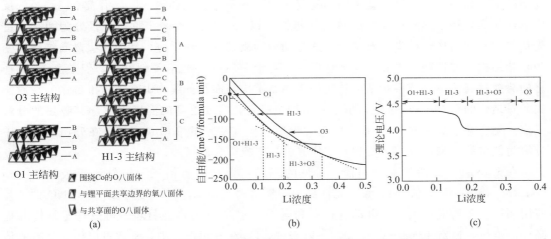

图 3-18　O3，O1 和 H1-3 三种主结构（a）；自由能随锂离子浓度变化的相图（b）；放电曲线计算图（c）[132]

以上所探讨的都是 O3 结构的 LiCoO$_2$，为热力学稳定状态，可以直接合成。相对的，另有 O2 结构的 LiCoO$_2$，为热力学亚稳态，通常只能利用离子交换的方法得到，氧离子堆积周期为 ABAB…[136,137]。与 O3 的结构相变不同，O2 结构所经历的相有：O2，T#2，T#2′，O6[138-140]。过去认为，O2 和 O3 结构一旦生成之后，其相变过程就是分离的，不会相互转变。2012 年，中科院物理所卢侠等[141]通过球差校正 ABF-STEM 技术首次从 O3 结构 LiCoO$_2$ 纳米颗粒的电化学循环中直接观测到 O2 结构的存在。另外，此实验表明当脱锂至 $x=0.5$ 时生成 O1 结构，与之前认为当 $x=0.15$ 时方能生成 O1 结构不同[132]。

金属钴价格昂贵、有毒、对环境污染大，而且钴酸锂电极在实际应用过程中由于仅有 0.5 个锂可以可逆脱嵌，因此过充和过放，均有可能降低性能，甚至引起安全性问题。Ni 与 Co 在元素周期表中属于同一周期，与 Co 相比，Ni 的价格只有 Co 的一半且毒性更小，LiNiO$_2$ 的实际放电容量可以达到 200mA·h/g[142]。2001 年，Ohzuku 和 Makimura 首次提出镍钴锰三元材料[143,144]。但其实在此之前，已有日本电池公司、日本三洋电机公司、韩国 SDI、韩国 LG 化学等开始布局三元材料方面的专利申请，其中核心的专利则为美国 3M 公司拥有，并在 2002 年开始实现 111 型三元材料的商品化应用。为了提高性能、降低成本，高镍掺杂的钴酸锂电极成为研究热点，其中 Ni 为主要活性成分，一般来说，更多的活性元素可以提高电池容量。然而，层状 LiNiO$_2$ 的 Li-Ni 混占位严重、相结构并不稳定，材料一致性和重现性差，这是因为：①Ni^{2+} 与 Li$^+$ 具有相似的离子半径，过多引入 Ni，将会导致 Li-Ni 混占位，破坏了循环稳定的结构，导致容量衰减；②当充电电压 >4.0V 时，反应不可逆、降低循环寿命；③充放电过程中形成的 Ni^{4+} 具有较高的氧化性，会导致电解液分解[145-147]。用第三种稳定元素与 Ni 进行共掺杂，常用于改善其相稳定性与安全性，掺杂元素的种类和组分对于循环性、热稳定性、结构稳定性等具有显著调控作用。从相的热力学稳定性来看，仅有 LiNiO$_2$ 和 LiCoO$_2$ 可以在任意比例形成固溶体，其他掺杂元素均为有限固溶体；而纳米化颗粒可以提高固溶体范围。Wang 等[148]利用第一性原理计算了锂离子电池中几种正极材料（Li_xNiO_2、Li_xCoO_2、$LiMn_2O_4$）在不同温度下的三元相图，并计算了相图中的稳定化合物和不稳定化合物，以及沿着不同路径的反应熵，从而解释了热降解反应的热力学机制。以 Li_xNiO_2 为例，脱锂路径分为两步，首先从层状转变为稳定的尖晶石结构，这是动力学驱动的放热反应；然后从尖晶石结构转变为岩盐结构，并伴随着失氧，为吸热反应。对于 Li_xCoO_2 也有类似性质，但是由于 Co 迁移的势垒过高，不会从层状结构转变为尖晶石结构，而是分解为 $LiCoO_2$ 和 Co_3O_4。对于尖晶石结构的 $LiMn_2O_4$ 来说，由于其结构更稳定，需要更高的温度来驱动其吸热的分解反应。完全充电的 $\lambda\text{-}Mn_2O_4$ 首先由动力学驱动生成稳定的 $\beta\text{-}MnO_2$ 相，在升温的过程中分解为低价的 $\alpha\text{-}Mn_2O_3$ 和 Mn_3O_4，分解热与实验符合良好。

在 Li-Ni-Co-O 的基础上调控，常见的三元正极材料包括三元镍钴锰正极材料 [$Li(Ni_xCo_yMn_z)O_2$，NCM] 和三元镍钴铝正极材料 [$LiNi_{1-x-y}Co_{0.x}Al_yO_2$，NCA]，其应用领域主要为电动工具、电动自行车和电动汽车。Al 为惰性材料，可以稳定结构并提高理论电压，通常 $1-x-y>0.9$。一般来说，Co 在三元正极中，常常可以提高材料的倍率性能，并且一定程度上增强材料的循环性能，那么，是否可以进一步降低 Co 含量，甚至发展无 Co 的层状正极呢？这里，首先要厘清 Ni 在层状正极循环过程中不稳定的原因[149]。Ni 因为有较强的磁矩，三个 Ni^{3+} 如果呈三角形排列，则常常有两种相反的磁矩，形成"磁族挫"，如图 3-19(a) 所示。因为 Li^+ 没有磁矩，其易于与 Ni^{3+} 交换，来缓解"磁族挫"，如图 3-19(b) 所示。当 Ni 进入 Li 层的时候，Li-O 层的厚度减小、阻塞锂离子输运，并导致电池衰减。而 Co^{3+} 阳离子没有磁矩，可以固定过渡金属层，并缓解"磁族挫"，如图 3-19(c) 所示。事实上，从热力学角度，并未有证据直接表明昂贵的 Co 在这个正极材料中是不可取代的，例如在 $Li_xNi_{0.95}M_{0.05}O_2$ 中，M 如果为 Al、Mn、Mg，甚至会比 Co 更容易抑制在 Li_xNiO_2 中观察到的多重相变（如图 3-20 所示），并降低提高温度时充电正极与电解质的反应活性；此时，$LiNi_{0.95}Al_{0.05}O_2$ 的循环容量保持率也不逊色于 15% Co 掺杂的商业 NCA 正极[150]。目前，特斯拉正在与宁德时代商讨无钴动力电池，这是下一代正极材料的发展方向之一。

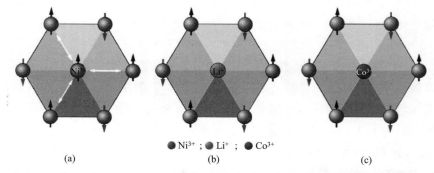

图 3-19 "磁族挫"（a）；Li-Ni 混排缓解"磁族挫"（b）；Co 加入缓解"磁族挫"（c）[149]

三元正极的重要组分 $LiMnO_2$ 具备成本优势，但是 Mn^{3+} 的电子组态为 $t_{2g}^3\text{-}e_g^1$，会导致 Jahn-Teller 畸变效应，体系由层状结构向尖晶石结构转变，造成容量的下降和电压的

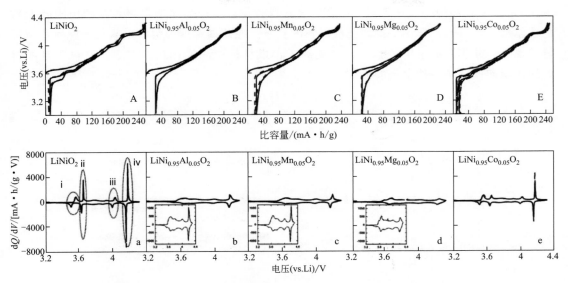

图 3-20 电压（V）与比容量（Q）的关系（A~E）；容量对电压的偏分（dQ/dV）与电压（V）的关系（a~e）[150]

降低。富锂层状 Li_2MnO_3 比层状 $LiMnO_2$ 热力学更稳定，但是在高电压下（4.5V）会产生电化学反应，不仅使电解质更容易分解，还会同时产生不可逆的晶格失氧。可以将具有层状结构的 Li_2MnO_3 [$Li(Li_{1/3}Mn_{2/3})O_2$] 引入层状固溶体中，可以循环至 4.8V，并可以得到超过 250mA·h/g 的可逆容量。此时在过渡金属层中含有不变价的 Li^+，且充放电过程中 Mn^{4+} 不活泼，氧化还原反应主要由 Ni^{2+}/Ni^{4+} 提供，所以这种材料的结构比较稳定。值得注意的是，高容量正极材料的阳离子氧化还原行为，仅能提供约为一半的实际放电容量。目前受到广泛接受的观点认为，对富锂三元材料中阴离子氧化还原的利用，是一种有希望将正极理论容量提高近两倍的新思路[152-155]。这一方法的基础在于可逆、稳定的阴离子氧化还原反应。其机理如图 3-21 所示[151]：在锂基过渡金属氧化物材料中，过渡金属的 d 轨道和氧的 p 轨道重叠，形成的成键（M—O）与反键（M—O）*能带分别具有强的配体特性和金属特性，如图 3-21（a）和（b）所示。成键（M—O）与反键（M—O）*的能量差值成为电荷转移项 Δ，取决于 M 和 O 之间的电负性差异。在高 O/M 比的 Li_2MO_3 中，存在非键的 O_{NB}，作为孤对电子独立存在于（M—O）成键之上，如图 3-21（c）所示。O_{NB} 提供了不同于反键（M—O）*的新的能带，为获取电子提供了可能性。相对的，传统正极 $LiMO_2$ 的氧化还原反应仅发生于反键（M—O）*，或者说 d 能带，此时对应着阳离子的氧化还原，两者区别如图 3-21（d）和（e）所示。接下来我们考虑 d-d 库仑相互作用项 U，这使反键（M—O）*劈裂为空的上哈伯德能带（Upper-Hubbard bands, UHB）和满的下哈伯德能带（Lower-Hubbard bands, LHB），如图 3-20（f）~（h）所示。当 $U \ll \Delta$ 时，如图 3-21（f）所示，电子的得失发生在 LHB，这和传统 $LiMO_2$ 比较类似。当 $U/2 \approx \Delta$ 时，如图 3-21（g）所示，LHB 和 O_{NB} 重叠，因此两者均可以得失电子，使费米能级稳定，反映在晶体结构上会造成 M—O 扭曲以及 O—O 距离缩短[156]；如果该过程可逆，则可以提供额外的容量。当 $U \gg \Delta$ 时，如图 3-21（h）所示，得失电子发生在 LHB 上，晶格不可逆失氧。因此，寻找或者设计 $U/2 \approx \Delta$ 的材料，并利用阴离子氧化还原反应，可以从理论上提高正极材料的容量。注意到，这种阴离子氧化还原导致高容量的现象，并非锂离子电池独有，在钠离子正极材料中同样存在[157,158]。这对于电池材料的认识和理解是一种全新的模式。

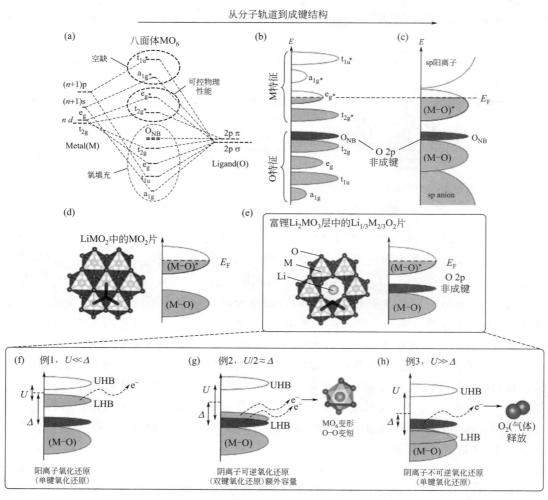

图 3-21　氧的能带结构以及阴离子氧化还原机制[151]

(a) 分子轨道；(b) 和 (c) 在 (a) 的基础上展宽；(d) $LiMO_2$；(e) Li_2MnO_3；

(f)～(h) 为 Li_2MnO_3 的能带结构，用 d-d 库仑斥力项 U 和电荷转移项 Δ 的相互作用分类

安全问题在锂离子电池研究中受到广泛关注，为了得到更高的能量密度，更多的锂离子从正极材料脱出，过渡金属成为高氧化态，这一相变通常伴随着热量和气体，使反应失控造成安全隐患。为了提高锂离子电池的热稳定性，需要理解正极材料不同充电态的热降解反应机制。Yang 等[159]及其合作者通过原位 XRD 与质谱联用研究了（过）充电正极材料 $Li_xNi_{0.8}Co_{0.15}Al_{0.05}O_2$ 在热分解过程中结构变化与气体释放之间的直接联系。在加热过程中，充电态 $Li_xNi_{0.8}Co_{0.15}Al_{0.05}O_2$ 正极材料从层状（$R\text{-}3m$）结构相变为无序排列的

尖晶石结构（$Fd3m$）以及岩盐结构（$Fm3m$）的混合物，热分解过程伴随着 O_2 和 CO_2 的析出，如图 3-22(a) 所示。充电的程度也严重影响加热过程中的结构变化和 O 的析出，如图 3-22(b) 所示。高温下脱锂态的正极材料与电解液之间的化学反应被认为是锂离子电池灾难性事故的主要原因，研究没有电解液和有不同电解液时加热到不同温度正极材料、负极材料结构稳定性及其演变，对于理解电池吸放热反应、安全性研究具有重要的意义。Khalil Amine[160] 及其合作者发展了一种高能量 XRD（high enrgy XRD，HEXRD）技术来研究升温过程中电极材料与添加了不同锂盐的电解质之间的化学反应，寻求更加稳定的匹配。研究发现，采用电解质质量比为 3:7 的 EC/EMC 的 $LiPF_6$ 电解质时，层状结构到尖晶石结构的相变发生在 197℃，结束于 231℃，继续升温则新生成的尖晶石结构分解。相变和尖晶石的分解会放出大量的热，可能会导致锂离子电池的热击穿，引起着火甚至爆炸。因为 HF 会导致脱锂正极的热分解，为了提高相变和热分解的起始温度需要减少电解质中的 HF[161]。用 $Li_2B_{12}F_{12}$ 代替 $LiPF_6$，两个反应的温度提升到 215℃ 和 270℃。相似地，$Li_2B_{12}F_{12}$ 基的电解质对应的反应温度分别为 214℃ 和 256℃。

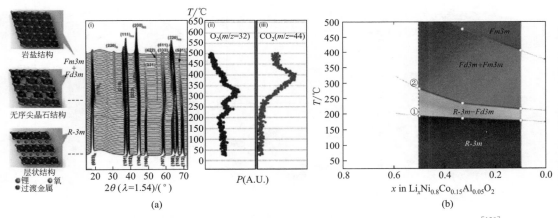

图 3-22　原位 XRD 与质谱联用以及 O_2、CO_2 析出 (a)；热分解过程的相变 (b)[159]

3.2.4.3　橄榄石 $LiFePO_4$ 电极充放电过程的相分离

橄榄石结构的磷酸铁锂（$LiFePO_4$）材料虽然在电压和容量方面不如 $LiCoO_2$，但是具有更高的安全性能，目前在商业上广泛应用。其充放电反应机制模型和发展如图 3-23 所示。1997 年，Goodenough 等提出 $LiFePO_4$。不同于 $LiCoO_2$ 在大范围内都是固溶体的"S"形曲线，其充放电曲线为两相反应的"L"形，同时提出"核壳结构"的两相反应模型[162]。2000 年，Andersson 等利用 XRD 和穆斯堡尔谱证实了锂离子脱嵌过程中形成了 $LiFePO_4$ 和 $FePO_4$ 两相[169]。在此基础上，于 2001 年提出的"半径模型（Radial model）"和"马赛克模型（Mosaic model）"，解释了脱嵌锂中的容量损失，同时指出脱嵌锂在指定颗粒的多个位置同时出现[163]。刘立君利用 X 射线光电子能谱分析（X-ray photoelectron spectroscopy analysis，XPS）测试材料表面的电子结构情况，认为充电过程符合"半径模型（Radial model）"，但是放电过程符合"马赛克模型（Mosaic model）"[164]。2004 年，Srinivasan 等[165] 通过推论两相反应区域之外还存在局部的固溶体区域，并提出

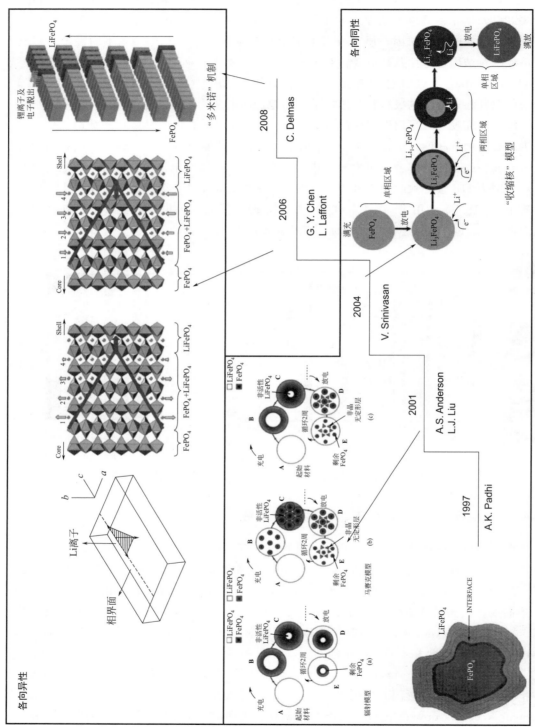

图 3-23 LiFePO$_4$ 电池充放电模型与反应机制[162-168]

"收缩核（Shrinking core）"模型，认为在放电初始整个颗粒形成贫锂固溶体相 Li_yFePO_4；随后，首先在表面形成富锂的固溶体相 $Li_{1-x}FePO_4$，并随着反应进行向内推进，直至全部形成富锂相 $Li_{1-x}FePO_4$；完全放电时，固溶体单相转变为 $LiFePO_4$ 单相（有关固溶体极限的相图在上一节中已经介绍）。以上的两相反应为各向同性。2006 年，Chen 等利用高分辨率透射电镜（high resolution transmission electron microscopy，HRTEM）研究 $LiFePO_4$-$FePO_4$ 片状晶体，观察到 a-c 平面内的无序区域，其平行于 c 轴，且沿着 a 轴移动，锂离子的嵌入脱嵌沿着 b 轴，如图 3-23 左上所示意[166]。同年 Laffont[167] 等通过电子能量损失谱（electron energy loss spectroscopy，EELS）分析单个 Li_xFePO_4 颗粒不同位置 $LiFePO_4$/$FePO_4$ 组分分布，其微观反应机制支持 Chen[166] 的结论，且没有观察到固溶体。但是，由于实验并非原位测量，是否存在暂态的固溶体，并没有得到明确结论。2008 年，Delmas 等提出"多米诺（Domino-cascade）"机制，试图解释在材料电子电导、离子电导均很低的情况下，纳米级别结晶完好的颗粒可以大电流充放电的原因[168]。以上的两相反应都具有方向性。

2011 年，谷林等利用先进的球差校正环形明场扫描透射成像技术（ABF-STEM），直接在正极材料 $LiFePO_4$ 中观察到锂离子，并首次在部分充电的直径为 65nm 的 $LiFePO_4$ 单晶纳米线中观测到了锂离子隔行脱出现象，如图 3-24 所示，其中黄色圈为锂离子存在的位置，橘色圈为锂离子脱出的位置。此现象可以类比于石墨中的"Staging-Ⅱ，2阶"的现象。这一发现与之前提出的各类两相反应模型均不一致，为一单相结构[170]。随后，索鎏敏等在部分脱锂的尺寸为 200nm 的 $Li_{0.90}Nb_{0.02}FePO_4$ 中首次直接观察到了 a 方向存在高度有序的具有 2 阶结构的 $LiFePO_4$/$FePO_4$ 界面，2 阶结构界面厚度大约 2nm。两相边界呈弧形（而不是像多米诺骨牌模型描述的直线形），并沿 c 方向移动。结合前面的实验，把界面的研究扩展到 $LiFePO_4$/Stage-Ⅱ/$FePO_4$ 三相共存[171]。

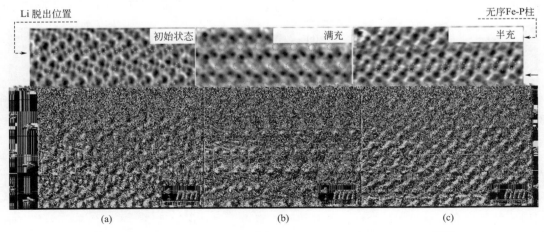

图 3-24　初始状态（a）、完全充电状态（b）和部分充电状态（c）[170]

实验发现单相中间结构的同时，2011 年 Malik 等[172] 通过 Monte Carlo 模拟计算证明了 $LiFePO_4$ 的脱嵌锂过程存在固溶体单相反应路径的可能性，如图 3-25(a)、（b）所示。研究发现，只需施加较低的过电位，就可能实现单相反应而无需发生两相分离，如图 3-25(c) 所示。而平缓的过电势曲线，导致尽管锂离子的迁移率很高[173-176]，但是由固溶体到相分离的弛豫过程十分缓慢[177-180]。受锂离子扩散速率的影响，小颗粒的两相分离发生在不同颗粒，大颗粒的两相分离在同一颗粒中出现。然而，Monte Carlo 模拟得到的几个能量较低的中间态，只有一个与 STEM 观察到的"2 阶"结构相似，其余几个低能量结构并没有实验证据；且该结果没有阐明阶结构出现的内在机理。2012 年，孙洋等通过密度泛函计算进一步研究了 $LiFePO_4$ 中 2 阶结构的形成机制[181]，如图 3-26 所示。计算结果表明，2 阶结构是由锂离子传输动力学导致的热力学亚稳态结构。锂离子之间除了直接静电相互作用，还存在着借助于 Fe^{2+}/Fe^{3+} 氧化还原电对的间接相互作用，因此锂离子传输动力学条件的限制使得充电时 $LiFePO_4$ 只能采取隔层脱锂而不是更为直观的顺序脱锂。隔层脱锂将产生"2 阶"结构，然而热力学能量最低原理却支持两相分离反应的发生。动力学与热力学条件的相互竞争导致单个 $LiFePO_4$ 颗粒脱锂的中间过程是三相共存的：整个颗粒主要由 $LiFePO_4$ 与 $FePO_4$（或富锂相与贫锂相）组成，而两相之间存在少量的由动力学限制导致的"2 阶"结构。需要指出的是这项研究中没有考虑尺寸效应，而小尺寸颗粒的高比表面积可能改变反应条件，特别是热力学条件，这些都有待于进一步的研究去证实。

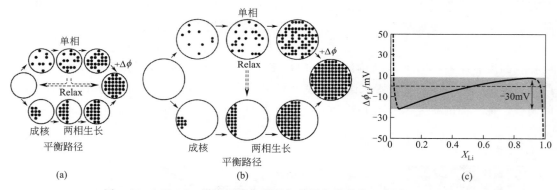

图 3-25 $LiFePO_4$ 脱嵌锂平衡路径与非平衡路径的比较 [(a) (b)]，其中 (a) 为小颗粒，(b) 为稍大的颗粒；(c) $0.05 < x_{Li} < 0.9$ 范围内单颗粒电势[172]

3.2.4.4 尖晶石结构高压电极

尖晶石材料 $LiMn_2O_4$ 与层状材料 $LiCoO_2$ 相比具有更高的工作电压和更低的价格。在 $LiCoO_2$ 中，只有接近一半的锂可以可逆地脱出，这限制了电极材料中锂的利用率。而从比例来看，降低一半的锂含量似乎可以充分利用过渡金属的氧化还原能力、提高锂离子利用比例。尖晶石材料和岩盐结构具有相同的面心立方氧化物晶格。不同的是，尖晶石结构中，阳离子仅仅占据八面体位点的一半，其余占据四面体位点；相对而言，岩盐结构中，阳离子仅占据八面体位点。如图 3-27(a) 所示，在 $LiMn_2O_4$ 中，Li 原子占 $8a$ 位，氧原子占 $32e$ 位，Mn 原子占 $16d$ 位，为 Fd-$3m$ 空间群；注意到，由于其结构中锂离子 $8a$ 位与空位 $16c$ 位连通成三维的锂离子扩散通道，且该通道不会被过渡金属阻塞，因此，

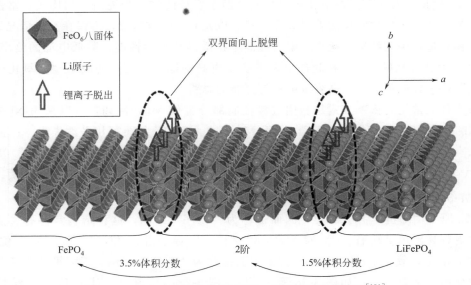

图 3-26 具有 2 阶结构的 LiFePO$_4$ 的双界面脱锂模型[181]

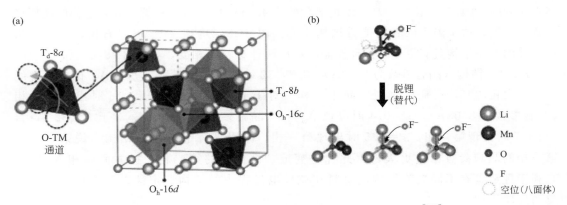

图 3-27 尖晶石结构（a）和 F 掺杂富锂尖晶石结构（b）[182]

尖晶石结构正极材料具有已知最快的倍率性能。在 LiMn$_2$O$_4$ 中，由于 Mn^{3+}：Mn^{4+} = 1：1，且 Mn^{3+} 会导致 Jahn-Teller 效应，因而当放电导致 Mn^{3+} 比例大于 50% 时发生畸变，循环性变差。当前最常见的改进循环性能的思路是用掺杂的方式减少 Mn^{3+} 的存在[8,183-188]。从富锂三元正极材料得到启发，可以用富锂的方式替代部分 Mn，占据 16c 位置，这种"过化学计量比"的方式可以提高 Mn 分布的无序度、从而抑制尖晶石结构充放电过程中的一级相变。为了补偿多余的 Li 所减少的电荷，可以用部分 F 替代 O，例如两种基于无序尖晶石结构的富锂正极材料，Li$_{1.68}$Mn$_{1.6}$O$_{3.7}$F$_{0.3}$ 和 Li$_{1.68}$Mn$_{1.6}$O$_{3.4}$F$_{0.6}$，如图 3-27(b) 所示；这里，虽然过量的 Li 部分占据了原来空着的 16c 的位置，但是当过渡层 Li/Mn 保持低比例时，局域的短程序依旧维持了锂离子通道的渗流连通。注意到，随着锂离子占据 16c 的位置，原先占据着锂的 8a 位以及占据 Mn 的 16d 位，占据率均降低，此时实际结构其实介于尖晶石结构和无序的岩盐结构之间[182]。基于尖晶石的富锂含氟正极材料，M—F 低能键的引入增加了局域的富锂结构，增加的 Li—O 键带来更多非杂化的氧 2p 轨道，其阴离子的氧化还原性能有助于容量的进一步提高。相比较，富锂三元化合物随着锂离子扩散常伴随着过渡金属迁移以及 O—O 键形成，不仅在充电情况下不稳定，在循环过程中其结构演化也会阻碍锂离子扩散的动力学性能、降低电池的倍率性能。富锂含 F 尖晶石材料，可以在超高倍率下实现更高的能量密度，挑战了"O 氧化还原本质缓慢"这一普遍观点，为高容量电极打开新的思路[189]。

另一种更常见的掺杂方式是利用过渡金属部分替代 Mn，例如，可以采用 Ni 掺杂形成 $LiNi_{0.5}Mn_{1.5}O_4$，充电到 4.5V，源于 Mn^{3+} 的 4.1V 平台确实几乎消失[190]。基于 Ni 可以从 +2 价氧化到 +4 价，Dahn 研究组把 $LiNi_{0.5}Mn_{1.5}O_4$ 的充电范围提高到 5.0V，发现 4.7V 有容量，成为新的高电压正极材料[191]。$LiNi_{0.5}Mn_{1.5}O_4$ 有两种晶体结构，空间群为 Fd-$3m$ 和 $P4_332$[192]。对于 Fd-$3m$ 结构，Li 原子占 $8a$ 位，氧原子占 $32e$ 位，Ni 原子和 Mn 原子"随机"占 $16d$ 位，因而被称为"无序"结构，这种结构常伴有岩盐杂质相，并更容易出现氧缺陷[190]。对于 $P4_332$ 结构，Li 原子占 $8c$ 位，Ni 原子占 $4a$ 位，Mn 原子占 $12d$ 位，氧原子占 $8c$ 和 $24e$ 位，因而被称为"有序"结构。王丽平等制备并测试了两种结构的化学性能，如图 3-28 所示[190]。图 3-28(a) 图中，对于 Fd-$3m$ 结构的样品，由于存在 Mn^{3+} 而在 4.0V 有少量的容量；图 3-28(b) 的微分曲线中，Fd-$3m$ 的氧化峰在 4.69V 和 4.75V 位置，$P4_332$ 的氧化峰在 4.74V 和 4.77V 位置，并且它们的还原峰位置也不同，这是因为两者的能带结构不同。图 3-28(c) 和 (d) 可以看出，空间群为 $P4_332$ 的有序结构具有更差的循环性能。对比两种材料的相变，如图 3-28(e) 和 (f) 所示，Fd-$3m$ 结构（晶胞参数为 8.12Å）和 $P4_332$ 结构（晶胞参数为 8.16Å）的样品都经历了 3 次相转变，只是转变的区间不一样。对于 Fd-$3m$ 结构，$x=0.6$ 时出现第二相（晶胞参数为 8.08Å），$x=0.4$ 时出现第三相（晶胞参数为 8.00Å）。对于 $P4_332$ 结构，$x=0.7$ 时出现第二相，$x=0.5$ 时出现第三相。原因是对于 $P4_332$ 结构来说第三相和第一相的晶胞参数相差更大，因此优先变成"缓冲"的第二相。这方面的相变研究，有利于理解具有不同缺陷结构的材料在充放电过程中的结构稳定性以及充放电曲线对应的反应机理。

3.2.4.5 MnO 的相转变反应

2000 年 Poizot 等发现了过渡金属氧化物的可逆储锂现象[193]；之后的研究发现大部分过渡金属硫化物、氟化物、氮化物、磷化物都有类似现象[194-203]。此现象可以表示为：$MX+Li \longleftrightarrow LiX+M(X=O，S，F，N，P，Se；M=Ti，V，Cr，Mn，Fe，Co，Ni，Cu，Zn，Sn，Sb，Ru，Mo)$。这一类反应称为相转变反应（conversion reaction）。过渡金属化合物材料的储锂容量可以达到 400~1100mA·h/g，化学惰性 Li_2O 和 LiF 在纳米复合结构中可以发生可逆的生成和分解，拓宽了人们对纳米尺寸效应对电极材料反应活性的影响认识，为寻找新的储能材料提供了理论指导，该方向吸引了大量的基础研究。热力学计算表明，MnO 负极材料具有较低的理论电位[201]。早期的 Poizot 小组报道，MnO 负极材料只有在恒电压间隙滴定（PITT）模式下用 1/300C 的极小电流才显示容量[195]。禹习谦等发现薄膜 MnO 材料表现出很高的体积容量（>3484mA·h/cm^3 0.125C），较优的倍

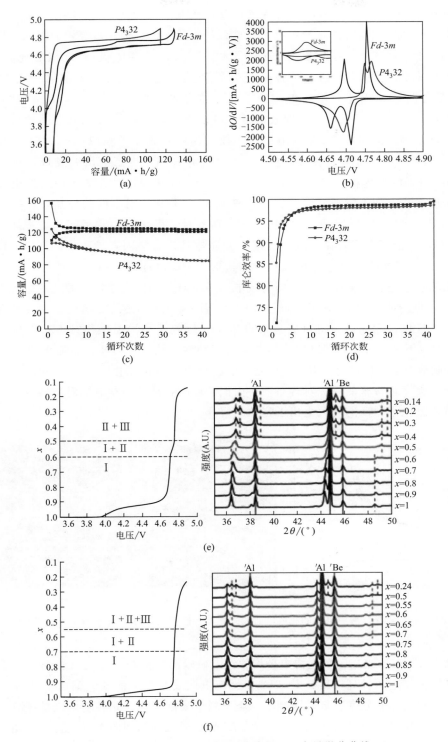

图 3-28　Fd-$3m$ 和 $P4_332$ 的充放电曲线（a），容量微分曲线（b），循环性能（c）以及库仑效率曲线（d）；Fd-$3m$ 结构（e）和 $P4_332$ 结构的充电曲线以及 XRD 图（f）[190]

率性能（在 5C 倍率下，55%容量保持率）以及较低的极化[204]。这些结果表明 MnO 负极材料动力学性能并没有像之前报道那样差，基于此原因，钟开富等较为深入地研究了反应机制和动力学特征[205,206]。图 3-29(a) 给出了 MnO 薄膜电极前 25 周的充放电曲线，从首周放电曲线上可以看到，样品存在类似于其他过渡金属氧化物的相转变反应平台[193,195,201,207-212]，且平台电位在 0.2V；从充放电曲线可以粗略看出，MnO 比其他过渡金属氧化物[211,213-217]极化小、首周效率低。可以看到，经过一次充放电循环后 Mn 的极化减小。这是因为循环后的 MnO 由初始的较大晶粒变成纳米颗粒[205]，其中 MnO 的晶粒尺寸大约在 5nm，如图 3-29(b) 所示[218]。钟开富和崔忠慧等研究发现，对于不同尺寸、结构的 MnO 粉体材料以及不同温度制备的 MnO 薄膜，其充放电曲线的电位差在不同电流密度下具有如下特征[206,219]：从图 3-30 可以看到，对不同的样品，在一定的电流密度下都能满足欧姆定律，同时当电流减小到零时，极化并不趋近于零，截距为 0.6~0.7V 左右，外推零电流时存在较大的电位差显然是热力学因素导致的。近来，Doe 等解释了 FeF_3 在 1/200C 充放电电位极化的来源时[220]，通过第一性原理计算表明，锂的嵌入和脱出的反应路径可能不一致，即充放电选择不对称的反应路径，相变的微观过程不是可逆、对称的。对于 MnO 来说，通过 TEM、XRD、EXAFS、XAS 的表征并没有发现中间相存在的证据，我们认为也可能是来自于尺寸效应及结构无序化对吉布斯自由能的影响[206]，但也不完全排除局部出现短程的非稳态结构导致的吉布斯自由能出现差异。Maier 等通过经验热力学分析讨论了 RuO_2 在充放电过程中的电位变化[221,222]，分析了尺寸效应中表面能、无序化带来的熵变对吉布斯自由能的影响，从热力学方面对相转变反应后放电平台比首周放电平台高的现象提出了定性的解释。

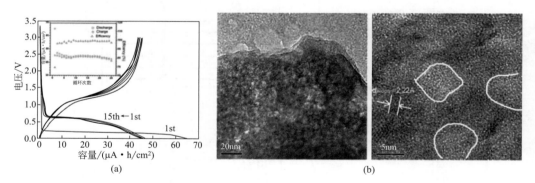

图 3-29 (a) MnO 薄膜电极的充放电曲线及循环特性；(b) MnO 电极充电后的 TEM[218]

虽然 MnO 在过渡金属氧化物中平均脱锂电位较高，但相转变类材料与石墨及硅负极相比，平均工作电压较高（1.0~1.8V），且充电曲线与放电曲线之间的极化较大（通常大于 1V），首次库仑效率较低，限制了该类材料的实际应用。相转变反应是锂离子电池中常见的反应，材料一般经历从高结晶度到低结晶度或无序结构的不可逆转变。相转变反应是否均存在热力学引起的充放电曲线之间的电位差，以及引起电位差的具体的热力学原因和定量分析仍需要大量的研究。而这些问题的理解和研究或许有助于解决该类材料的极化问题，并对解决金属空气电池（对应 MO、M_2O、M_2O_2、MO_2 的分解）的极化也有一

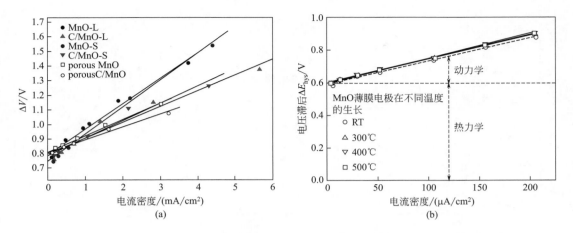

图 3-30　不同 MnO 粉体材料电位极化与电流密度的关系（a）[206]；不同温度下
制备的 MnO 薄膜电极电位极化与电流密度的关系（b）[219]

定的指导意义，因此值得深入研究，但该类材料在室温下不适合作为锂离子电池的负极材料。

3.3　相　　图

3.3.1　相图与相律

相图也叫相平衡图，又叫状态图，它描述处于平衡状态下物质的组分、物相和外界条件的相互关系，是一个物质体系相平衡图示的总称[223]。利用相图可以得到以下信息：体系是由哪些平衡相构成；平衡相的百分比与相对量；条件变化后，相变过程的进行倾向等。其中图 3-31(a) 是典型的温-压相图，绿色线代表凝固点随压力的变化；蓝色线表示沸点随压力的变化；红色线表示升华点随压力的变化。图 3-31(b) 表示温度-压力-比容三维典型相图。

吉布斯相律是相图的基本原理[224,225]，它是由约西亚·威拉德·吉布斯于1875年首先提出的，其表达式为：$F=C-P+n$。式中，P 为相态数目；F 为系统的自由度，是指在不产生新相的前提下，可以在一定范围内变动的独立变量的数目；C 为系统的独立组元数，也就是在一定的温度、压力等条件下的平衡体系中，在数量上可以连续、无限制变动而不受其他物质数量制约的物质的数目；n 为外界因素。一般来说 $n=2$，代表压力和温度；而在固体系统中，由于蒸气压影响很小，通常取 $n=1$；同时，外界因素也包括电场、磁场的影响系统平衡的因素。除了相律以外，相图还要遵循其他的基本原理和规则[226]，包括连续性原理、相应原理、化学变化的统一性原理、相区接触规则、溶解度规则、相线交点规则等。具体见参考文献 [1] 及其引用文献。传统绘制相图是基于实验的方法，例

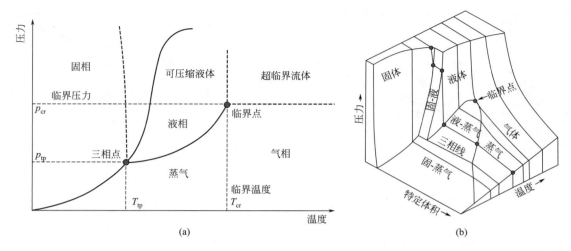

图 3-31 固定量的纯净物质的典型相图
(a) p-T 相图；(b) p-V-T 3D 相图（维基百科）

如热分析、差热分析、化学分析、电子探针微量分析、显微方法、X 射线结构分析、膨胀计方法、电导率方法、磁性分析法、能谱分析等[227]。

无论是合成制备还是电化学反应，需要知道电极材料的相与材料中锂含量之间的关系，即能够绘制出电极材料的锂含量相图。由于锂电池中电极材料的相主要与锂/锂空位在晶体结构中的占位有关，而锂在固体中室温下扩散系数或离子电导率有可能较低，因此在实验中电极材料是否处于完全的平衡态往往较难判断。因此，通过原位实验技术获得的相变信息，可能存在一定的误差和一定范围的不确定性。即便如此，相图对于理解电极材料的结构稳定性、电位曲线的变化以及电极过程动力学等基础科学问题仍然是必不可少的研究内容，以下举几个典型材料的相图。

3.3.2 典型材料的相图

3.3.2.1 锂硅合金相图

电极材料在连续的脱嵌锂过程中发生一系列的相变是一种常见的锂存储模式，以合金型反应为例。在图 3-32(a) Li-Si 合金相图中，我们看到，在室温下，Li 与 Si 可以形成 $Li_{22}Si_5$、$Li_{13}Si_4$、Li_7Si_3 和 $Li_{12}Si_7$ 等合金相，简单计算表明 Li 嵌入 Si 中形成这些合金相的理论容量分别为 4200mA·h/g，3100mA·h/g，2200mA·h/g，1600mA·h/g，这对早期理解硅负极嵌锂机制具有重要的参考价值。通过第一性原理密度泛函计算，不同 Li-Si 合金的生成能可以计算如图 3-32(b) 所示，在系列 Li-Si 合金中，锂原子转移了 0.73 个电子，与合金相和周围 Si 原子数无关[228]。需要注意的是，上述合金相图是在高温下获得的。在室温下，锂嵌入硅引起结构的无定形化，但短程序仍然保持；NMR 的研究表明，反应仍然按照系列合金化反应发生[229,230]。较为复杂的是在嵌锂深度达到 $Li_{15}Si_4$ 时，出现了从无定形向晶态转化的现象[231]。

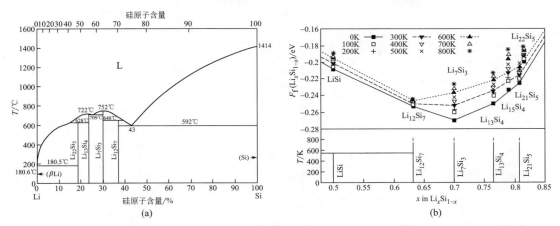

图 3-32　Li-Si 合金相图（a）[232]；第一性原理计算的 Li-Si 合金生成能与组成关系图（b）[228]

3.3.2.2　LiFePO$_4$ 相图

1997 年 Padhi 等报道了 LiFePO$_4$ 作为锂离子电池正极材料的研究[162]。2005 年 Delacourt 等测定了不同比例 LiFePO$_4$ 与 FePO$_4$ 的变温 XRD 发现[177]，如图 3-33(a) 和 (b) 所示，在 450℃时，存在 Li$_x$FePO$_4$ 固溶体。当降温时，Li$_x$FePO$_4$ 分解成两种非橄榄石型的混合物，它们的组成取决于温度与 x 的初始值。当温度低于 140℃±20℃时，两相体系变为更加复杂的体系，其中，LiFePO$_4$ 与 FePO$_4$ 跟另外两种橄榄石型的化合物 Li$_{x1}$FePO$_4$ 与 Li$_{x2}$FePO$_4$ 共存于这一体系，将这一体系的混合物置于室温下老化，四相体系将逐渐转变为 LiFePO$_4$ 与 FePO$_4$ 两相体系。2006 年 Dodd 进一步测定了不同锂含量及不同温度下的 LiFePO$_4$-FePO$_4$ 的 XRD 图谱[180]，确定了其相图，如图 3-33(c) 所示，室温下 LiFePO$_4$ 和 FePO$_4$ 是独立的两相，当温度升至 300℃以上时，其转变为无序的固溶体相，在 200℃附近，其表现为共析体系，其转变焓约为 700J/mol。从报道的相图来看，体相材料 LiFePO$_4$ 与 FePO$_4$ 的固溶体范围较窄，两端均小于 5%（原子百分数）。MIT 的 Y. M. Chiang 等报道认为[233]，材料尺寸较小后，固溶体范围发生显著变化，掺杂也因此更容易发生，即使是 12%（原子百分数）的 Zr 也可以掺杂进入 LiFePO$_4$，如图 3-33(d) 所示。

3.3.2.3　LiMn$_2$O$_4$ 组成相图

锂锰氧化合物是锂离子电池广泛研究的正极材料，该体系的三元组成相图以及扩展的相图参见图 3-34。上述在 1994～1997 年发表的组成相图包括了所有迄今为止的在研究的 Li-Mn-O 材料。通过解读上述相图，有助于准确理解嵌脱锂可能引起的相变，以及可能发展的锰酸锂正极材料。

3.3.2.4　Li-Ni-Co-Mn-O 多元组成相图

富锂正极材料是当前锂离子电池广泛研究的高容量正极材料，该体系的三元组成相图以及扩展的相图参见图 3-35，由 Thackeray 在 2004～2005 年发表，张联齐等也丰富了该相图的内容。虽然上述相图并非直接通过热学或其他手段测量获得，但图 3-35 包括了迄今为止多

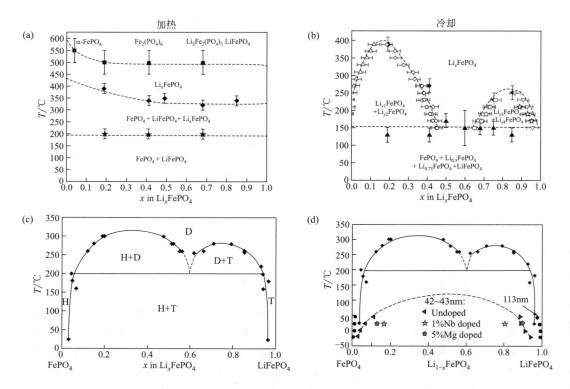

图 3-33 升温（a）和降温（b）时，Li_xFePO_4（$0 \leqslant x \leqslant 1$）的温度-相态分布曲线[177]；$LiFePO_4$（Triphylite，T）与 $FePO_4$（Heterosite，H）的相图（c）[180]；纳米尺寸 $LiFePO_4$ 材料相图（d）[233]

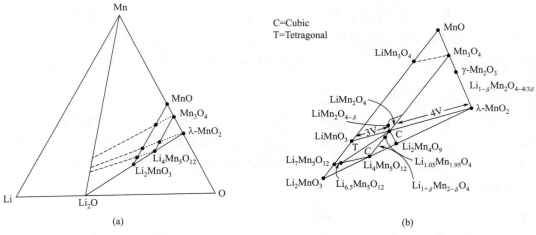

图 3-34 Li-Mn-O 三元组成相图（a）以及拓展的 MnO-Li_2MnO_3-MnO_2 三元组成相图（b）[234]

数在研究的富锂正极材料，此类相图的绘制对于设计该系列材料的组成具有非常重要的指导意义。在锂离子电池的研究中，通常关注组分相图（三元或四元组成相图）。基于前期对于 Li_2MnO_3 电化学反应活性的研究[236]，高容量的富锂三元材料受到关注，图 3-36 是其三元相图[236]。其中，Li_2MnO_3 和 $LiMO_2$ 为固溶体正极材料二元组分的两端，而另一个三角顶点 MO_2 用来体现电池充放电的电化学过程。以 $xLi_2MnO_3·(1-x)LiMO_2$，$x=0.3$ 为例，其充放电电压在小于 4.4V 的时候，其反应沿着相图中的虚线箭头进行。然而，当充电电压大于 4.4V 的时候，表现出额外的容量。从相图来看，高电压下的氧化还原反应不能由 M 和 Mn 贡献。也就是说，这部分额外的容量不来源于过渡金属阳离子，脱锂以 Li_2O 的方式进行。该工作为基于阴离子氧化还原的高容量富锂三元正极材料奠定了理论基础。

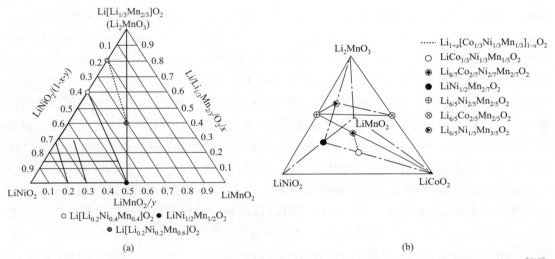

图 3-35　$LiNiO_2$-$LiMnO_2$-Li_2MnO_3 相图（a）；$LiNiO_2$-$LiCoO_2$-$LiMnO_2$-Li_2MnO_3 四面体相图（b）[235]

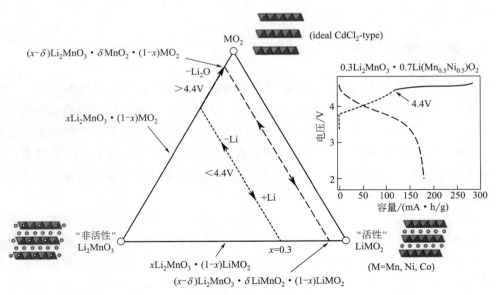

图 3-36　$xLi_2MnO_3·(1-x)LiMO_2$ 电极的组分相图以及电化学反应路径[237]

3.4 相图的计算

3.4.1 CALPHAD

传统绘制相图采用实验的方法，存在一些局限性。例如，测得一个相图需要庞大的数据；对于高温、低温条件很难控制实验条件；高温等条件导致物质挥发，组分变化；化学分析时偏析造成的成分不均匀；测量也可能会干扰平衡条件[238]。将热力学理论与计算机手段相结合，产生了新的交叉分支学科——CALPHAD（CALculation of PHAse Diagram, Computer Coupling of Phase Diagrams and Thermochemistry）。这是一种比较便捷地绘制相图的手段。1908 年，van Laar 将吉布斯自由能的概念应用于相平衡中，进行了数学分析；1957 年 Meijering 尝试计算真实三元系统的完整相图；Kaufman 在 1959 年用晶格稳定性和基于原子相互作用参数的正规溶液模型开始发展 Ti 和 Zr 二元系统相图的数据库，1967 年引入理想溶液模型研究非磁性过渡金属，又在 1968 年引入更多相互作用参数实现亚正规溶液模型；1970 年，Kaufman 和 Bernstein 出版了第一本关于相图定量计算的教科书，标志着 CALPHAD 学科的建立；1973 年举办了第一届 CALPHAD 会议；1975 年 CALPHAD 有限公司在马萨诸塞州成立；1977 年，美国国家标准局组织了关于合金和陶瓷相图研究的研讨会，同年，CALPHAD 期刊由派格蒙出版公司出版，当年发表的文献包括 Pettifor 的第一性原理和 Lukas 等的热统计学对相图计算的应用；1981 年 Sundman 和 Agren 发展了 Hillert 和 Staffansson 的亚点阵模型，并发展出 Thermo-Calc 软件；1981 年以后，更多的模型和计算软件提了出来，并建立了相图热力学数据库，具体见文献[239]。

3.4.2 第一性原理计算

第一性原理可以广泛地用来预测锂离子电池的多种性质，包括嵌锂电位[240]、锂-空位的有序性[132]、锂离子扩散[173,241]、过渡金属有序性[242]、电子结构和热力学化学势平衡[243]等。将这种方法应用于锂离子电池相图与相变的研究，已经取得一些成果，包括新的电极材料的筛选和开发[175,244-247]、多相材料的结构优化[175]、对于不同实验条件或使用条件下的相稳定性和热稳定性[248,249]、反应路径和分解路径[250]等，其中一些应用在前面章节已经提到。

Doe 等[250]利用第一性原理计算 Li-Fe-F 相图。图 3-37(a) 为其三元相图，并标识出不同三角相区的化合物相对于锂金属的电位；Doe 等对相图中超过 100 个点进行计算，探索不同化学计量比化合物的可能结构以及相稳定性，并预测了一些新型稳定化合物，其中实心圆点为稳定相化合物，"×"是亚稳相化合物。图 3-37(b) 为 FeF_3 的扭曲的钙钛矿结构（ABO_3，A 为空位），图 3-37(c) 为 FeF_2 的金红石结构。图 3-37(d) 为图 3-37(a)

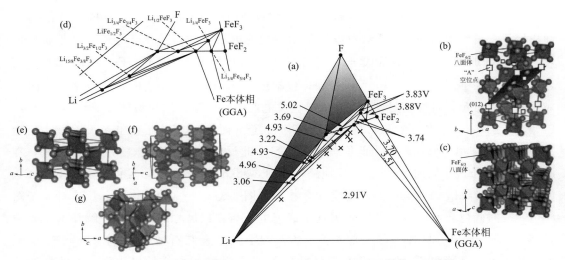

图 3-37 Li-Fe-F 三元相图（a）；FeF_3（b），FeF_2（c）结构；相稳定点所代表的化合物（d）；$Li_{3/4}Fe_{3/4}F_3$（e），$Li_{3/2}Fe_{1/2}F_3$（f），$Li_{15/8}Fe_{3/8}F_3$（g）结构[220]

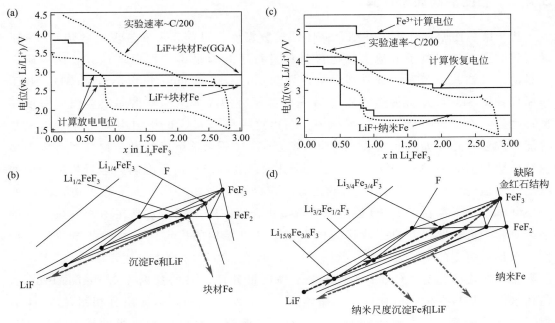

图 3-38 Fe 块材（a）(b)、Fe 纳米颗粒（c）(d) 的电压曲线和反应路径[220]

的局部相图，并明确标出每一个稳定相的具体化合物，其中红色为钙钛矿结构，蓝色为金红石结构，绿色为钛铁矿结构，灰色为尖晶石结构，LiF 为岩盐结构。图 3-37（e）~（g）分别为 $Li_{3/4}Fe_{3/4}F_3$（Li-Fe 离子交换的金红石结构）、$Li_{3/2}Fe_{1/2}F_3$（刚玉/钛铁矿结构）、$Li_{15/8}Fe_{3/8}F_3$（尖晶石结构）的结构。图 3-38(a) 和（b）为 Fe 块材的电压曲线和反应路径，由于反应有金属 Fe 的析出，且 Fe 的氟化物和 Fe 金属电子态变化很大（分别为局域的 3d 电子和非局域的电子）。将金属 Fe 的化学势利用反应 $FeF_2 + 2Li \longrightarrow Fe + 2LiF$ 进行简单修正，得到如图 3-38（a）红色虚线所示的理论电压曲线，其化学势依然高于实验所得。用 1nm 的球形 Fe 颗粒替代块材，其能量增加，因而 FeF_2 优先于 Fe 生成；假设反应过程中生成除 FeF_2 以外的 Fe^{2+} 亚稳相，包括 $LiFeF_3$ 和 $Li_{2/3}Fe_{3/4}F_3$，反应路径如图 3-38（d）绿线所示，对应于图 3-38（c）的绿色放电曲线。以互扩散模型解释亚稳相的形成：Fe 比 Li 更难扩散，为了补偿嵌锂过程中 Li 扩散至（而 Fe 来不及扩散出）化合物中所增加的电荷，Fe 离子倾向于成为低价态（+2 价）。此反应动力学上更容易发生。同理，在脱锂过程中 Fe 离子倾向于 +3 价，反应路径如图 3-38（d）蓝线所示，且得到 $Li_{3/4}Fe_{3/4}F_3$、$Li_{3/2}Fe_{1/2}F_3$、$Li_{15/8}Fe_{3/8}F_3$ 三种稳定化合物〔如图 3-37（e）~（g）所示〕；反应对应于图 3-38（d）的蓝色充电曲线。此结果展现了嵌锂电位与颗粒尺寸的关系，并解释了实验中"电势滞后"现象。另外，在研究过程中还发现了其中金红石结构的 $Li_{3/4}Fe_{3/4}F_3$ 脱锂后容易得到"三金红石（trirutile）"结构的 FeF_3，此结构的能量仅比稳定态的钙钛矿结构略高，且纳米尺度样品的 XRD 谱图与金红石结构的 FeF_2 类似，这解释了实验中在完全脱锂后会形成类似 FeF_2 结构的原因。

3.4.3 相图的高通量计算

基于第一性原理计算，并结合晶体结构数据库，目前，由麻省理工学院 G. Ceder 教授研究组对于锂离子电池大量电极材料进行了高通量第一性原理计算，这种研究材料的方式被其命名为"材料基因组计划"（material genome initiative，MGI）。其工作原理如图 3-39（a）所示[251,252]。首先根据已有的晶体结构数据库建立概率密度模型；之后，在任意一个 A-B-C 系统中，利用这一模型来认识相图中某一化学组成是否能形成复合物〔如图 3-39（a）中红点所示〕；这一化学组成的复合物可能对应着多种结构〔如图 3-39（a）中 1，2，3 三种结构〕；最后，利用更准确的能量模型，例如密度泛函理论（DFT），选出复合物的最稳定结构。图 3-39（b）是利用此方法预测的 A-B-O 体系中不存在于 ICSD 数据库的新型化合物。图 3-39（c）是在图 3-39（a）基础上，对元素进行取代并根据取代的概率模型计算结构的稳定性，并判断电荷的平衡性，来预测新型的取代复合物。图 3-39（d）标识了离子对的相关性程度 g_{ab}，即一对元素相互取代的趋势。Ceder 团队建立了高通量计算结果的数据库 Materials Project，包括软件（材料探索，电池探索，晶体工具包，结构预测，多元组分相图，Pourbaix 相图，反应计算，热力学数据等 15 种功能），超级计算机以及功能材料的筛选。截至 2020 年 2 月 21 日，已计算 124515 种无机材料，52827 个能带结构，4401 种嵌入电池以及 16128 种相转变反应电池材料体系。数据均可以通过网站 http://www.materialsproject.org/查询，且数据库在不断更新。在国内，北京大学深圳研究生院的潘锋教授，也建立了材料基因数据库 http//www.pkusam.com/database.html，并不断完善内容；该数据库的计算利用了基于图形处理器（graphics processing unit，GPU）的 PWmat，其相对于中央处理器（central processing unit，CPU）具有更优良的并行计算性能，在计算电子能带结构等方面具有优势。

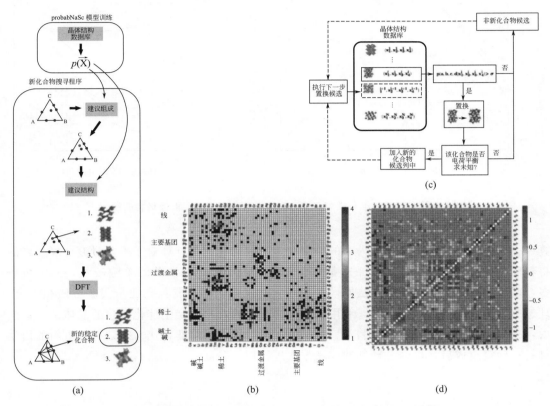

图 3-39　复合物开发程序的数据挖掘机制（a）；预测的新型 A-B-O 复合物（b）[251]；
取代的概率模型预测新型复合物（c）；离子对相关性程度 g_{ab}（d）[252]

利用数据库中材料的能量以及多元组分相图，可以进一步挖掘锂离子电池中无机固体电解质的电化学窗口以及对于正负极的稳定性[253-255]。此外，还可以进一步预测界面修饰，例如正极包覆材料，文献［256］对不同正极的稳定包覆相进行了计算、筛选和预测。对于电化学窗口的计算，可以通过调整研究对象中锂的比例，并在相图数据库中搜寻，是否存在多种化合物的混合，其能量低于改变了锂含量的研究对象。当反应能量低于 0meV/原子的时候，则反应发生，所对应的化学势超出了稳定的电化学窗口，如图 3-40(a) 所示。对于电解质/正极材料之间稳定性的计算，可以通过调整正极材料的脱锂量，并搜寻是否存在多种化合物的混合，其组分总和等于脱锂正极材料与研究对象所构成的组分总和，使得混合物的能量低于电解质材料与电极材料的能量和。当反应发生时，此时该脱锂量的正极材料与固体电解质研究对象之间不稳定，如图 3-40(b) 所示。

相对于实验，高通量计算基础上形成的相图有利于判断实验相图的完整性，挖掘在一类材料体系中迄今未知的化合物。但目前由于第一性原理计算方法对强关联材料体系的计

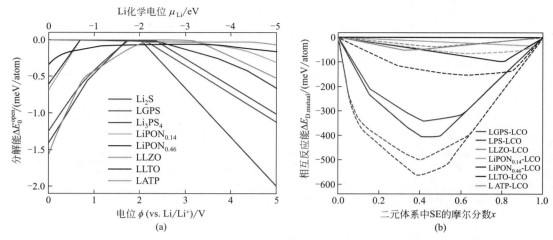

图 3-40 固体电解质材料的电化学窗口（a），以及相对 LiCoO$_2$ 正极材料的稳定性（b）[254]

算还存在较大的困难，计算时人为的因素较多，因此高通量计算对于不含过渡金属元素的材料体系更加准确。此外，对于非整比化合物，由于计算时涉及的单胞较大、原子数较多，基于第一性原理的计算方法同样存在较大的困难，仍需要与高通量实验方法相结合。理想的处理过程可能是：首先，利用第一性原理计算，来确定主体化合物的晶体结构；然后，在该成分附近通过高通量制备、测试、表征，进而了解非整比、掺杂等产生新的衍生材料；确定组成后，再结合精细的计算方法进行理论验证。

3.5 相、相变与相图的实验研究方法

由于锂离子电池材料的多样性、复杂性、非均相、有序与无序结构共存并相互转化，相结构的确认往往并不是一件容易的事情。同其他材料领域的相与相变研究一样，几乎所有在结构研究中应用到的研究方法在锂离子电池相变研究中都有应用，同时由于锂离子电池中嵌脱锂对结构的影响，在锂离子电池研究中，还发展了一些在惰性气氛下的原位相变研究实验技术。图 3-41 列举了具有不同空间分辨率的表征方法，相与相变的研究可以综合采用这些方法来研究。

X 射线衍射（XRD）是检验物相最直接也最常用的方法，其被广泛用于相分析及相变的研究。基于同步辐射的 XRD 及原位 XRD，由于其高通量和高时间分辨的特点，被广泛应用于锂离子电池电极材料的结构和相变原位研究。由于轻元素对于 X 射线散射截面较小，XRD 衍射图谱获得的信息主要来自于较重的元素，因此对材料中 Li 的精确占位信息的确定存在一定的困难。锂的中子散射截面较大，常被用来研究含锂材料的精细结构，如 Li 及过渡金属的混占位、锂在固体中的扩散通道等。

一些相结构是依靠衍射方法难以区分的，如此前提到的 LiNi$_{0.5}$Mn$_{1.5}$O$_4$ 材料有两种

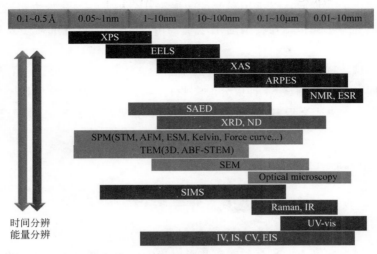

图 3-41 具有不同空间分辨率的表征方法

空间群结构 Fd-$3m$ 和 $P4_332$。XRD 很难区分这两种结构，但是由于两种样品的结构对称性不同，导致其拉曼散射光谱不一致，故可用拉曼光谱来对其进行区分。

衍射和散射技术获得的是平均结构或者在材料中具有较高结晶度与较高散射截面的结构单元的信息，一般来说，当衍射能够识别新相的形成时，新相至少占有 2%～5% 的体积比例。能够在新相形成的早期观察，有利于准确理解电极材料的相变机理及其对电化学性能的影响。透射电子显微镜（transmission electron microscope，TEM）是高空间分辨下观察材料微观结构的有力手段之一，适合于局部结构的分析。而且透射电镜技术还可以区分表面和近表面结构。

在锂离子电池材料中，锂在结构中的占位是材料研究关注的重点，但一般的透射电镜技术空间分辨率高于 0.1nm，无法观察锂。在 20 世纪末首台球差校正器问世之后，球差校正电子显微技术得到了广泛应用，使得在亚埃尺度的高空间分辨率下直接获取结构信息成为可能。通过环形明场像技术，Li 也能在电镜下直接成像，这使得许多涉及局部结构的问题能够获得原子尺度的了解[170]。

如前所述，高通量计算可以获得非强关联体系、整比化合物的相图，对于复杂体系、强关联材料、非整比化合物，需要借助于实验的方法，目前逐渐兴起的组合化学方法，对于材料相图的研究，能够大大提高效率。代表性的组合化学方法包括"组合材料芯片"技术（combinatorial material processing）全称为"材料组合生长与表征平台技术"，由项晓东博士和 Peter Schultz 博士于 1993～2000 年期间在美国劳伦斯伯克利国家实验室（Lawrence Berkeley National Laboratory）共同发展和完善。该技术可在一块基片上同时集成生长和表征上千种不同组分的新材料，并专门开发了自动化的微区表征平台，实现高通量的材料筛选及"材料相图"的系统描绘工作。该技术的核心特征包括：①通过精妙的材料芯片设计，在同一基片上、在相同或不同的热力学参数下，一次性地合成覆盖大范围组分或整个二元/三元"相图"组分的样品，降低由多次实验所带来的数据离散性；②利用自动化、高速度、综合性微区表征平台的强大分析测试能力，对样品进行多参数微区分析表征，建立完整周密的数据库；③采用科学的数据统计分析方法，找出大量数据中隐含的趋势与规律。类似的方法还包括在美国通用电气公司（GE）的赵继成博士发展的"扩散多元节"方法等。

在锂电池领域，Dahn 小组较早通过高通量制备方法，系统研究了 Sn、Si 基合金负极材料，发展了同时制备并测试 64 种成分的实验技术，这种测试技术加快了电极材料筛选的速度[257]。此类方法适合于多组元的合金类负极，也被用于 $LiFe_{1-x}Mn_xPO_4$ 固溶体正极材料研究[258]。

3.6 本章结语

研究电池材料在制备中的相与相变，获得准确完备的相图对于开发、设计、优化电池材料具有十分重要的意义。锂离子电池材料合成制备中的相变与相图方面的知识虽然还远未完善，但正日渐积累。获得除了组成相图之外，包括温压相图及各类物理性质的相图，相信也为期不远，这些努力对于材料基础科学的发展以及储能材料的开发一定具有积极的推动作用。在锂离子电池相图相变研究中同样重要的，是在充放电过程中由于嵌脱锂引起的材料的相变与相图。高通量的计算、制备、表征技术已经开始在锂离子电池材料研究中获得应用，普及后将会大大加快新相材料开发及相图绘制的速度。高空间分辨率、时间分辨率、能量分辨率的技术也被广泛地应用于电池材料的体相、表面相、界面相结构、组成及其演化的研究。锂离子电池中的相变与相图方面的知识虽然还远未完善，但正日渐积累。从原子尺度到宏观尺度了解相变过程、相变的驱动力、相的稳定性、相变对电化学性能的影响；获得除了组成相图之外，包括温压相图及各类物理性质的相图。这些努力对于材料基础科学的发展以及储能材料的开发一定具有积极的推动作用。

本章需要进一步思考的基础科学问题：

1. 锂电池中的正负极材料的相及相变主要与锂和锂空位的排布有关，形成有序结构的主要原因有哪些？
2. 哪些正负极材料在嵌脱锂时形成的相边界是相干边界（coherent）？哪些无法形成？能形成相干边界的材料体系是否显示了优异的循环性？锂电池中都有哪些种类的相边界？
3. $Li_4Ti_5O_{12}$ 与 $Li_7Ti_5O_{12}$ 为何能形成原子尺度的相边界？$LiFePO_4$ 和 $FePO_4$ 的相边界具有哪些结构特征？
4. 相变发生的主要原因是受电子输运驱动还是离子输运驱动还是都有关系？
5. 在设计和寻找电极材料时，倾向于选择哪些种类的相变特征的材料？哪些材料相变时会发生较大体积形变？为什么？
6. 在 8 种可逆储锂机制中，每一种都具备哪些相变特征？
7. 堆垛层错在充放电过程中发生的变化属于一级相变吗？

8. 一直以来快离子导体的高温无序相具备较高的离子电导率，能否通过掺杂、淬火等手段将高温快离子导电相稳定到电池的工作温度？
9. 层状化合物中嵌脱锂导致的 Staging（阶）结构是有意思的现象，为什么会形成高阶结构？其热力学本质是什么？从 3 阶结构过渡到 2 阶结构的微观机制是什么？
10. 相转变过程中，嵌锂和脱锂过程是否会选择不同的反应路径？反应路径中产物的结构差异性与嵌锂和脱锂方向有关的热力学原因是什么？

参考文献

[1] 高健. 锂离子导体的基础研究 [D]. 北京：中国科学院，2015.
[2] 高健，吕迎春，李泓. 锂电池基础科学问题（Ⅲ）——相图与相变 [J]. 储能科学与技术，2013，2（3）：250-266.
[3] 高健，吕迎春，李泓. 锂电池基础科学问题（Ⅳ）——相图与相变（2）[J]. 储能科学与技术，2013，2（4）：383-401.
[4] Grey C P, Dupre N. NMR studies of cathode materials for lithium-ion rechargeable batteries [J]. Chem Rev, 2004, 104（10）：4493-4512.
[5] Sugiyama J, Mukai K, Nozaki H, et al. Antiferromagnetic spin structure and lithium ion diffusion in Li_2MnO_3 probed by μ^+ SR [J]. Phys Rev B, 2013, 87（2）：024409.
[6] Yao J, Konstantinov K, Wang G X, et al. Electrochemical and magnetic characterization of $LiFePO_4$ and $Li_{0.95}Mg_{0.05}FePO_4$ cathode materials [J]. Journal of Solid State Electrochemistry, 2007, 11（2）：177-185.
[7] Ramzan M, Ahuja R. Ferromagnetism in the potential cathode material $LiNaFePO_4F$ [J]. Epl-Europhys Lett, 2009, 87（1）：
[8] Li G H, Ikuta H, Uchida T, et al. The spinel phases $LiM_yMn_{2-y}O_4$ （M＝Co, Cr, Ni）as the cathode for rechargeable lithium batteries [J]. J Electrochem Soc, 1996, 143（1）：178-182.
[9] Ivancevic V G. Complex nonlinearity：Chaos, phase transitions, topology change and path integrals（understanding complex systems）[J]. Berlin：Springer, 2008.
[10] 于渌，郝柏林. 相变和临界现象 [M]. 北京：科学出版社，1984.
[11] 程晓农，戴起勋，邵红红. 材料固态相变与扩散 [M]. 北京：化学工业出版社，2006.
[12] Charpin J, Botter F, Briec M, et al. Investigation of γ lithium aluminate as tritium breeding material for a fusion reactor blanket [J]. Fusion Engineering and Design, 1989, 8（4）：07-13.
[13] Okuno K, Kudo H. Tritium diffusivity in lithium-based ceramic breeders irradiated with neutrons [J]. Fusion Engineering and Design, 1989（8）355-358.
[14] Johnson C E, Hollenberg G W, Roux N, et al. Current experimental activities for solid breeder development [J]. Fusion Engineering and Design, 1989（8）145-153.
[15] Nishikawa M, Baba A, Kawamura Y. Tritium inventory in a $LiAlO_2$ blanket [J]. Journal of nuclear materials, 1997, 246（1）：1-8.
[16] Luo T, Oda T, Oya Y, et al. Existence states of deuterium irradiated into $LiAlO_2$ [J]. Journal of Nuclear Materials, 2008, 372（1）：53-58.
[17] Patil K Y, Yoon S P, Han J, et al. The effect of lithium addition on aluminum-reinforced alpha-$LiAlO_2$ matrices for molten carbonate fuel cells [J]. Int J Hydrogen Energ, 2011, 36（10）：6237-6247.
[18] Terada S, Nagashima I, Higaki K, et al. Stability of $LiAlO_2$ as electrolyte matrix for molten carbonate fuel cells [J]. J Power Sources, 1998, 75（2）：223-229.
[19] Zhou L, Lin H X, Yi B L. Sintering behavior of porous alpha-lithium aluminate matrices in molten carbonate fuel cells at high temperature [J]. J Power Sources, 2007, 164（1）：24-32.
[20] Choi H J, Lee J J, Hyun S H, et al. Phase and Microstructural Stability of Electrolyte Matrix Materials for Molten Carbonate Fuel Cells [J]. Fuel Cells, 2010, 10（4）：613-618.
[21] Appetecchi G B, Dautzenberg G, Scrosati B. A new class of advanced polymer electrolytes and their relevance in plastic-like, rechargeable lithium batteries [J]. J Electrochem Soc, 1996, 143（1）：6-12.
[22] Ulihin A S, Slobodyuk A B, Uvarov N F, et al. Conductivity and NMR study of composite solid electrolytes

based on lithium perchlorate [J]. Solid State Ionics, 2008, 179 (27-32): 1740-1744.

[23] Capuano F, Croce F, Scrosati B. Composite polymer electrolytes [J]. J Electrochem Soc, 1991, 138 (7): 1918-1922.

[24] Wen Z Y, Itoh T, Ikeda M, et al. Characterization of composite electrolytes based on a hyper-branched polymer [J]. J Power Sources, 2000, 90 (1): 20-26.

[25] Ceder G, Chiang Y M, Sadoway D R, et al. Identification of cathode materials for lithium batteries guided by first-principles calculations [J]. Nature, 1998, 392 (6677): 694-696.

[26] Cao H, Xia B J, Zhang Y, et al. $LiAlO_2$-coated $LiCoO_2$ as cathode material for lithium ion batteries [J]. Solid State Ionics, 2005, 176 (9-10): 911-914.

[27] Kim H S, Kim Y, Kim S I, et al. Enhanced electrochemical properties of $LiNi_{1/3}Co_{1/3}Mn_{1/3}O_2$ cathode material by coating with $LiAlO_2$ nanoparticles [J]. J Power Sources, 2006, 161 (1): 623-627.

[28] Cheng F Q, Xin Y L, Huang Y Y, et al. Enhanced electrochemical performances of 5 V spinel $LiMn_{1.58}Ni_{0.42}O_4$ cathode materials by coating with $LiAlO_2$ [J]. J Power Sources, 2013, 239: 181-188.

[29] Park J S, Meng X B, Elam J W, et al. Ultrathin lithium-ion conducting coatings for increased interfacial stability in high voltage lithium-ion batteries [J]. Chem Mater, 2014, 26 (10): 3128-3134.

[30] Chen Z, Qin Y, Amine K, et al. Role of surface coating on cathode materials for lithium-ion batteries [J]. J Mater Chem, 2010, 20 (36): 7606-7612.

[31] Lei L, He D, Zou Y, et al. Phase transitions of $LiAlO_2$ at high pressure and high temperature [J]. Journal of Solid State Chemistry, 2008, 181 (8): 1810-1815.

[32] Marezio M, Remeika J P. High-pressure synthesis and crystal structure of alpha-$LiAlO_2$ [J]. J Chem Phys, 1966, 44 (8): 3143.

[33] Li X, Kobayashi T, Zhang F, et al. A new high-pressure phase of $LiAlO_2$ [J]. Journal of Solid State Chemistry, 2004, 177 (6): 1939-1943.

[34] Marezio M, Remeika J P. Polymorphism of $LiMO_2$ compounds and high-pressure single-crystal synthesis of $LiBO_2$ [J]. The Journal of Chemical Physics, 1966, 44 (9): 3348-3353.

[35] Chang C-H, Margrave J L. High-pressure-high-temperature syntheses. Ⅲ. Direct syntheses of new high-pressure forms of lithium aluminum oxide and lithium gallium oxide and polymorphism in $LiMO_2$ compounds (M=boron, aluminum, gallium) [J]. Journal of the American Chemical Society, 1968, 90 (8): 2020-2022.

[36] Danek V, Tarniowy M, Suski L. Kinetics of the α → γ phase transformation in $LiAlO_2$ under various atmospheres within the 1073 — 1173 K temperatures range [J]. Journal of Materials Science, 2004, 39 (7): 2429-2435.

[37] Rasneur B, Charpin J. Chemical properties of lithium ceramics: Reactivity with water and water vapour [J]. Journal of Nuclear Materials, 1988, 155-157: 461-465.

[38] Finn P A. The effects of different environments on the thermal stability of powdered samples of $LiAlO_2$ [J]. J Electrochem Soc, 1980, 127 (1): 236.

[39] Ribeiro R A, Silva G G, Mohallem N D S. The influences of heat treatment on the structural properties of lithium aluminates [J]. J Phys Chem Solids, 2001, 62 (5): 857-864.

[40] Byker H J, Eliezer I, Eliezer N, et al. Calculation of a phase diagram for the lithium oxide-aluminum oxide ($LiO_{0.5}$-$AlO_{1.5}$) system [J]. The Journal of Physical Chemistry, 1979, 83 (18): 2349-2355.

[41] Byker H J, Eliezer I, Eliezer N, et al. Calculation of a phase-diagram for the $LiO_{0.5}$-$AlO_{1.5}$ system [J]. J Phys Chem, 1979, 83 (18): 2349-2355.

[42] 雷力, 贺端威. 高温高压法合成六方相偏铝酸锂晶体 [J]. 中国科技论文在线精品论文, 2012, 5 (3): 227-231.

[43] Kang Y C, Park S B, Kwon S W. Preparation of submicron size gamma lithium aluminate particles from the mixture of alumina sol and lithium salt by ultrasonic spray pyrolysis [J]. J Colloid Interf Sci, 1996, 182 (1): 59-62.

[44] Wen Z Y, Gu Z H, Xu X H, et al. Research on the preparation, electrical and mechanical properties of gamma-$LiAlO_2$ ceramics [J]. Journal of Nuclear Materials, 2004, 329: 1283-1286.

[45] Shimura T, Murahashi D, Iwahara H, et al. Lithium ionic conduction in $LiAlO_2$-based oxides at elevated temperatures [J]. Solid State Ionics: Trends in the New Millennium, Proceedings, 2002, 613-620.

[46] Indris S, Heitjans P, Uecker R, et al. Li Ion dynamics in a $LiAlO_2$ single crystal studied by Li-7 NMR spectroscopy and conductivity measurements [J]. J Phys Chem C, 2012, 116 (27): 14243-14247.

[47] Gao J, Shi S, Xiao R, et al. Synthesis and ionic transport mechanisms of α-$LiAlO_2$ [J]. Solid State Ionics,

2016, 286: 122-134.

[48] Lehmann H A, Hesselbarth H. Zur Kenntnis Der Lithiumaluminate . Ⅰ. Über Eine Neue Modifikation Des LiAlO$_2$ [J]. Z Anorg Allg Chem, 1961, 313 (1-2): 117-120.

[49] Choi H J, Lee J J, Hyun S H, et al. Cost-effective synthesis of alpha-LiAlO$_2$ powders for molten carbonate fuel cell matrices [J]. Fuel Cells, 2009, 9 (5): 605-612.

[50] Kharlamova O A, Mitrofanova R P, Isupov V P. Mechanochemical synthesis of fine-particle gamma-LiAlO$_2$ [J]. Inorg Mater, 2007, 43 (6): 645-650.

[51] Isupov V P, Kharlamova O A, Chupakhina L E, et al. Mechanochemical synthesis of gamma-LiAlO$_2$ studied by Li-6 and Al-27 NMR and synchrotron X-Ray diffraction [J]. Inorg Mater, 2011, 47 (7): 763-767.

[52] Isupov V P, Eremina N V. Effect of mechanical activation of Al(OH)$_3$ on its reaction with Li$_2$CO$_3$ [J]. Inorg Mater, 2012, 48 (9): 918-924.

[53] Nakagawa K, Ohzu H, Akasaka Y, et al. Allotropic phase transformation of lithium aluminate in MCFC electrolyte plates [J]. Denki Kagaku Oyobi Kogyo Butsuri Kagaku, 1997, 65 (3): 231-235.

[54] Kinoshita K, Sim J W, Ackerman J P. Preparation and characterization of lithium aluminate [J]. Mater Res Bull, 1978, 13 (5): 445-455.

[55] Kinoshita K, Sim J, Kucera G. Synthesis of fine particle size lithium aluminate for application in molten carbonate fuel cells [J]. Mater Res Bull, 1979, 14 (10): 1357-1368.

[56] Poeppelmeier K R, Kipp D O. Cation Replacement in Alpha-LiAlO$_2$ [J]. Inorganic Chemistry, 1988, 27 (5): 766-767.

[57] Poeppelmeier K R, Kipp D O. Cation Replacement in Alpha-LiAlO$_2$ [J]. Solid State Ionics, 1988, 26 (2): 170.

[58] Poeppelmeier K R, Chiang C K, Kipp D O. Synthesis of High-Surface-Area Alpha-LiAlO$_2$ [J]. Inorganic Chemistry, 1988, 27 (25): 4523-4524.

[59] Takizawa K, Hagiwara A. The transformation of LiAlO$_2$ crystal structure in molten Li/K carbonate [J]. J Power Sources, 2002, 109 (1): 127-135.

[60] Luo C, Martin M. Stability and defect structure of spinels Li$_{1+x}$Mn$_{2-x}$O$_{4-\delta}$: Ⅰ. In situ investigations on the stability field of the spinel phase [J]. Journal of Materials Science, 2007, 42 (6): 1955-1964.

[61] Kelder E, Jak M, Schoonman J, et al. Quality control of Li$_{1+\delta}$Mn$_{2-\delta}$O$_4$ spinels with their impurity phases by Jaeger and Vetter titration [J]. J Power Sources, 1997, 68 (2): 590-592.

[62] Thackeray M, Mansuetto M, Dees D, et al. The thermal stability of lithium-manganese-oxide spinel phases [J]. Mater Res Bull, 1996, 31 (2): 133-140.

[63] Boulineau A, Croguennec L, Delmas C, et al. Thermal stability of Li$_2$MnO$_3$: From localized defects to the spinel phase [J]. Dalton Transactions, 2012, 41 (5): 1574-1581.

[64] 谢斌. 锂离子电池新型电解质材料的研究 [D]. 北京: 中国科学院, 2008.

[65] Liang H, Li H, Wang Z, et al. New binary room-temperature molten salt electrolyte based on urea and LiTFSI [J]. The Journal of Physical Chemistry B, 2001, 105 (41): 9966-9969.

[66] Xie B, Li L, Li H, et al. A preliminary study on a new LiBOB/acetamide solid phase transition electrolyte [J]. Solid State Ionics, 2009, 180 (9): 688-692.

[67] He X M, Pu W H, Wang L, et al. Plastic crystals: An effective ambient temperature all-solid-state electrolyte for lithium batteries [J]. Prog Chem, 2006, 18 (1): 24-29.

[68] Timmermans J. Plastic crystals-a historical review [J]. J Phys Chem Solids, 1961, 18 (1): 1-8.

[69] Post B. The cubic form of carbon tetrachloride [J]. Acta Crystallogr, 1959, 12 (4): 349.

[70] Staveley L A. Phase transitions in plastic crystals [J]. Annu Rev Phys Chem, 1962, 13: 351.

[71] Cooper E I, Angell C A. Ambient-temperature plastic crystal fast ion conductors (plicfics) [J]. Solid State Ionics, 1986, 18 (9): 570-576.

[72] Chandra D, Helms J H, Majumdar A. Ionic conductivity in ordered and disordered phases of plastic crystals [J]. Journal of The Electrochemical Society, 1994, 141 (7): 1921-1927.

[73] Hattori M, Fukada S-I, Nakamura D, et al. Studies of the anisotropic self-diffusion and reorientation of butylammonium cations in the rotator phase of butylammonium chloride using 1H magnetic resonance, electrical conductivity and thermal measurements [J]. Journal of the Chemical Society, Faraday Transactions, 1990, 86 (22): 3777-3783.

[74] Ishida H, Furukawa Y, Kashino S, et al. Phase transitions and ionic motions in solid trimethylethylammonium

[74] iodide studied by ^{1}H and ^{127}I NMR, electrical conductivity, X-ray diffraction, and thermal analysis [J]. Bunsenges Phys Chem, 1996, 100 (4): 433-439.

[75] Tanabe T, Nakamura D, Ikeda R. Novel ionic plastic phase of $[(CH_3)_4N]$SCN obtainable above 455 K studied by proton magnetic resonance, electrical conductivity and thermal measurements [J]. Journal of the Chemical Society, Faraday Transactions, 1991, 87 (7): 987-990.

[76] Macfarlane D R, Meakin P, Sun J, et al. Pyrrolidinium imides: A new family of molten salts and conductive plastic crystal phases [J]. The Journal of Physical Chemistry B, 1999, 103 (20): 4164-4170.

[77] Long S, Macfarlane D R, Forsyth M. Fast ion conduction in molecular plastic crystals [J]. Solid State Ionics, 2003, 161 (1): 105-112.

[78] Long S, Macfarlane D R, Forsyth M. Ionic conduction in doped succinonitrile [J]. Solid State Ionics, 2004, 175 (1): 733-738.

[79] Alarco P-J, Abu-Lebdeh Y, Abouimrane A, et al. The plastic-crystalline phase of succinonitrile as a universal matrix for solid-state ionic conductors [J]. Nat Mater, 2004, 3 (7): 476-481.

[80] Macfarlane D R, Huang J H, Forsyth M. Lithium-doped plastic crystal electrolytes exhibiting fast ion conduction for secondary batteries [J]. Nature, 1999, 402 (6763): 792-794.

[81] Casciola M, Costantino U, Merlini L, et al. Preparation, structural characterization and conductivity of $LiZr_2(PO_4)_3$ [J]. Solid State Ionics, 1988, 26 (3): 229-235.

[82] Zhao Y S, Daemen L L. Superionic conductivity in lithium-rich anti-perovskites [J]. Journal of the American Chemical Society, 2012, 134 (36): 15042-15047.

[83] Braga M H, Ferreira J A, Stockhausen V, et al. Novel Li_3ClO based glasses with superionic properties for lithium batteries [J]. J Mater Chem A, 2014, 2 (15): 5470-5480.

[84] Matsuo M, Nakamori Y, Orimo S-I, et al. Lithium superionic conduction in lithium borohydride accompanied by structural transition [J]. Applied Physics Letters, 2007, 91 (22).

[85] Maekawa H, Matsuo M, Takamura H, et al. Halide-stabilized $LiBH_4$, a room-temperature lithium fast-ion conductor [J]. Journal of the American Chemical Society, 2009, 131 (3): 894-895.

[86] Tang W S, Matsuo M, Wu H, et al. Liquid-like ionic conduction in solid lithium and sodium monocarba-closo-decaborates near or at room temperature [J]. Advanced Energy Materials, 2016, 6 (8).

[87] Hagenmuller P, Van Gool W. Solid electrolytes: General principles, characterization, materials, applications [M]. New York: Academic Press, 1978.

[88] Salamon M B. Physics of superionic conductors [M]. Berlin: Springer-Verlag GmbH, 1979.

[89] Johnston W V, Wiedersi H, Lindberg G W. Heat capacity, transformations, and thermal disorder in solid electrolyte $RbAg_4I_5$ [J]. J Chem Phys, 1969, 51 (9): 3739.

[90] Pardee W J, Mahan G D. Second-order phase-transition in super ionic conductor $RbAg_4I_5$-Comment [J]. J Chem Phys, 1974, 61 (5): 2173-2174.

[91] Lederman F L, Salamon M B, Peisl H. Evidence for an order-disorder transformation in solid electrolyte $RbAg_4I_5$ [J]. Solid State Communications, 1976, 19 (2): 147-150.

[92] Graham L J, Chang R. Temperature and pressure-dependence of elastic properties of $RbAg_4I_5$ [J]. J Appl Phys, 1975, 46 (6): 2433-2438.

[93] Nagao M, Kaneda T. Ultrasonic-attenuation of silver ions in $RbAg_4I_5$ single-crystals [J]. Phys Rev B, 1975, 11 (8): 2711-2716.

[94] Leung K M, Huber D L. Ultrasonic-attenuation near the 208-K phase-transition of $RbAg_4I_5$ [J]. Phys Rev Lett, 1979, 42 (7): 452-456.

[95] Udovic T J, Matsuo M, Unemoto A, et al. Sodium superionic conduction in $Na_2B_{12}H_{12}$ [J]. Chem Commun, 2014, 50 (28): 3750-3752.

[96] Udovic T J, Matsuo M, Tang W S, et al. Exceptional superionic conductivity in disordered sodium decahydro-closo-decaborate [J]. Adv Mater, 2014, 26 (45): 7622-7626.

[97] Tang W S, Unemoto A, Zhou W, et al. Unparalleled lithium and sodium superionic conduction in solid electrolytes with large monovalent cage-like anions [J]. Energy & Environmental Science, 2015, 8 (12): 3637-3645.

[98] Soloninin A V, Dimitrievska M, Skoryunov R V, et al. Comparison of anion reorientational dynamics in MCB_9H_{10} and $M_2B_{10}H_{10}$ (M=Li, Na) via nuclear magnetic resonance and quasielastic neutron scattering studies [J]. The Journal of Physical Chemistry C, 2017, 121 (2): 1000-1012.

[99] Varley J B, Kweon K, Mehta P, et al. Understanding ionic conductivity trends in polyborane solid electrolytes from Ab initio molecular dynamics [J]. ACS Energy Letters, 2017, 2 (1): 250-255.

[100] Lu Z, Ciucci F. Structural origin of the superionic Na conduction in $Na_2B_{10}H_{10}$ closo-borates and enhanced conductivity by Na deficiency for high performance solid electrolytes [J]. J Mater Chem A, 2016, 4 (45): 17740-17748.

[101] Goodenough J B, Hong H Y P, Kafalas J A. Fast Na^+-ion transport in skeleton structures [J]. Mater Res Bull, 1976, 11 (2): 203-220.

[102] Hong H Y P. Crystal structures and crystal chemistry in the system $Na_{1+x}Zr_2Si_xP_{3-x}O_{12}$ [J]. Mater Res Bull, 1976, 11 (2): 173-182.

[103] Shannon R D, Taylor B E, English A D, et al. New Li solid electrolytes [C] //Armstrong R D. International Symposium on Solid Ionic and Ionic-Electronic Conductors. Pergamon. 1977: 783-796.

[104] Taylor B E, English A D, Berzins T. New solid ionic conductors [J]. Mater Res Bull, 1977, 12 (2): 171-181.

[105] Arbi K, Hoelzel M, Kuhn A, et al. Structural factors that enhance lithium mobility in fast-ion $Li_{1+x}Ti_{2-x}Al_x(PO_4)_3$ ($0 \leqslant x \leqslant 0.4$) conductors investigated by neutron diffraction in the temperature range 100-500 K [J]. Inorganic Chemistry, 2013, 52 (16): 9290-9296.

[106] Catti M, Stramare S, Ibberson R. Lithium location in NASICON-type Li^+ conductors by neutron diffraction. Ⅰ. Triclinic alpha′-$LiZr_2(PO_4)_3$ [J]. Solid State Ionics, 1999, 123 (1-4): 173-180.

[107] Catti M, Stramare S. Lithium location in NASICON-type Li^+ conductors by neutron diffraction: Ⅱ. Rhombohedral alpha-$LiZr_2(PO_4)_3$ at $T=423$ K [J]. Solid State Ionics, 2000, 136: 489-494.

[108] Petit D, Colomban P, Collin G, et al. Fast ion-transport in $LiZr_2(PO_4)_3$-Structure and conductivity [J]. Mater Res Bull, 1986, 21 (3): 365-371.

[109] Catti M, Stramare S, Ibberson R. Lithium location in NASICON-type Li^+ conductors by neutron diffraction. I. Triclinic α′-$LiZr_2(PO_4)_3$ [J]. Solid State Ionics, 1999, 123 (1): 173-180.

[110] Petit D, Colomban P, Collin G, et al. Fast ion transport in $LiZr_2(PO_4)_3$: Structure and conductivity [J]. Mater Res Bull, 1986, 21 (3): 365-371.

[111] Sudreau F, Petit D, et al. Dimorphism, phase-transitions, and transport-properties in $LiZr_2(PO_4)_3$ [J]. Journal of Solid State Chemistry, 1989, 83 (1): 78-90.

[112] Roy S, Padma Kumar P. Framework flexibility of sodium zirconium phosphate: Role of disorder, and polyhedral distortions from Monte Carlo investigation [J]. Journal of Materials Science, 2012, 47 (12): 4946-4954.

[113] Anantharamulu N, Koteswara Rao K, Rambabu G, et al. A wide-ranging review on Nasicon type materials [J]. Journal of Materials Science, 2011, 46 (9): 2821-2837.

[114] Aono H, Sugimoto E, Sadaaka Y, et al. Ionic conductivity of the lithium titanium phosphate $Li_{1+x}M_xTi_{2-x}(PO_4)_3$, M=Al, Sc, Y, and La systems [J]. Journal of the Electrochemical Society, 1989, 136 (2): 590.

[115] Lenain G E, Mckinstry H A, Alamo J, et al. Structural model for thermal expansion in $MZr_2P_3O_{12}$ (M=Li, Na, K, Rb, Cs) [J]. Journal of Materials Science, 1987, 22 (1): 17-22.

[116] Derosa P A, Balbuena P B. A lattice-gas model study of lithium intercalation in graphite [J]. J Electrochem Soc, 1999, 146 (10): 3630-3638.

[117] Yamaki J, Egashira M, Okada S. Potential and thermodynamics of graphite anodes in Li-ion cells [J]. J Electrochem Soc, 2000, 147 (2): 460-465.

[118] Marquez A, Vargas A, Balbuena P B. Computational studies of lithium intercalation in model graphite in the presence of tetrahydrofuran [J]. J Electrochem Soc, 1998, 145 (10): 3328-3334.

[119] Nalimova V A, Guerard D, Lelaurain M, et al. X-ray-investigation of highly saturated Li-graphite intercalation compound [J]. Carbon, 1995, 33 (2): 177-181.

[120] Woo K C, Mertwoy H, Fischer J E, et al. Experimental phase-diagram of lithium-intercalated graphite [J]. Phys Rev B, 1983, 27 (12): 7831-7834.

[121] Guerard D, Herold A. Intercalation of lithium into graphite and other carbons [J]. Carbon, 1975, 13 (4): 337-345.

[122] Dahn J R. Phase-Diagram of Li_xC_6 [J]. Phys Rev B, 1991, 44 (17): 9170-9177.

[123] 胡进. 锂离子电池纳米结构负极材料储锂性能研究 [D]. 北京: 中国科学院物理研究所, 2005.

[124] Gavilán Arriazu E M, López De Mishima B A, Oviedo O A, et al. Criticality of the phase transition on stage two in a lattice-gas model of a graphite anode in a lithium-ion battery [J]. Phys Chem Chem Phys, 2017, 19

(34): 23138-23145.

[125] Liu Q, Li S, Wang S, et al. Kinetically determined phase transition from stage Ⅱ (LiC_{12}) to stage Ⅰ (LiC_6) in a graphite anode for Li-ion batteries [J]. The Journal of Physical Chemistry Letters, 2018, 9 (18): 5567-5573.

[126] Reimers J N, Dahn J R. Electrochemical and Insitu X-Ray-Diffraction Studies of Lithium Intercalation in Li_xCoO_2 [J]. J Electrochem Soc, 1992, 139 (8): 2091-2097.

[127] Reimers J N, Dahn J R, Vonsacken U. Effects of impurities on the electrochemical properties of $LiCoO_2$ [J]. J Electrochem Soc, 1993, 140 (10): 2752-2754.

[128] Ohzuku T, Ueda A. Solid-state redox reactions of $LiCoO_2$ (R (3) over-Bar-M) for 4 volt secondary lithium Cells [J]. J Electrochem Soc, 1994, 141 (11): 2972-2977.

[129] Shao-Horn Y, Levasseur S, Weill F, et al. Probing lithium and vacancy ordering in O3 layered Li_xCoO_2 (x approximate to 0.5) -An electron diffraction study [J]. J Electrochem Soc, 2003, 150 (3): A366-A373.

[130] Menetrier M, Saadoune I, Levasseur S, et al. The insulator-metal transition upon lithium deintercalation from $LiCoO_2$: electronic properties and Li-7 NMR study [J]. J Mater Chem, 1999, 9 (5): 1135-1140.

[131] Marianetti C A, Kotliar G, Ceder G. A first-order Mott transition in Li_xCoO_2 [J]. Nat Mater, 2004, 3 (9): 627-631.

[132] Van der Ven A, Aydinol M K, Ceder G. First-principles evidence for stage ordering in Li_xCoO_2 [J]. J Electrochem Soc, 1998, 145 (6): 2149-2155.

[133] Mizushima K, Jones P C, Wiseman P J, et al. Li_xCoO_2: a new cathode material for batteries of high-energy density [J]. Mater Res Bull, 1980, 15 (6): 783-789.

[134] Amatucci G G, Tarascon J M, Klein L C. CoO_2, the end member of the Li_xCoO_2 solid solution [J]. J Electrochem Soc, 1996, 143 (3): 1114-1123.

[135] Li W, Currie C. Morphology effects on the electrochemical performance of $LiNi_{1-x}Co_xO_2$ [J]. J Electrochem Soc, 1997, 144 (8): 2773-2779.

[136] Delmas C, Braconnier J J, Hagenmuller P. A new variety of $LiCoO_2$ with an unusual oxygen packing obtained by exchange-reaction [J]. Mater Res Bull, 1982, 17 (1): 117-123.

[137] Carlier D, Saadoune I, Croguennec L, et al. On the metastable O2-type $LiCoO_2$ [J]. Solid State Ionics, 2001, 144 (3-4): 263-276.

[138] Mendiboure A, Delmas C, Hagenmuller P. New layered structure obtained by electrochemical deintercalation of the metastable $LiCoO_2$ (O2) Variety [J]. Mater Res Bull, 1984, 19 (10): 1383-1392.

[139] Carlier D, Saadoune I, Menetrier M, et al. Lithium electrochemical deintercalation from $O2-LiCoO_2$-Structure and physical properties [J]. J Electrochem Soc, 2002, 149 (10): A1310-A1320.

[140] Carlier D, Van Der Ven A, Delmas C, et al. First-principles investigation of phase stability in the $O-2-LiCoO_2$ system [J]. Chem Mater, 2003, 15 (13): 2651-2660.

[141] Lu X, Sun Y, Jian Z, et al. New Insight into the Atomic Structure of Electrochemically Delithiated O3-$Li_{1-x}CoO_2$ ($0 \leqslant x \leqslant 0.5$) Nanoparticles [J]. Nano Letters, 2012, 12 (12): 6192-6197.

[142] Liu Z, Zhen H, Kim Y, et al. Synthesis of $LiNiO_2$ cathode materials with homogeneous Al doping at the atomic level [J]. J Power Sources, 2011, 196 (23): 10201-10206.

[143] Ohzuku T, Makimura Y. Layered lithium insertion material of $LiNi_{1/2}Mn_{1/2}O_2$: A possible alternative to $LiCoO_2$ for advanced lithium-ion batteries [J]. Chemistry Letters, 2001, 30 (8): 744-745.

[144] Ohzuku T, Makimura Y. Layered Lithium Insertion Material of $LiCo_{1/3}Ni_{1/3}Mn_{1/3}O_2$ for Lithium-Ion Batteries [J]. Chemistry Letters, 2001, 30 (7): 642-643.

[145] Kalyani P, Kalaiselvi N. Various aspects of $LiNiO_2$ chemistry: A review [J]. Science and Technology of Advanced Materials, 2005, 6 (6): 689.

[146] Chen Z, Zou H, Zhu X, et al. First-principle investigation of Jahn-Teller distortion and topological analysis of chemical bonds in $LiNiO_2$ [J]. Journal of Solid State Chemistry, 2011, 184 (7): 1784-1790.

[147] Han C H, Kim J H, Paeng S H, et al. Electrochemical characteristics of $LiNiO_2$ films prepared for charge storable electrode application [J]. Thin Solid Films, 2009, 517 (14): 4215-4217.

[148] Wang L, Maxisch T, Ceder G. A first-principles approach to studying the thermal stability of oxide cathode materials [J]. Chem Mater, 2007, 19 (3): 543-552.

[149] Li M, Lu J. Cobalt in lithium-ion batteries [J]. Science, 2020, 367 (6481): 979-980.

[150] Li H, Cormier M, Zhang N, et al. Is cobalt needed in Ni-rich positive electrode materials for lithium ion batteries? [J]. Journal of The Electrochemical Society, 2019, 166 (4): A429-A439.

[151] Assat G, Tarascon J-M. Fundamental understanding and practical challenges of anionic redox activity in Li-ion batteries [J]. Nature Energy, 2018, 3 (5): 373-86.

[152] Luo K, Roberts M R, Guerrini N, et al. Anion redox chemistry in the cobalt free 3d transition metal oxide intercalation electrode Li [Li$_{0.2}$Ni$_{0.2}$Mn$_{0.6}$] O$_2$ [J]. Journal of the American Chemical Society, 2016, 138 (35): 11211-11218.

[153] Grimaud A, Hong W T, Shao-Horn Y, et al. Anionic redox processes for electrochemical devices [J]. Nat Mater, 2016, 15 (2): 121-126.

[154] Luo K, Roberts M R, Hao R, et al. Charge-compensation in 3D-transition-metal-oxide intercalation cathodes through the generation of localized electron holes on oxygen [J]. Nature Chemistry, 2016, 8 (7): 684-691.

[155] Assat G, Foix D, Delacourt C, et al. Fundamental interplay between anionic/cationic redox governing the kinetics and thermodynamics of lithium-rich cathodes [J]. Nat Commun, 2017, 8 (1): 2219.

[156] Zhao E, Zhang M, Wang X, et al. Local structure adaptability through multi cations for oxygen redox accommodation in Li-rich layered oxides [J]. Energy Storage Materials, 2020, 24: 384-393.

[157] Rong X, Hu E, Lu Y, et al. Anionic redox reaction-induced high-capacity and low-strain cathode with suppressed phase transition [J]. Joule, 2019, 3 (2): 503-517.

[158] Vergnet J, Saubanère M, Doublet M-L, et al. The structural stability of P2-layered Na-based electrodes during anionic redox [J]. Joule, 2020, 4 (2): 420-434.

[159] Bak S M, Nam K W, Chang W, et al. Correlating structural changes and gas evolution during the thermal decomposition of charged Li$_x$Ni$_{0.8}$Co$_{0.15}$Al$_{0.05}$O$_2$ cathode materials [J]. Chem Mater, 2013, 25 (3): 337-351.

[160] Chen Z H, Ren Y, Jansen A N, et al. New class of nonaqueous electrolytes for long-life and safe lithium-ion batteries [J]. Nat Commun, 2013, 4.

[161] Benedek R, Thackeray M M, Van De Walle A. Free energy for protonation reaction in lithium-ion battery cathode materials [J]. Chem Mater, 2008, 20 (17): 5485-5490.

[162] Padhi A K, Nanjundaswamy K S, Goodenough J B. Phospho-olivines as positive-electrode materials for rechargeable lithium batteries [J]. J Electrochem Soc, 1997, 144 (4): 1188-1194.

[163] Andersson A S, Thomas J O. The source of first-cycle capacity loss in LiFePO$_4$ [J]. J Power Sources, 2001, 97 (8): 498-502.

[164] 刘立君. 锂离子电池正极材料研究 [D]. 北京: 中国科学院, 2003.

[165] Srinivasan V, Newman J. Discharge model for the lithium iron-phosphate electrode [J]. J Electrochem Soc, 2004, 151 (10): A1517-A1529.

[166] Chen G Y, Song X Y, Richardson T J. Electron microscopy study of the LiFePO$_4$ to FePO$_4$ phase transition [J]. Electrochem Solid St, 2006, 9 (6): A295-A298.

[167] Laffont L, Delacourt C, Gibot P, et al. Study of the LiFePO$_4$/FePO$_4$ two-phase system by high-resolution electron energy loss spectroscopy [J]. Chem Mater, 2006, 18 (23): 5520-5529.

[168] Delmas C, Maccario M, Croguennec L, et al. Lithium deintercalation in LiFePO$_4$ nanoparticles via a domino-cascade model [J]. Nat Mater, 2008, 7 (8): 665-671.

[169] Andersson A S, Kalska B, Haggstrom L, et al. Lithium extraction/insertion in LiFePO$_4$: An X-ray diffraction and Mossbauer spectroscopy study [J]. Solid State Ionics, 2000, 130 (1-2): 41-52.

[170] Gu L, Zhu C, Li H, et al. Direct observation of lithium staging in partially delithiated LiFePO$_4$ at atomic resolution [J]. Journal of the American Chemical Society, 2011, 133 (13): 4661-4663.

[171] Suo L M, Han W Z, Lu X, et al. Highly ordered staging structural interface between LiFePO$_4$ and FePO$_4$ [J]. Phys Chem Chem Phys, 2012, 14 (16): 5363-5367.

[172] Malik R, Zhou F, Ceder G. Kinetics of non-equilibrium lithium incorporation in LiFePO$_4$ [J]. Nat Mater, 2011, 10 (8): 587-590.

[173] Morgan D, Van Der Ven A, Ceder G. Li conductivity in Li$_x$MPO$_4$ (M=Mn, Fe, Co, Ni) olivine materials [J]. Electrochem Solid St, 2004, 7 (2): A30-A32.

[174] Chung S Y, Bloking J T, Chiang Y M. Electronically conductive phospho-olivines as lithium storage electrodes [J]. Nat Mater, 2002, 1 (2): 123-8.

[175] Kang B, Ceder G. Battery materials for ultrafast charging and discharging [J]. Nature, 2009, 458 (7235):

190-193.

[176] Kim D H, Kim J. Synthesis of LiFePO$_4$ nanoparticles in polyol medium and their electrochemical properties [J]. Electrochem Solid St, 2006, 9 (9): A439-A442.

[177] Delacourt C, Poizot P, Tarascon J M, et al. The existence of a temperature-driven solid solution in Li$_x$FePO$_4$ for $0 \leqslant x \leqslant 1$ [J]. Nat Mater, 2005, 4 (3): 254-260.

[178] Delacourt C, Rodriguez-Carvajal J, Schmitt B, et al. Crystal chemistry of the olivine-type Li$_x$FePO$_4$ system ($0 \leqslant x \leqslant 1$) between 25 and 370 degrees C [J]. Solid State Sci, 2005, 7 (12): 1506-1516.

[179] Chen G Y, Song X Y, Richardson T J. Metastable solid-solution phases in the LiFePO$_4$/FePO$_4$ system [J]. J Electrochem Soc, 2007, 154 (7): A627-A632.

[180] Dodd J L, Yazami R, Fultz B. Phase diagram of Li$_x$FePO$_4$ [J]. Electrochem Solid St, 2006, 9 (3): A151-A155.

[181] Sun Y, Lu X, Xiao R J, et al. Kinetically controlled lithium-staging in delithiated LiFePO$_4$ driven by the Fe center mediated interlayer Li-Li interactions [J]. Chem Mater, 2012, 24 (24): 4693-4703.

[182] Ji H, Wu J, Cai Z, et al. Ultrahigh power and energy density in partially ordered lithium-ion cathode materials [J]. Nature Energy, 2020, 5: 213-221.

[183] Tarascon J M, Wang E, Shokoohi F K, et al. The spinel phase of LiMn$_2$O$_4$ as a cathode in secondary lithium cells [J]. J Electrochem Soc, 1991, 138 (10): 2859-2864.

[184] Bittihn R, Herr R, Hoge D. The swing system, a nonaqueous rechargeable carbon metal-oxide cell [J]. J Power Sources, 1993, 43 (1-3): 223-231.

[185] Sigala C, Guyomard D, Verbaere A, et al. Positive electrode materials with high operating voltage for lithium batteries: LiCr$_y$Mn$_{2-y}$O$_4$ ($0 \leqslant y \leqslant 1$) [J]. Solid State Ionics, 1995, 81 (3-4): 167-170.

[186] Amine K, Tukamoto H, Yasuda H, et al. Preparation and electrochemical investigation of LiMn$_{2-x}$Me$_x$O$_4$ (Me: Ni, Fe, and $x=0.5$, 1) cathode materials for secondary lithium batteries [J]. J Power Sources, 1997, 68 (2): 604-608.

[187] Kawai H, Nagata M, Kageyama H, et al. 5 V lithium cathodes based on spinel solid solutions Li$_2$Co$_{1+X}$Mn$_{3-X}$O$_8$: $-1 \leqslant X \leqslant 1$ [J]. Electrochim Acta, 1999, 45 (1-2): 315-327.

[188] Ein-Eli Y, Howard W F, Lu S H, et al. LiMn$_{2-x}$Cu$_x$O$_4$ spinels ($0.1 \leqslant x \leqslant 0.5$): A new class of 5 V cathode materials for Li batteries-I. Electrochemical, structural, and spectroscopic studies [J]. J Electrochem Soc, 1998, 145 (4): 1238-1244.

[189] House R A, Bruce P G. Lightning fast conduction [J]. Nature Energy, 2020, 5 (3): 191-192.

[190] 王丽平. 锂离子电池高电压正极材料 LiNi$_{0.5}$Mn$_{1.5}$O$_4$ 的研究 [D]. 北京: 中国科学院, 2011.

[191] Zhong Q M, Bonakdarpour A, Zhang M J, et al. Synthesis and electrochemistry of LiNi$_x$Mn$_{2-x}$O$_4$ [J]. J Electrochem Soc, 1997, 144 (1): 205-213.

[192] Kim J H, Myung S T, Yoon C S, et al. Comparative study of LiNi$_{0.5}$Mn$_{1.5}$O$_4$-delta and LiNi$_{0.5}$Mn$_{1.5}$O$_4$ cathodes having two crystallographic structures: Fd_3 over-bar-m and $P4_3$32 [J]. Chem Mater, 2004, 16 (5): 906-914.

[193] Poizot P, Laruelle S, Grugeon S, et al. Nano-sized transition-metaloxides as negative-electrode materials for lithium-ion batteries [J]. Nature, 2000, 407 (6803): 496-499.

[194] Debart A, Dupont L, Poizot P, et al. A transmission electron microscopy study of the reactivity mechanism of tailor-made CuO particles toward lithium [J]. J Electrochem Soc, 2001, 148 (11): A1266-A1274.

[195] Poizot P, Laruelle S, Grugeon S, et al. Rationalization of the low-potential reactivity of 3d-metal-based inorganic compounds toward Li [J]. J Electrochem Soc, 2002, 149 (9): A1212-A1217.

[196] Larcher D, Sudant G, Leriche J B, et al. The electrochemical reduction of Co$_3$O$_4$ in a lithium cell [J]. J Electrochem Soc, 2002, 149 (3): A234-A241.

[197] Li H, Richter G, Maier J. Reversible formation and decomposition of LiF clusters using transition metal fluorides as precursors and their application in rechargeable Li batteries [J]. Adv Mater, 2003, 15 (9): 736-739.

[198] Badway F, Pereira N, Cosandey F, et al. Carbon-metal fluoride nanocomposites-Structure and electrochemistry of FeF$_3$: C [J]. J Electrochem Soc, 2003, 150 (9): A1209-A1218.

[199] Balaya P, Li H, Kienle L, et al. Fully reversible homogeneous and heterogeneous Li storage in RuO$_2$ with high capacity [J]. Adv Funct Mater, 2003, 13 (8): 621-625.

[200] Silva D C C, Crosnier O, Ouvrard G, et al. Reversible lithium uptake by FeP$_2$ [J]. Electrochem Solid St,

2003, 6 (8): A162-A165.

[201] Li H, Balaya P, Maier J. Li-storage via heterogeneous reaction in selected binary metal fluorides and oxides [J]. J Electrochem Soc, 2004, 151 (11): A1878-A1885.

[202] Fu Z W, Wang Y, Yue X L, et al. Electrochemical reactions of lithium with transition metal nitride electrodes [J]. J Phys Chem B, 2004, 108 (7): 2236-2244.

[203] Fu Z W, Li C L, Liu W Y, et al. Electrochemical reaction of lithium with cobalt fluoride thin film electrode [J]. J Electrochem Soc, 2005, 152 (2): E50-E55.

[204] Yu X Q, He Y, Sun J P, et al. Nanocrystalline MnO thin film anode for lithium ion batteries with low overpotential [J]. Electrochem Commun, 2009, 11 (4): 791-794.

[205] Zhong K F, Xia X, Zhang B, et al. MnO powder as anode active materials for lithium ion batteries [J]. J Power Sources, 2010, 195 (10): 3300-3308.

[206] Zhong K F, Zhang B, Luo S H, et al. Investigation on porous MnO microsphere anode for lithium ion batteries [J]. J Power Sources, 2011, 196 (16): 6802-6808.

[207] Hu J, Li H, Huang X J. Cr_2O_3-based anode materials for Li-ion batteries [J]. Electrochem Solid St, 2005, 8 (1): A66-A69.

[208] Grugeon S, Laruelle S, Dupont L, et al. Combining electrochemistry and metallurgy for new electrode designs in Li-ion batteries [J]. Chem Mater, 2005, 17 (20): 5041-5047.

[209] Hu J, Li H, Huang X J. Influence of micropore structure on Li-storage capacity in hard carbon spherules [J]. Solid State Ionics, 2005, 176 (11-12): 1151-1159.

[210] Dupont L, Grugeon S, Laruelle S, et al. Structure, texture and reactivity versus lithium of chromium-based oxides films as revealed by TEM investigations [J]. J Power Sources, 2007, 164 (2): 839-848.

[211] Dupont L, Laruelle S, Grugeon S, et al. Mesoporous Cr_2O_3 as negative electrode in lithium batteries: TEM study of the texture effect on the polymeric layer formation [J]. J Power Sources, 2008, 175 (1): 502-509.

[212] Sun J P, Tang K, Yu X Q, et al. Overpotential and electrochemical impedance analysis on Cr_2O_3 thin film and powder electrode in rechargeable lithium batteries [J]. Solid State Ionics, 2008, 179 (40): 2390-2395.

[213] Grugeon S, Laruelle S, Herrera-Urbina R, et al. Particle size effects on the electrochemical performance of copper oxides toward lithium [J]. J Electrochem Soc, 2001, 148 (4): A285-A292.

[214] Luo J Y, Zhang J J, Xia Y Y. Highly electrochemical reaction of lithium in the ordered mesoporous beta-MnO_2 [J]. Chem Mater, 2006, 18 (23): 5618-5623.

[215] Jiao F, Harrison A, Bruce P G. Ordered three-dimensional arrays of monodispersed Mn_3O_4 nanoparticles with a core-shell structure and spin-glass behavior [J]. Angew Chem Int Edit, 2007, 46 (21): 3946-3950.

[216] Hu J, Li H, Huang X J, et al. Improve the electrochemical performances of Cr_2O_3 anode for lithium ion batteries [J]. Solid State Ionics, 2006, 177 (26-32): 2791-2799.

[217] Gireaud L, Grugeon S, Pilard S, et al. Mass spectrometry investigations on electrolyte degradation products for the development of nanocomposite electrodes in lithium ion batteries [J]. Anal Chem, 2006, 78 (11): 3688-3698.

[218] 禹习谦. 基于薄膜技术的锂离子电池新材料研究 [D]. 北京: 中国科学院, 2010.

[219] Cui Z H, Guo X X, Li H. Improved electrochemical properties of MnO thin film anodes by elevated deposition temperatures: Study of conversion reactions [J]. Electrochim Acta, 2013, 89: 229-238.

[220] Doe R E, Persson K A, Meng Y S, et al. First-Principles investigation of the Li-Fe-F phase diagram and equilibrium and nonequilibrium conversion reactions of iron fluorides with lithium [J]. Chem Mater, 2008, 20 (16): 5274-5283.

[221] Delmer O, Balaya P, Kienle L, et al. Enhanced potential of amorphous electrode materials: Case study of RuO_2 [J]. Adv Mater, 2008, 20 (3): 501.

[222] Delmer O, Maier J. On the chemical potential of a component in a metastable phase-application to Li-storage in the RuO_2-Li system [J]. Phys Chem Chem Phys, 2009, 11 (30): 6424-6429.

[223] 梁敬魁. 相图与相结构: 相图的理论、实践和应用 [M]. 北京: 科学出版社, 1993.

[224] Atkins P W. de Paula, Atkins' Physical chemistry [M]. 8th edition. Oxford: Oxford University Press, 2006.

[225] 曾燕伟. 无机材料科学基础 [M]. 武汉: 武汉理工大学出版社, 2012.

[226] 刘长俊. 相律及相图热力学 [M]. 北京: 高等教育出版社, 1995.

[227] 陆学善. 相图与相变 [M]. 合肥: 中国科学技术大学出版社, 1990.

[228] Chevrier V L, Zwanziger J W, Dahn J R. First principles study of Li-Si crystalline phases: Charge transfer, electronic structure, and lattice vibrations [J]. Journal of Alloys and Compounds, 2010, 496 (1): 25-36.

[229] Li H, Huang X, Chen L, et al. The crystal structural evolution of nano-Si anode caused by lithium insertion and extraction at room temperature [J]. Solid State Ionics, 2000, 135 (1): 181-191.

[230] Zhou G W, Li H, Sun H P, et al. Controlled Li doping of Si nanowires by electrochemical insertion method [J]. Appl Phys Lett, 1999, 75 (16): 2447-2449.

[231] Hatchard T D, Dahn J R. In situ XRD and electrochemical study of the reaction of lithium with amorphous silicon [J]. J Electrochem Soc, 2004, 151 (6): A838-A842.

[232] Vandermarel C, Vinke G J B, Vanderlugt W. The phase-diagram of the system lithium-silicon [J]. Solid State Communications, 1985, 54 (11): 917-919.

[233] Meethong N, Kao Y H, Speakman S A, et al. Aliovalent substitutions in olivine lithium iron phosphate and impact on structure and properties [J]. Adv Funct Mater, 2009, 19 (7): 1060-1070.

[234] Thackeray M M. Manganese oxides for lithium batteries [J]. Prog Solid State Ch, 1997, 25 (1-2): 1-71.

[235] 张联齐, 肖成伟, 杨瑞娟. 有序/无序岩盐结构的 $Li_{1+x}M_{1-x}O_2$ 锂离子电池正极材料 [J]. 化学进展, 2011, 23 (2/3).

[236] Robertson A D, Bruce P G. Mechanism of electrochemical activity in Li_2MnO_3 [J]. Chem Mater, 2003, 15 (10): 1984-1992.

[237] Thackeray M M, Kang S-H, Johnson C S, et al. Li_2MnO_3-stabilized $LiMO_2$ (M=Mn, Ni, Co) electrodes for lithium-ion batteries [J]. J Mater Chem, 2007, 17 (30): 3112-25.

[238] 陆学善. 相图与相变 [M]. 合肥: 中国科学技术大学出版社, 1990.

[239] Saunders N M A P. CALPHAD (calculation of phase diagrams): A comprehensive guide [M]. Oxford: Oxford University Press, 1998.

[240] Aydinol M K, Kohan A F, Ceder G. Ab initio calculation of the intercalation voltage of lithium transition metal oxide electrodes for rechargeable batteries [J]. J Power Sources, 1997, 68 (2): 664-668.

[241] Van Der Ven A, Ceder G. Lithium diffusion in layered Li_xCoO_2 [J]. Electrochem Solid St, 2000, 3 (7): 301-304.

[242] Meng Y S, Van Der Ven A, Chan M K Y, et al. Ab initio study of sodium ordering in $Na_{0.75}CoO_2$ and its relation to Co^{3+}/Co^{4+} charge ordering [J]. Phys Rev B, 2005, 72 (17).

[243] Maxisch T, Zhou F, Ceder G. Ab initio study of the migration of small polarons in olivine Li_xFePO_4 and their association with lithium ions and vacancies [J]. Phys Rev B, 2006, 73 (10).

[244] Ong S P, Chevrier V L, Hautier G, et al. Voltage, stability and diffusion barrier differences between sodium-ion and lithium-ion intercalation materials [J]. Energy & Environmental Science, 2011, 4 (9): 3680.

[245] Hautier G, Jain A, Chen H, et al. Novel mixed polyanions lithium-ion battery cathode materials predicted by high-throughput ab initio computations [J]. J Mater Chem, 2011, 21 (43): 17147-17153.

[246] Jain A, Hautier G, Moore C, et al. A computational investigation of $Li_9M_3(P_2O_7)_3(PO_4)_2$ (M=V, Mo) as cathodes for Li ion batteries [J]. J Electrochem Soc, 2012, 159 (5): A622-A633.

[247] Chen H, Hautier G, Jain A, et al. Carbonophosphates: A new family of cathode materials for Li-ion batteries identified computationally [J]. Chem Mater, 2012, 24 (11): 2009-2016.

[248] Ping Ong S, Wang L, Kang B, et al. Li-Fe-P-O_2 phase diagram from first principles calculations [J]. Chem Mater, 2008, 20 (5): 1798-1807.

[249] Ong S P, Jain A, Hautier G, et al. Thermal stabilities of delithiated olivine MPO_4 (M=Fe, Mn) cathodes investigated using first principles calculations [J]. Electrochem Commun, 2010, 12 (3): 427-430.

[250] Doe R E, Persson K A, Meng Y S, et al. First-principles investigation of the Li-Fe-F phase diagram and equilibrium and nonequilibrium conversion reactions of iron fluorides with lithium [J]. Chem Mater, 2008, 20 (16): 5274-5283.

[251] Hautier G, Fischer C C, Jain A, et al. Finding nature's missing ternary oxide compounds using machine learning and density functional theory [J]. Chem Mater, 2010, 22 (12): 3762-3767.

[252] Hautier G, Fischer C, Ehrlacher V, et al. Data mined ionic substitutions for the discovery of new compounds [J]. Inorganic Chemistry, 2011, 50 (2): 656.

[253] Richards W D, Miara L J, Wang Y, et al. Interface stability in solid-state batteries [J]. Chem Mater, 2016, 28 (1): 266-273.

[254] Zhu Y, He X, Mo Y. First principles study on electrochemical and chemical stability of solid electrolyte-electrode interfaces in all-solid-state Li-ion batteries [J]. J Mater Chem A, 2016, 4 (9): 3253-3266.

[255] Zhu Y, He X, Mo Y. Origin of outstanding stability in the lithium solid electrolyte materials: Insights from thermodynamic analyses based on first-principles calculations [J]. ACS Applied Materials & Interfaces, 2015, 7 (42): 23685-23693.

[256] Xiao Y, Miara L J, Wang Y, et al. Computational screening of cathode coatings for solid-state batteries [J]. Joule, 2019, 3 (5): 1252-1275.

[257] Fleischauer M D, Hatchard T D, Rockwell G P, et al. Design and testing of a 64-channel combinatorial electrochemical cell [J]. J Electrochem Soc, 2003, 150 (11): A1465-A1469.

[258] Roberts M R, Vitins G, Denuault G, et al. High throughput electrochemical observation of structural phase changes in $LiFe_{1-x}Mn_xPO_4$ during charge and discharge [J]. J Electrochem Soc, 2010, 157 (4): A381-A386.

第4章

电池界面问题

电池中固液界面的性质对锂离子电池充放电效率、能量效率、能量密度、功率密度、循环性、服役寿命、安全性、自放电等特性具有重要的影响。对界面问题的研究是锂离子电池基础研究的核心。本章小结了对锂离子电池电极表面固体电解质中间相（SEI）形成机理及其组成结构的认识，介绍了近年来对锂离子输运机制，SEI 膜改性研究，以及用透射电镜（TEM）及原子力显微镜（AFM）中力曲线等实验技术来分析 SEI 膜的形貌、厚度、覆盖度及力学性能等。

4.1 锂离子电池界面问题

锂离子电池具备优越的综合电化学性能，广泛应用于消费电子领域。电动汽车，大型储能设备等的发展迫切需要更高功率密度，更高能量密度，更长循环寿命，更好安全性的锂离子电池[1-3]。目前的研究主要集中在开发新型高容量正负极电极材料，新型电解液体系等。在锂离子电池的研究和开发中，已经认识到，界面特性对锂离子电池的各方面性能均会产生重要的影响。

电池中常见的界面类型有固固界面，包括电极材料在脱嵌锂过程中产生的两相界面（$LiFePO_4/FePO_4$，$Li_4Ti_5O_{12}/Li_7Ti_5O_{12}$），多晶结构的电极材料晶粒与晶粒之间形成的晶界，电极材料、导电添加剂、粘接剂、集流体之间相处的多个固固界面等。固固界面一般存在空间电荷层以及缺陷结构，其物理化学特性会影响离子与电子的输运、电极结构的稳定性、电荷转移的速率。如果电极材料中存在大量的晶界，晶界处也可储存少量的额外的锂。

锂离子电池中更为重要的界面是固液界面。现有的锂离子电池多采用非水液态有机溶剂电解质。当充放电电位范围较宽时，在正负极表面，会形成一层或多层固体电解质膜（solid electrolyte interphase，SEI），其示意图见图 4-1。

锂离子电池工作电位范围为 2～4.3V。其中，石墨类负极工作电位范围在 0～1.0V（vs. Li^+/Li），正极工作电位范围一般在 2.5～4.3V（vs. Li^+/Li）。而目前商用电解液不发生氧化还原反应的电化学窗口一般为 1.2~3.7V（vs. Li^+/Li）。目前已知的非水有机溶剂电解质还没有一个体系的电化学窗口能够超过 0~5.0V（vs. Li^+/Li）。

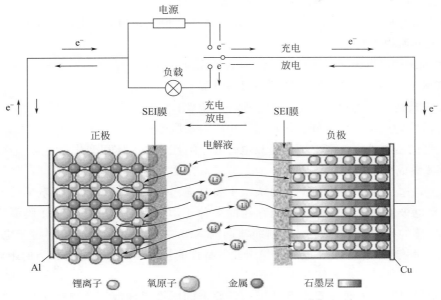

图 4-1　锂离子电池及 SEI 膜示意图[4]

Goodenough 等描绘了锂离子电池中电极 Fermi 能级与电解质中 HOMO（最高占据分子轨道），LUMO（最低未占据分子轨道）的示意图（非严格测量或计算结果，便于理解），参见图 4-2。可以看出，当有机溶剂或锂盐的 LUMO 低于负极的 Fermi 能级时，负极中的电子将注入到 LUMO，导致溶剂或锂盐被还原；而当 HOMO 高于正极的 Fermi 能级时，电子将注入到正极，导致溶剂或锂盐被氧化。在电池充电过程中，溶剂或锂盐在电极表面被还原或氧化，产生的物质中其不能溶解的部分将沉积覆盖在负极或正极表面上。通常这些物质含有锂离子，可以导通锂离子，但是对电子绝缘，因此电极表面膜被认为是固体电解质中间相（SEI）膜。如果 SEI 膜不能致密覆盖在电极表面，或者 SEI 膜不

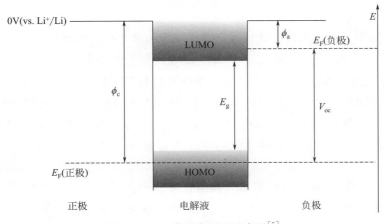

图 4-2　SEI 膜形成原理示意图[5]

是电子绝缘体，则溶剂或锂盐可以持续从电极上得失电子，发生氧化还原副反应，消耗正极的锂源，会降低充放电效率。如果 SEI 膜可以有效地组织后续的溶剂或锂盐的还原，则具有钝化膜的性质，有时被称为表面钝化膜（surface passivating film）。

采用非水有机溶剂的锂离子电池中，从有利于获得优异的综合电化学性能考虑，一般希望 SEI 膜具有以下特征[6]：

① 在 SEI 膜厚度超过电子隧穿长度时表现完全的电子绝缘；
② 高离子电导，使得锂离子顺利通过 SEI 膜；
③ 形貌及化学结构稳定，不随电池循环而改变；
④ 与活性物质结合性良好，循环过程中不容易脱落；
⑤ 良好的机械性能，能够适应充放电过程中活性物质的体积变化；
⑥ SEI 膜成分不溶于电解液。

事实上，迄今为止，对 SEI 膜的认识主要是通过多种分析手段，积累了典型电解质体系在典型电极上，不同电位下可能存在的多种化学物质的知识，对 SEI 膜厚度通过对局部结构敏感的 TEM、SIMS、Ar 刻蚀辅助的 XPS 获得了一定的了解，对于电子电导、力学行为、覆盖度、溶解度、电化学稳定性等特性的精确了解目前还有相当的距离，亟须新的研究手段对其进一步系统深入地研究。

当正负极材料的表面覆盖了 SEI 膜后，原来的固液界面演变为电极活性颗粒和其他材料与 SEI 膜形成的固固界面，SEI 膜与电解质形成的固液界面。

需要说明的是，如果电极材料的充放电电位范围较窄，例如负极的嵌锂电位高于 1.2V（vs. Li^+/Li），正极的脱锂电位低于 3.5V（vs. Li^+/Li），则正负极表面可以不发生电解质的氧化还原反应，不会形成 SEI 膜。此外，如果正负极表面预先生长了类似固体电解质的表面修饰层，能够起到防止电解质在电极上发生氧化还原反应的作用，则可以不再形成 SEI 膜。

4.2 SEI 膜结构及生长机理

对 SEI 膜结构及生长机理的认识是逐步加深和细致化的，重要的模型如图 4-3 所示。

早在 1970 年，A. N. Dey 就发现金属锂长时间浸泡在有机溶剂中会形成表面膜[7]。Peled 发现非水电池中碱金属及碱土金属与电解液接触时会形成一层表面膜[8]，是金属与电解液的一个中间相，具有电解质的特点，故命名为固态电解质中间相（SEI），作者还指出 SEI 膜的厚度预计在 1.5～2.5nm，因为厚度受电子隧穿距离限制。SEI 膜的成分和结构通常认为是靠近电极材料的为无机物层，主要包含 Li_2CO_3、LiF、Li_2O 等；中间层为有机物层，包括 $ROCO_2Li$、ROLi、$RCOO_2Li$（R 为有机基团）等；最外面为聚合物层，例如 PEO-Li 等。Peled 在研究 SEI 膜阻抗时发现其成分中微观颗粒相之间的晶界电阻 R_{gb} 要比体相的离子电阻大，因此提出 SEI 膜各成分颗粒是相互堆砌而成，类似马赛克结构（Masaic model）[9]。关于 SEI 膜的形成机理和组成分析，Aurbach 等利用红外光

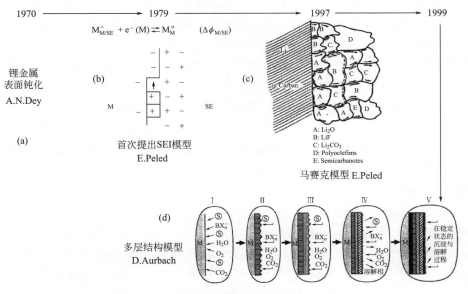

图 4-3 SEI 膜生长机理模型发展示意图[7-10]

(a) Dey 1970 年首次发现锂金属表面钝化现象；(b) Peled 1979 年首次提出 SEI 模型；
(c) 1997 年 Peled 提出的马赛克模型；(d) Aurbach 提出的多层结构模型

谱、Raman 光谱、电化学阻抗谱、XPS 等做了大量工作[10-13]，提出多层结构模型［见图 4-3（d）］，以金属锂为例，新鲜的金属锂浸泡于电解液，由于其活泼的金属性，锂金属与电解液成分发生反应形成一层表面膜，这种反应可认为是自发的，选择性低（low selectivity）；之后的电化学过程，电解液会继续得电子，此时的反应选择性高（high selectivity），产物与第一步不同；后续长循环过程，形成的表面膜可部分溶解于电解液，再者电极材料体积形变、电解液中痕量水也将影响 SEI 膜的成分与结构。SEI 膜的多层结构通过分子动力学模拟也可以构建[14]，如图 4-4 所示。

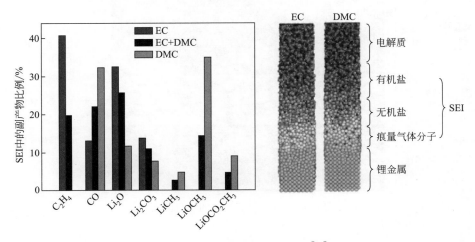

图 4-4 SEI 膜成分及其分布图[14]

分别计算了溶剂碳酸乙烯酯（EC）、碳酸二甲酯（DMC）以及两者共存，锂金属表面 SEI 膜的生成情况，图 4-4 中可以看出只有 EC 时，SEI 的成分为 C_2H_4，CO，Li_2O，Li_2CO_3 等气体和无机物；当只有 DMC 时，SEI 主要成分为 $LiOCH_3$，$LiOCO_2CH_3$ 有机物，当两者共存时，产物为靠近锂金属表面的无机物内层和靠近电解液的有机物外层组成，与上文提到的多层模型基本吻合。

Tang 等[15]使用核磁共振（NMR）探针原位研究了恒电流循环过程中 rGO 中 SEI 钝化层的生长和演化。结果表明，大部分 SEI 层在第一个周期内形成，并在第二个充电周期结束时变得稳定。同时，提出了一种定量方法来确定与 SEI 形成相关的不可逆 Li^+ 的数量，该方法使用了垂直于平坦锂金属表面的纳米/微米结构的锂金属纤维所产生良好可分离的共振峰。根据 NMR 的研究，在第一个循环中的插层反应 Li^+ 的量计算为 36%，这与实际的电池测试结果一致。

Michael 等[16]提出 SEI 层内部有无机无孔层以及由有机、无机和聚合物电解质还原产物组成的外部多孔层（图 4-5），其运输过程包括表面反应、通过内部 SEI 的离子迁移、内部-外部 SEI 界面的溶剂化以及通过外部多孔 SEI 的电解质迁移。而在实际研究过程中，通常并未观察到 SEI 的外部多孔层，这可能是在做 TEM 等表征前往往会进行清洗、晾干等操作导致该多孔层被破坏[17]。Zhou 等[18]通过原位液相质谱观察也证实了 SEI 外部多孔层的存在。

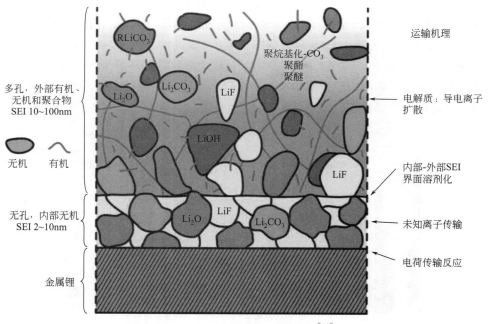

图 4-5　SEI 和相关过程的示意图[16]

4.3 SEI 膜表征手段

由于 SEI 膜具有多层无序结构，成分复杂，SEI 膜的成长受电极材料、电解液成分、温度、充放电条件、电极中的黏结剂和导电添加剂、集流体等多种因素影响，SEI 的成分与微观结构基本不存在普适性的规律，针对特定的电极与电解质体系需要具体问题具体分析，且不易表征。Novák 等对 SEI 膜表征方法做了总结[19]，包括：俄歇电子能谱（AES），飞行时间二次离子质谱（ToF-SIMS），扫描探针显微镜（SPM），扫描电子显微镜（SEM），透射电子显微镜（TEM），红外吸收光谱（IRAS），拉曼光谱（RS），X 射线衍射（XRD），电子能量损失谱（ELLS），X 射线近边吸收谱（XANES），电化学阻抗谱（EIS），差分扫描量热仪（DSC），程序升温脱附仪（TPD），核磁共振仪（NMR），原子吸收光谱（AAS），电化学石英晶体微天平（EQCM），离子色谱（IC），二次电子聚焦离子束与元素线性扫描分析仪（FIB-ELSA），傅立叶变换红外光谱（FTIR），X 射线光电子能谱（XPS）等。笔者研究团队自 1998 年开始对系列正负极材料的 SEI 膜的形貌、稳定性、力学特性、覆盖度等问题展开了研究[20-27]，在此简介如下。

4.3.1 SEI 膜形貌

TEM 适合探测纳米材料的微观结构，常用来研究电极材料的 SEI 膜形貌。图 4-6(a) 显示 SnO 负极放电（嵌锂）至 0V 时球形颗粒表面有一个完整的半透明壳层介质，推断为 SEI 膜，厚度为几个纳米。这是锂离子电池领域首张 SEI 膜的透射电镜照片。过渡金属氧化物 Cr_2O_3 放电至 0V 时[图 4-6(b)]活性颗粒表面存在一层不均匀的 SEI 膜，厚度在 30～90nm 之间，SEI 膜在被电子束照射后会收缩甚至消失。碳纳米管（CNTs）放电至 0V 时表面也出现一层均匀的无定形 SEI 膜，厚度在 5～7nm [图 4-6(c)]。图 4-6(d) 显示硬碳球（HCS）在经历 20 周循环后，再放电至 20mV，表面的 SEI 膜厚度为 10～20nm。图 4-6(e) 是 nano-silicon 首周放电至 0V 并短路 24h 后的形貌，由于晶体硅嵌锂后晶格被破坏，形成无定形的 Li-Si 合金，因此区分表面的物质与 Li-Si 合金必须借助于其他手段，对颗粒表面进行电子能量损失谱（ELLS）发现 Li、Si、P 元素的存在，如图 4-6(e) 圈内区域所示，这里 P 元素的存在证实了 nano-silicon 在放电至 0V 时有 SEI 膜的生成，而 P 元素来自于 $LiPF_6$，Li 元素来自 SEI 膜，Si 元素可能为 SEI 膜的成分，也有可能来自原先存在于 nano-silicon 表面的 SiO_x。$Li_4Ti_5O_{12}$ 的嵌锂平台在 1.5V，研究表明在 1.0～3.0V 区间里循环时，颗粒表面不会生成 SEI 膜[28]，当嵌锂至 0V[图 4-6(f)]，此时可以清楚地观察到表面的无定形的 SEI 膜，说明在低电位时电解液还是无法避免被分解生成 SEI 膜，SEI 膜厚度达 20nm。图 4-6(g) 显示 $LiCoO_2$ 颗粒在循环 10 周后表面也会形成 SEI 膜，厚度大概在 5nm 左右。图 4-6(h) 显示 $LiNi_{0.5}Mn_{1.5}O_{4-\delta}$ 在充电至 4.9V 有 SEI 膜生成，其厚度在 2～10nm。

这里讨论 SEI 膜的厚度问题。上节提到 SEI 膜因其电子绝缘的属性使得其生长厚度

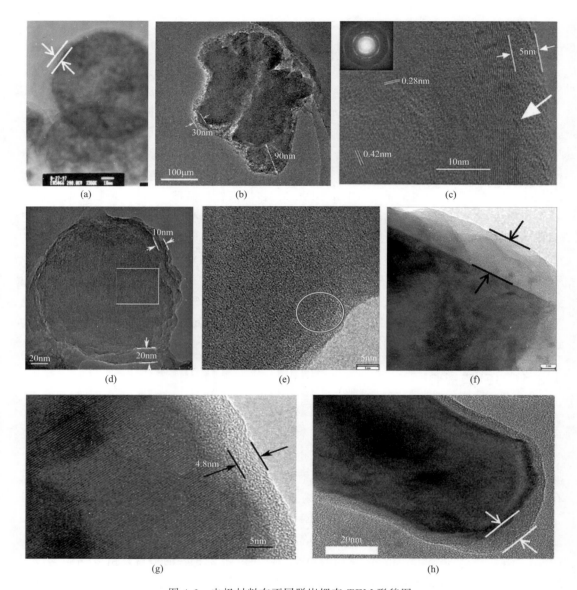

图 4-6 电极材料在不同脱嵌锂态 TEM 形貌图

(a) SnO 嵌锂至 0V；(b) Cr_2O_3 嵌锂至 0V；(c) CNTs 嵌锂至 0V；(d) HCS 20 周循环，嵌锂至 20mV；(e) nano-silicon 嵌锂至 0V；(f) $Li_4Ti_5O_{12}$ 首周嵌锂至 0V；(g) $LiCoO_2$ 循环 10 周；(h) $LiNi_{0.5}Mn_{1.5}O_{4-\delta}$ 充电至 4.9V，4.9V 恒压充电 24h

受限于电子隧穿距离（约 2nm）。然而通过 TEM 观察表明 SEI 膜厚度可以超过，甚至远远大于 2nm。过渡金属氧化物负极材料的 SEI 膜比石墨、碳纳米管、硬碳球大，可达 100nm 以上。另有文献报道 SEI 膜厚度在 100nm 以上[29]。厚的 SEI 膜必将带来大的不可逆容量，目前尚没有文献对如此厚的 SEI 膜的生长机制做出详细明确的解释。SEI 膜如果无法致密地覆盖电极层、SEI 膜的结构为多孔不均匀结构、SEI 膜中存在利于电子传输的缺陷，都有可能使得其厚度超过一般的电子隧穿距离。

由于 SEI 膜对电子束很敏感，在常规透射电镜下难以保持原有的化学状态，无法实现纳米尺度的原位观测。崔屹等[30,31]通过冷冻电镜技术在 2017 年首次拍摄出了具有原子级分辨率的 SEI 膜透射电镜照片，研究了 Li 原子及 SEI 膜界面处原子的分布情况（图 4-7）。作者还发现了枝晶的生长状况并揭示了不同电解液中 SEI 膜的结构。他们进一步确认了具有不同结构 SEI 膜的金属锂在脱锂过程中的形貌变化。通过研究发现，具有马赛克结构 SEI 的金属锂脱锂不均匀，从而形成大量失去与电极接触的金属锂，俗称"死锂"导致电池循环效率降低。与之相比，具有层状结构 SEI 的金属锂脱锂均匀，残留的"死锂"较少，所以循环效率也较高。经过对 SEI 中无机物（Li_2O、Li_2CO_3 等）纳米颗粒分布的分析，他们发现造成这种现象的原因是：在马赛克结构 SEI 中无机物纳米颗粒密度不均匀，无机物纳米颗粒含量高的区域锂离子传导速度快、脱锂速度快，当该处的金属锂被脱光后剩余的金属锂无法维持与电极的通路进而成为死锂。在层状结构 SEI 中无机物纳米颗粒的密度较为均匀，各处锂离子传导速率相当，从而可以均匀脱锂。

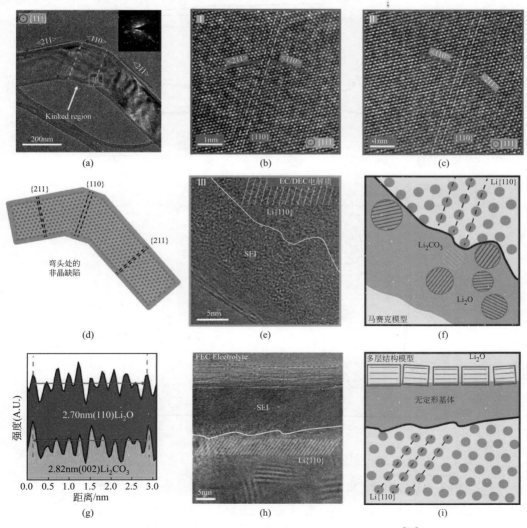

图 4-7　Li 金属枝晶和 SEI 界面的原子分辨率 TEM 分析[30]

4.3.2 SEI 膜的组成分析

拉曼光谱（RS）常用来研究界面区域的化学组成及其微结构，而 SEI 膜较薄，导致拉曼信号太弱而难以测量。中科院物理所团队最早利用表面增强拉曼光谱（SERS）研究了 Li/Ag 电池体系中 Ag 表面的 SEI 膜的形成，电解液为 1mol/L $LiPF_6$（EC：DEC＝1：1）及 1mol/L $LiClO_4$（PC：DMC＝1：1），其结果见图 4-8[32]。相对于普通的电极材料，可以观察到大量不属于电极材料的振动模式。

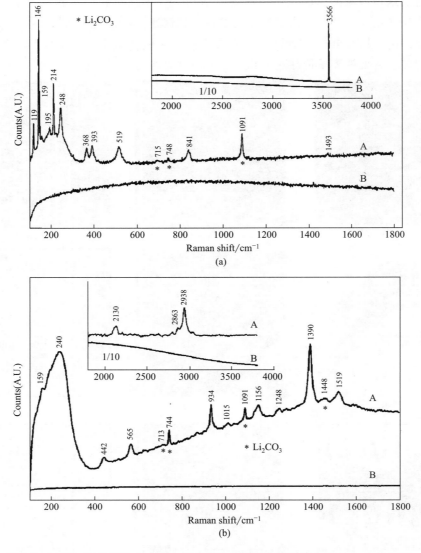

图 4-8　$LiPF_6$ 体系电极 Ag(A)，Al(B) 放电至 0V 的拉曼光谱（a）和 $LiClO_4$ 体系 Ag(A)，Al(B) 放电至 0V 的拉曼光谱，插图为长波数部分（b）[32]

图 4-8(a) 中 715cm^{-1}、748cm^{-1}、1091cm^{-1} 处的振动峰被指认为 Li_2CO_3，为 SEI 膜的主要组成成分；119cm^{-1}、146cm^{-1}、214cm^{-1}、248cm^{-1}、368cm^{-1}、393cm^{-1}、519cm^{-1}、841cm^{-1}、1493cm^{-1} 为有机物 $ROCO_2Li$。在 EC-DEC 体系中，其分解产物分别为 $(LiOCO_2CH_2)_2$ 与 $CH_3CH_2OCO_2Li$。图 4-8(b) 中同样存在 713cm^{-1}、744cm^{-1}、1091cm^{-1} 为 Li_2CO_3 的振动峰，而 240cm^{-1}、442cm^{-1}、565cm^{-1}、1015cm^{-1}、1248cm^{-1}、1390cm^{-1}、1519cm^{-1}、2130cm^{-1}、2863cm^{-1} 处的振动峰来自 PC 的分解产物 $CH_3CH(OCO_2Li)CH_2OCO_2Li$，及 DMC 的分解产物 CH_3OCO_2Li。对应的 B 曲线 Li/Al 体系中却没有观察到拉曼信号，说明证实放电态的 Ag 对 SEI 膜的拉曼信号有增强作用而 Al 没有。

Nanda 等[33]采用尖端增强拉曼光谱（TERS）研究了恒电流循环不同圈数的非晶硅（a-Si）薄膜阳极，并提出了一种在 a-Si 阳极上的纳米马赛克-多层混合 SEI 模型（图 4-9）。对于 1 倍循环的 a-Si，TERS 在固体电解质相间（SEI）形貌与化学图谱之间显示出良好的相关性，对应于碳酸二亚乙酯锂（LEDC）和类似聚环氧乙烷（PEO）低聚物物质的分布。随后的电化学循环使 SEI 层变得相对较厚且粗糙，对于 5 倍循环 a-Si，其化学成分主要由 LEDC 单体二聚体主导。对于 20 倍循环 a-Si，TERS 信号主要由各种构型和氟化物的羧酸盐（RCO_2Li）化合物（$Li_xPO_yF_z$）产生。

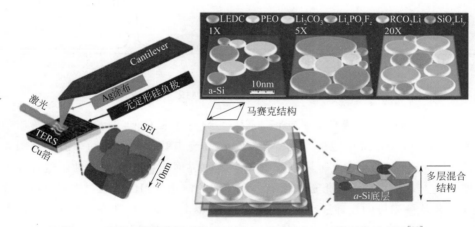

图 4-9　TERS 实验装置和纳米马赛克-多层混合 SEI 模型的示意图[33]

物理所曾艳等[34]利用 TG-DSC-MS 研究 Cr_2O_3 在首周放电至 0.005V，之后充电至 1.1V、2.0V、3.0V 时 SEI 膜的变化，温度范围为 40~500℃。有机物在高温下分解有其独特的键断裂方式，形成离子碎片，离子碎片被质谱仪检测到，显示出不同的荷质比响应峰，通过分析该温度范围内 SEI 膜的热分解产物，比对参考化合物的 TG-DSC-MS 曲线，可以初步分析 SEI 膜中热分解的产物包括 $ROCO_2Li$、$ROLi$、含（CH_2—CH_2—O）的低聚物和 PEO 等。

Wang 等[35]通过液态核磁共振光谱（NMR）对石墨阳极上生长的 SEI 进行直接比较

（图 4-10），其研究结果表明碳酸亚乙酯锂（LEMC）可能是 SEI 的主要成分，而不是人们通常认为的碳酸二亚乙酯锂（LEDC）。LEMC 和碳酸锂（LMC）的单晶 X 射线衍射研究显示了不寻常的层状结构和 Li^+ 协调环境。LEMC 的 Li^+ 电导率 $>1\times10^{-6}$ S/cm，而 LEDC 几乎是离子绝缘体。

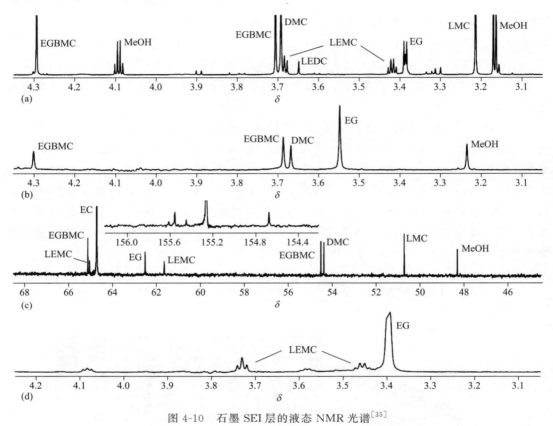

图 4-10 石墨 SEI 层的液态 NMR 光谱[35]

Zhang 等[36]通过 X 射线光电子能谱（XPS）研究了 $LiCoO_2$ 上的阴极电解质中间相（CEI）层。在 $LiCoO_2$/Li 电池中，观察到 CEI 层在充电和放电时的动态演变（图 4-11）。基于定量 XPS 分析，已经建立了阴极和锂阳极之间界面产物的强相关性。这种相关性主要源于锂阳极上 SEI 层的可逆形成和溶解。CEI 层的演化可归因于通过电解质的顺序反应以及 SEI 碎片从锂阳极的物理迁移。在 $LiCoO_2$/石墨电池中，CEI 在 $LiCoO_2$ 上的变化由于在石墨阳极上形成的相对稳定的固体电解质中间相（SEI）层而变得不太重要，这进一步支持阴极上的 CEI 与锂阳极上的固体电解质中间相（SEI）之间的强相关性。这些结果揭示了在 $LiCoO_2$ 上 CEI 的动力学演变的起源，并强调指出，在解释 CEI 在阴极上时，尤其是在使用锂作为阳极的可充电锂电池中，应考虑阳极的影响。

4.3.3 SEI 膜热电化学稳定性

稳定的 SEI 膜可改善电极材料界面特性，有利于提高电池长期循环性能。Aurbach 等[11]在研究金属锂和石墨负极材料的 SEI 膜问题时，发现锂金属充放电过程中 SEI 膜不

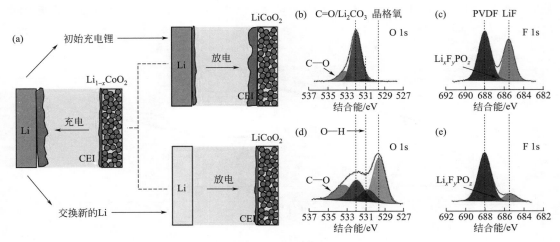

图 4-11 验证实验程序的示意图

稳定,锂金属在充放电过程中体积变化较大,容易破坏 SEI 膜,导致新鲜的锂表面暴露于电解液而生成更厚的 SEI 膜;而对石墨负极来说,锂离子在石墨层间的脱嵌引起的体积变化较小,故其 SEI 膜形成且在后续的循环中较稳定,因此石墨负极比锂金属具备更好的循环性。SEI 膜的电化学不稳定性还体现在其他负极。前面提到的 CNTs[22]在放电末态其 SEI 膜厚度为 5~7nm,而充电至 3.5V 后最大厚度仅为 3nm,说明 SEI 膜在充电至高电位时可分解。HCS[23]的 SEI 膜在放电末态比充电态明显要厚,同样说明 SEI 膜在充电时不稳定,易分解。Cr_2O_3[21]、MnO[27]在放电时能生成非常厚的 SEI 膜,充电时也观察到其厚度减薄现象。利用 TG-DSC-MS[34]对不同状态 Cr_2O_3 的 SEI 膜成分进行热分析,表明 $ROCO_2Li$、ROLi 是电化学稳定的物质;而电化学不稳定的物质为含(CH_2—CH_2—O)的低聚物和 PEO 等。

Andersson 等[37,38]认为在高温时 SEI 膜中有机成分如烷基酯锂不稳定,可分解生成 Li_2CO_3。曾艳的结果也证实 SEI 膜中有机与聚合物成分的热稳定性较差,会在不同的温度分解[34]。

4.3.4 SEI 膜力学特性及覆盖度分析

SEI 膜的力学特性较少被研究,在此介绍一种基于 AFM 的方法及其在探测 SEI 膜方面的应用。AFM 具有较高的分辨率,其工作基本原理是利用尖端尖锐的探针,在扫描成像过程中通过控制探针与样品之间的作用力以得到材料的形貌分布。其工作模式主要有两种:接触模式(contact mode)和轻敲模式(tapping mode)。通过对针尖的修饰,可实现对样品除形貌以外的微区物理化学属性测量,如电学及力学测量。导电原子力显微镜(C-AFM)利用导电针尖,通过在针尖施加直流或交流激励,探测样品微区导电特性。电化学原子力显微镜(EC-AFM)是针对电化学测量而设计的,可原位或非原位探测电极表面在锂离子脱嵌过程的形貌及 SEI 膜的变化[39-41]。而利用 AFM 来探测样品力学属性,在锂离子领域中应用较少,生物医学领域里用该方法来探测细胞是否发生病变[42]。由于探针固定于有一定弹性的悬臂梁,探针与样品相互作用过程中悬臂将发生弹性形变,因而悬臂梁的形变量大小可用来反映样品的力学属性,将施加于样品上力大小与样品形变之间的关系称为力曲线。原理见图 4-12[43]。

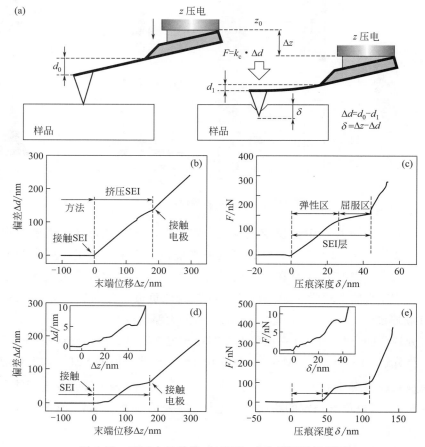

图 4-12 原子力显微镜(AFM)力曲线测量示意图

图 4-12(a) 显示的是当针尖靠近样品而未接触样品时,悬臂梁处于自然状态(d_0),当针尖继续向样品方向移动并且扎入样品内部,此时针尖与样品都将发生形变,分别为 $\Delta d = d_0 - d_1$,δ;由几何关系可知 $\delta = \Delta z - \Delta d$,式中,$\Delta z$ 为带动针尖移动的压电陶瓷在 z 轴的位移,此过程中样品受到的力可被实时测量:$F = k_c \cdot \Delta d$(k_c 为探针悬臂力学常数),由此得到力曲线 F vs.δ。根据样品的软硬程度不同,可探测 SEI 膜的分层结构。图 4-12(c) 为单层 SEI 膜结构,由图(b)经过变换得到,$\delta = 0$ 代表针尖接触样品表面,开始相互作用,在 $0 < \delta < 45$nm,为 SEI 膜的力学响应区域,呈现弹性区(elastic region)和屈服区(yield region),SEI 膜的厚度为该两区域的厚度总和;在 $\delta > 45$nm 区域,力曲线斜率较大,说明相互作用力持续增大的同时样品的形变量很小,由此推断针尖已经扎穿单层 SEI 膜并接触底端电极部分,由于电极硬度较针尖悬臂梁大很多,施加在样品上的力不足以引起电极明显的形变;图 4-12(d) 和(e) 显示 SEI 膜两层结构。$0 < \delta < 45$nm 为一层[弹性区,屈服区见图 4-12(e) 插图],然而针尖扎穿第一层,并没有图(c)显示的接触到底端电极,而是在 45nm $< \delta <$ 110nm 区域出现另外一组弹性区;屈服区,$\delta >$

110nm 的响应部分来自电极，由此可见此处探测得到的 SEI 膜为两层结构，其厚度分别为 45nm、65nm。另外，如果力曲线中不包含弹性区和屈服区，只存在近似于图 4-12(c) 中 $\delta>43$nm 的部分，那么认为该区域没有被 SEI 膜覆盖。

以磁控溅射制备的非晶硅薄膜作为工作电极，金属锂箔作对电极，电解液是 1mol/L LiPF$_6$（EC：DMC 体积比为 1∶1）含有 2%的碳酸亚乙烯酯（VC），组装 Swaglock 型模拟电池。选择首周室温恒电流放电至 1V，接着 1V 恒电位放电 48h，放电结束后在手套箱里拆除电池，用 DMC 清洗极片上残留的电解液溶质，真空抽干 DMC 后将样品转移到惰性气氛保护的 AFM（Bruker，MultiMode 8），利用力曲线方法观察硅薄膜经循环后表面的 SEI 膜生长情况。我们在样品上任意选取 $45\mu m\times 45\mu m$ 的区域进行 225 条力曲线实验测量，通过分析每条力曲线上所得 SEI 膜的厚度，可以得到 SEI 膜厚度的统计信息［见图 4-13(a)］，图 4-13(a) 可以看出 SEI 膜厚度在 0~30nm 之间，而主要集中于 0~10nm。利用力曲线信息还可以得到 SEI 膜的杨氏模量（Young Modulus）[43]。实验所采用的 AFM 探针针尖为锥形，针尖半径 $r=2$nm，可采用 Sneddon 模型：$F=(2/\pi)[E/(1-\nu^2)]\delta^2\tan(\alpha)$，式中 F 为针尖样品相互作用力；E 为杨氏模量；ν 为泊松系数（这里设为 0.5）；δ 为力曲线上弹性区的厚度；$\alpha=20°$ 为探针针尖半角。SEI 膜厚度-杨氏模量统计信息见图 4-13(b)，图中可看出此时的 SEI 膜都是单层结构，杨氏模量几乎全部大于 200MPa。结合厚度和杨氏模量信息，可以将硅表面的 SEI 膜三维结构呈现出来，具体过程如下：选择 $45\mu m\times 45\mu m$ 的区域，在该区域上均匀地作 225 条力曲线，因此每条力曲

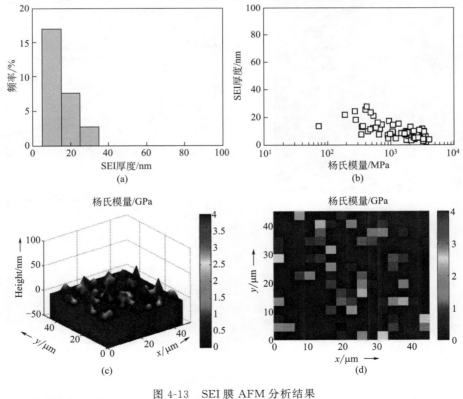

图 4-13 SEI 膜 AFM 分析结果

(a) SEI 膜厚度；(b) SEI 膜厚 vs. 杨氏模量；(c) SEI 膜三维成像；(d) SEI 膜二维投影图

线的 x，y 轴坐标可以随之确定，作为 SEI 膜的 x-y 平面坐标；再者，力曲线方法测量得到 SEI 膜的厚度，作为 SEI 膜的 z 轴坐标；将 SEI 膜的杨氏模量用不同颜色表示（0～4GPa），三维成像图见图 4-13(c)，图 4-13(d) 表示图 (c) 在 x-y 平面的投影图，说明在首周放电至 1V，1V 恒电位放电 48h 后 SEI 膜并没有全部覆盖电极表面，覆盖度只有 28%，生长过程是非均匀的。此方法针对电极上一个较大的随机区域对 SEI 膜进行测量，因此可较精确地得到 SEI 膜的覆盖度信息，结果表明 SEI 膜是持续生成的过程。

4.3.5 锂离子在 SEI 膜中的输运

锂电池充电放电过程中，锂离子要往返从活性物质一侧穿过 SEI 膜到电解液一侧，由于 SEI 膜结构不均匀，成分也不单一，故研究锂离子在 SEI 膜中的传输机理较为不易。有研究指出锂离子在 SEI 膜中的输运步骤是速率控制步骤[8,44]。施思齐等通过 TOF-SIMS、TEM 并结合密度泛函理论（DFT）研究锂离子在 SEI 膜中的输运问题并提出 "two layer/two mechanism"[45]。如图 4-14 所示。

将 SEI 膜沉积在铜箔上，透射电镜观察到 SEI 膜 [图 4-14(a)] 为两层结构，铜箔左边为浅灰色的有机物层和深灰色的无机物层，总厚度为 10～20nm。同位素交换实验见图

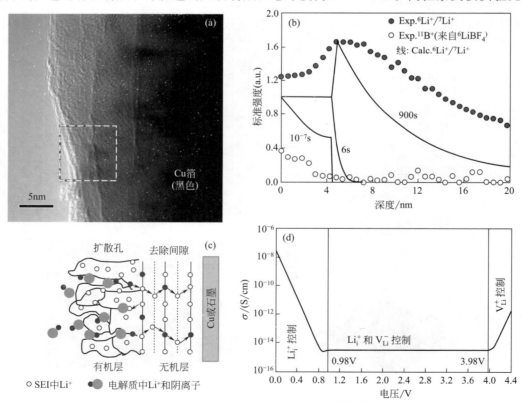

图 4-14　锂离子在 SEI 膜中的输运[45,46]

(a) SEI 膜 TEM 形貌；(b) TOF-SIMS，元素在剖面的分布；(c) SEI 中锂离子传输机制；
(d) 不同电位下锂离子在 Li_2CO_3 中的传输机制

4-14(b),实验中将含^{7}Li$^+$的SEI膜浸泡在^{6}Li^{11}BF$_4$电解液中900s,从图中看出^{6}Li$^+$/^{7}Li$^+$比例的峰值出现在距电解液/SEI膜5nm处,而不是电解液/SEI膜界面处,由此证实锂离子在SEI膜从电解液一侧输运到电极一侧,且其在SEI两不同成分中输运机制不同,而5nm近似为靠近电解液有机物SEI膜的厚度。DFT计算结果如图4-14(c)所示:锂离子在多孔的有机物SEI层通过孔扩散机制输运,而在较为致密的无机物层中则通过锂离子去除孔(interstitial knock-off)机制输运。施思齐等[46]又通过DFT计算研究0~4.4V区间锂离子在Li$_2$CO$_3$中的输运[见图4-14(d)],发现在0.98V以下,锂离子传输主要依赖间隙位锂离子(Li$_i^+$),在3.98V以上依赖于锂离子空位(V$_{Li}^-$)传输,而在0.98~3.98V区间内为两者共同作用的传输机制。

4.3.6 SEI膜的动态生长过程

电解液在高低电位(超过电化学窗口)不稳定被分解形成SEI膜。该机理在负极表面得到大量实验和理论的验证;而正极表面的SEI膜形成机理与负极并不相同,仍存在不清楚的地方。目前普遍接受关于正极SEI膜的形成机理有以下三种[47]:①漂移机理,即认为正极表面SEI膜是有机电解液在负极上的还原产物在电解液中饱和后,扩散并沉积到正极材料的表面形成的;②亲核反应机理,即认为可能是由有机电解液中亲电性的溶剂分子(如EC,DMC)与带负电的正极材料发生亲核反应或在正极上被氧化后的不溶产物沉积在电极表面形成的;③自发反应机理,即认为是由于正极材料与电解液之间自发反应的产物。

笔者团队结合相关研究,提出了SEI膜的动态生长过程。如图4-15所示。以负极为例,有机溶剂在负极表面在不同电位下得到电子后还原形成不同的自由基或阴离子,与受

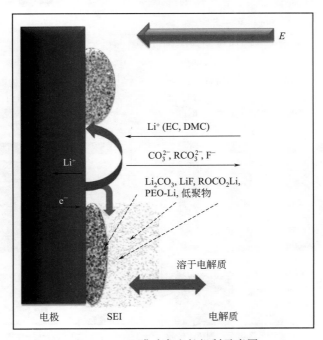

图4-15 SEI膜动态生长机制示意图

电场驱动扩散而来的锂离子相遇先后形成多种化合物,这些化合物在电极表面具有较高的浓度,形成了不溶于电解液或可溶于电解液的物质。不溶于电解液的物质在电极的表面能较高的地方优先沉积出来,形成岛状成长的早期的 SEI 膜沉积物,进一步还原产物或者沉积在裸露的电极表面,或者沉积覆盖在早期的沉积物上。这取决于裸露表面还是 SEI 膜早期沉积物表面的表面能以及集合形状。从图 4-13(c)的研究结果看,SEI 膜应该是动态生长的过程,且确实存在累积生长的情况,导致 SEI 膜不均匀覆盖在电极表面,造成 SEI 膜厚度超过电子隧穿距离,因为始终裸露的表面能够不断提供可漂移可沉积的 SEI 还原产物,进行后续的非均匀生长。

Michan 等[48]通过 ^{7}Li、^{19}F 和 ^{13}C 固态核磁共振光谱法监测了 Si 负极的 SEI 在多个电化学循环中的动态变化(图 4-16)。通过使用聚焦离子束和扫描电子显微镜,确证 SEI 的增长与电极弯曲度的增加相关。建立了锂化能力损失的两阶段模型:最初,锂化能力稳步下降,Li$^+$ 稳定地不可逆地消耗,并且可以看到明显的 SEI 增长。后来,低于初始锂化能力的 50% 时,去锂化的 Si 较少,从而导致体积膨胀和收缩的减少;Li$^+$ 的不可逆消耗率降低,Si 的 SEI 层厚度变得稳定。锂化能力的下降主要归因于动力学,电极弯曲度的增加严重限制了 Li$^+$ 扩散通过整个电极。在电化学容量曲线中看到的锂化过程中的最终变化归因于不均匀的锂化,该反应开始于隔膜附近/颗粒表面上。

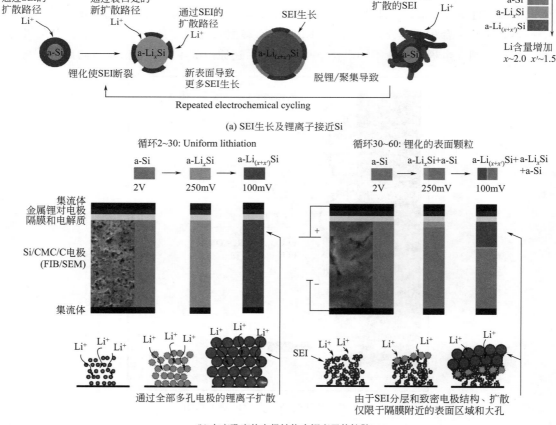

(a) SEI生长及锂离子接近Si

(b) 在变致密的电极结构中锂离子的扩散

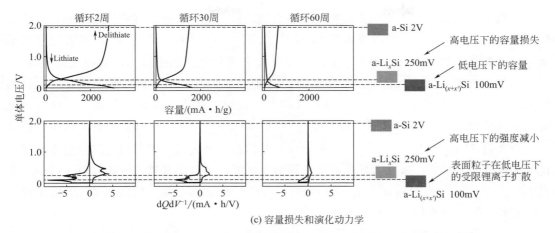

(c) 容量损失和演化动力学

图 4-16　作为 Si 复合电极失效机理的 SEI 生长和 Li^+ 扩散的多尺度模型[48]

4.4　界面改性

由于不稳定的 SEI 膜，裸露的表面会消耗锂离子电池中正极的锂源，影响循环性、充放电效率，增大电池内阻，产生气体，降低安全性，为了稳定电极表面，通常会在电极材料或电极表面进行表面修饰，或者在电解质中添加能形成更稳定 SEI 膜的添加剂。这些措施对于采用超细粉体、纳米电极材料尤为重要。图 4-17 给出了几种常见锂电池正负极材料脱嵌锂电位及相应的界面问题，处在电解液电化学窗口（stable window）外的电极材料充放电过程电解液将不可避免会发生副反应。为了使电极材料，特别是纳米材料在宽电位范围能够工作，需要形成核壳结构等来拓展可应用的电位范围。

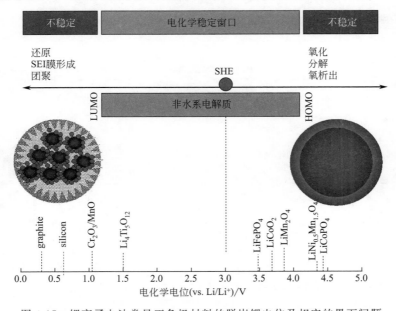

图 4-17　锂离子电池常见正负极材料的脱嵌锂电位及相应的界面问题

材料的表面修饰对于提高界面特性是显著的。Koch 等[49]通过理论计算显示，相比于 O_2、CO_2、F_2 和 SO_2 气体，考虑到 Li_3N 中的高离子电导率，经过 N_2 处理的层可以形成最具有弹性的钝化层，可最大程度地减少裂纹形成和锂枝晶生长的可能性。何宇等[50]通过原子层沉积（ALD）在图形化硅薄膜上包覆 2nm 的 Al_2O_3，硅薄膜的循环性、每周库仑效率能得到显著改善，见图 4-18。每周库仑效率的提高，反映了界面稳定性的显著提高。Hu 等[51]基于 α-氰基丙烯酸乙酯（ECA）前驱体与 $LiNO_3$ 添加剂的原位聚合设计了一个人造 SEI 层（图 4-19）。CN^- 和 NO_3^- 基团在循环过程中可与锂金属反应形成具有优异力学性能的聚（α-氰基丙烯酸乙酯）界面无机层，从而促进离子传导并阻止进一步不良的界面反应。使得含有该人造 SEI 膜的 $LiFePO_4$ 锂金属电池，既使以 2C 的倍率经过 500 次循环后，仍具有 93% 的容量保持率。

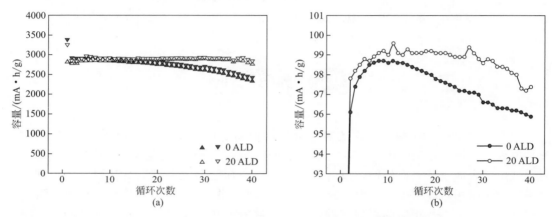

图 4-18　0ALD（未包覆）与 20ALD（2nm Al_2O_3）包覆硅薄膜循环性（a）与库仑效率（b）对比（电压范围为 0.005~2V）[50]

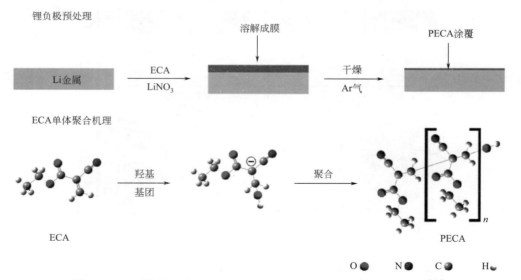

图 4-19　金属锂的预处理过程示意图以及 ECA 单体的相应聚合机理[51]

实际上，Wang 等[52]将人造 SEI 层材料分为完全锂化和非完全锂化的材料。Li_2O、Li_2CO_3、LiF、Li_3N 等已完全锂化，在电池运行期间不会被锂化。但是，由于 Li 具有最低的还原电势，任何非完全锂化涂层如氧化物、硫化物、碳等都可以被锂化，如 Li 可以代替 M_xO_y 中的 M 形成 Li_2O+M，并且 M 也可以被连续地锂化。如 Kim 等[53]使用 DFT 计算表明 SiO_2 和 Al_2O_3 开始在 Si（约 0.4eV Si 锂化）之前被锂化，分别在 0.68V 和 0.9eV 左右。通过 ReaxFF MD 对 Si@氧化物核壳进行模拟，进一步证明当 Si 核完全锂化时，SiO_2 和 Al_2O_3 壳的组成将高度锂化[54]。Choi 等[55]将约 10nm 的二维 MoS_2 包覆在锂金属负极外并进行锂化，层状 MoS_2 纳米结构在锂化过程中的相变反应减少了负极材料的界面电阻，实现了包覆层紧密接触和较高的 Li^+ 传输效率（图 4-20）。以上这些研究表明，当我们使用非完全锂化的材料修饰材料表面时，还需要进一步考虑锂化过程对人造 SEI 层形貌、电子及离子传输、力学性能等因素的影响。

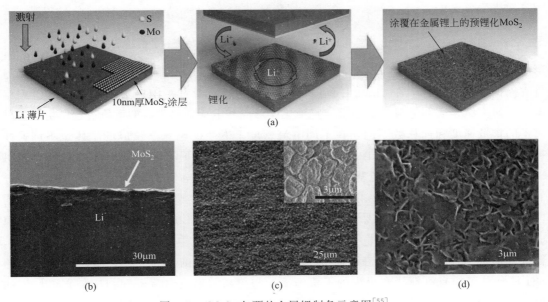

图 4-20　MoS_2 包覆的金属锂制备示意图[55]

采用电解液成膜添加剂，例如碳酸亚乙烯酯（VC）能显著改善石墨、硅负极的循环性能[13,56]，氟代碳酸乙烯酯（FEC）改善硅负极性能[57,58]，双草酸基硼酸锂（LiBOB）可改善 $LiNiO_2$-graphite 全电池性能[59,60]。少量添加剂的使用不会改变主体电解液的性质，其作用体现在对 SEI 膜的改性，期望得到均匀且更加致密，力学性能良好足以抵抗材料体积形变，电池长循环过程结构稳定的 SEI 膜。如 Soto 等[61]根据 ab initio MD（AIMD）计算得出 EC 溶剂易形成不稳定的 SEI 物质（例如 Li_2EDC 和低聚物），它们容易受到自由基的攻击，导致 SEI 持续增长，而添加剂（例如 VC 或 FEC）可导致生成更稳定的聚合物，从而使 SEI 外层更加紧凑和实现受控的 SEI 增长（图 4-21）。

此外，温度、电流密度、压力等因素对形成 SEI 的质量影响很大。Rodrigues 等[62]发现使用离子液体电解质并将形成温度提高到 90℃ 将使在石墨表面上形成的 SEI 变厚，具有更好的热稳定性和自放电性能。另外，电流密度和压力对形成 SEI 的质量影响很大。

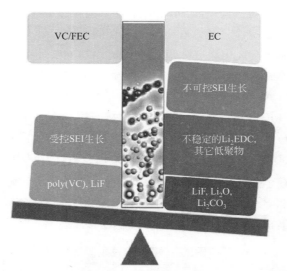

图 4-21　基于 AIMD 计算比较 SEI 的生长原理指出了还原型物种的化学稳定性[61]

形成电压主要影响成膜反应路径，而形成电流密度主要影响成膜反应速率。Ota 等[63]通过表面探针技术分析了在 1mol/L LiPF$_6$ PC/ES 电解液中石墨阳极上形成 SEI 的过程组合。其研究结果表明电极表面的反应是钝化膜形成和电荷转移之间的竞争过程。由于各种离子的扩散速度和迁移数不同，在不同电流密度下发生的电化学反应的主体也不同，并且所产生的 SEI 具有不同的性质。在高电流密度的情况下，首先在高电势下形成无机 SEI，然后在发生锂嵌入的电势附近形成有机 SEI；在低电流密度下，只含有有机组分的 SEI 在起始电势（1.5V vs. Li/Li$^+$就开始形成。因此，操纵形成电流密度对于获得均匀、紧凑和薄的 SEI 十分重要。Zhu 等[64]通过调节 SEI 形成电流密度，调节 SEI 的组成和结构以优化石墨电极的电化学性能。结果表明，SEI 主要形成在 1.1～1.4V 之间，较低的形成电流密度有利于在石墨电极表面上形成优异的 SEI。在这种条件下形成的 SEI 具有更多的有机锂盐和更少的无机锂盐，与更高的形成电流密度相比，它更均匀地包裹在石墨阳极的表面上。同时，衍生的 SEI 更稳定和更厚，可以有效地稳定电极/电解质的界面，从而在形成步骤之后增强石墨负极材料的循环稳定性，从而缓冲其在循环中的体积变化。

目前，表面修饰、成膜添加剂、形成温度、形成电流等界面改性的效果通过半电池或全电池得到了验证，但其机制还不是非常清楚，有待综合多种表征手段来澄清，以利于电极材料与电极的进一步开发和优化。

4.5　本章结语

采用非水有机溶剂的锂离子电池中固固界面和固液界面对输运、储存、反应均有显著的影响。由于界面层的结构为纳米尺度，局部组成和结构不均匀，组成复杂，影响组成与结构的条件众多，导致目前对电极材料界面特性的控制主要是根据大量的实验获得。目前，高通量的计算已开始应用，计算各类溶剂的 HOMO、LUMO 轨道，预测还原或氧化产物，分析和模拟固固、固液界面；具有高空间分辨率、时间分辨率、能量分辨率、化学信息监测的原位、非原位表征手段也在迅速发展，这些均有利于对界面问题的深入认识。尽管还不能提供明确的理论指导，新的表面修饰方法、更稳定多功能的表面修饰材料、更好的添加剂预计在未来的几年内仍然是当前锂离子电池技术发展的核心。从更长远考虑，全固态电池会显著改变目前由于使用液体电解质带来的复杂的科学与技术问题，也可能会产生全新的固固界面问题，还有待进一步研究[65]。

本章需要进一步思考的基础科学问题：

1. 电荷转移过程与 SEI 膜形成存在什么样的关系？
2. SEI 成膜均匀性与正负极表面电子电导率的关系是什么？
3. SEI 膜成膜均匀性、覆盖度与电流密度的关系是什么？
4. 高温下 SEI 膜的溶解、分解行为和产物怎样？
5. 如何确定 SEI 膜厚度？
6. SEI 膜的分解电位如何确定？哪些正极表面存在 SEI 膜？生长机制如何？是否会电化学分解？
7. 哪些离子导电、电子绝缘材料在 4.9V 不分解？
8. 如何实现人造 SEI 膜？
9. SEI 离子电导和电子电导如何测量？
10. SEI 膜热分解温度多高？热分解产物是什么？如何形成热稳定性好的 SEI 膜？

参考文献

[1] Tarascon J, Armand M. Issues and challenges facing rechargeable lithium batteries [J]. Nature, 2001, 414: 359.
[2] Armand M, Tarascon J M. Building better batteries [J]. Nature, 2008, 451: 652-657.
[3] Li H, Wang Z, Chen L, Huang X. Research on advanced materials for li-ion batteries [J]. Advanced Materials, 2009, 21: 4593-4607.
[4] Alper J. The Battery: Not yet a Terminal Case [J]. Science, 2002, 296: 1224-1226.
[5] Goodenough J B, Kim Y. Challenges for rechargeable Li batteries [J]. Chemistry of Materials, 2010, 22: 587-603.
[6] Xu K. Nonaqueous liquid electrolytes for lithium-based rechargeable batteries [J]. Chemical Reviews-Columbus, 2004, 104: 4303-4418.
[7] Dey A N. Film formation on lithium anode in propylene carbonate [J]. Journal of the Electrochemical Society, 1970, 117: C248.
[8] Peled E. The Electrochemical behavior of alkali and alkaline earth metals in nonaqueous battery systems—the solid electrolyte interphase model [J]. Journal of the Electrochemical Society, 1979, 126: 2047.
[9] Peled E, Golodnitsky D, Ardel G. Advanced model for solid electrolyte interphase electrodes in liquid and polymer electrolytes [J]. Journal of the Electrochemical Society, 1997, 144: L208-L210.
[10] Aurbach D, Markovsky B, Levi M, et al. New Insights into the Interactions between Electrode Materials and Electrolyte Solutions for Advanced Nonaqueous Batteries [J]. Journal of Power Sources, 1999, 81: 95-111.

[11] Aurbach D. Review of selected electrode—solution interactions which determine the performance of Li and Li ion batteries [J]. Journal of Power Sources, 2000, 89: 206-218.

[12] Aurbach D, Zinigrad E, Cohen Y, et al. A short review of failure mechanisms of lithium metal and lithiated graphite anodes in liquid electrolyte solutions [J]. Solid State Ionics, 2002, 148: 405-416.

[13] Aurbach D, Gamolsky K, Markovsky B, et al. On the use of vinylene carbonate (vc) as an additive to electrolyte solutions for Li-ion batteries [J]. Electrochimica Acta, 2002, 47: 1423-1439.

[14] Kim S-P, Duin A C T V, Shenoy V B. Effect of electrolytes on the structure and evolution of the solid electrolyte interphase (SEI) in Li-ion batteries: A molecular dynamics study [J]. Journal of Power Sources, 2011, 196: 8590-8597.

[15] Tang W, Goh B-M, Hu M Y, et al. In situ Raman and nuclear magnetic resonance study of trapped lithium in the solid electrolyte interface of reduced graphene oxide [J]. The Journal of Physical Chemistry C, 2016, 120: 2600-2608.

[16] Hess M. Non-linearity of the solid-electrolyte-interphase overpotential [J]. Electrochimica Acta, 2017, 244: 69-76.

[17] Fears T M, Doucet M, Browning J F, et al. Evaluating the solid electrolyte interphase formed on silicon electrodes: A comparison of Ex Situ X-Ray photoelectron spectroscopy and in situ neutron reflectometry [J]. Phys Chem Chem Phys, 2016, 18: 13927-40.

[18] Zhou Y, et al. Real-time mass spectrometric characterization of the solid-electrolyte interphase of a lithium-ion battery [J]. Nat Nanotechnol, 2020.

[19] Verma P, Maire P, Novák P. A review of the features and analyses of the solid electrolyte interphase in Li-ion batteries [J]. Electrochimica Acta, 2010, 55: 6332-6341.

[20] Li H, Huang X, Chen L. Direct imaging of the passivating film and microstructure of nanometer-scale sno anodes in lithium rechargeable batteries [J]. Electrochemical and solid-state letters, 1998, 1: 241-243.

[21] Hu J, Li H, Huang X, et al. Improve the electrochemical performances of Cr_2O_3 anode for lithium ion batteries [J]. Solid State Ionics, 2006, 177: 2791-2799.

[22] Wang Q, Li H, Chen L, et al. Investigation of lithium storage in bamboo-like cnts by hrtem [J]. Journal of the Electrochemical Society, 2003, 150: A1281-A1286.

[23] Hu J, Li H, Huang X. Electrochemical behavior and microstructure variation of hard carbon nano-spherules as anode material for Li-ion batteries [J]. Solid State Ionics, 2007, 178: 265-271.

[24] 舒杰. 锂离子电池界面及负极材料相关问题研究 [D]. 北京: 中国科学院物理研究所, 2007.

[25] Liu N, Li H, Wang Z, et al. Origin of solid electrolyte interphase on nanosized $LiCoO_2$ [J]. Electrochemical and solid-state letters, 2006, 9: A328-A331.

[26] Wang L P, Li H, Huang X J. Electrochemical properties and interfacial reactions of $LiNi_{0.5}Mn_{1.5}O_4$-Delta nanorods [J]. Progress in Natural Science-Materials International, 2012, 22: 207-212.

[27] Zhong K, Xia X, Zhang B, et al. Mno powder as anode active materials for lithium ion batteries [J]. Journal of Power Sources, 2010, 195: 3300-3308.

[28] Ariyoshi K, Yamato R, Ohzuku T. Zero-strain insertion mechanism of Li [$Li_{1/3}Ti_{5/3}$] O_4 for advanced lithium-ion (shuttlecock) batteries [J]. Electrochimica Acta, 2005, 51: 1125-1129.

[29] Schwager F, Geronov Y, Muller R. Ellipsometer studies of surface layers on lithium [J]. Journal of the Electrochemical Society, 1985, 132: 285-289.

[30] Li Y, et al. Atomic structure of sensitive battery materials and interfaces revealed by cryo-electron microscopy [J]. Science, 2017, 358: 506-510.

[31] Li Y, Huang W, Li Y, et al. Correlating structure and function of battery interphases at atomic resolution using cryoelectron microscopy [J]. Joule, 2018, 2: 2167-2177.

[32] Li H, Mo Y, Pei N, et al. Surface-enhanced raman scattering study on passivating films of ag electrodes in lithium batteries [J]. The Journal of Physical Chemistry B, 2000, 104: 8477-8480.

[33] Nanda J, Yang G, Hou T, et al. Unraveling the nanoscale heterogeneity of solid electrolyte interphase using tip-enhanced raman spectroscopy [J]. Joule, 2019, 3: 2001-2019.

[34] Zeng Y, Li L, Li H, et al. TG-MS Analysis on thermal decomposable components in the SEI film on Cr_2O_3 powder anode in Li-ion batteries [J]. Ionics, 2009, 15: 91-96.

[35] Wang L, et al. Identifying the components of the solid-electrolyte interphase in Li-ion batteries [J]. Nat Chem,

2019, 11: 789-796.

[36] Zhang J-N, Li Q, Wang Y, et al. Dynamic evolution of cathode electrolyte interphase (CEI) on high voltage Li-CoO$_2$ cathode and its interaction with Li anode [J]. Energy Storage Materials, 2018, 14: 1-7.

[37] Andersson A M, Edstroöm K. Chemical composition and morphology of the elevated temperature SEI on graphite [J]. Journal of the Electrochemical Society, 2001, 148: A1100.

[38] Andersson A M, Abraham D P, Haasch R, et al. Surface characterization of electrodes from high power lithium-ion batteries [J]. Journal of the Electrochemical Society, 2002, 149: A1358.

[39] Jeong S-K, Inaba M, Iriyama Y, et al. AFM study of surface film formation on a composite graphite electrode in lithium-ion batteries [J]. Journal of Power Sources, 2003, 119-121: 555-560.

[40] Inaba M, Tomiyasu H, Tasaka A, et al. Atomic force microscopy study on the stability of a surface film formed on a graphite negative electrode at elevated temperatures [J]. Langmuir: the ACS Journal of Surfaces and Colloids, 2004, 20: 1348-1355.

[41] Lucas I T, Pollak E, Kostecki R. In Situ AFM Studies of SEI formation at a Sn electrode [J]. Electrochemistry Communications, 2009, 11: 2157-2160.

[42] Plodinec M, Loparic M, Monnier C A, et al. The nanomechanical signature of breast cancer [J]. Nature Nanotechnology, 2012, 7: 757-765.

[43] Zhang J, Wang R, Yang X, et al. Direct observation of inhomogeneous solid electrolyte interphase on mno anode with atomic force microscopy and spectroscopy [J]. Nano Letters, 2012, 12: 2153-2157.

[44] Xu K, von Cresce A, Lee U. Differentiating contributions to "ion transfer" barrier from interphasial resistance and Li$^+$ desolvation at electrolyte/graphite interface [J]. Langmuir: the ACS Journal of Surfaces and Colloids, 2010, 26: 11538-11543.

[45] Shi S, Lu P, Liu Z, et al. Direct calculation of Li-ion transport in the solid electrolyte interphase [J]. Journal of the American Chemical Society, 2012, 134 (37): 15476-15487.

[46] Shi S, Q Y, Li H, et al. Defect thermodynamics and diffusion mechanisms in Li$_2$Co$_3$ and implications for the solid electrolyte interphase in Li-ion batteries [J]. The Journal of Physical Chemistry B, 2013, 117 (7): 8579-8593.

[47] 柳娜. 锂离子电池正极材料的界面及改性研究 [D]. 北京: 中国科学院物理研究所, 2006.

[48] Michan A L, Divitini G, Pell A J, et al. Solid electrolyte interphase growth and capacity loss in silicon electrodes [J]. J Am Chem Soc, 2016, 138: 7918-7931.

[49] Koch S L, Morgan B J, Passerini S, et al. Density functional theory screening of gas-treatment strategies for stabilization of high energy-density lithium metal anodes [J]. Journal of Power Sources, 2015, 296: 150-161.

[50] He Y, Yu X, Wang Y, et al. Alumina-coated patterned amorphous silicon as the anode for a lithium-ion battery with high coulombic efficiency [J]. Advanced Materials, 2011, 23: 4938-4941.

[51] Hu Z, Zhang S, Dong S, et al. Poly (Ethyl α-cyanoacrylate)-based artificial solid electrolyte interphase layer for enhanced interface stability of Li metal anodes [J]. Chemistry of Materials, 2017, 29: 4682-4689.

[52] Wang A, Kadam S, Li H, et al. Review on modeling of the anode solid electrolyte interphase (SEI) for lithium-ion batteries [J]. NPJ Computational Materials, 2018, 4.

[53] Kim S-Y, Qi Y. Property evolution of Al$_2$O$_3$ coated and uncoated Si electrodes: A first principles investigation [J]. Journal of The Electrochemical Society, 2014, 161: F3137-F3143.

[54] Kim S Y, Ostadhossein A, van Duin A C, et al. Self-generated concentration and modulus gradient coating design to protect Si nano-wire electrodes during lithiation [J]. Phys Chem Chem Phys, 2016, 18: 3706-3715.

[55] Cha E, Patel M D, Park J, et al. 2D MoS$_2$ as an efficient protective layer for lithium metal anodes in high-performance Li-S batteries [J]. Nat Nanotechnol, 2018, 13: 337-344.

[56] Martin L, Martinez H, Ulldemolins M, et al. Evolution of the Si electrode/electrolyte interface in lithium batteries characterized by XPS and AFM techniques: The influence of vinylene carbonate additive [J]. Solid State Ionics, 2012, 215: 36-44.

[57] Etacheri V, Haik O, Goffer Y, et al. Effect of fluoroethylene carbonate (FEC) on the performance and surface chemistry of si-nanowire Li-ion battery anodes [J]. Langmuir: the ACS journal of surfaces and colloids, 2012, 28: 965-76.

[58] Lin Y M, Klavetter K C, Abel P R, et al. High performance silicon nanoparticle anode in fluoroethylene carbonate-based electrolyte for Li-ion batteries [J]. Chem Commun (Camb), 2012, 48: 7268-7270.

[59] Xu K, Zhang S, Jow T R. Libob as additive in LiPF$_6$-based lithium ion electrolytes [J]. Electrochemical and solid-state letters, 2005, 8: A365-A368.

[60] Zhang S, Xu K, Jow T. Enhanced performance of Li-ion cell with LiBF$_4$-Pc based electrolyte by addition of small amount of LiBOB [J]. Journal of Power Sources, 2006, 156: 629-633.

[61] Soto F A, Ma Y, Martinez de la Hoz J M, et al. Formation and growth mechanisms of solid-electrolyte interphase layers in rechargeable batteries [J]. Chemistry of Materials, 2015, 27: 7990-8000.

[62] Rodrigues M-T F, Sayed F N, Gullapalli H, et al. High-temperature solid electrolyte interphases (SEI) in graphite electrodes [J]. Journal of Power Sources, 2018, 381: 107-115.

[63] Ota H, Sato T, Suzuki H, et al. TPD-GC/MS analysis of the solid electrolyte interface (SEI) on a graphite anode in the propylene carbonate/ethylene sulfite electrolyte system for lithium batteries [J]. Journal of Power Sources, 2001, 97-98: 107-113.

[64] Zhu T, Hu Q, Yan G, et al. Manipulating the composition and structure of solid electrolyte interphase at graphite anode by adjusting the formation condition [J]. Energy Technology, 2019. 7 (9): doi: 10.1002/ente.201900273.

[65] He Y, Hu H, Zhang K, et al. Mechanical insights into the stability of heterogeneous solid electrolyte interphase on an electrode particle [J]. Journal of Materials Science, 2016, 52: 2836-2848.

第 5 章
离子在固体中的输运

锂离子电池充放电过程中，锂离子需要在固体中输运，包括：电极活性材料、电极/液态电解质接触界面产生的固体电解质层、全固态电池中的固体电解质，以及导电添加剂/粘接剂/活性颗粒形成的固固界面。一般而言，固相内部及固相之间的离子传输是电池动力学过程中相对较慢的步骤，因此离子在固体中的传输是锂电池材料研究的重要基础科学问题。本章小结了固态离子学基础知识中，关于离子在固体中的传输机制及其扩散的驱动力，并讨论了影响离子电导率的几种关键因素，简介了锂离子在正极、负极以及固态电解质中的输运特性，以及界面输运特性，讨论了内源锂和外源锂输运特性的差异以及尺寸效应对于离子输运性质的影响，简单介绍了离子输运的表征方法。

锂离子电池中的电化学反应包括电荷转移和物质扩散，例如电子、空穴、锂离子、其他阳离子、阴离子、带电粒子基团等。其中，锂离子在电池内部、电子在外电路的正极和负极之间的输运过程，可以为外部电器供电。由于离子在固相中的传输一般是电池工作中最慢的步骤，因此，提高电池的实际输出能量密度/倍率特性/能量效率，以及控制自放电率，需要准确了解和调控离子在固体中的输运特性。

依照 Funke 的分类，完全有序的晶体被定义为是第一级（level one），此时离子不能离开晶格位[1,2]。当晶格位点无序、伴随热力学导致的点缺陷，定义为第二级（level two），离子输运伴随着缺陷从晶格位到晶格位的随机运动；事实上，现代材料学通常基于第二级概念。对于无序结构来说，情况将完全不同，点缺陷不再是独立的结构元素，甚至不再是通常意义的"缺陷"，因为整个结构是无序的；因此，离子输运不能被描述成单独缺陷在静态能面的随机行走，相反地，需要考虑复杂的多体问题，包括载流子的相互作用，以及载流子与周围晶格（surrounding matrix）的相互作用；并将这种情况定义为第三级（level three），其中包括结构无序的晶体（例如常见快离子导体中锂离子在超量位点无序分布）、玻璃、聚合物以及纳米化复合系统。锂离子电池中的电极材料多为粉末材料，一般不采用大块单晶，个别厂家的粉末材料产品单个颗粒结晶度很高，为单晶小颗粒。多数的商品电极材料为包含缺陷的多晶粉末材料。在采用液态电解质的锂离子电池中，一般在正负极与电解质接触的表面还生长一层或多层固体电解质中间相（solid electrolyte interphase，SEI）膜，在充放电过程中，锂离子必须通过这层 SEI 膜。SEI 膜层是由多种成分形成的无定形结构。全固态锂电池使用固态材料作为电解质。固体电解质材料（solid state electrolytes，SSE）包括无机固体、聚合物固体及其复合物，其中既有多晶材料，也有无定形材料，也有复合了晶态和非晶态的材料。因此，离子在锂离子电池材料中的输运涉及在单晶固体、多晶固体和无序固体中的输运问题。

离子在单晶固体中一般通过晶格（lattice）中的格点空位（vacancy site）或间隙空位（interstitial site）输运。对于快离子导体，通常情况下离子在大量位点协同输运，这通常是其高离子电导率的关键原因。对于多晶固体，离子既可以在晶格中传输，也可以在晶粒之间的晶界（grain boundary）或者颗粒的固液界面（interface）以及固气表面（surface）传输；界面失配导致的无序结构，常常提供了额外的高离子电导率。离子在实际材料中的传输路径（trajectory）是重要的基础科学问题。锂离子电池中电解质材料为纯离子导体，电极材料为混合离子导体。由于电中性的要求，充放电过程中离子和电子会同时嵌入或脱出电极材料（是否严格时间同步是另外的基础科学问题）。由于离子与电子之间存在着相互作用，混合离子导体中离子的输运会受到电子输运特性的影响。无论是混合离子导体还是纯离子导体，在化学势梯度和电场梯度的驱动下，内在和外在离子可以在固体中或沿着/穿过固体传输。发生传输的离子既可以是来自于晶格内部的内在离子，也可以是外来离子（例如，对于电解质、负极来说，来自正极材料的离子可以视为外来离子）。

上面提到的离子在固体中的输运问题是固体离子学的核心研究内容，对这部分内容，有兴趣的读者可以参看相关的专著和综述类文献[3-6]。本文首先简要总结了关于离子输运的基础知识，之后总结了锂离子电池电极材料、电解质材料的离子输运特性，主要从锂离子电池与电化学实用性方面对于实际材料中的离子输运问题进行阐述。其中更多的物理相关的理论性内容，以及离子输运的实验研究方法，详情可见文献[5，6]。计算是一种研究微观机理的有效方法，关于离子输运的计算研究方法，详情可见文献[7，8]。

5.1 离子输运的相关参数

对于锂离子在固体中的输运，主要使用扩散系数（diffusion coefficient，D）和离子电导率（ionic conductivity，σ）来进行描述。考虑无外力情况，可以由涨落理论和能量均分定理得到爱因斯坦关系[式(5-1)]，具体推导不赘述。这里，u 为迁移率，在很多文献和书籍中常用 μ 来表示，这里为和化学势区分，用 u 来表示。玻尔兹曼常数 k_B 与理想气体常数 R 存在正比关系 $R=k_B N_A$（N_A 为阿伏伽德罗常数），因此为了描述宏观现象，我们选择用 R 作为下文参数。

$$D = u k_B T \tag{5-1}$$

实际情况中的扩散，其本质是各种力所驱动下的各种流。以锂离子电池为例，锂离子的输运是在各种梯度力的作用下（例如浓度/化学势梯度和电场梯度）所产生的宏观的扩散或者迁移行为。对于理想体系，在浓度梯度 ∇c_i 驱动下，物质 i 的扩散过程可以由菲克第一定律和菲克第二定律来描述：

$$j_i = -D_i \nabla c_i \tag{5-2}$$

$$\frac{\partial c_i}{\partial t} = \nabla(D \nabla c_i) \tag{5-3}$$

菲克第一定律（Fick first law）描述了浓度梯度驱动下，空间中的物质流，物质 i 将沿其浓度场决定的负梯度方向进行扩散，其扩散流大小与浓度梯度成正比。扩散系数 D_i 反映了物质 i 扩散的能力，单位是 cm^2/s。菲克第二定律描述了物质 i 在介质中的浓度分布随时间变化的扩散。菲克定律是一种宏观现象的描述，它将浓度以外的一切影响物质扩散的因素都包括在扩散系数之中。菲克第二定律是由菲克第一定律以及连续性方程式(5-3)所得：

$$\frac{\partial c_i}{\partial t} + \nabla \cdot j_i = 0 \tag{5-4}$$

对于非理想体系，如果选择用化学势梯度 $\nabla \mu_i$ 来代替浓度梯度 ∇c_i。设某多组分体系中，i 组分的质点沿 x 方向扩散所受到的力 F_i 应等于该组分化学势（μ_i）在 x 方向上梯度的负值：

$$F_i = -\frac{\partial \mu_i}{\partial x} \tag{5-5}$$

相应的质点运动平均速度 V_i 正比于作用力 F_i：

$$V_i = u_i F_i = -u_i \frac{\partial \mu_i}{\partial x} \tag{5-6}$$

其中 u_i 为单位力作用下，组分 i 质点的平均速率，或称为淌度 [单位为 $m^2/(s \cdot V)$]。则 x 方向的电流为：

$$j_i = c_i V_i = -c_i u_i \frac{\partial \mu_i}{\partial x} = -c_i u_i \frac{\partial \mu_i}{\partial c_i} \times \frac{\partial c_i}{x} \tag{5-7}$$

其中 c_i 为该组分的浓度。与菲克第一定律比较可得：

$$D_i = c_i u_i \frac{\partial \mu_i}{\partial c_i} = u_i \frac{\partial \mu_i}{\partial \ln c_i} \tag{5-8}$$

定义组分 $N_i = c_i/c$，则有 $d\ln c_i = d\ln N_i$，
因此：

$$D_i = u_i \frac{\partial \mu_i}{\partial \ln N_i} \tag{5-9}$$

考虑组分 i 的活度系数 γ_i，则 i 的组分活度为 $N_i \gamma_i$。此时，对于非理想体系物质 i 的化学势：

$$\mu_i = \mu_i^0 + RT(\ln N_i + \ln \gamma_i) \tag{5-10}$$

两边对 $\ln N_i$ 做偏分，有：

$$\frac{\partial \mu_i}{\partial \ln N_i} = RT\left(1 + \frac{\partial \ln \gamma_i}{\partial \ln N_i}\right) \tag{5-11}$$

比较式(5-8) 和式(5-10) 可以得到扩散系数的一般热力学关系式：

$$D_i = u_i RT \left(1 + \frac{\partial \ln \gamma_i}{\partial \ln N_i}\right) \tag{5-12}$$

其中 $1 + \frac{\partial \ln \gamma_i}{\partial \ln N_i}$，为扩散系数的热力学因子。

① 对于理想混合体系：活度系数 $\gamma_i = 1$，则有 $D_i = D_i^* = u_i RT$，通常称 D_i^* 为自扩散系数；D_i 为本征扩散系数。

② 对于非理想混合体系：

a. 当 $1 + \frac{\partial \ln \gamma_i}{\partial \ln N_i} > 0$，则 $D_i > 0$，称为正常扩散，即物质流将从高浓度处流向低浓度处，扩散的结果使溶质趋于均匀化；

b. 当 $1 + \frac{\partial \ln \gamma_i}{\partial \ln N_i} < 0$，则 $D_i < 0$，称为反常扩散或逆扩散，扩散的结果使溶质偏聚或分相，如固溶体中的某些元素偏聚、玻璃在旋节区分相和晶界上选择性吸附过程等。

当外电场加到材料上时，电流或快或慢地会达到一个稳态直流值。我们可以通过在电场存在下出现的带电粒子数和它们的迁移速度来表示稳态过程。设带电量为 q_i，则电流密度 j_i 与漂移速度 V_i 的关系为：

$$j_i = n_i q_i V_i \tag{5-13}$$

考虑漂移速度受电场驱动，电导率 σ_i 的定义为：

$$\sigma_i = j_i / E = n_i q_i V_i / \nabla \varphi \tag{5-14}$$

迁移速度正比于局部作用的电场强度，迁移率由以下比值确定：

$$u_i = V_i / (q_i \cdot \nabla \varphi) \tag{5-15}$$

因此，电导率是载流子浓度和迁移率的乘积：

$$\sigma_i = n_i q_i^2 u_i \tag{5-16}$$

带入公式(5-1)的爱因斯坦关系，则得到能斯特-爱因斯坦方程：

$$\sigma_i = n_i q_i^2 D_i / (RT) \tag{5-17}$$

由于材料的实际电导率由多种载流子贡献而成：

$$\sigma = \sum \sigma_i \tag{5-18}$$

由此定义迁移数（transference number）：

$$t_i = \frac{\sigma_i}{\sigma} \tag{5-19}$$

迁移数是指各种可动的导电粒子在导电过程中的导电份额。对于电极材料而言，希望电子和离子的输运速度都比较高，对于电子和离子迁移数的比值没有严格的要求。对于电解质材料而言，则希望工作离子的输运对电流起主要贡献，且对于电子是绝缘体，电子的迁移数应小于 1%，以防止内部短路和自放电。对于固态锂离子导体而言，这一要求不难实现。对于液态的电解质，一般锂离子的迁移数在 0.2～0.4 之间，阴离子的迁移对离子电流产生了较大贡献，这会引起在电极侧的极化，提高了界面传输的电阻。

考虑到化学势与电势的共同驱动，离子的漂移速度与离子所受到的所有作用力的总和成正比，在同时存在化学势梯度与电场梯度驱动力时：(离子漂移速度)＝(离子迁移率)×(化学势作用力＋电场作用力)，也就是：

$$V_i = -u_i(\nabla \mu_i + \nabla \varphi) \tag{5-20}$$

在稀溶液体系中，化学势梯度与浓度梯度有如下关系：

$$\nabla \mu_i = \frac{RT}{c_i} \nabla c_i \tag{5-21}$$

由式(5-20) 和式(5-21) 得：

$$V_i = -\frac{u_i RT}{c_i}\left(\nabla c_i + c_i \frac{\nabla \varphi}{RT}\right) \tag{5-22}$$

由此可以得到离子的流量表达式：

$$j_i = c_i V_i = -u_i RT\left(\nabla c_i + c_i \frac{\nabla \varphi}{RT}\right) = -D_i\left(\nabla c_i + c_i \frac{\nabla \varphi}{RT}\right) \tag{5-23}$$

式(5-23) 也叫做能斯特-普朗克方程。其中扩散系数和迁移率的关系为 $D_i = u_i RT$，其来源于式(5-1) 中的爱因斯坦方程。

需要指出的是，以上为一般性的关于浓度梯度、化学势梯度及电场梯度驱动下带电粒子输运的讨论，未考虑固体材料的结构特点。在固体中，离子的输运机制与结构有关，存在不同的输运机制。固体中浓度梯度、电场梯度的建立与材料的结构、电子及离子的电导率有关。具体到微观的原子尺度，驱动力如何作用于内在和外来的可迁移离子，浓度梯度与电场梯度在空间分布的非线性和受结构因素的制约，都使得问题更为复杂，需要深入的理论分析和实验研究。正常固体中，粒子的扩散过程基于热激活的缺陷在周期场中的布朗运动，且跃迁过程符合随机行走模型，详见文献 [9-13]：

$$D_r = \langle R_n^2 \rangle/(6t_n) = 1/6(\nu a^2 c) \tag{5-24}$$

式中，D_r 为无序随机扩散系数，R_n 为 n 次跃迁后的净位移；$\langle \ \rangle$ 表示所有过程的平均值；t_n 为时间；a 为跃迁间距或自由行程；c 为可供跃迁的缺陷浓度；ν 为平均跃迁频率。定义 f 为相关因子：$f = D^*/D_r$。其中，D^* 为示踪原子扩散系数。f 与几何结构和输运机制有关[14]。

假设粒子的振动服从于热活化过程，则跃迁频率满足：

$$\nu = \nu_0 \exp\left(-\frac{\Delta G}{k_B T}\right) = \nu_0 \exp\frac{\Delta S}{k_B} \exp\left(-\frac{\Delta H}{k_B T}\right) \tag{5-25}$$

其中，ν_0 为试跳频率；ΔG、ΔS、ΔH 分别为粒子从基态激活到过渡态的自由能、熵和焓的变化；k_B 为玻尔兹曼常数；T 为开氏温度。由式(5-24)、式(5-25) 可得：

$$D^* = \frac{1}{6}fa^2 c\nu_0 \exp\frac{\Delta S}{k_B} \exp\left(-\frac{\Delta H}{k_B T}\right) \tag{5-26}$$

注意到，D^* 所描述的是单种粒子扩散，也称为自扩散系数。

多元系统往往存在几种粒子同时进行扩散，称为互扩散。常利用 Darken 方程描述[15]，以二元系统为例：

$$\widetilde{D} = (N_2 D_1^* + N_1 D_2^*)\left(1 + \frac{\partial \ln \gamma_1}{\partial \ln C_1}\right) \tag{5-27}$$

其中 $\tilde{D}$ 为互扩散系数，或化学扩散系数。然而该方程不适用于离子的互扩散过程，需要进一步考虑体系电中性等复杂因素。值得注意的是，与载流子为单一的锂离子的锂离子固体电解质不同，电极材料还需要考虑电子、空穴和带电缺陷等因素。因此，对于电极材料的测量，循环伏安法（cyclic voltammetry，CV）[16,17]、恒电流间歇滴定法（galvanostatic intermittent titration technique，GITT）[18,19]、恒电位间歇滴定法（potentiostatic intermittent titration technique，PITT）[17,20-22]、电化学阻抗法（electrochemical impedance spectroscopy，EIS）[16,17,19,22-24] 所给出的通常是化学扩散系数[25]。本章仅对此术语进行辨析，不再详细介绍。

如前文所述，锂离子电池中的离子在固体中的输运常受电势的驱动，无序扩散将转变为有序过程，在这些情形下定义 D^σ 为电导扩散系数（这里采取了常见的表示方法，其中 $k_B = R/N_A$，$c = n/N_A$）：

$$D^\sigma = \frac{\sigma_{dc} k_B T}{c q^2} \tag{5-28}$$

注意到，D^σ 并非从 Fick 第一定律推出（因此并非精确意义的"扩散系数"），仅是通过能斯特-爱因斯坦方程[式(5-17)]，从形式得到直流离子电导率 σ_{dc} 的导出值（因此具有"扩散系数"的量纲）。为了建立其与 D^* 的关系，定义式(5-26)与式(5-28)的比值，

$$H_R = D^*/D^\sigma \tag{5-29}$$

称其为"Haven ratio"，体现了扩散进程中粒子（particles）的移位效应（displacement effects），此时假设一个离子不仅可以跃迁至最近邻空位，也可以跃迁至已被占据位置、并导致该位置离子向同方向跃迁，直至最后一个离子跃迁入空位位置；若载流子的离子-离子相互作用导致协同效应（cooperative effects），行为将更加复杂，详见文献[26]。简单而不精确地说，若离子存在运动关联、或者电子对电导存在贡献的时候，$H_R < 1$；当类似空位对或掺杂-空位对这类缺陷出现，可以对扩散做出贡献、却不体现在电导率上的时候，$H_R > 1$[27]。其精确值可以通过测量 σ_{dc} 和 D^* 得出，关于其解读在文献[26-28]中详尽介绍。在具有熔融亚晶格结构的快离子导体，以及结构无序的导电玻璃或聚合物电解质中，Haven ratio 值对于确认扩散机制具有重要的意义，然而，对其的解读也相应地将更加困难[14]。

以上讨论均为固体中缺陷扩散的随机行走模型，其局限性在于它要求载流子的传递过程不因其他缺陷的存在而受影响。除此之外，考虑到多体效应，另有晶格气体模型（Lattice-gas models）[29-33]，假设可移动的离子与离子占据位可比拟，在跃迁过程中有相互作用。对于快离子导体，假设不做迁移运动的离子束缚在平衡位置附近作有限幅度的振动，运动离子既能在空位网络的格点上做振荡，又能跨越（两平衡位置间的）势垒进行扩散，甚至在势垒上停留；以此为基础，提出跳跃-弛豫模型（jump-relaxation models），Funke[31,34] 和 Ngai[35] 等均对其进行了深入研究。事实上，对于快离子导体而言，其静态势能面中的能谷位置（载流子的等效位置）数往往多于其实际占位（实际载流子）数，造成了载流子亚晶格的无序，从有序态相变成为无序态，使得电导率显著上升。固体电解质存在刚性的骨架结构，其变化相对很小；而液体中载流子的环境是通过周围壳层的相对电荷（counter ions）的数量和位置进行调整的，这种调整中的塑性是液体的重要特征。离

子的协同运动常常导致更低的活化能，此时，静态势能面不再使用，需要利用动态势能面理解这一进程。处理方法，详见文献[5]。

5.2 离子在晶格内的输运机制

微观上看，在一定的温度下，粒子在凝聚态物质（包括液体和固体）的平衡位置存在着随机跳跃，进而与周围的粒子交换位置。在一定的驱动力作用下，粒子将偏离平衡位置，形成净的宏观扩散现象。离子在晶体中扩散的微观机制主要包括 Schottky 类型的空位传输机制以及 Frenkel 类型的间隙位传输机制。对于锂离子电池的实际体系，扩散机制更为复杂，例如 β-Al_2O_3 的推填子机制[36]，$LiCoO_2$ 的双空位传输机制[37]，Li_3N 的集体（collective）输运机制[38,39]，Li_2CO_3 中间隙位-空位交换类型的"knock-off"机制[40,41]等。常见的锂电池固体中的扩散机制如表 5-1[42]。

表 5-1 固体中的扩散机制[42]

扩散机制		描述
非空位机制	间隙机制	间隙原子的尺寸小于点阵原子，间隙原子占据间隙位，与点阵原子构成间隙固溶体
	集体(collective)输运机制	间隙原子与点阵原子的尺寸相当，原子的扩散涉及一个以上原子的集体运动，通常形成置换式固溶体
	推填子(interstitialcy)机制	集体输运机制的一种，扩散过程中至少有两个原子同时运动。这种机制在热扩散中可以忽略
	间隙位-格点位交换机制(interstitial-substitutional exchange)	间隙原子同时占据间隙位和格点位，通过间隙位和格点位的交换来实现扩散
空位机制	空位	适用于金属或置换式固溶体中的自扩散
	双空位	通过空位的聚集实现的扩散

间隙位扩散机制适用于间隙固溶体中间隙原子的扩散。在间隙固溶体中，尺寸较大的骨架原子构成了固定的晶体点阵，而尺寸较小的间隙原子处在点阵的间隙中。由于间隙固溶体中间隙数目较多，而间隙原子数量又很少，这就意味着在任何一个间隙原子周围几乎都存在间隙位置，这就为间隙原子向周围的扩散提供了必要的结构条件。尺寸较小的间隙原子在固溶体中的扩散就是按照从一个间隙位置跳动到其近邻的另一个间隙位置的方式进行的。这种方式也叫直接间隙扩散机制（图 5-1），也是最简单的一种扩散机制。

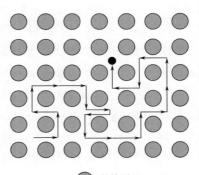

图 5-1 直接间隙扩散机制[43]

空位扩散机制适用于置换式固溶体的扩散，原子通过跳跃到邻近的空位实现扩散（图 5-2）。晶格中的结点并非完全被原子所占据，而是存在一定比例的空位。空位的数量随温度的升高而增加，在一定温度下对应着一定的空位浓度。由于熵的增加，在一定温度下存在一定浓度空位的晶体热力学能量更低。在置换式固溶体（或纯金属）中，由于原子尺寸相差不太大（或者相等），因此不能进行间隙扩散。空位扩散机制在这类固体中起到了重要的作用。固体电极与电解质材料设计时，通过掺杂产生空位，通常是提高离子电导率的重要方式。当空位团聚的时候，还可能存在多空位机制，如图 5-2（b）所示的双空位机制；其中 $LiCoO_2$ 就是遵循双空位机制，如图 5-2（c）所示[37]。

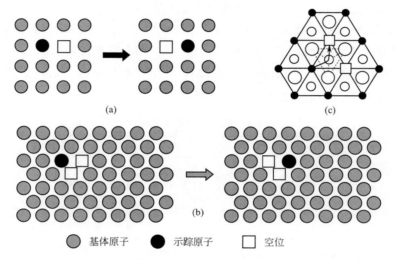

图 5-2　扩散的空位机制（a）[43]和双空位机制（b）以及在 $LiCoO_2$ 中的实例（c）

当间隙原子同时占据间隙位和格点位的时候，原子可以通过间隙位-格点位交换的形式输运，如图 5-3 所示。间隙方式的扩散系数通常要远高于取代方式的扩散系数，然而，间隙位"溶质"原子的浓度却小于"取代位"原子的浓度。在这种情况下，输运为间隙位-取代位共同作用机制。如果这种输运是通过空位来完成的，则称为解离机制；如果输运仅通过自间隙原子来完成，则称为"踢出"机制（knock-off）。

除了前述的两种主要机制，还可能存在集体输运机制，即几个原子同时运动的机制，原子的集体运动方式类似于链状或者履带状。这种机制适用于无定形体系，图 5-4（a）为无定形态 Zr-Ni 合金。固体电解质 Li_3N 中锂离子输运亦遵循此机制，如图 5-4（b）所示[38,39]。另外，此机制对于碱金属离子在氧化物离子导电玻璃里面的输运也起到重要作用。推填子机制（interstitialcy mechanisms）[图 5-4（c）]和自间隙位机制（self-interstitials）同样属于集体输运机制，因为离子跃迁过程需要不止一个原子同时运动。注意到，快离子导体中超高的电导率常来源于协同输运，这已经越来越被广泛认知，并作为快离子导体判断和设计的依据。

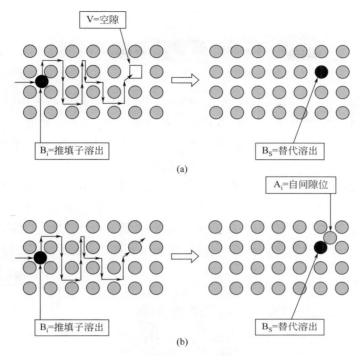

图 5-3 间隙-取代交换机制
(a) 解离机制；(b) "踢出" 机制[43]

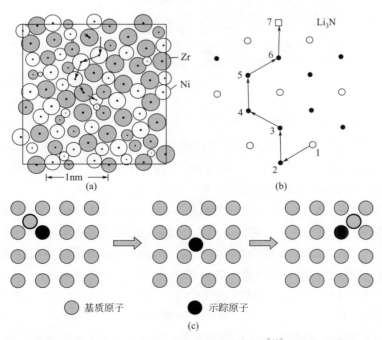

图 5-4 合金（a）和 Li_3N（b）中的集体输运机制[43]及推填子机制（c）

5.3 离子在晶界处的输运机制

晶界是由结构不同或者取向不同的晶粒互相接触而形成的，如图 5-5(a) 所示。它与晶粒的取向、成分、成键状态以及形貌大小等有很大关系[44]。各类纳米复合物容易形成不同形式的晶界。图 5-5(b) 为纳米复合物的分类，上行是三个维度均为纳米尺度的情况，下行是只有两个或一个维度为纳米尺度的情况，分别导致了层状或棒状结构。第一列为同类型晶粒的纳米复合物，第二列为不同类型晶粒混合的纳米复合物，第三列为孤立的纳米晶粒分散到主体基质（host matrix）的情况。主体基质可能是晶体，也可能是无定形态（包括玻璃态，聚合物等）。对于图 5-5(a) 这种多晶结构，通常可以用阻抗谱来研究体相和晶界的电导以及电子电导，如图 5-5(c) 所示，为其对应的等效电路以及阻抗谱拟合。

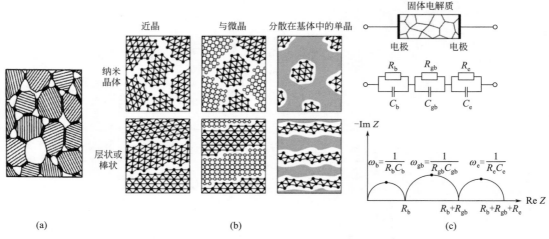

图 5-5　纳米复合物的结构（a）、分类（b）和等效电路表示（c）[27]

为了描述所测得的两相混合物导电性质的阻抗谱，最早提出的模型是串联模型和并联模型（平行层模型），如图 5-6(a) 所示。这两个模型描述了两种极端情况，其中相均是非连续的，因此可能与一些材料的微结构相差较大。后来，Beekmans 和 Heyne 提出了砖层模型，将两种极端情况融为一体。如图 5-6(b) 所示它是由立方形晶粒堆砌而成，晶粒之间由上平面的晶粒间界分开。假定电流是一维的，电流在晶粒角上的弯曲忽略不计，这样电流只能沿着两条途径进行；通过晶粒并穿过晶粒间界 [图 5-6(b) 左]，或者沿着晶粒间界 [图 5-6(b) 右]。砖层模型假定在个别晶粒之间有一连续的晶粒间界，但是很多情况下，晶粒间界上一些区域晶粒间接触良好，Bauerle 将这些区域称为"捷径"（short-circuit），如图 5-6(c_1) 所示。其等效电路如图 5-6(c_2) 和 (c_3) 所示，两者是等价的。除了上述的砖层模型，若考虑电流分布的实际情况，还提出了有效介质模型。Maxwell 首先提出球状粒子分散到连续相中的有效介质模型，随后 Brailsford 和 Hohnke 提出了同心球模型。Fricke 对椭球状粒子建立了类似 Maxwell 模型的表达式，引入形状因素，可使阻抗弧变形[45]。

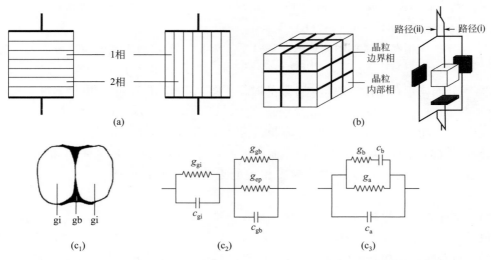

图 5-6 串联层和并联层模型（a）、砖层模型（b）和有效介质模型（c）[45]

为了解释复合物电解质体系电导率的整体提高，Dudney 还提出了晶界、体相和表面相串并联的电阻网络模型（resistor-network model）[46]，如图 5-7（a）所示。整体的电导率为：

$$\sigma = (1-x)\sigma_b + x\sigma_A + 2\left[\frac{1}{r_b} + x\left(\frac{1}{r_A} - \frac{1}{r_b}\right)\right] \times \frac{\sigma_{b/b}(1-x^2)r_A^2 + 2\sigma_{b/A}(1-x)xr_Ar_B + \sigma_{A/A}x^2r_b^2}{[(1-x)r_A + xr_b]^2}$$

(5-30)

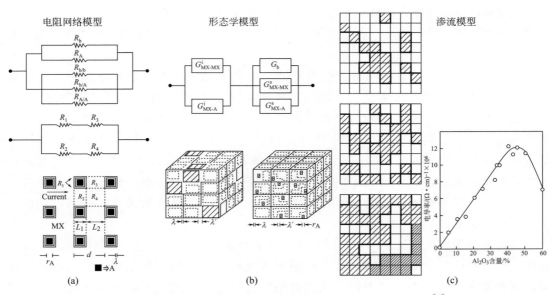

图 5-7 电阻网络模型（a）、形态学模型（b）和渗流模型（c）[4]

式中，x 是分散体在整个体系中的体积分数；r_b 和 r_A 分别是主体相和分散体相的晶粒半径；$\sigma_{b/b}$、$\sigma_{b/A}$、$\sigma_{A/A}$ 是相应界面的电导率。这一模型解释了主体相与分散体相颗粒大小，以及分散体相在整个体系中所占的体积分数对电导率的影响。

考虑到分散体团聚或均匀分散的不同情况，Uvarov等[47]提出了改进的形态学模型（morphological model），如图5-7(b)所示。

① 当分散体相在主体相中团聚的时候［图5-7(b) 左］，则交流电导率可以表示为：
$$\sigma_{ac}=\sigma_S(\lambda/r_b)(\beta/\gamma)(1-f)^2+\sigma'_S(\lambda'/r_b)(\beta/\gamma)f(1-f) \tag{5-31}$$

② 当分散体相在主体相中均匀分散的时候［图5-7(b) 右］，交流电导率可以表示为：
$$\sigma_{ac}=\sigma_S(\lambda/r_b)(\beta/\gamma)(1-f)^2+\sigma'_S(\lambda'/r_A)(\beta'/\gamma')f(1-f) \tag{5-32}$$

式中，σ_S 和 σ'_S 分别为主体相-主体相、主体相-分散体相界面的电导率；λ 和 λ' 分别为主体相-主体相、主体相-分散体相界面的厚度；r_b 和 r_A 分别是主体相和分散体相的晶粒半径；β，β'，γ，γ' 是和样品形貌相关的无量纲的几何因子；f 为分散体的体积分数。

当分散体相为绝缘体的时候，电导率随着分散体的体积分数的提高而显著增大，超过临界值后又会减小，如图5-7(c) 右图[48]所示。分散体相（深色方块）在主体相（白色方块）中随机分布，两相分界面为高电导率的界面相。分散体体积分数小于临界值，界面相随分散体增多而增多；当超过临界值时，分散体继续增多，导致分散体团聚在一起，界面相反而减少。这是典型的渗流现象[4]，宏观的渗流模型可以解释[49]。

上述唯象的模型便于理解两相复合材料的电导行为，但不涉及离子在界面处传输性质的变化。为了分析复合材料由于界面引起的不同于体相的离子电导行为，把晶界附近的结构分成主体基质、分散体、表面三部分，对于电导率惰性的分散体，分散体自身的电导率贡献通常可以忽略。因此，表面部分电导率对总电导率的提高做出主要贡献。在"Jow 和 Wagner 模型"[4,50]中，假设在主体基质和分散体之间存在空间电荷的区域分布，这一区域产生过量的缺陷浓度。缺陷浓度的径向分布函数采用不同模型来描述（如图5-8所示），具体推导和描述详见综述［4］。径向分布函数表述了单一分散体附近载流子浓度、电导率的变化。

图5-9给出了通过示踪原子实验测量的晶格内和晶界处的等浓度示意图，可以看到，晶界处原子的扩散距离长，表明该实验中示踪原子通过晶界，原子迁移速度快。这是由于晶界处原子排列不规则，点阵畸变严重，空位密度比晶内高的原因。分析平行于晶界及垂直于晶界处离子电导行为是固体离子学研究的一个重要课题。Joachim Maier 在 Carl Wagner（图5-10）等人的研究基础上，进一步深入发展了空间电荷层对离子传输的影响理论，假设载流子的迁移率不受空间电荷层的影响，分析了平衡时载流子在空间电荷层两侧的浓度变化，解释了晶界处离子电导率显著提高的机制，详细分析参见参考文献［4］。

空间电荷层理论方法不考虑实际晶体结构对载流子在空间中分布的限制，也不考虑局部应力对离子迁移率的影响，也无法预测出现导电增强的体积比例。实际晶界还有可能存在着杂质聚集偏析和不同形式的缺陷。这使得晶界既有可能有利于传输，也有可能阻塞离子传输，需要具体问题具体分析，不易准确预测。虽然存在着上述困难，前述的各种界面模型及空间电荷层模型还是提供了从不同的角度思考界面离子传输行为的理论指导。

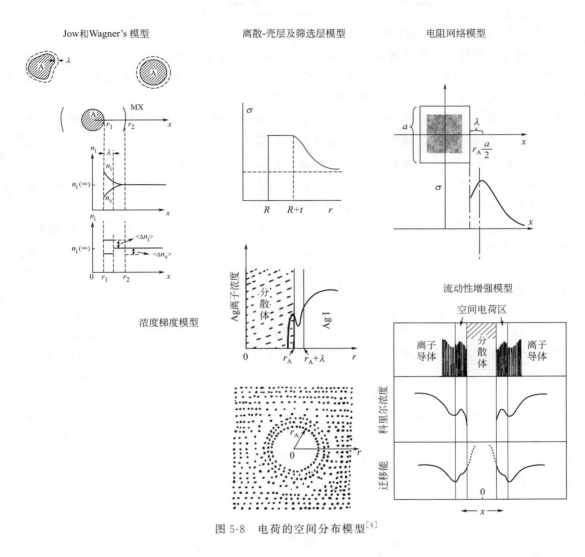

图 5-8 电荷的空间分布模型[4]

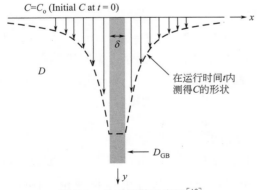

图 5-9 晶界扩散示意图[42]

图 5-10 Carl Wagner

Carl Wagner 自 20 世纪 30 年代建立的离子输运的理论和图像主要是基于固体内部的可以移动的离子[51,52]。在锂离子电池中，离子可以从外部进入。因此存在一个问题，离子是如何穿过固体的，是通过连续的间隙位或空位从一侧传输到另一侧，还是通过与晶格内锂离子的交换传输，需要进一步澄清。不同的传输方式，与固体的微观缺陷结构有关。在讨论具体的扩散机制中，准确判断到底是采用哪种扩散机制并不容易。迄今为止的一个仔细研究过的例子是研究锂离子通过 Li_2CO_3 时的传输机制，利用同位素交换，结合二次离子质谱，通过第一性原理分子动力学计算对实验数据的拟合，推测在 Li_2CO_3 中通过"踢出"（knock-off）交换机制传输[6,7]，要比直接传输活化能更低[38,39]。

所以，当考虑锂离子在固体中的输运和扩散的问题时，对于锂离子的来源，我们认为有两种，一种是内源锂，这是我们最常见的一种，也就是本身就在晶格位点中存在的锂离子，而在充放电过程中，最初的锂离子也是由这些晶格位点中的锂离子提供的。另外一种是外源锂，也就是通过预锂化的方式（化学锂化或者电化学锂化）从外界向晶格位点、间隙位，非晶材料或者纳米复合物的界面处引入锂离子。外源锂的一个典型例子是纳米复合物体系 LiF/Ti。在这个体系中，LiF 中含有一定的锂元素，但是 LiF 的离子键很强，LiF 晶格中的锂在电场作用下基本上是不可移动的。通过对刚制备出来的 LiF/Ti 体系进行直流极化测试，也就是在体系两端加一个恒定电场，发现在电流响应中并没有与离子有关的迁移电流。而通过化学锂化的方式在 LiF/Ti 的体系中引入额外的锂元素，并再次对这个体系进行直流极化测试，发现在电流响应曲线的开始阶段，出现了一段明显对应于离子迁移的衰减电流。这个例子说明，外源的锂离子可以沿着 LiF 和 Ti 的界面进行传输。从以上对于外源锂的简单介绍可以看出，一种本身不含锂元素或者含有锂元素但是本身的锂离子电导率不太高的材料和体系，可能由于存在界面而传导外来的锂离子。因此在具体的离子输运特性的研究中，需要仔细分析材料中的内源锂与外源锂的输运特性。

5.4 无序态与电导率

电导测试发现，不同于有序晶体的 Arrhenius 行为 $\sigma_{DC}T = A_\sigma \exp[-E_\sigma/(k_B T)]$，对于无序的玻璃态和聚合物，也包括某些锂离子与"空位"无序占据的固体电解质晶态，直流电导率的对数与温度倒数的关系可能偏离直线，可以由 Vogel-Fulcher-Tammann（VFT）方程或 Williams-Landel-Ferry（WLF）方程比较好的拟合：

$$\sigma_{DC}T \propto \exp\left(-\frac{E}{k_B(T-T_0)}\right) \tag{5-33}$$

当 $T_0 = 0$ 时，式（5-33）退化为（reduce to）正常的（normal）Arrhenius 方程。该方程通常用构型熵理论（configurational entropy theory）[53]和自由体积理论（free-volume theory）[54,55]简单地予以解释。构型熵理论认为，原子（或离子）的相关涨落导致的不同构型是过渡态形成的原因，过渡态的构型熵 $S_C \approx \Delta C_p/T + \text{const}$，则跃迁概率可写作 $\nu_i =$

$\nu_i^0 \exp \frac{b}{TS_C}$，可整理为式(5-33)。自由体积理论则基于这样一个图像：载流子的跃迁仅发生在体积 V 比临界体积 V^* 大的空位中，与温度相关的平均自由体积近似为：$V^f = V_0 \alpha^f (T - T_0)$，这里 α^f 为液态中自由体积的热膨胀系数；T_0 是自由体积消失的温度；V_0 是温度 T_0 时液体的体积。扩散系数可写作：$D = D_0 \frac{\gamma V^*}{V^f}$；亦可整理为式(5-33)。若加以考虑能垒效应[56]，仅当原子能量高于能垒 E_V 的时候跃迁才得以发生，则扩散系数可改写为 $D = D_0 \left(-\frac{\gamma V^*}{V^f} + \frac{E_V}{kT} \right)$，可较好地拟合聚合物[57,58]和玻璃态[59,60]中锂离子扩散系数（或电导率）随温度变化的阿伦尼乌斯曲线。上述努力，均在通过修正扩散系数方程来解释造成阿伦尼乌斯曲线偏离线性的原因。事实上，不同材料的输运机理通常差异很大。以锂离子电解质为例，无机玻璃中锂离子的迁移率通常接近1，而聚合物中阳离子和阴离子是同时输运的；更加重要的区别在于，聚合物中离子的输运通常是通过链段的蠕动和骨架离子的涨落，随机地提供离子通道。Moynihan 和 Angell 利用"耦合"和"非耦合"的概念来分别区分聚合物和无机玻璃中输运机制的不同[61-65]，并解释了无机固体电解质可在玻璃转变温度（T_g）以下工作。对于无机玻璃，在载流子扩散的时间尺度，骨架离子被认为是具有高稳定性的，然而骨架离子和载流子的相互作用却不能忽略。相对于式(5-33)，无机玻璃中的阿伦尼乌斯方程通常采取 Rasch-Hinrichsen 关系（Rasch and Hinrichsen relationship）来更好地拟合阿伦尼乌斯曲线：

$$\sigma(T) = T^m \sigma_0 \exp\left(-\frac{\Delta E}{kT} \right) (m = 0 \text{ 或 } -1) \tag{5-34}$$

可以更好地拟合，被称为 Rasch-Hinrichsen 关系（Rasch and Hinrichsen relationship）[66,67]。对于无序固体中输运问题的解析解，基本是不可能得到的，这是因为：

① 从非平衡热力学的观点来看，玻璃转变相当于热涨落的冻结，源于外部参数（例如温度）的迅速变化；遍历的和非遍历的（ergodic to non-ergodic）相转变现象，在非平衡热力学的统计力学领域，同样是前沿问题；临界区域的弛豫为非指数、非线性的[61]。

② 离子在随机势垒中的局域振动和扩散现象，可以用随机行走模型和渗流模型唯象地描述；然而，无定形并非没有结构，虽然没有晶体中的长程序，仍具有局域结构的短程序；而且，除了玻璃态类骨架原子的短程序结构，对于晶态快离子导体，包括钙钛矿体系和石榴石体系，为锂离子亚晶格的半无序结构，此时，同一位置所对应的概率椭圆中，锂离子占据随半径范围的增大而连续减小。

③ 运动离子-运动离子以及运动离子-骨架离子的关联效应是最富有挑战性的部分，涉及离子的多体相互作用[61]；包括由于跃迁位被占据、或者所有其他离子对载流子的库仑作用而引起的"向后的效应（backward correlation）"（可由全频分析中的高频响应得到）[34,68]，以及多个载流子运动的"协同效应"（通常只能通过计算模拟的方法预测）[69-71]，但是在同位素示踪实验中得以间接验证[40,41]。

④ 银离子和铜离子化合物通常比锂离子导体具有更高的离子电导率，或许要从量子力学和化学键分析的角度，重新审视和理解。

理论并不能给出最终的答案，常常仅利用唯象的方法来描述所观察到的现象；例如一些现象学和半微观的方法，包括晶格气体模型、Ngai 提出的耦合机制（coupling scheme）[72]、Funke 提出的跃迁-弛豫机制（jump relaxation model）[73]、Elliott 和 Owens 提出的扩散控制的弛豫模型（diffusion-controlled relaxation model）[74]等。实验表明一些材料的无定形态具有更高的电导率。这或许源于玻璃态无机材料中存在大量的缺陷；或许由于玻璃态制备有利于去除孔隙、消除晶界电阻；或许在玻璃态形成过程中，由于热涨落的冻结，本征地形成高离子导电的异质结构[75-77]；此外，在玻璃态中混入、或者直接析出晶态材料，得到微晶/纳晶玻璃，通常可以进一步提高电导率，这或许源于形成了新的缺陷的富集的有序/无序界面，此时，除以上描述中离子在均匀体系中的简单跃迁，还需利用渗流模型考虑离子偏向于在特定区域的输运[78-80]。

图 5-11 给出了现阶段各种锂离子电解质材料的离子电导率与温度的关系曲线，可以看出，无机电解质多服从阿伦尼乌斯关系，聚合物电解质与低温时的有机电解质则需要通过各种修正方程，来进行非线性部分的拟合。

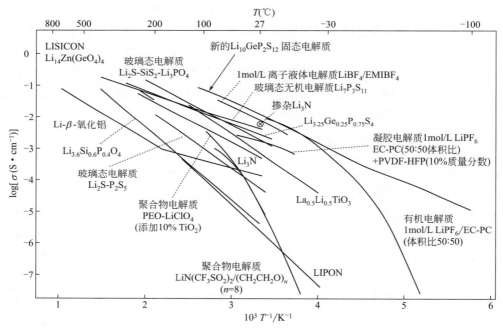

图 5-11　锂离子电解质材料的离子电导率与温度的关系曲线（离子电导的对数与温度的倒数作图）[81]

5.5　锂离子在电极材料中的输运

5.5.1　锂离子在正极材料中的输运

正极材料在充电时，锂离子从晶格中脱出，扩散或迁移到了对电极，同时过渡金属离子的价态也发生了相应的变化，以图 5-12 中 $LiFePO_4$ 电极为例，充电过程 Fe^{2+} 被氧化成了 Fe^{3+}，锂离子需要在正极材料的晶格中扩散和迁移，这就要求正极材料需要有一定的离子电导率，同时这个过程还伴随着过渡金属离子氧化反应，这就要求正极材料同时也要有一定的电子电导。与电解质材料不同，电极材料为混合离子导体。在电极材料中，电子和锂离子都会在电场的作用下运动，在这个过程中，电子与离子的相互作用将会对离子电导产生影响。比如 $LiAlO_2$、$LiNiO_2$ 和 $LiCoO_2$ 三种材料具有类似的晶格结构，但是后两者具有较高的电子电导和离子电导，而 $LiAlO_2$ 的电子电导和离子电导都较低[82]。在微观尺度上更为准确地分析离子的输运特性需要借助于量子力学的方法。根据目前的研究成果，认为锂离子在橄榄石结构的 $LiFePO_4$ 中具有一维的离子扩散通道[84,85]，在层状结构的 $LiCoO_2$ 中具有二维的离子扩散通道[86-89]，且充电态空位提高，锂离子输运遵循双空位机制，以降低跃迁能垒；在实验中测得的扩散系数差别较大[37,84]。在尖晶石结构的 $LiMn_2O_4$ 中具有三维的离子扩散通道[83,90]，这是尖晶石型电极材料通常具有高倍率性能的原因。通过对 $LiFePO_4$ 的计算，阐明了 $LiFePO_4$ 的（0 1 0）面是最稳定的取向，同时也是锂离子扩散最快的方向[91]；Ceder 小组分析认为 $LiFePO_4$ 的导电机理为小极化子理论[92]；孙洋等采用第一性原理分子动力学方法模拟 $LiFePO_4$ 脱锂结构相变时，认为锂离子在一维离子通道输运时与周围阳离子、阴离子、电子的相互作用，导致 $LiFePO_4$/$FePO_4$ 相界面形成更为复杂的"阶结构"[93]。可以通过掺杂使其在更高的维度实现锂离子通道连通，并提高倍率性能。在尖晶石结构的 $LiMn_2O_4$ 中具有三维的离子扩散通道。当材料体系确定后，其本征的电子电导率、离子电导率、扩散系数就确定下来。表 5-2 给出了特定条件下测试的三种正极材料的锂离子扩散系数和电子电导率，具体的参数和测量方法，以及不同掺杂材料，详见文献［42］。

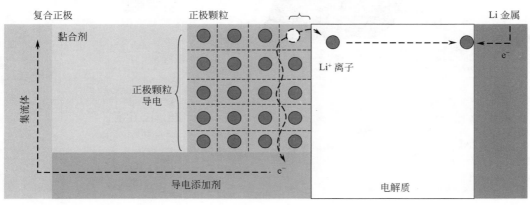

图 5-12　锂离子电池正极材料 $LiFePO_4$ 充电过程中的离子和电子输运过程示意图[42]

表 5-2　正极材料中锂离子扩散系数和电子电导率[42]

正极材料	$D_{Li}/(cm^2/s)$	$\sigma/(S/cm)$	参考文献
$LiCoO_2$	$10^{-13} \sim 10^{-11}$	$10^{-4} \sim 10^{-1}$	[94-96]
$LiMn_2O_4$	$10^{-11} \sim 10^{-8}$	$10^{-6} \sim 10^{-4}$	[97-100]
$LiFePO_4$	$10^{-18} \sim 10^{-15}$	$10^{-9} \sim 10^{-7}$	[101-103]

在锂离子电池体系中，由于电子电导和离子电导决定了整个器件的关键性能，包括倍率性能和循环性能[104,105]，在设计电池体系时需要综合考虑两者的关系和影响。以超高容量富锂正极为例，其额外的容量来自于阴离子得失电子。对于富锂层状材料来说，由于金属锂部分占据过渡金属层，充放电过程中不仅包含锂离子迁移，同时过渡金属也需要迁移；而电子在氧上面的转移伴随着O—O键的变化。因此，离子和电子的输运过程中，不仅发生了预期之外的扩散进程，导致正极材料的循环性变差；同时 O 的氧化还原进程较慢，制约了反应的动力学速度。相对而言，对于富锂含氟尖晶石材料而言，如图 5-13(a) 所示，它具有三维的锂离子通道、高的锂离子电导率、快的倍率性能。F 的掺杂降低了锰的氧化还原态，如图 5-13(b) 所示，并通过调节短程序、使其局域结构为尖晶石型，促进了超化学计量比阳离子条件下锂离子通道逾渗。实验证明额外的电导率同样来自于 O 的氧化还原，且这一速度更快。

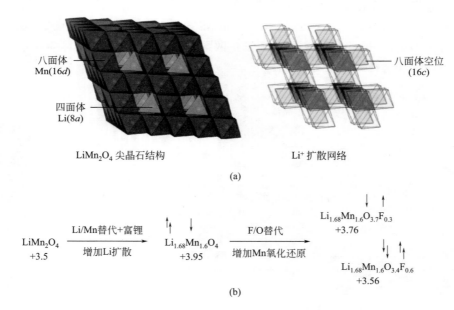

图 5-13　尖晶石型局域结构提高锂离子电导率[105]
（a）$LiMn_2O_4$ 尖晶石结构以及锂离子扩散通道；（b）$LiMn_2O_4$ 尖晶石结构富锂-含氟修饰

当材料的离子电导较低时，在不影响其他电化学性能的前提下，可以通过前述影响离子电导率的因素对材料进行结构上的调整。对于正极材料而言，最为有效的降低离子输运阻抗的办法是减少材料的尺寸，减小锂离子从正极颗粒扩散到电解液的路径。材料的电子电导较低时可以通过表面修饰、掺杂来改善。最有效的例子是在 $LiFePO_4$ 上的碳包覆方法。目前，由于可重复的高纯样品制备上的困难，以及各类测量方法存在的局限性，正极材料本征输运性质，包括扩散系数、离子电导率、迁移率、载流子浓度、离子在晶格中的跃迁频率、平均跳跃距离、昂萨格系数等物理量并没有精确获得，也没有公认的权威数据。此外，正极材料在充放电过程中，锂离子在晶格中的浓度不断发生变化，甚至引起相变，这也导致了输运特性随嵌锂量的不同而发生变化。材料改性调控离子-电子混合输运特性，是后续研究应该关注的科学问题。

5.5.2 锂离子在负极材料中的输运

目前在商业上广泛使用的负极材料基本上都是各种碳材料，比如天然石墨、人造石墨、中间相碳微球（MCMB）或非石墨化的软碳、硬碳材料[113,114]。在不同的碳材料中，测量获得的锂扩散系数也不同，如表 5-3 所示，测量结果均为锂的化学扩散系数；具体测试方法和测试条件，详见文献［42］。锂离子在石墨类材料中的输运涉及连续的相转变，从而在开路电压曲线中形成明显的电压平台[115]。由于石墨的层间结合力远比层内小，且层间距离大，因此在石墨层间容易嵌入一些其他原子、基团或离子，形成石墨层间化合物（GIC），具体见"第 3 章相、相变与相图"。在 GIC 中，每层中嵌入一些其他原子基团或离子称为一阶 GIC，每隔 $n-1$ 层插入 1 层的称为 n 阶 GIC。Li-GIC 主要有 4 阶、3 阶、2 阶（LiC_{18}）、1 阶（LiC_6）4 种化合物[116]，不同阶结构中，锂的扩散系数不同[117]。

表 5-3 锂在各种碳材料中的化学扩散系数[42]

负极材料	$D_{Li}/(cm^2/s)$	参考文献
天然石墨	$10^{-11} \sim 10^{-9}$	[106-108]
MCMB	$10^{-11} \sim 10^{-8}$	[109]
石墨化 MCMB	$10^{-10} \sim 10^{-8}$	[110]
石墨	$10^{-11} \sim 10^{-7}$	[111]
HOPG	$10^{-12} \sim 10^{-11}$	[112]

值得注意的是，具有高倍率性能的锂离子尖晶石 $Li_4Ti_5O_{12}$ 负极中锂离子输运的特殊机制。该材料脱嵌锂过程遵循两相反应，锂离子在这两个相中的输运均较慢。然而，在两相之间，存在包含扭曲的锂多面体的亚稳相边界。在亚稳态相边界时，锂离子跃迁过程中产生共享 Li-O 多面体面的中间构型，此时锂离子跃迁活化能低于两相，其机理如图 5-14 所示[118]，其中图 5-14(a) 中的蓝线是富锂相，绿线为贫锂相，红线为高锂离子电导率的中间态亚稳相。

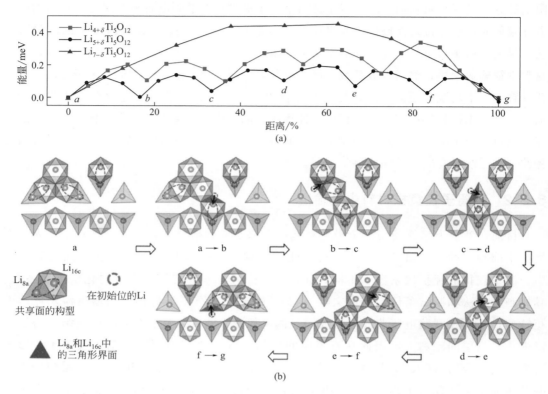

图 5-14 锂离子输运活化能（a）和中间相的锂离子输运通道（b）[118]

5.6 锂离子在固体电解质中的输运

为了提高锂离子电池的安全性，采用固体电解质的全固态锂离子电池引起了广泛的关注。固体电解质又称快离子导体，具有许多优点，包括较宽的电化学窗口、更好的安全性、更好的抗碰撞性能[119-121]。常温下接近 10^{-3} S/cm 的离子电导率，是作为良好锂离子电池电解质的判断依据。目前已经研究的固体电解质材料有几百种，包括无机电解质、聚合物电解质、无机与聚合物复合型电解质。对于离子在不同固态电解质材料中的输运机制、离子电导率等问题，在以往的综述文章[4,5,122,123]里已有较为详尽的介绍。目前不含液态增塑剂的纯聚合物固体电解质室温离子电导率可以达到 10^{-3} S/cm，100℃时离子电导率达到 5×10^{-2} S/cm。聚合物电解质导电，主要通过聚合物链段上的官能团与锂盐相互作用，使得锂盐发生解离，锂离子与聚合物链上的阴离子基团相互作用，并跟随聚合物链段蠕动，被阴离子基团释放后与下一个官能团复合，从而在空间发生位移。锂离子电导率的提高依赖于链段蠕动的速度以及聚合物与锂盐的相互作用。较高的锂盐解离能将阻碍锂盐的解离，不易形成能自由移动的锂离子。广泛研究的无机固体电解质按阴离子种类区分主要包括氧化物、硫化物以及含卤素类[81,124,125]，目前的研究范围包括了 Li_4SiO_4 及相关相、Li_5AlO_4 及相关相、LISICONS、NASICON、钙钛矿、反钙钛矿、石榴石、$Li_{10}GeP_2S_{12}$、Li-B-H 型硼酸盐固体电解质以及各种含卤族元素固体电解质等。在综述［7，126］中已经详细阐述了固体电解质的发展、种类以及不同材料的计算方法和结论。一般而言，氧化物类固体电解质材料的离子电导率常常受晶界限制，要使电解质的总体电导率提高，必须调控并改善材料的晶界，制备出致密的产物。由于 Li-S 的相互作用比 Li-O 的相互作用弱，一般硫系电解质具有更高的离子电导率。玻璃态硫化物固体电解质室温下大约为 10^{-4} S/cm。2011 年，日本东京工业大学制备出了室温离子电导率接近 10^{-2} S/cm 的 Li_2S-GeS_2-P_2S_5 系硫化物固体电解质材料[81]。但是，硫化物固态电解质在应用时存在很多问题，比如硫化物材料在室温条件下在空气中不稳定，遇到水汽易发生反应生成有毒的 H_2S 气体，高温遇到氧气会发生氧化燃烧反应等。

为了研究结构与离子电导率的关系，可以通过高通量计算，得到大量结构的锂离子电导率的相关参数，并学习现有高离子电导率材料的结构特征，用以预测新型高离子电导率材料。例如，可以利用快速的键价和方法，以分钟为单位，搜索现有的晶体学数据库，计算含锂材料的锂离子通道，并通过锂离子通道是否连通来进行判断和筛选[127-129]。此外，通过建立阴离子堆垛方式与锂离子输运的关系，有效预测了硫化物中锂离子导体高电导率通常符合体心立方结构，因其可以使锂离子在邻近的四面体之间以较低的活化能跃迁，如图 5-15 所示[130]。

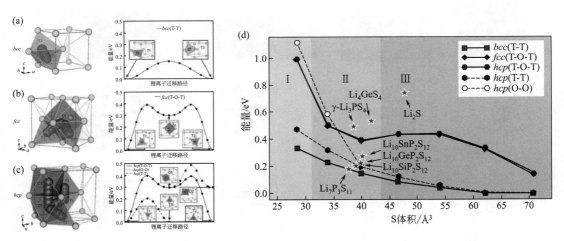

图 5-15　锂离子在 bcc（a），fcc（b），hcp（c）型阴离子晶格中的扩散通道以及活化能与晶格体积的关系（d）[130]

需要注意的是，文献［5］预测了无序与协同运动是无机固体电解质高电导率的关键。目前在这两方面已经逐渐被理论所证实[131]。常见的无机固体电解质，其锂离子可以占据的位置通常高于锂离子的实际数目。而从宏观来讲，局域的跃迁之所以可以连成宏观的锂离子连续通道，是锂离子-空位的定向随机跃迁导致的。对于快离子导体，锂离子在可占据位置周围具有比较大的占据空间，锂离子在热力学驱动下在占据位附近振动，沿扩散方向扩展为"运动椭球",这种扩展占据,使锂离子可以在多余的位置以比较低的活化能扩散,如图 5-16 所示[132]。当锂离子-空位的随机跃迁,在时间标度上有因果性或者具有时间关联性的时候,锂离子的跃迁将不完全随机,而是有一定的协同作用。协同作用下,锂离子扩散的动态活化能(活化能在宏观晶体的空间尺度,以及运动关联所覆盖的时间尺度的平均)将进一步降低。可以利用 Van Hove 关联函数研究锂离子跃迁在时间尺度上的关系,如图 5-17 所示[133]。在快离子导体中,活化能常常低于 0.5eV,远远低于普通晶体中的缺陷扩散。

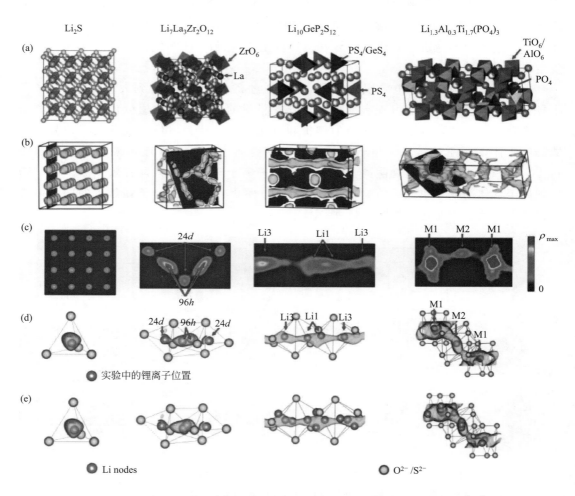

图 5-16 非超离子导体与超离子导体中,锂离子的概率密度和锂的位置[132]
(a) 晶体结构;(b) 锂离子的密度;(c) 截面;(d) 实验所得锂离子位置;(e) 拓扑分析预测锂离子节点

另外,对于未来的全固态电池,有机-无机复合电解质是最有希望真正达成应用的。而对于复合电解质来说,总电导率不仅依赖于复合材料本身的电导率,也备受界面电导率、组分比例、不同组分形貌(颗粒大小,形状,分散情况等等)所影响。

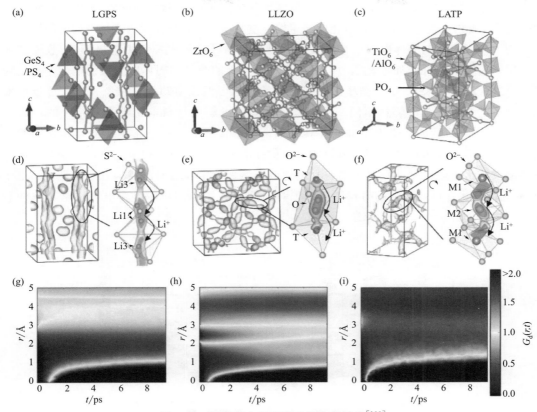

图 5-17 锂离子在超离子导体中的扩散[133]

(a)~(c) 晶体结构；(d)~(f) 锂离子的概率密度；(g)~(i) 对于指定锂离子，其动力学的 Van Hove 关联函数

5.7 离子在电极/固体电解质界面

在全固态电池的应用中，电解质材料与正负极的匹配带来了界面问题，可能会显著提高全电池电阻，甚至成为电池电阻的最大制约因素，减小界面电阻可以有效地改善全电池电导率。固体电解质与电极材料接触，最大的问题是固固界面物理接触面积通常较小，甚至为点接触。薄膜电池利用射频溅射制备，通过严格控制制备工艺通常可以得到致密且基本无晶界的电解质层和电极层，且两层之间以原子的尺度紧密接触。对于三明治结构全固态电池，如果利用氧化物作为固体电解质，则通常需要将电极材料与固体电解质材料混合作为电极，并通过烧结的方式增大接触面积、减小界面电阻。对于硫化物而言，其具有良好的延展性，杨氏模量介于有机聚合物和氧化物陶瓷之间[134,135]，通常通过冷压的方法就可以达到较小的晶界电阻[134-136]，并与其他材料（包括电极和复合电解质中的另一成分等）有比较好的接触[137]；控制材料形貌也是提高固相-固相之间物理接触的手段之

一[138]。当负极为锂或者锂铟合金，由于金属具有比较好的延展性，物理接触就不是最主要的问题了。此时，化学稳定性、电化学稳定性、空间电荷层，转而成为必须考虑的问题。

为了确保电极与固体电解质接触良好，以及在循环过程中抑制由电极膨胀/收缩造成的分层，常需要额外施加堆叠压力。堆压对合金负极的容量、固体电解质的电化学稳定窗口以及锂金属在ASSBs中的循环性都有很大的影响，但该值却常常被忽视。表征堆压对于电导率的影响，可以解释不同文献中，同一固体电解质材料电化学测量结果的巨大差异。如图5-18所示的一个固态电池夹具和一个重量传感器，可以精确控制堆压。实验表明，高的制造压强可以降低固体电解质材料的孔隙率，当压强达到一定值，其相对密度将不会继续增加，电导率达到最大值，且电池的容量保持率和倍率性能均有所提高。当电解质-集流体已建立良好接触的时候，电池循环过程中继续施加堆压，对电解质电导率的影响可以忽略不计。

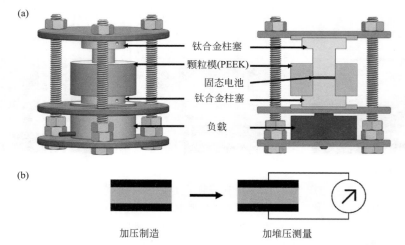

图5-18 带压力传感器的固态电池夹具（a）；加压制造及加堆压测量电导率和电池循环（b）[138]

5.7.1 固体电解质/负极与混合输运

固体电解质体相锂离子迁移率通常为1，为电子绝缘体；然而，在界面处，可能由于电子隧穿、化学反应、电化学反应等原因，导致复杂的混合输运。此种效应在负极一侧的固体电解质中尤其明显。锂离子在各类碳材料中的输运过程，还伴随着SEI膜的形成和穿过SEI膜的动力学过程。SEI膜主要是由电解液还原分解产生，组成成分包括碳酸锂、烷基氧锂、烷基碳酸酯锂和氟化锂等[139]，而SEI膜的结构与诸多因素有关，比如充放电电流密度、温度、电解液中的添加剂、溶剂和所使用的锂盐[139-142]。离子在SEI膜中的输运受到了膜的结构和厚度的影响。10nm厚的SEI膜的体相电阻约为$10 \sim 1000 \Omega \cdot cm^2$，而晶界电阻则有$10 \sim 100 \Omega \cdot cm^2$[143]，因此晶界对于离子在SEI膜中的输运起到了重要的影响。但是目前对于锂离子在SEI膜中的输运机理并不十分清楚，还需要更加细致的研究。同时，SEI膜的存在使通过减小电极材料颗粒尺寸来提高锂离子扩散动力学的方法不一定具有优势，因为减小颗粒尺寸后增加的表面积会促进SEI膜的生成，导致界面电阻的增加。为了研究 SEI 膜形成的微观机理,利用第一性原理得到电子隧穿速率,认为非绝热条件下,溶剂分子重组能使电子转移减慢[144]。SEI 膜的成分很复杂[145],输运机制不容易研究。其中主要的无机成分为 Li_2CO_3,电导率极低[40,41,146,147];利用第一性原理计算[40,41,148,149],认为是间隙位锂离子主要以"knock-off"机制输运[40,41,148]。此外,Li 还可能以原子形式输运[41],涉及离子与电子的混合输运,并可能导致 SEI 膜的不断增厚。

目前广泛研究的负极材料还包括尖晶石结构的钛酸锂负极($Li_4Ti_5O_{12}$)[150]以及硅负极[151],前者为三维锂离子导体,具有优异的离子输运特性。后者在嵌脱锂后会形成无定形 Li-Si 合金[152]。离子的输运在无定形相中发生。硅负极在充放电过程中具有比石墨负极更大的极化[153],目前还不清楚是否与离子输运性质有关。

氧化物固体电解质中,钙钛矿体系和 NASICON 体系通常具有较大的电导率,然而两者的高电导率组分均含有可变价的 Ti,导致其可能被还原;实验表明,两者均可以发生嵌锂反应[154,155]。基于此原因,以 NASICON 为固体电解质,可以通过首周放电,在正极对侧原位析出 Li 负极。Ti 的变价对应着电子的转移。对于一些硫化物固体电解质,特别对于含 Ge 的材料(包括含 Ge 的玻璃陶瓷[156]和晶体 $Li_{10}GeP_2S_{12}$[157]),以及一些含 Si 材料[156,158,159],其还原电位甚至更高,即使和石墨接触也会反应。通常需要选择和电极材料电化学窗口匹配的固体电解质材料,或者在正负极侧选用不同的固体电解质材料,使电解质同时满足不容易被负极还原,且不容易被正极氧化[156,157,159]。值得注意的是,对于通常认为与金属锂稳定的石榴石型固体电解质,其与锂接触会造成非化学计量比嵌锂、并导致电子绝缘体-电子导体相变,这是其循环过程中容易被锂枝晶穿透的重要原因[160]。

5.7.2 固体电解质/正极与空间电荷层

固体电解质与正极材料的接触和晶界,是其应用的最大难点之一。此外界面的相稳定性也是巨大的问题[161-163],在之前章节已经介绍,不再赘言。得益于硫化物固体电解质,晶界问题基本解决。Takada 组首次提出,限制高电导率硫化物基固体电解质全固态电池的倍率性能的,不再是无机固体电解质部分,而是固体电解质与正极的界面。由于电解质层的硫比电极层的氧对于锂离子的束缚能力更弱,导致电解质侧界面的锂离子耗尽,界面电阻增大[164]。Takada 组从循环伏安法(cyclic voltammetry,CV)的充电初发现类似于电容现象的"斜坡",并对电极层用不同厚度的多种材料(包括 $Li_4Ti_5O_{12}$[164,165]、$LiTaO_3$[165]、和 $LiNbO_3$[165,166]等)包覆,利用交流阻抗谱详细研究了包覆材料与包覆厚度对于界面电阻的抑制效果。为了深入了解微观机理,Takata 与 Huruyama 等合作,计算了 $LiCoO_2/\beta\text{-}Li_3PS_4$ 和 $LiCoO_2/LiNbO_3/\beta\text{-}Li_3PS_4$ 界面空间电荷层。结果显示,界面处 LPS 侧锂离子耗损,LCO 附近锂离子富集(这一点与传统平衡态在 LCO 内富集略有差别),导致热平衡态的空间电荷层;在充电初态,LCO 侧锂离子由于空间电荷层的阻碍作用,富集在界面处,导致放电前段电容现象的产生;$LiNbO_3$ 缓冲层有效缓解了空间电荷层的生成,并大大抑制充电初阶段空间电荷层的增强现象[167]。然而,该工作人为将两个晶格参数差异较大的材料,以异质结的方式构造出几个原子层的界面,这种方法过于简化了。事实上，界面可能有多种形式，与界面两侧材料的晶体结构、晶胞参数以及构成元素的性质等，均有很大关系，不能一概而论[168]。此外，得益于实验手段的发展，利用三维电子全息成像，从实验上直接观察到了空间电荷层的存在，其在空间尺度上延伸至几百纳米深度，远远大于传统液体中的空间电荷层[169-171]。

5.8 影响离子输运的因素

通常而言，我们研究所关注的，是对于确定体系优化/提高电导率。其中，更高的载流子浓度、更多的锂离子可占据位、两者满足渗流模型使通道连通、且锂离子在通道内以较低的能量迁移，通常是提高电导率的主要关注点。在一定温度下，迁移率及扩散系数与载流子浓度、离子脱离平衡位置势垒的高度、跳跃频率、平均跳跃距离有关，可以利用缺陷化学的方法对目标材料进行掺杂改性。阿伦尼乌斯公式中指前因子主要取决于可移动的离子的浓度，可以通过异价元素替代来调控空位浓度和可移动离子浓度，或通过构造异质界面/界面修饰来调控界面处空间电荷层的性质、增加缺陷浓度。阿伦尼乌斯公式中活化能的大小反映了离子迁移势垒的高低，它与晶体结构有关，即离子与周围点阵的库仑相互作用的大小。掺杂/取代的元素不同，导致电解质中骨架元素对于锂离子的作用会有差别；更小的库仑相互作用通常带来更低的跃迁活化能，可以使锂离子更自由地流动。从这一点出发，在近似结构的情况下，离子导体电导率高低大体的趋势是氟化物＜氧化物＜硫化物＜氮化物＜磷化物。为了降低骨架阴离子对阳离子的束缚，将F，O用S，N，Cl等替代是经常考虑的方法。此外，掺杂取代对于固溶体范围内晶胞参数的微调，通常会影响电导率；合适的晶胞参数以及锂离子通过的"瓶颈"的大小，可以优化得到更高的电导率。

对于混合离子导体，离子的输运还与电子的输运性质密切相关。Wagner对此进行了详细分析。输运是由于物体内部某种不均匀引起的，在平衡态附近，流量 J 与不均匀因素 X 成正比（$J=LX$），比值 L 称为动理系数（coefficient）。当不均匀因素不止一种的时候，有 $J_k=\sum_l L_{kl}X_l$，其中 L_{kl} 为交叉动理系数，这样就把力学量耦合起来。对于多种离子和电子的共同作用来说，相互作用包括电子-电子、电子-离子、离子-离子相互作用，动理系数分别为 L_{ee}、L_{ei}（$=L_{ie}$，昂萨格关系）和 L_{ii}，通过这种方式，就把离子和电子输运通过动理系数的概念联系在一起了。Lee等通过动理系数的比值来推断相互作用的强弱，并指出在 TiO_2 中（电子-电子相互作用）≫（电子-离子相互作用）＞（离子-离子相互作用）[172]。

改性的策略主要有两种：形貌和组分。改变形貌主要应用于复合材料，其中有机-无机复合材料，可能是未来固态电池电解质的最终解决方案。当尺寸减小的时候，改变形貌和改变组分的改性方法，两者的界线很模糊。复合材料有希望提高离子电导率，不同种类的复合材料导致了迥异的相界，并使导电提高机制有所不同，例如空间电荷层的形成，位错密度增加，或者形成新相[173]。其中Maier提出的空间电荷层理论受到最广泛的接受[168,174-177]，他认为固体电解质体相为电中性平衡，带正、负电荷的缺陷局域相等；然而在两相界面处，电化学势驱动导致不同电荷的缺陷浓度会发生变化，并引起空间电荷层。缺陷浓度的不平衡，随着从界面至体相方向深度的增加而衰减，并定义德拜屏蔽长度 (Debye screening length，λ)：$\lambda = \sqrt{\dfrac{\varepsilon_0 \varepsilon_r RT}{2(zF)^2 C_b}}$。这里，$\varepsilon_0$ 和 ε_r 分别是真空介电常数和相对介电常数；C_b 为载流子浓度；z 为电荷数；F 为法拉第常数。在界面控制的材料中，尺寸减小可以是界面区域比例提高，即纳米粒子学（nanoionics）。Maier 将尺寸效应分为两类[178,179]：显性尺寸效应是指材料的电阻（与输运有关）或电容（与储能有关）性能直接受到几何结构的影响；而隐形尺寸效应与材料的有效参数有关，尤其对于异质对象，甚至可以利用尺寸效应，通过调节的方法改变输运或者储能机制。如果局域的影响和孤立界面相同，则称为"平凡尺寸效应（trivial size effects）"；如果界面之间相互影响（perceive each other），则称为"非平凡尺寸效应（true size effects）"。后者源于尺寸减小导致非局域电子的增多，是一种量子效应[180]。

尺寸效应对于全电池其他环节的离子输运，也具有显著影响。离子在介质中输运的最短时间与扩散距离的平方成正比，与扩散系数成反比 $\left(\tau \propto \dfrac{L^2}{D}\right)$，离子在电极固相中的扩散一般是电极反应的最慢步骤。当材料的尺寸显著降低时，比如电极颗粒的直径由 $10\mu m$ 降低到 100nm，离子输运的最短时间将缩短 4 个数量级，这将显著提高电极的电子、离子的输运、存储、反应的动力学速率，进而提高电池的功率密度。同时，考虑到单位质量的物质，其比表面积与尺寸成反比关系，小尺寸材料具有较大的比表面积，这使得在同样质量的情况下，小尺寸材料允许更高的单位面积电流密度。纳米材料具有较高的缺陷密度，对于纳米尺寸的颗粒而言，具有较高的表面能，因此热力学不稳定。研究表明，在充放电过程中，纳米颗粒之间由于离子的迁移、扩散会出现显著的团聚、融合，发生电化学团聚现象[152,181,182]。同时，纳米碳负极材料由于表面能较高，会促进 SEI 膜的生成，从而消耗正极的锂元素。但是，小尺寸材料又显示出了显著的动力学优势[183,184]。研究发现，不同于纳米正极材料 $LiCoO_2$ 在电解液中自发浸泡会形成表面电子绝缘层[185]，$LiFePO_4$ 和 $Li_4Ti_5O_{12}$ 的嵌脱锂电极电位避开了电解质的氧化还原分解电位，并且化学性质稳定。因此 $LiFePO_4$[186] 和 $Li_4Ti_5O_{12}$ 由于尺寸效应具有的动力学优势得以显示出来。总而言之，电极材料的纳米化虽然可以显著改善离子的输运特性，但必须同时考虑到纳米化后引起的热力学不稳定，及其带来的一系列问题。

5.9 实验表征方法

为了研究动力学过程的微观细节，可以利用电磁波或探测粒子，通过吸收、反射、散射的方法，研究其与材料的作用。现象和对应的研究方法如表 5-4 所示，不同频率的动态模式如图 5-19 所示。对于电导率的交流阻抗谱测量，可以从直流到 10MHz 量级；高于此频率可以通过低温阻抗谱，间接得到高频得不到的信息。10MHz～GHz 可以用无线电波段探测。GHz 以上利用微波技术。更高频率，可以利用远红外、红外和可见光波段的吸收和反射，以及准弹性散射（包括基于 γ 射线的穆斯堡尔谱和基于中子的准弹性中子散射）。声子谱的光学模式一般在 10^{13} Hz，体现了离子的振动；而对于高离子电导率对应的声子谱模式软化，振动频率降低，对应着离子扩散。在更高的频率，光子的相互作用与离子振荡相比很快，因此 X 射线或中子衍射谱被离子的瞬时组态决定，即得到结构信息。详情见文献 [2, 187]。

表 5-4 现象和所用方法的频率范围[187]

辐射		λ^{-1}/cm^{-1}	λ	频率/Hz		现象和方法
			3Å	10^{18}		
	X 射线		30Å	10^{17}		光子相互作用
	紫外光		300Å	10^{16}		
			3000Å	10^{15}		
	可见光		3μm	10^{14}		
	红外	1000	30μm	10^{13}		离子振动
		100	300μm	10^{12}		傅里叶光谱
	远红外	10		10^{11}		中子散射
		1		10^{10}	40GHz	光散射
	微波		30cm	10^{9}	2GHz	（拉曼，布里渊）
			3m	10^{8}		声学声子
	短波		30m	10^{7}		
无线电	中波		300m	10^{6}	1MHz	
	长波		3000m	10^{5}		
				10^{4}		
				10^{3}	1kHz	交流阻抗
				10^{2}		测量
交流				10		
				1		
				10^{-1}		电极扩散
				10^{-2}		现象
				10^{-3}	1mHz	
				H		
				L		
直流				$10^{-\infty}$		直流测量

5.9.1 晶体结构与锂原子占位

快离子导体相变的很多问题均源于结构，例如：①载流子亚晶格和骨架离子亚晶格；②快离子相变导致的结构无序化；③声子模式软化导致的锂离子的偏移等，与高电导率的

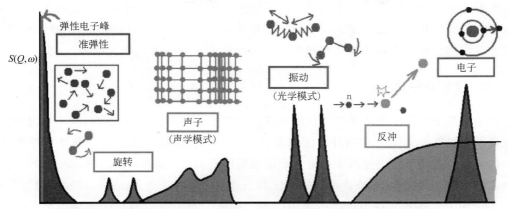

图 5-19 动态模式图（来自 2013 Oxford school on Neutron Scattering 讲座）

关系。X 射线衍射为结构确认的常用手段，然而在包含重元素的化合物中观测轻元素十分困难（目前，单晶的 X 射线衍射确定锂的占位已成为可能[189,190]）；相较而言，由于热中子束的动能很低，因而利用中子衍射有可能探测周期表中几乎所有元素的位置和动态特性，适合于锂离子位置的确认，如图 5-20 所示。衍射谱中既包含长程周期结构信息（布喇格峰/德拜线），也包含来自短程关联的类液体的漫散射；然而当系统在化学上和结构上更加复杂时，所能提取的局域结构信息的价值则大大减弱。可采用小角 X 射线（small-angle-X-ray scattering，SAX）以及小角中子衍射（small-angle neutron scattering，SANS）、低角度衍射谱第一衍射锐锋（first-sharp diffraction peak，FSDP），分辨约为 1～10nm 尺度的中程序结构（intermediate range order）。短程序（short range order）方面，扩展 X 射线吸收谱（extended X-ray absorption fine structure，EXAFS）可以给出受激原子与近邻原子配位的短程结构信息、受激原子最近邻的类型和数目信息，以及这种信息与材料高离子电导率之间的结构关系。利用短程有序（short-range order，SRO）模型解释：吸收原

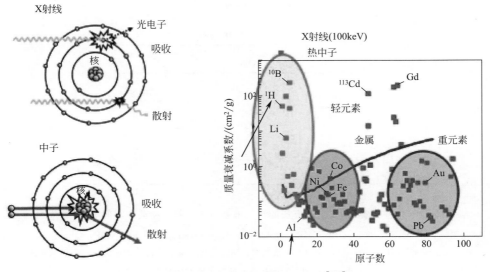

图 5-20 中子衍射与 X 射线的互补[188]

子的出射光电子波受到近邻原子的散射而形成散射光电子波,出射光电子波和散射光电子波在吸收原子处相互干涉,使吸收系数 μ 发生变化。当两者位相相同时,出现干涉极大,而当两者位相相差 π 时,出现干涉极小,从而形成了 XAFS 谱的振荡结构。

5.9.2 锂扩散通道

对于快离子导体中锂离子扩散过程而言,锂离子在非晶格位的弛豫时间不可忽略。此时锂离子在温度驱动下,常离开固定的晶格位点,并在晶格位附近呈"椭球"形统计分布,在更高的温度下连成锂离子通道;利用变温中子衍射(neutron diffraction,ND),以及最大熵模型(MEM)精修,可以探测锂离子通道。此方法由 Yamada 首次运用到 LiFePO$_4$ 中[191,192],并在固体电解质中得到应用。例如,在石榴石型固体电解质的发展历史中,首先利用 X 射线衍射对整体结构确认[194],通过粉末中子衍射可以精确定出 Li 在晶体结构中的位置和占有率,提供晶体材料结构方面的关键信息[193];随后通过变温粉末中子衍射结合最大熵拟合方法,研究锂离子在无机晶体材料中的输运路径[195]。

5.9.3 电导率和扩散系数

测量扩散系数可分为直接法和间接法;图 5-21 以测试手段的时间/空间尺度进行归纳[27]。

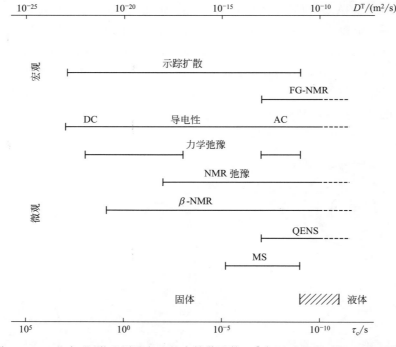

图 5-21 一些宏观/微观测试方法的自扩散系数 D^* 与运动关联时间 τ_c 典型范围
(FG-NMR:field gradient NMR,场梯度 NMR;β-NMR:β-辐射探测 NMR;QENS:quasi-elastic neutron scattering,准弹性中子散射;MS:moessbauer spectroscopy,莫斯堡尔光谱);
阴影部分表示固相到液相的相变,此时关联时间减小两个量级[27]

5.9.3.1 直接法

同位素示踪的方法是测量固体中的扩散系数（包括自扩散和杂质扩散）的最直接、最准确的方法。由于测试中仅需要痕量的同位素，所以不会影响目标材料的化学成分。示踪元素可以是放射性同位素，也可以是稳定同位素。确定浓度-深度图的最好方法是对样品连续切片，并逐一测试每部分的同位素含量。如果扩散系数过小、扩散深度小于 $1\mu m$，精确控制切片和精确测量含量将很困难。可以利用其他确定浓度-深度图的方法，例如二次离子质谱（secondary ion mass spectrometry，SIMS），用一次离子（primary ions）连续炮轰（bombard）样品表面原子，并利用质谱分析溅射出来的二次离子。其他方法还包括：微区电子探针分析（electron microprobe analysis，EMPA）、俄歇电子能谱（auger electron spectroscopy，AES）、卢瑟福背散射谱（rutherford backscattering spectrometry，RBS）、核反应分析（nuclear reaction analysis，NRA）以及（脉冲）梯度场核磁共振[field gradient nuclear magnetic resonance，（P）FG NMR][196]。具体请见参考文献 [197] 以及相关引用文献。

5.9.3.2 间接法

（1）阻抗谱

可以通过测试电导扩散系数 D^σ，并通过常数 H_R 求得实际的自扩散系数，但是对于电极材料而言，其涉及化学势、浓度和离子-电子的交叉关联（cross ionic/electronic phenomenological coefficients），更为复杂[198]。交流阻抗谱最为常见，其优点之一是可以分辨多晶固体电解质的体相和晶界电阻，即可以研究微结构对于整体电导率的影响[45]。其原理是，施加一个交流电压 $U(\omega)=U_0 e^{i\omega t}$，样品响应电流为 $I(\omega)=I_0 e^{i(\omega t+\Phi)}$，其与外加电压具有相同的频率，但是产生一个相位 Φ。则所得的复阻抗为：$Z(\omega)=\dfrac{U(\omega)}{I(\omega)}=Z_0 e^{-i\Phi}=Z_0\cos\Phi-iZ_0\sin\Phi$，可以通过电阻和电容或电感的等效电路来拟合，如图 5-22 所示，并给出微观结构的信息[27]。

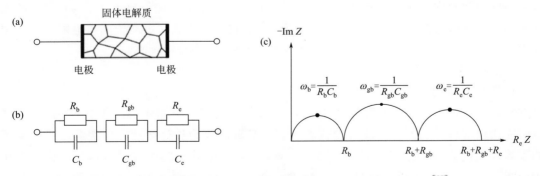

图 5-22 多晶固体电解质（a）；等效电路（b）和 Nyquist 图（c）[27]

此外，可以利用微电极或纳米尺度电极，直接扫描材料中各"局域"点（包括晶界和体相）的电导率[199-202]。例如在材料表面各处施加微电极，如图 5-23 所示；或利用导电原子力显微镜（conducting atomic force microscopy）作为纳米尺度电极。也可以利用霍尔效应研究电导特性[203-206]。注意，所研究对象均为具有更高导电率、基于典型亚晶格熔化模型的银离子导体，然而就笔者目前的认识来看，Hall 效应中霍尔迁移率和电导迁移率的关系或许并不确切。这是由于 Hall 效应基于准液态模型，而对于大多数锂离子导体而言运动离子之间、运动离子与晶格之间通常有更复杂的长程相互作用，预计将会导致更大的偏差。

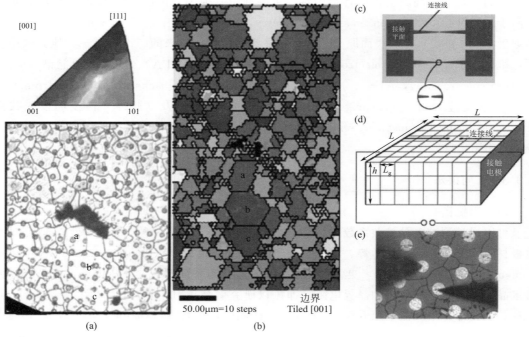

图 5-23　多晶 SrTiO$_3$ 和顶端微电极的光学显微镜照片（a）；晶界取向（b）；微电极（c）；探测模型示意图（d）；利用针尖圆形使微电极接触材料照片（e）[199]

（2）力或磁的弛豫

力的弛豫利用了材料中原子的运动受外部作用影响，而铁磁材料的磁矩与局域序的相互作用可以导致类似于磁滞回线的弛豫现象[197]。这两种方法在锂离子固体电解质中应用较少。

（3）核的方法[207]

包括核磁共振（NMR）[196,208,209]、穆斯堡尔谱（MS）[210]和准弹性中子散射（QENS）[197,211]。不同方法和原理在文献 [207] 中进行了详细介绍。其中 NMR 对于电池研究比较常用[27,187,208]，既可以检测短程、又可以检验长程运动，也常常能区分原子的不等效位置。NMR 有多种，其中 FG NMR 常用于介孔材料中粒子输运与结构相互作用的测量[196]，由于其具有毫秒量级的时间尺度，也可测量锂离子固体电解质中的长程关联。而电泳 NMR（electrophoretic NMR）主要应用于液体中带电粒子的测量[209]。除了这些宏观的应用，NMR 可以用来测试微观性质，最常见的包括瞬态研究（transient studies）和宽线研究（wideline studies）。在瞬态测量中，是给外加静磁场中的样品施加一个短而强的射频辐射脉冲,通过这个脉冲,就把附加的能量传导到核上,从而把相位关联施加于磁场中核的简正运动上。撤去脉冲,核把附加能量传递给周围环境,相位关联以两个特征时间(自旋-晶格弛豫时间 T_1 和自旋-自旋弛豫时间 T_2)衰减。T_1 和 T_2 的大小,反映了核与环境的相互作用强度。瞬态测量的扩展方法,包括在旋转坐标系中测试自旋-点阵弛豫时间 $T_{1\rho}$,性质介于 T_1 和 T_2 之间。在宽线研究中,把样品暴露在外加静磁场当中,并连续地施加小的射频场。核所吸收的射频功率的位置、宽度和形状,就反映了它们与微观的局部环境(邻近核的磁矩、电场梯度和化合物的其他组分,例如顺磁电子和传导电子)的相互作用。快离子导体会导致其极限运动变窄,该方法常用于检测异质结构导致的界面高电导新相生成[212,213],可以分辨痕量高离子电导相。

5.9.4　全频电导分析[2, 187]

在全频分析中,低频部分可以通过直流/交流阻抗谱完成。高频测量主要分为准弹性散射和高频电磁谱测量;这里,仅关注和电导率有关的测量方式,如图 5-24 所示,并根据测量方式所依赖的频率不同,将其统称为全频电导分析。虽然对于不同频段,所用测试手段不同,但是原理是一致的:复电导率通过测量外场诱导的某量的振幅和相位来确定。对于阻抗谱来说,"某量"是指电流和电压;对于其他技术,是没有电极的,相关的"量"则为透射或反射电磁波的复场(complex field)的强度,并通过解决满足界面边界条件的 Maxwell 方程(保证电磁场的连续性)来得到复电导率。由于是速度关联函数的傅里叶变换,因此它体现出速度关联函数的一些特征。全频谱的电导率测试方法跨越了 17 个数量级(17 decades),不仅提供了测量带电颗粒在极大时间跨度运动的技术,更通过高频段的测量方法观测到极小的时间尺度,从而试图解释基本跃迁过程。注意到,热激活缺陷、强电解质溶液以及无序晶格中,全谱的特征完全不同。按照 Funke 的分类,热激活缺陷扩散行为遵循随机行走模型,因此点缺陷相互之间没有影响,导致所有关联函数的交叉项都消失;且随机行走不存在记忆效应,导致电流的关联函数仅包括自关联部分,为 $\delta(t)$,傅里叶变换后电导率为常数,如图 5-25(a) 所示。而对于无序晶格离子导体,如图 5-25(c) 所示,关联函数的绝对值在 $t=0$ 处有一个陡峰,同样为自关联效应;并随着 $t>0$ 逐渐衰减,表示向后跃迁的概率逐渐减少,其原理如图 5-25(d) 所示。与如图 5-25(b) 所示的具有 Debye-Hückel-Onsager-Falkenhagen 效应的强电解质稀溶液比较,当溶液中的离子在初始偏离平衡位置,两个效应使其有恢复平衡的倾向:离子向后跃迁,以及带负电的

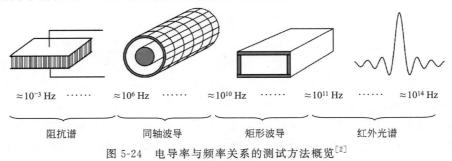

≈10^{-3} Hz　……　≈10^6 Hz　　　≈10^{10} Hz　　　≈10^{11} Hz　……　≈10^{14} Hz

阻抗谱　　同轴波导　　矩形波导　　红外光谱

图 5-24　电导率与频率关系的测试方法概览[2]

"云"向前跃迁,均导致了与初始位移相反的电荷流,傅里叶变换表现为电导率随频率略有升高。而对于晶格无序的固体电解质来说,运动以不同晶格位势阱间跃迁的方式进行,附近的离子扮演"离子云"的角色,类似地,也存在两个不同的弛豫过程:离子向后跃迁,或另外的"离子云"向前跃迁,而图5-25(b)和图5-25(c)如此大的区别,其原因在于,两种竞争机制的比例是不同的:固体中,离子向后的概率要大得多,导致更强的色散。

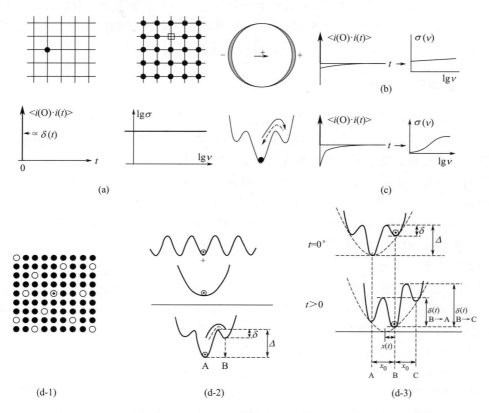

图 5-25 热激活缺陷(a);强电解质稀溶液(b);结构无序固体电解质的电流自关联函数和电导率色散比较(c)[2];跃迁-弛豫模型(d)[73]

5.10 本章结语

输运问题是锂离子电池的核心问题之一,包括锂离子在电极材料、电解质及其界面中的输运,离子的输运特性与锂离子电池的功率密度密切相关,掌握离子在固体中的输运特性对于设计和开发材料、提高电池性能具有十分重要的意义。固体电解质的发展历史表明,似乎材料体系的诞生驱动着对于离子输运机制的理解。离子(而不是电子)可以导电,导致了缺陷化学理论的建立;为了解释高电导率,引入亚晶格熔化的图像;为了解释表征过程中发现的反常现象，例如准弹性谱展宽、高频区电导与频率的相关性等，提出了各种动态模型将多体相互作用纳入考量。物理图像随着表征手段的改进而不断修正。然而，由于当前的理论模型无法得到解析解，所以如何高效地简化理论模型，使计算和实验的精确性和时间成本得以平衡，对科研工作者来说是一个挑战。为了研究离子输运机制，现有的方法可分为直接方法和间接方法。直接方法主要是基于示踪原子检测，目前，无论是切片、或者利用 SIMS 分析逐层溅射出来的二次离子，均是非原位、破坏性的方法；间接方法通常得到的是微观机制的平均，对于常见的多晶固体电解质，不具有空间分辨率。微探针测量电导率的方法，可以对全空间扫描并统计不同微结构处电导率-结构的关系[202]。目前，环形明场球差校正扫描透射电子显微镜（annular bright field imaging using aberration-corrected scanning transmission electron microscopy，ABF-STEM）是迄今为止直接观察锂元素的最好的工具。然而，对于固体电解质而言，其困难在于样品电子电导低，造成辐照损伤[214]。希望接下来的技术可以控制辐照损伤，获得原子在空间运动的图像。更进一步，由于 4D 超快电子显微技术（4D ultrafast electron microscopy）引入了时间的维度[215]，已成为继球差校正电镜之后人们所关注的一个新的电镜技术发展方向，它通过将原子分辨率的透射电镜与超快激光技术及原位技术结合，来同时实现 pm 级的空间分辨率及 ps 级的时间分辨率。此外，利用扫描隧道显微镜（scanning tunnel microscope，STM）可以原位观察锂离子在表面的运动，并研究电场诱导下的运动，设计实验进行原位的动力学研究，对于理解锂离子输运机制将有很大帮助；其困难同样在于样品导电性。由于技术极限，以及时间成本和经济成本的原因，很多重要的问题依然无法通过现有的实验手段获得答案。相对地，随着计算机科学和信息学的飞速发展，理论、模型、计算和模拟，已成为实验的有力补充。"材料基因组计划（MGI）"旨在将实验、计算和数据库三者结合，其中，计算为重要的一环[216-218]。注意到，当材料尺寸缩小为纳米量级的时候，材料性质往往会伴随反常的尺寸效应，这一效应的值取决于尺寸大小和材料形状，这就为构-性关系增加了两个维度。Qian 等[219]提出纳米材料基因工程的概念，将尺寸-结构的维度纳入考量，为材料的快速研发应用，以及已有材料的纳米尺度的修饰设计打开了新思路。

目前，离子在固体中输运的诸多科学问题需要进一步阐述，比如：

① 实际体系里更复杂的各种驱动力和各种载流体的相互作用，例如，离子和电子的相互作用、离子与骨架原子的相互作用、离子与离子之间的相互作用如何影响离子的输运特性；我们不仅要考虑其空间的关联特性，还要考虑运动的时间相关性，例如，载流子之间，以及载流子与骨架之间的协同输运，可能是高离子电导率的关键；

② 实际体系中电化学势的分布，以及驱动力的来源与大小，例如，固体中浓度梯度和电场梯度建立的微观机制，如何驱动内在和外来离子在固体中运动，在垂直于及平行于表面和界面的离子输运行为；

③ 固固界面的锂离子输运和相变与动力学亚稳态问题：离子从一相进入另一相，其微观过程不仅仅包含热力学稳定的两相，通常还需要考虑离子输运过程中热力学不稳定的亚稳相；亚稳相不仅对于界面稳定性具有重要影响，甚至有可能决定了锂离子输运的微观机理；

④ 具有高离子电导率材料的结构和组成特点：虽然硫化物固体电解质中，bcc 型结构与高离子电导率具有较好的对应关系。然而，对于已知的绝大部分氧化物固体电解质而言，并不存在简单的对应关系。较为宽敞、连贯的锂离子通道，是高离子电导率的必要非充分条件。已知的快离子导体氧化物通常具有结构的无序性，更多的锂离子可占据位置，以及锂离子在其所在位置具有较"强烈"的热振动。然而，如何定量化"无序性"特征，并可以用这一特征去反向地利用这些特性筛选、设计固体电解质，仍是需要研究的问题；

⑤ 已知利用材料的尺寸效应，是除了掺杂调控和界面调控之外，另一种有效的锂离子输运调控方式。这种调控方式并不是"独立"的，事实上，在更微观的尺度，掺杂和界面调控实现了统一。然而，尺寸减小，通常意味着表面和界面的增加，此时大尺度下热力学稳定参数将不再适用。小尺寸的相-相变-物理特性，需要重新来研究。例如，在利用小尺寸动力学优势的同时，来减小热力学不稳定性带来的影响。

本章需要进一步思考的基础科学问题：

1. 关于离子电导率，难道只有 Arrhenius 和 VTF 方程两种温度依赖关系？
2. 同一种材料，固相离子电导率一定比液相低吗？
3. 界面应力、界面中间反应层、界面缺陷对离子输运有何影响？
4. 离子在界面富集的最大允许浓度是多少？
5. 空间电荷层存在对于离子垂直和平行界面的输运有什么影响？外在电场下有何影响？
6. 外场作用下，载流子有效质量的大小会影响其动力学，为什么银离子导体、铜离子导体的室温电导率很高？
7. 离子在正负极材料颗粒表面和内部的输运驱动力主要是浓度梯度还是电场梯度还是兼而有之？如何设计实验判断？
8. 离子在颗粒表面的传输速率如何测量？表面离子电导率如何定义？
9. 聚合物电解质能否参考无机化合物的缺陷化学的思路调控离子电导率？
10. 聚合物无机复合的电解质中，如何设计获得最高的表观离子电导率？

参考文献

[1] Funke K. Solid State Ionics：from michael faraday to green energy-the european dimension [J]. Sci Technol Adv Mat，2013，14 (4).

[2] Funke K，Cramer C，Wilmer D. Concept of mismatch and relaxation for self-diffusion and conduction in ionic materials with disordered structures [M] //Heitjans P，Kärger J. Diffusion in Condensed Matter. Berlin：Springer Berlin Heidelberg. 2005：857-893.

[3] Maier J. Physical Chemistry of ionic materials：ions and electrons in solids [M]. New York：Wiley，2004.

[4] Agrawal R，Gupta R. Superionic solid：Composite electrolyte phase-an overview [J]. Journal of Materials Science，1999，34 (6)：1131-1162.

[5] Gao J，Zhao Y S，Shi S Q，Li H. Lithium-ion transport in inorganic solid state electrolyte [J]. 2015，25 (1)：018211.

[6] 高健. 锂离子导体的基础研究 [D]. 北京：中国科学院，2015.

[7] 高健，何冰，施思齐. 锂离子电池无机固体电解质的计算[J]. 自然杂志，2016，38（5）：334-343.

[8] Shi S，Gao J，Liu Y，et al. Multi-scale computation methods：Their applications in lithium-ion battery research and development[J]. 2015，25（1）：018212.

[9] Barber M N，Ninham B W. Random and restricted walks：Theory and applications[M]. Washington D. C：Gordon and Breach，1970.

[10] Weiss G H. Aspects and applications of the random walk[M]. Amsterdam：North-Holland，1994.

[11] Van Kampen N G. Stochastic Processes in Physics and Chemistry[M]. Amsterdam：Elsevier Science，2011.

[12] Spitzer F. Principles of Random Walk[M]. Berlin：Springer，2001.

[13] Spohn H. Large Scale Dynamics of Interacting Particles[M]. Berlin：Springer Berlin Heidelberg，2011.

[14] Mehrer H. Diffusion in Solids：Fundamentals，Methods，Materials，Diffusion-Controlled Processes[M]. Berlin：Springer，2007.

[15] Kingery W. Introduction to Ceramics[M]. London：John Wiley & Sons，Incorporated，2013.

[16] Shin H-C，Pyun S-I. Investigation of lithium transport through lithium cobalt dioxide thin film sputter-deposited by analysis of cyclic voltammogram[J]. Electrochimica Acta，2001，46（16）：2477-2485.

[17] Tang S B，Lai M O，Lu L. Li-ion diffusion in highly（003）oriented $LiCoO_2$ thin film cathode prepared by pulsed laser deposition[J]. Journal of Alloys and Compounds，2008，449（1-2）：300-303.

[18] Striebel K A，Deng C Z，Wen S J，Cairns E J. Electrochemical Behavior of $LiMn_2O_4$ and $LiCoO_2$ Thin Films Produced with Pulsed Laser Deposition[J]. J Electrochem Soc，1996，143（6）：1821-1827.

[19] Julien C，Camacho-Lopez M A，Escobar-Alarcon L，et al. Fabrication of $LiCoO_2$ thin-film cathodes for rechargeable lithium microbatteries[J]. Materials Chemistry and Physics，2001，68（1-3）：210-216.

[20] Bouwman P J，Boukamp B A，Bouwmeester H J M，et al. Influence of diffusion plane orientation on electrochemical properties of thin film $LiCoO_2$ electrodes[J]. J Electrochem Soc，2002，149（6）：A699-A709.

[21] Jang Y-I，Neudecker B J，Dudney N J. Lithium diffusion in $Li_x CoO_2$（$0.45 < x < 0.7$）intercalation cathodes[J]. Electrochemical and Solid-State Letters，2001，4（6）：A74-A77.

[22] Xia H，Lu L，Ceder G. Li diffusion in $LiCoO_2$ thin films prepared by pulsed laser deposition[J]. Journal of Power Sources，2006，159（2）：1422-1427.

[23] Iriyama Y，Inaba M，Abe T，Ogumi Z. Preparation of c-axis oriented thin films of $LiCoO_2$ by pulsed laser deposition and their electrochemical properties[J]. Journal of Power Sources，2001，94（2）：175-182.

[24] Wang B，Bates J B，Hart F X，et al. Characterization of Thin-Film Rechargeable Lithium Batteries with Lithium Cobalt Oxide Cathodes[J]. J Electrochem Soc，1996，143（10）：3203-3213.

[25] Xie J，Imanishi N，Matsumura T，et al. Orientation dependence of Li-ion diffusion kinetics in $LiCoO_2$ thin films prepared by RF magnetron sputtering[J]. Solid State Ionics，2008，179（9-10）：362-370.

[26] Murch G E. The Haven Ratio in Fast Ionic Conductors[J]. Solid State Ionics，1982，7（3）：177-198.

[27] Heitjans P，Indris S. Diffusion and ionic conduction in nanocrystalline ceramics[J]. J Phys-Condens Mat，2003，15（30）：R1257-R1289.

[28] Murch G E. Diffusion Kinetics in Solids[M]. Phase Transformations in Materials. Berlin：Wiley-VCH Verlag GmbH & Co KgaA，2005：171-238.

[29] Sato H，Kikuchi R. Cation Diffusion and Conductivity in Solid Electrolytes.1.[J]. J Chem Phys，1971，55（2）：677.

[30] Sato H. Theoretical Background for the Mixed Alkali Effect[J]. Solid State Ionics，1990，40（1）：725-733.

[31] Sato H，Ishikawa T，Funke K. Frequency-Dependence of Ionic-Conductivity in Interacting Lattice Gas Systems[J]. Solid State Ionics，1992，53：907-923.

[32] Sato H，Datta A. Frequency-Dependence of Ionic-Conductivity in Lattice-Gas Models[J]. Solid State Ionics，1994，72：19-28.

[33] Sato H，Datta A，Ishikawa T. Kinetics of relaxation process of hopping ionic conduction in lattice gas models[J]. Solid State Ionics，1996，86（8）：1319-1323.

[34] Funke K. Jump Relaxation in Solid Electrolytes[J]. Progress in Solid State Chemistry，1993，22（2）：111-195.

[35] Ngai K L，Jain H. Conductivity Relaxation and Spin-Lattice Relaxation in Lithium and Mixed Alkali Borate Glasses-Activation Enthalpies，Anomalous Isotope-Mass Effect and Mixed Alkali Effect[J]. Solid State Ionics，1986，18（9）：362-367.

[36] Wang J C，Gaffari M，Choi S. Ionic-Conduction in Beta-Alumina-Potential-Energy Curves and Conduction Mecha-

nism [J]. J Chem Phys, 1975, 63 (2): 772-778.

[37] Van Der Ven A, Ceder G. Lithium diffusion in layered Li_xCoO_2 [J]. Electrochem Solid St, 2000, 3 (7): 301-304.

[38] Wolf M L. Observation of Solitary-Wave Conduction in a Molecular-Dynamics Simulation of the Superionic Conductor Li_3N [J]. J Phys C Solid State, 1984, 17 (10): L285-L288.

[39] Ihara S, Suzuki K. Molecular-Dynamics Study of Li_3N [J]. Phys Lett A, 1985, 110 (5): 265-268.

[40] Shi S Q, Lu P, Liu Z Y, et al. Direct Calculation of Li-Ion Transport in the Solid Electrolyte Interphase [J]. Journal of the American Chemical Society, 2012, 134 (37): 15476-15487.

[41] Shi S Q, Qi Y, Li H, et al. Defect Thermodynamics and Diffusion Mechanisms in Li_2CO_3 and Implications for the Solid Electrolyte Interphase in Li-Ion Batteries [J]. J Phys Chem C, 2013, 117 (17): 8579-8593.

[42] Park M, Zhang X, Chung M, et al. A review of conduction phenomena in Li-ion batteries [J]. Journal of Power Sources, 2010, 195 (24): 7904-7929.

[43] Mehrer H. Diffusion in Solids: Fundamentals, Methods, Materials, Diffusion-Controlled Processes [M]. Berlin: Springer, 2010.

[44] Mclean D. Grain boundaries in metals [J]. London: Oxford University Press, 1957.

[45] Barsoukov E, Macdonald J R. Impedance spectroscopy: theory, experiment, and applications [M]. New York: Wileycom, 2005.

[46] Dudney N J. Effect of Interfacial Space-Charge Polarization on the Ionic-Conductivity of Composite Electrolytes [J]. J Am Ceram Soc, 1985, 68 (10): 538-545.

[47] Uvarov N F, Isupov V P, Sharma V, Shukla A K. Effect of Morphology and Particle-Size on the Ionic Conductivities of Composite Solid Electrolytes [J]. Solid State Ionics, 1992, 51 (1-2): 41-52.

[48] Liang C C. Conduction Characteristics of Lithium Iodide Aluminum Oxide Solid Electrolytes [J]. J Electrochem Soc, 1973, 120 (10): 1289-1292.

[49] Bunde A, Dieterich W, Roman E. Dispersed Ionic Conductors and Percolation Theory [J]. Phys Rev Lett, 1985, 55 (1): 5-8.

[50] Jow T. The Effect of Dispersed Alumina Particles on the Electrical Conductivity of Cuprous Chloride [J]. Journal of The Electrochemical Society, 1979, 126 (11): 1963.

[51] Wagner C. The theory of the warm-up process [J]. Z Phys Chem B, 1933, 21 (25).

[52] Wagner C. Equations for transport in solid oxides and sulfides of transition metals [J]. Prog Solid State Ch, 1975, 10 (3): 16.

[53] Adam G, Gibbs J H. On the temperature dependence of cooperative relaxation properties in glass-forming liquids [J]. The Journal of Chemical Physics, 1965, 43 (1): 139-146.

[54] Cohen M H, Turnbull D. Molecular transport in liquids and glasses [J]. The Journal of Chemical Physics, 1959, 31 (5): 1164-1169.

[55] Turnbull D, Cohen M H. On the Free-Volume Model of the Liquid-Glass Transition [J]. The Journal of Chemical Physics, 1970, 52 (6): 3038-3041.

[56] Macedo P, Litovitz T. On the relative roles of free volume and activation energy in the viscosity of liquids [J]. The Journal of Chemical Physics, 1965, 42 (1): 245-256.

[57] Miyamoto T, Shibayama K. Free-volume model for ionic conductivity in polymers [J]. J Appl Phys, 1973, 44 (12): 5372-5376.

[58] Watanabe M, Sanui K, Ogata N, et al. Ionic conductivity and mobility in network polymers from poly (propylene oxide) containing lithium perchlorate [J]. J Appl Phys, 1985, 57 (1): 123-128.

[59] Souquet J L, Levy M, Duclot M. A single microscopic approach for ionic transport in glassy and polymer electrolytes [J]. Solid State Ionics, 1994, 70: 337-345.

[60] Souquet J L, Duclot M, Levy M. Ionic transport mechanisms in oxide based glasses in the supercooled and glassy states [J]. Solid State Ionics, 1998, 105 (1): 237-242.

[61] Angell C A. Phenomenology of Fast Ion Conducting Glasses: Facts and Confusions [M]. High Conductivity Solid Ionic Conductors, Singapore: World Scientific Publishing, 1989: 89-113.

[62] Angell C A. Fast ion motion in glassy and amorphous materials [J]. Solid State Ionics, 1983, 9: 3-16.

[63] Angell C A. Recent developments in fast ion transport in glassy and amorphous materials [J]. Solid State Ionics, 1986, 18-19, Part 1 (0): 72-88.

［64］ Zhang S，Chang Z，Xu K，Angell C A. Molecular and anionic polymer and oligomer systems with microdecoupled conductivities［J］. Electrochimica acta，2000，45（8）：1229-1236.

［65］ Angell C A. Mobile ions in amorphous solids［J］. Annual Review of Physical Chemistry，1992，43（1）：693-717.

［66］ Varshneya A K. Fundamentals of inorganic glasses［M］. Amsterdam：Elsevier，1993.

［67］ Rawson H. Inorganic glass-forming systems［M］. London：Academic press，1967.

［68］ Bunde A，Maass P，Roman H，et al. Transport of disordered structures：Effect of long range interactions［J］. Solid State Ionics，1990，40：187-191.

［69］ Habasaki J，Hiwatari Y. Molecular dynamics study of single and mixed alkali metasilicates-spatial and temporal characterization of the dynamics in the supercooled liquid and glassy states［J］. Journal of Non-Crystalline Solids，2002，307-310（0）：930-938.

［70］ Vogel M. Complex lithium ion dynamics in simulated LiPO_3 glass studied by means of multitime correlation functions［J］. Phys Rev B，2003，68（18）：184301.

［71］ Adams S，Swenson J. Structure conductivity correlation in reverse Monte Carlo models of single and mixed alkali glasses［J］. Solid State Ionics，2004，175（1-4）：665-669.

［72］ Ngai K L，Kanert O. Comparisons between the Coupling Model Predictions，Monte-Carlo Simulations and Some Recent Experimental-Data of Conductivity Relaxations in Glassy Ionics［J］. Solid State Ionics，1992，53：936-946.

［73］ Funke K. Ion transport in fast ion conductors-Spectra and models［J］. Solid State Ionics，1997，94（1-4）：27-33.

［74］ Elliott S R，Owens A P. Nuclear-Spin Relaxation in Ionically Conducting Glasses-Application of the Diffusion-Controlled Relaxation Model［J］. Phys Rev B，1991，44（1）：47-59.

［75］ Richert R. Evidence for Dynamic Heterogeneity near Tg from the Time-Resolved Inhomogeneous Broadening of Optical Line Shapes［J］. The Journal of Physical Chemistry B，1997，101（33）：6323-6326.

［76］ Angell C A. Relaxation in liquids, polymers and plastic crystals — strong/fragile patterns and problems［J］. Journal of Non-Crystalline Solids，1991，131-133，Part 1（0）：13-31.

［77］ Roland B，Gerald H，Thomas J，et al. Dynamical heterogeneity in α-and β-relaxations of glass forming liquids as seen by deuteron NMR［J］. Journal of Physics：Condensed Matter，2000，12（8A）：A383.

［78］ Kirkpatrick S. Percolation and Conduction［J］. Reviews of Modern Physics，1973，45（4）：574-88.

［79］ Bunde A，Dieterich W，Maass P，Meyer M. Ionic Transport in Disordered Materials［M］//Heitjans P，Kärger J. Diffusion in Condensed Matter. Berlin：Springer Berlin Heidelberg，2005：813-856.

［80］ Bunde A，Kantelhardt J. Diffusion and Conduction in Percolation Systems［M］//Heitjans P，Kärger J. Diffusion in Condensed Matter. Berlin：Springer Berlin Heidelberg. 2005：895-914.

［81］ Kamaya N，Homma K，Yamakawa Y，et al. A lithium superionic conductor［J］. Nature Materials，2011，10（9）：682-686.

［82］ Gao J，Shi S，Xiao R，Li H. Synthesis and ionic transport mechanisms of α-LiAlO$_2$［J］. Solid State Ionics，2016，286：122-134.

［83］ Ammundsen B，Rozière J，Islam M S. Atomistic simulation studies of lithium and proton insertion in spinel lithium manganates［J］. The Journal of Physical Chemistry B，1997，101（41）：8156-8163.

［84］ Morgan D，Van Der Ven A，Ceder G. Li conductivity in Li$_x$MPO$_4$（M＝Mn，Fe，Co，Ni）olivine materials［J］. Electrochemical and Solid-State Letters，2004，7（2）：A30-A32.

［85］ Ouyang C，Shi S，Wang Z，et al. First-principles study of Li ion diffusion in LiFePO$_4$［J］. Physical Review B，2004，69（10）：104303.

［86］ Van Der Ven A，Ceder G. Lithium diffusion mechanisms in layered intercalation compounds［J］. Journal of power sources，2001，97：529-531.

［87］ Koyama Y，Tanaka I，Adachi H，et al. First Principles Calculations of Formation Energies and Electronic Structures of Defects in Oxygen-Deficient LiMn$_2$O$_4$［J］. J Electrochem Soc，2003，150（1）：A63-A67.

［88］ Suzuki K，Oumi Y，Takami S，et al. Structural Properties of Li$_x$Mn$_2$O$_4$ as Investigated by Molecular Dynamics and Density Functional Theory［J］. Japanese Journal of Applied Physics，2000，39：4318.

［89］ Wolverton C，Zunger A. Cation and vacancy ordering in Li$_x$CoO$_2$［J］. Physical Review B，1998，57（4）：2242.

［90］ Goodenough J B. Design considerations［J］. Solid State Ionics，1994，69（3）：184-198.

［91］ Ouyang X，Lei M，Shi S，et al. First-principles studies on surface electronic structure and stability of LiFePO$_4$［J］. Journal of Alloys and Compounds，2009，476（1）：462-465.

[92] Maxisch T, Zhou F, Ceder G. Ab initio study of the migration of small polarons in olivine $Li_x FePO_4$ and their association with lithium ions and vacancies [J]. Physical review B, 2006, 73 (10): 104301.

[93] Sun Y, Lu X, Xiao R, et al. Kinetically Controlled Lithium-Staging in Delithiated $LiFePO_4$ Driven by the Fe Center Mediated Interlayer Li-Li Interactions [J]. Chemistry of Materials, 2012, 24 (24): 4693-4703.

[94] Dokko K, Mohamedi M, Fujita Y, et al. Kinetic characterization of single particles of $LiCoO_2$ by AC impedance and potential step methods [J]. J Electrochem Soc, 2001, 148 (5): A422-A426.

[95] Barker J, Pynenburg R, Koksbang R, et al. An electrochemical investigation into the lithium insertion properties of $Li_x CoO_2$ [J]. Electrochimica acta, 1996, 41 (15): 2481-2488.

[96] Levasseur S, Ménétrier M, Delmas C. On the dual effect of mg doping in $LiCoO_2$ and $Li_{1+\delta}CoO_2$: Structural, electronic properties, and ^7Li MAS NMR Studies [J]. Chemistry of materials, 2002, 14 (8): 3584-3590.

[97] Marzec J, Świerczek K, Przewoznik J, et al. Conduction mechanism in operating a $LiMn_2O_4$ cathode [J]. Solid State Ionics, 2002, 146 (3): 225-237.

[98] Molenda J, Kucza W. Transport properties of $LiMn_2O_4$ [J]. Solid State Ionics, 1999, 117 (1): 41-46.

[99] Saidi M, Barker J, Koksbang R. Thermodynamic and kinetic investigation of lithium insertion in the $Li_{1-x}Mn_2O_4$ spinel phase [J]. Journal of Solid State Chemistry, 1996, 122 (1): 195-199.

[100] Cao F, Prakash J. A comparative electrochemical study of $LiMn_2O_4$ spinel thin-film and porous laminate [J]. Electrochimica acta, 2002, 47 (10): 1607-1613.

[101] Shi S, Liu L, Ouyang C, et al. Enhancement of electronic conductivity of $LiFePO_4$ by Cr doping and its identification by first-principles calculations [J]. Physical Review B, 2003, 68 (19): 195108.

[102] Xu Y-N, Chung S-Y, Bloking J T, et al. Electronic structure and electrical conductivity of undoped $LiFePO_4$ [J]. Electrochemical and Solid-State Letters, 2004, 7 (6): A131-A134.

[103] Prosini P P, Lisi M, Zane D, et al. Determination of the chemical diffusion coefficient of lithium in $LiFePO_4$ [J]. Solid State Ionics, 2002, 148 (1): 45-51.

[104] Ma J, Wang C, Wroblewski S. Kinetic characteristics of mixed conductive electrodes for lithium ion batteries [J]. Journal of power sources, 2007, 164 (2): 849-856.

[105] House R A, Bruce P G. Lightning fast conduction [J]. Nature Energy, 2020, 5 (3): 191-192

[106] Zhang S, Xu K, Jow T. Low temperature performance of graphite electrode in Li-ion cells [J]. Electrochimica acta, 2002, 48 (3): 241-246.

[107] Funabiki A, Inaba M, Ogumi Z, et al. Impedance study on the electrochemical lithium intercalation into natural graphite powder [J]. J Electrochem Soc, 1998, 145 (1): 172-178.

[108] Tang X-C, Pan C-Y, He L-P, et al. A novel technique based on the ratio of potentio-charge capacity to galvano-charge capacity (RPG) for determination of the diffusion coefficient of intercalary species within insertion-host materials: theories and experiments [J]. Electrochimica acta, 2004, 49 (19): 3113-3119.

[109] Yang H, Bang H J, Prakash J. Evaluation of electrochemical interface area and lithium diffusion coefficient for a composite graphite anode [J]. J Electrochem Soc, 2004, 151 (8): A1247-A1250.

[110] Wang Q, Li H, Huang X, et al. Determination of chemical diffusion coefficient of lithium ion in graphitized mesocarbon microbeads with potential relaxation technique [J]. J Electrochem Soc, 2001, 148 (7): A737-A741.

[111] Levi M, Markevich E, Aurbach D. The effect of slow interfacial kinetics on the chronoamperometric response of composite lithiated graphite electrodes and on the calculation of the chemical diffusion coefficient of Li ions in graphite [J]. The Journal of Physical Chemistry B, 2005, 109 (15): 7420-7427.

[112] Nuli Y, Yang J, Jiang Z. Intercalation of lithium ions into bulk and powder highly oriented pyrolytic graphite [J]. Journal of Physics and Chemistry of Solids, 2006, 67 (4): 882-886.

[113] Wu Y, Rahm E, Holze R. Carbon anode materials for lithium ion batteries [J]. Journal of power sources, 2003, 114 (2): 228-236.

[114] Noel M, Suryanarayanan V. Role of carbon host lattices in Li-ion intercalation/de-intercalation processes [J]. Journal of Power Sources, 2002, 111 (2): 193-209.

[115] Flandrois S, Simon B. Carbon materials for lithium-ion rechargeable batteries [J]. Carbon, 1999, 37 (2): 165-180.

[116] Dahn J. Phase diagram of $Li_x C_6$ [J]. Physical Review B, 1991, 44 (17): 9170.

[117] Levi M, Aurbach D. Diffusion coefficients of lithium ions during intercalation into graphite derived from the simultaneous measurements and modeling of electrochemical impedance and potentiostatic intermittent titration char-

acteristics of thin graphite electrodes [J]. The Journal of Physical Chemistry B, 1997, 101 (23): 4641-4647.

[118] Zhang W, Seo D-H, Chen T, et al. Kinetic pathways of ionic transport in fast-charging lithium titanate [J]. Science, 2020, 367 (6481): 1030-1034.

[119] Akridge J R, Balkanski M. Solid state microbatteries [M]. New York: Plenum Publishing Corporation, 1988.

[120] Wakihara M. Recent developments in lithium ion batteries [J]. Materials Science and Engineering: R: Reports, 2001, 33 (4): 109-134.

[121] Zaghib K, Charest P, Guerfi A, et al. Safe Li-ion polymer batteries for HEV applications [J]. Journal of power sources, 2004, 134 (1): 124-129.

[122] Knauth P. Ionic conductor composites: theory and materials [J]. Journal of electroceramics, 2000, 5 (2): 111-125.

[123] Hull S. Superionics: crystal structures and conduction processes [J]. Reports on Progress in Physics, 2004, 67 (7): 1233-1314.

[124] Armand M, Tarascon J M. Building better batteries [J]. Nature, 2008, 451 (7179): 652-657.

[125] Quartarone E, Mustarelli P. Electrolytes for solid-state lithium rechargeable batteries: recent advances and perspectives [J]. Chemical Society Reviews, 2011, 40 (5): 2525-2540.

[126] Zhang B, Tan R, Yang L, et al. Mechanisms and properties of ion-transport in inorganic solid electrolytes [J]. Energy Storage Materials, 2018, 10: 139-159.

[127] Xiao R, Li H, Chen L. High-throughput design and optimization of fast lithium ion conductors by the combination of bond-valence method and density functional theory [J]. Scientific Reports, 2015, 5 (1): 14227.

[128] Xiao R, Li H, Chen L. Candidate structures for inorganic lithium solid-state electrolytes identified by high-throughput bond-valence calculations [J]. Journal of Materiomics, 2015, 1 (4): 325-332.

[129] Gao J, Chu G, He M, et al. Screening possible solid electrolytes by calculating the conduction pathways using Bond Valence method [J]. Science China Physics, Mechanics & Astronomy, 2014, 57 (8): 1526-1536.

[130] Wang Y, Richards W D, Ong S P, et al. Design principles for solid-state lithium superionic conductors [J]. Nat Mater, 2015, 14 (10): 1026-1031.

[131] Nolan A M, Zhu Y, He X, et al. Computation-Accelerated Design of Materials and Interfaces for All-Solid-State Lithium-Ion Batteries [J]. Joule, 2018, 2 (10): 2016-2046.

[132] He X M, Pu W H, Wang L, et al. Plastic crystals: An effective ambient temperature all-solid-state electrolyte for lithium batteries [J]. Prog Chem, 2006, 18 (1): 24-29.

[133] He X, Zhu Y, Mo Y. Origin of fast ion diffusion in super-ionic conductors [J]. Nature Communications, 2017, 8 (1): 15893.

[134] Hayashi A, Noi K, Sakuda A, Tatsumisago M. Superionic glass-ceramic electrolytes for room-temperature rechargeable sodium batteries [J]. Nat Commun, 2012, 3: 856.

[135] Sakuda A, Hayashi A, Tatsumisago M. Sulfide Solid Electrolyte with Favorable Mechanical Property for All-Solid-State Lithium Battery [J]. Sci Rep-Uk, 2013, 3.

[136] Rangasamy E, Sahu G, Keum J K, et al. A high conductivity oxide-sulfide composite lithium superionic conductor [J]. Journal of Materials Chemistry A, 2014, 2 (12): 4111-4116.

[137] Trevey J E, Stoldt C R, Lee S-H. High Power Nanocomposite TiS2 Cathodes for All-Solid-State Lithium Batteries [J]. J Electrochem Soc, 2011, 158 (12): A1282-A1289.

[138] Doux J-M, Yang Y, Tan D H S, et al. Pressure effects on sulfide electrolytes for all solid-state batteries [J]. J Mater Chem A, 2020, 8 (10): 5049-5055.

[139] Zhang S S. A review on electrolyte additives for lithium-ion batteries [J]. Journal of Power Sources, 2006, 162 (2): 1379-1394.

[140] Abe K, Miyoshi K, Hattori T, et al. Functional electrolytes: Synergetic effect of electrolyte additives for lithium-ion battery [J]. Journal of Power Sources, 2008, 184 (2): 449-455.

[141] Ota H, Sato T, Suzuki H, et al. TPD-GC/MS analysis of the solid electrolyte interface (SEI) on a graphite anode in the propylene carbonate/ethylene sulfite electrolyte system for lithium batteries [J]. Journal of power sources, 2001, 97: 107-113.

[142] Aurbach D, Ein-Eli Y, Markovsky B, et al. The Study of Electrolyte Solutions Based on Ethylene and Diethyl Carbonates for Rechargeable Li Batteries Ⅱ. Graphite Electrodes [J]. J Electrochem Soc, 1995, 142 (9): 2882-2890.

[143] Peled E, Golodnitsky D, Ardel G. Advanced model for solid electrolyte interphase electrodes in liquid and polymer electrolytes [J]. J Electrochem Soc, 1997, 144 (8): L208-L210.

[144] Leung K, Qi Y, Zavadil K R, et al. Using Atomic Layer Deposition to Hinder Solvent Decomposition in Lithium Ion Batteries: First-Principles Modeling and Experimental Studies [J]. Journal of the American Chemical Society, 2011, 133 (37): 14741-14754.

[145] Aurbach D. Review of selected electrode-solution interactions which determine the performance of Li and Li ion batteries [J]. Journal of Power Sources, 2000, 89 (2): 206-218.

[146] Dissanayake M A K L, Mellander B E. Phase-Diagram and Electrical-Conductivity of the Li_2SO_4-Li_2CO_3 System [J]. Solid State Ionics, 1986, 21 (4): 279-285.

[147] Shannon R D, Taylor B E, English A D, et al. New Li Solid Electrolytes [J]. J Electrochem Soc, 1977, 124 (8): C271.

[148] Qi Y. Direct calculation of Li-ion transport in the solid electrolyte interphase (SEI) [J]. Abstr Pap Am Chem S, 2013, 246.

[149] Iddir H, Curtiss L A. Li Ion Diffusion Mechanisms in Bulk Monoclinic Li_2CO_3 Crystals from Density Functional Studies [J]. J Phys Chem C, 2010, 114 (48): 20903-20906.

[150] Zaghib K, Simoneau M, Armand M, et al. Electrochemical study of $Li_4Ti_5O_{12}$ as negative electrode for Li-ion polymer rechargeable batteries [J]. Journal of Power Sources, 1999, 81: 300-305.

[151] Kasavajjula U, Wang C, Appleby A J. Nano-and bulk-silicon-based insertion anodes for lithium-ion secondary cells [J]. Journal of Power Sources, 2007, 163 (2): 1003-1039.

[152] Li H, Huang X, Chen L, et al. The crystal structural evolution of nano-Si anode caused by lithium insertion and extraction at room temperature [J]. Solid State Ionics, 2000, 135 (1): 181-191.

[153] Sun J, Tang K, Yu X, et al. Overpotential and electrochemical impedance analysis on Cr_2O_3 thin film and powder electrode in rechargeable lithium batteries [J]. Solid State Ionics, 2008, 179 (40): 2390-2395.

[154] Delmas C, Nadiri A, Soubeyroux J L. The nasicon-type titanium phosphates $ATi_2(PO_4)_3$ (A = Li, Na) as electrode materials [J]. Solid State Ionics, 1988, 28-30, Part 1 (0): 419-423.

[155] Shan Y J, Inaguma Y, Itoh M. The Effect of Electrostatic Potentials on Lithium Insertion for Perovskite Oxides [J]. Solid State Ionics, 1995, 79: 245-251.

[156] Takada K, Inada T, Kajiyama A, et al. Solid-state lithium battery with graphite anode [J]. Solid State Ionics, 2003, 158 (3-4): 269-274.

[157] Shin B R, Nam Y J, Oh D Y, et al. Comparative Study of TiS2/Li-In All-Solid-State Lithium Batteries Using Glass-Ceramic Li_3PS_4 and $Li_{10}GeP_2S_{12}$ Solid Electrolytes [J]. Electrochimica Acta, 2014, 146: 395-402.

[158] Kennedy J H, Zhang Z M. Improved Stability for the SiS_2-P_2S_5-Li_2S-LiI Glass System [J]. Solid State Ionics, 1988, 28: 726-728.

[159] Takada K, Nakano S, Inada T, et al. Compatibility of Lithium Ion Conductive Sulfide Glass with Carbon-Lithium Electrode [J]. J Electrochem Soc, 2003, 150 (3): A274-A277.

[160] Gao J, Guo X, Li Y, et al. The Ab Initio Calculations on the Areal Specific Resistance of Li-Metal/$Li_7La_3Zr_2O_{12}$ Interphase [J]. Advanced theory and simulations, 2019, 2 (6): 1900028.

[161] Yamamoto K, Yoshida R, Sato T, et al. Nano-scale simultaneous observation of Li-concentration profile and Ti-, O electronic structure changes in an all-solid-state Li-ion battery by spatially-resolved electron energy-loss spectroscopy [J]. Journal of Power Sources, 2014, 266: 414-421.

[162] Zhu Y, He X, Mo Y. Origin of Outstanding Stability in the Lithium Solid Electrolyte Materials: Insights from Thermodynamic Analyses Based on First-Principles Calculations [J]. ACS Applied Materials & Interfaces, 2015, 7 (42): 23685-23693.

[163] Zhu Y, He X, Mo Y. First principles study on electrochemical and chemical stability of solid electrolyte-electrode interfaces in all-solid-state Li-ion batteries [J]. J Mater Chem A, 2016, 4 (9): 3253-3266.

[164] Ohta N, Takada K, Zhang L Q, et al. Enhancement of the high-rate capability of solid-state lithium batteries by nanoscale interfacial modification [J]. Advanced Materials, 2006, 18 (17): 2226.

[165] Takada K, Ohta N, Zhang L Q, et al. Interfacial modification for high-power solid-state lithium batteries [J]. Solid State Ionics, 2008, 179 (27-32): 1333-1337.

[166] Takada K, Ohta N, Zhang L Q, et al. Interfacial phenomena in solid-state lithium battery with sulfide solid electrolyte [J]. Solid State Ionics, 2012, 225: 594-597.

[167] Haruyama J, Sodeyama K, Han L Y, et al. Space-Charge Layer Effect at Interface between Oxide Cathode and Sulfide Electrolyte in All-Solid-State Lithium-Ion Battery [J]. Chem Mater, 2014, 26 (14): 4248-4255.

[168] Maier J. Ionic-Conduction in-Space Charge Regions [J]. Progress in Solid State Chemistry, 1995, 23 (3): 171-263.

[169] Yamamoto K, Iriyama Y, Asaka T, et al. Dynamic Visualization of the Electric Potential in an All-Solid-State Rechargeable Lithium Battery [J]. Angew Chem Int Edit, 2010, 49 (26): 4414-4417.

[170] Yamamoto K, Iriyama Y, Asaka T, et al. Direct observation of lithium-ion movement around an in-situ-formed-negative-electrode/solid-state-electrolyte interface during initial charge-discharge reaction [J]. Electrochemistry Communications, 2012, 20: 113-116.

[171] Yamamoto K, Hirayama T, Tanji T. Development of advanced electron holographic techniques and application to industrial materials and devices [J]. Microscopy, 2013, 62: S29-S41.

[172] Lee D-K, Yoo H-I. Electron-Ion Interference and Onsager Reciprocity in Mixed Ionic-Electronic Transport in TiO_2 [J]. Phys Rev Lett, 2006, 97 (25): 255901.

[173] Wagner J B. Composite Solid Ion Conductors [M] //High Conductivity Solid Ionic Conductors. Singapore: World Scientific, 1989: 146-165.

[174] Maier J. Space-Charge Regions in Solid 2-Phase Systems and Their Conduction Contribution. 3. Defect Chemistry and Ionic-Conductivity in Thin-Films [J]. Solid State Ionics, 1987, 23 (1-2): 59-67.

[175] Maier J. Composite Electrolytes [J]. Materials Chemistry and Physics, 1987, 17 (5): 485-498.

[176] Maier J. Defect Chemistry and Conductivity Effects in Heterogeneous Solid Electrolytes [J]. J Electrochem Soc, 1987, 134 (6): 1524-1535.

[177] Maier J. Defect Chemistry in Heterogeneous Systems [J]. Solid State Ionics, 1995, 75: 139-145.

[178] Maier J. Pushing Nanoionics to the Limits: Charge Carrier Chemistry in Extremely Small Systems [J]. Chem Mater, 2014, 26 (1): 348-360.

[179] Maier J. Control parameters for electrochemically relevant materials: the significance of size and complexity [J]. Faraday Discussions, 2014, 176 (0): 17-29.

[180] Maier J. Nanoionics: ion transport and electrochemical storage in confined systems [J]. Nat Mater, 2005, 4 (11): 805-815.

[181] Li H, Shi L, Lu W, et al. Studies on capacity loss and capacity fading of nanosized SnSb alloy anode for Li-ion batteries [J]. J Electrochem Soc, 2001, 148 (8): A915-A922.

[182] Li H, Shi L, Wang Q, et al. Nano-alloy anode for lithium ion batteries [J]. Solid State Ionics, 2002, 148 (3): 247-258.

[183] Maier J. Nanoionics: ion transport and electrochemical storage in confined systems [J]. Nature materials, 2005, 4 (11): 805-815.

[184] Aricò A S, Bruce P, Scrosati B, et al. Nanostructured materials for advanced energy conversion and storage devices [J]. Nature materials, 2005, 4 (5): 366-377.

[185] Liu N, Li H, Wang Z, et al. Origin of solid electrolyte interphase on nanosized $LiCoO_2$ [J]. Electrochemical and solid-state letters, 2006, 9 (7): A328-A331.

[186] Gibot P, Casas-Cabanas M, Laffont L, et al. Room-temperature single-phase Li insertion/extraction in nanoscale Li(x) $FePO_4$ [J]. Nature materials, 2008, 7 (9): 741-747.

[187] Hagenmuller P, Van Gool W. Solid electrolytes: general principles, characterization, materials, applications [M]. New York: Academic Press, 1978: 549.

[188] Tang W S, Matsuo M, Wu H, et al. Liquid-Like Ionic Conduction in Solid Lithium and Sodium Monocarba-closo-Decaborates Near or at Room Temperature [J]. Advanced Energy Materials, 2016, 6 (8).

[189] Bron P, Johansson S, Zick K, et al. $Li_{10}SnP_2S_{12}$: An Affordable Lithium Superionic Conductor [J]. Journal of the American Chemical Society, 2013, 135 (42): 15694-15697.

[190] Kuhn A, Koehler J, Lotsch B V. Single-crystal X-ray structure analysis of the superionic conductor $Li_{10}GeP_2S_{12}$ [J]. Physical Chemistry Chemical Physics, 2013, 15 (28): 11620-11622.

[191] Yamada A, Yashima M. Experimental visualization of lithium diffusion in Li_xFePO_4 [J]. J Crystallogr Soc Jpn, 2009, 51 (2): 175-181.

[192] Kwon O, Hirayama M, Suzuki K, Kato Y, Saito T, Yonemura M, Kamiyama T, Kanno R. Synthesis, structure, and conduction mechanism of the lithium superionic conductor Li10＋delta Ge1＋delta P2-delta S12

[J]. Journal of Materials Chemistry A, 2015, 3 (1): 438-446.

[193] Cussen E J. The structure of lithium garnets: cation disorder and clustering in a new family of fast Li^+ conductors [J]. Chem Commun, 2006, (4): 412-413.

[194] Mazza D. Remarks on a Ternary Phase in the La_2O_3-Nb_2O_5-Li_2O, La_2O_3-Ta_2O_5-Li_2O System [J]. Mater Lett, 1988, 7 (5-6): 205-207.

[195] Han J T, Zhu J L, Li Y T, et al. Experimental visualization of lithium conduction pathways in garnet-type $Li_7La_3Zr_2O_{12}$ [J]. Chem Commun, 2012, 48 (79): 9840-9842.

[196] Kärger J, Stallmach F. PFG NMR Studies of Anomalous Diffusion [M] //Heitjans P, Kärger J. Diffusion in Condensed Matter. Berlin: Springer Berlin Heidelberg, 2005: 417-459.

[197] Mehrer H. Diffusion: Introduction and Case Studies in Metals and Binary Alloys [M] //Heitjans P, Kärger J. Diffusion in Condensed Matter. Berlin: Springer Berlin Heidelberg, 2005: 3-63.

[198] Heitjans P. Use of Beta-Radiation-Detected Nmr to Study Ionic Motion in Solids [J]. Solid State Ionics, 1986, 18-9: 50-64.

[199] Fleig J, Rahmati B, Rodewald S, et al. On the localized impedance spectroscopic characterization of grain boundaries: General aspects and experiments on undoped $SrTiO_3$ [J]. J Eur Ceram Soc, 2010, 30 (2): 215-220.

[200] Fleig J, Maier J. Microcontact impedance measurements of individual highly conductive grain boundaries: General aspects and application to AgCl [J]. Physical Chemistry Chemical Physics, 1999, 1 (14): 3315-3320.

[201] Lee W, Prinz F B, Chen X, et al. Nanoscale impedance and complex properties in energy-related systems [J]. Mrs Bull, 2012, 37 (7): 659-667.

[202] Louie M W, Hightower A, Haile S M. Nanoscale Electrodes by Conducting Atomic Force Microscopy: Oxygen Reduction Kinetics at the Pt vertical bar $CsHSO_4$ Interface [J]. Acs Nano, 2010, 4 (5): 2811-2821.

[203] Read P L, Katz E. Ionic Hall Effect in Sodium Chloride [J]. Phys Rev Lett, 1960, 5 (10): 466-468.

[204] Kaneda T, Mizuki E. Hall-Effect of Silver Ions in $RbAg_4I_5$ Single-Crystals [J]. Phys Rev Lett, 1972, 29 (4): 937.

[205] Liou Y J, Hudson R A, Wonnell S K, et al. Ionic Hall-Effect in Crystals-Independent Versus Cooperative Hopping in AgBr and Alpha-AgI [J]. Phys Rev B, 1990, 41 (15): 10481-10485.

[206] Stuhrmann C H J, Kreiterling H, Funke K. Ionic Hall effect measured in rubidium silver iodide [J]. Solid State Ionics, 2002, 154: 109-112.

[207] Schatz G, Weidinger A. Nuclear condensed matter physics: Nuclear methods and applications [M]. Berlin: Wiley, 1996.

[208] Heitjans P, Schirmer A, Indris S. NMR and β-NMR Studies of Diffusion in Interface-Dominated and Disordered Solids [M] //Heitjans P, Kärger J. Diffusion in Condensed Matter. Berlin: Springer Berlin Heidelberg, 2005: 367-415.

[209] Holz M. Field-Assisted Diffusion Studied by Electrophoretic NMR [M] //Heitjans P, Kärger J. Diffusion in Condensed Matter. Berlin: Springer Berlin Heidelberg, 2005: 717-742.

[210] Vogl G, Sepiol B. The Elementary Diffusion Step in Metals Studied by the Interference of Gamma-Rays, X-Rays and Neutrons [M] //Heitjans P, Kärger J. Diffusion in Condensed Matter. Berlin: Springer Berlin Heidelberg, 2005: 65-91.

[211] Springer T, Lechner R. Diffusion Studies of Solids by Quasielastic Neutron Scattering [M] //Heitjans P, Kärger J. Diffusion in Condensed Matter. Berlin: Springer Berlin Heidelberg, 2005: 93-164.

[212] Stockmann H J. A Stochastic Relaxation Model for Interstitial Diffusion [J]. J Phys-Condens Mat, 1989, 1 (31): 5101-5114.

[213] Borgs P, Kehr K W, Heitjans P. Longitudinal Spin Relaxation in Simple Stochastic-Models for Disordered-Systems [J]. Phys Rev B, 1995, 52 (9): 6668-6683.

[214] Santhanagopalan D, Qian D, Mcgilvray T, et al. Interface Limited Lithium Transport in Solid-State Batteries [J]. J Phys Chem Lett, 2014, 5 (2): 298-303.

[215] Mohammed O F, Yang D-S, Pal S K, et al. 4D Scanning Ultrafast Electron Microscopy: Visualization of Materials Surface Dynamics [J]. Journal of the American Chemical Society, 2011, 133 (20): 7708-7711.

[216] Hautier G, Fischer C C, Jain A, et al. Finding Nature's Missing Ternary Oxide Compounds Using Machine Learning and Density Functional Theory [J]. Chem Mater, 2010, 22 (12): 3762-3767.

[217] Ceder G. Opportunities and challenges for first-principles materials design and applications to Li battery materials [J]. Mrs Bull, 2010, 35 (9): 693-701.
[218] Jain A, Hautier G, Moore C J, et al. A high-throughput infrastructure for density functional theory calculations [J]. Comp Mater Sci, 2011, 50 (8): 2295-2310.
[219] Qian C, Siler T, Ozin G A. Exploring the Possibilities and Limitations of a Nanomaterials Genome [J]. Small, 2015, 11 (1): 64-69.

第 6 章

锂离子电池正极材料

 6.1　正极材料概述

提高锂离子电池正极材料的综合性能以满足其对能量存储日益提高的要求，一直是锂离子电池领域最重要的研究方向。目前的正极材料主要基于层状结构、尖晶石结构以及橄榄石结构，采用这些材料的锂离子电池可以基本满足消费电子、电动车辆、规模储能等要求。本节将对目前广泛使用的锂离子电池正极材料的性能特点和发展状况进行概述。

1980 年，Armand 等提出了摇椅式电池（rocking chair battery）的概念，在充放电过程中，Li^+ 在正负极层状化合物之间来回不停穿梭。鉴于含 Li 的负极材料在空气中一般不稳定、安全性较差，目前开发的锂离子电池均以正极材料作为锂源。

为了使锂离子电池具有较高的能量密度、功率密度，较好的循环性能及可靠的安全性能，对正极材料的选择应满足以下条件：①正极材料起到锂源的作用，它不仅要提供在可逆的充放电过程中往返于正负极之间的 Li^+，而且还要提供首次充放电过程中在负极表面形成 SEI 膜时所消耗的 Li^+；②提供较高的电极电位，这样电池输出电压才可能高；③整个电极过程中，电压平台稳定，以保证电极输出电位的平稳；④为使正极材料具有较高的能量密度，要求正极活性物质的电化学当量小，并且能够可逆脱嵌的 Li^+ 量要大；⑤Li^+ 在材料中的化学扩散系数高，电极界面稳定，具有高功率密度，使锂电池可适用于较高的充放电倍率，满足动力型电源的需求；⑥充放电过程中结构稳定，可逆性好，保证电池的循环性能良好；⑦具有比较高的电子和离子电导率；⑧化学稳定性好，无毒，资源丰富，制备成本低。

能全面满足上述要求的正极材料体系并不多，目前也没有明确的理论可以指导正极材料的选择，锂离子电池的正极材料研究主要是在固体化学与固体物理的基础上，由个别研究者提出材料体系，然后经过长期的研究开发使材料逐渐获得应用。几个标志性的研究有：1981 年，Goodenough 等提出层状 $LiCoO_2$ 材料可以用作锂离子电池的正极材料。1983 年，Thackeray 等发现 $LiMnO_4$ 尖晶石是优良的正极材料，具有低价、稳定和优良的导电、导锂性能，其分解温度高，且氧化性远低于 $LiCoO_2$，即使出现短路、过充电，

也能够避免燃烧、爆炸的危险。1991 年，Sony 公司率先解决了已有材料的集成技术，推出了最早的商业化锂离子电池，他们采用的体系是以无序非石墨化石油焦炭为负极，$LiCoO_2$ 为正极，$LiPF_6$ 溶于碳酸丙烯酯（PC）和乙烯碳酸酯（EC）为电解液，这种电池作为新一代的高效便携式储能设备进入市场后，在无线通信、笔记本电脑等方面得到了广泛应用。$LiFePO_4$ 的研发开始于 1997 年 Goodenough 等的开创性工作，由于 $LiFePO_4$ 结构稳定、安全性能好、高温性能好、循环寿命长，同时又具有无毒、无污染、原材料来源广泛、价格便宜等优点，目前已开始应用于电动汽车和大容量储能电池。

6.2 典型的锂离子电池正极材料

在目前的锂离子电池体系中，整个电池的比容量受限于正极材料的容量。在电池的生产中，正极材料的成本占材料总成本的 30% 以上。因此，制备成本低、同时具有高能量密度的正极材料是目前锂离子电池研究与生产的重要目标。

目前商业化使用的锂离子电池正极材料按结构主要分为以下三类：①六方层状晶体结构的 $LiCoO_2$；②立方尖晶石晶体结构的 $LiMn_2O_4$；③正交橄榄石晶体结构的 $LiFePO_4$。其晶体结构如图 6-1 所示，目前已经应用的锂离子电池正极材料的 Li^+ 扩散系数及理论容量等信息见表 6-1。

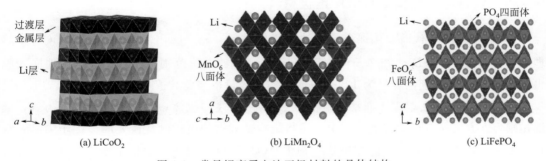

(a) $LiCoO_2$ (b) $LiMn_2O_4$ (c) $LiFePO_4$

图 6-1 常见锂离子电池正极材料的晶体结构

表 6-1 常见锂离子电池正极材料及其性能

项目	磷酸铁锂	锰酸锂	钴酸锂	三元镍钴锰
化学式	$LiFePO_4$	$LiMn_2O_4$	$LiCoO_2$	$Li(Ni_xCo_yMn_z)O_2$
晶体结构	橄榄石结构	尖晶石	层状	层状
空间点群	$Pmnb$	$Fd\text{-}3m$	$R\text{-}3m$	$R\text{-}3m$
晶胞参数/Å	$a=4.692, b=10.332, c=6.011$	$a=b=c=8.231$	$a=2.82, c=14.06$	—
锂离子表观扩散系数/(cm^2/s)	$1.8\times10^{-16} \sim 2.2\times10^{-14}$	$10^{-14} \sim 10^{-12}$	$10^{-11} \sim 10^{-12}$	$10^{-10} \sim 10^{-11}$
理论密度/(g/cm^3)	3.6	4.2	5.1	—

续表

项目	磷酸铁锂	锰酸锂	钴酸锂	三元镍钴锰
振实密度/(g/cm³)	0.80~1.10	2.2~2.4	2.8~3.0	2.6~2.8
压实密度/(g/cm³)	2.20~2.30	>3.0	3.6~4.2	>3.40
理论容量/(mA·h/g)	170	148	274	273~285
实际容量/(mA·h/g)	130~140	100~120	150~180	155~220
相应电池电芯的质量比能量/(W·h/kg)	130~160	130~180	180~240	180~240
平均电压/V	3.4	3.8	3.8	3.6
电压范围/V	3.2~3.7	3.0~4.3	3.0~4.5	2.5~4.6
循环性/次	2000~6000	500~2000	500~1000	800~2000
环保性	无毒	无毒	钴有毒性	镍、钴有毒
安全性能	好	良好	差	中
适用温度/℃	20~75	>50 快速衰退	20~55	20~55
价格/(万元/吨)	15~20	9~15	26~30	15.5~16.5
主要应用领域	电动汽车及大规模储能	电动工具、电动自行车、电动汽车及大规模储能	传统3C电子产品	电动工具、电动自行车、电动汽车及大规模储能

6.2.1 六方层状结构 $LiCoO_2$ 正极材料

$LiCoO_2$ 是第一代商业化锂离子电池的正极材料。完全脱出 1mol Li 需要 $LiCoO_2$ 的理论容量为 274mA·h/g，在 2.5~4.25V vs. Li^+/Li 的电位范围内一般能够可逆地嵌入脱出 0.5 个 Li，对应理论容量为 138mA·h/g，实际容量也与此数值相当。

$LiCoO_2$ 具备多种晶型，通常高温相呈层状结构，主要包含 O3 及 O2 两种排列方式，低温相则为立方尖晶石结构排列。其中，层状结构钴酸锂的电化学性能要明显优于尖晶石结构，因此受到了更多的关注。在层状结构中，O3-$LiCoO_2$ 为热力学稳定结构，O 沿（001）方向的排布式为 ABCABC…，在亚稳态的 O2-$LiCoO_2$ 中 O 沿（001）方向的排布式为 ABACABAC…。此外，钴酸锂还包含 O4 相及完全脱锂后形成的 O1-CO_2 相。在不同的层状结构中，电化学循环过程随着 Li 含量的不断变化（Li 离子和空位的相互作用），都会发生 Co 和 O 阵列的重排，导致新相的出现。目前的研究及工业应用主要集中在 O3 相 $LiCoO_2$。

在 O3 相 $LiCoO_2$ 中，随着 Li^+ 的脱出，材料会经历三个相变过程。以 $Li_{1-x}CoO_2$ 表示，第一个相变过程发生在锂脱出量 $x=0.07~0.25$ 过程中，由 H1→H2 转变，c 轴伸长约 2%，Co—Co 间距明显缩短，引起能带分散，造成价带与导带重叠，电导率提高，使材料由原来的半导体向金属导体转变。在脱锂量 $x=0.25~0.5$ 过程中，$Li_{1-x}CoO_2$ 的结构和金属性的电子电导保持不变。第二个相变发生在 $x=0.5$ 附近，这一相变过程伴随着 Li^+ 无序与有序的转变以及材料由六方相向单斜相转变。第三个相变则出现在 4.5V 附近，在脱锂量 $x=0.6~0.9$ 过程中，出现 O3 相向 O1 相的转变，其间还会形成过渡相

H1-3。早期认为 O3-LiCoO$_2$ 的电化学循环过程中体现固溶体的行为，2003 年，Yang 等利用 HRTEM 给出了 O3-LiCoO$_2$ 的晶格原子相，随后 STEM 给出了更清晰的 O3-LiCoO$_2$ 原子相。2012 年，Lu 等通过球差矫正 ABF-STEM 技术首次在脱锂态 O3 结构的 Li$_{1-x}$CoO$_2$ 中直接观察到 O2 结构，认为在 $0.07 \leqslant x \leqslant 0.25$ 过程中 O3 向 O2 转变，$0.25 \leqslant x \leqslant 0.43$ 过程中 O2 向 O1 转变，在 $0.43 \leqslant x \leqslant 0.52$ 过程中 O2 向 O1 转变完成。构建了 O1、O2 和 O3 三个相在电化学循环过程中的有效转换关系。

为了能将更多的锂离子从晶体结构中可逆脱出，广泛采用掺杂、包覆等方法来对其进行改性。Cho 等报道了 ZrO$_2$ 等金属氧化物包覆改善 LiCoO$_2$ 在高截止电压内（2.75～4.4V）的循环稳定性并提高其可逆容量（170mA·h/g）的实验现象，他们认为，所包覆的氧化物与 LiCoO$_2$ 发生反应，在 LiCoO$_2$ 表面形成一层非常薄的固溶体层 LiCo$_{1-x}$M$_x$O$_2$（M＝Zr、Al、Ti、B），从而抑制了 Li$_x$CoO$_2$ 在充放电循环过程中（$0.5 < x < 1$）晶胞沿 c 轴方向的膨胀。Dahn 等也通过实验证明了 ZrO$_2$ 包覆后 LiCoO$_2$ 在高电压下容量保持率的提高。他们通过原位 XRD 研究认为，表面包覆并不能阻止和抑制 LiCoO$_2$ 在循环过程中晶格常数的变化。Wang 等通过原位 XRD 研究在宽电位范围内（3.0～5.2V vs. Li$^+$/Li）多次循环后 LiCoO$_2$ 表面包覆 Al$_2$O$_3$ 前后结构的变化，结果表明，表面包覆并没有抑制充放电过程中 LiCoO$_2$ 正极材料本应发生的结构相变；相反，表面包覆使这种相变能够可逆地进行。向电解液中添加纳米 Al$_2$O$_3$ 同样表明材料电化学性能的提高并非由于在充放电循环过程中本应发生的结构相变被抑制，而是使这种相变变得更加顺畅，使材料的结构变化更加可逆。在包覆方面较成功的还有 AlPO$_4$、Al$_2$O$_3$ 和 MgO 等，在掺杂方面较成功的有 Mn 掺杂、Al 掺杂以及 Ti、Mg 共掺杂等。

近年来，随着机理研究与技术发展的不断进步，钴酸锂材料也取得了长足的发展。2013 年，4.35V 钴酸锂成功产业化，突破了传统钴酸锂只能脱出 50％Li 的限制。随后，在 2014 年，4.40V 钴酸锂也在产品中取得应用。2019 年，4.45V 钴酸锂进入供货阶段，并有望在 2020～2021 年进入市场。至此，钴酸锂材料实际发挥容量已经超过 180mA·h/g，压实密度超过 4.2g/cm^3，体积能量密度超过 700W·h/L。在此基础上，科研工作者对高电压钴酸锂进行了更加深入的基础研究，试图探索钴酸锂真正的应用极限。2018 年，Lu 等人通过表面 Li-Al-F 处理获得了 4.6V 稳定循环的钴酸锂，他们认为通过表面处理能够隔绝电极表面与电解液的直接接触、抑制活性 Co 损失，同时，表面形成的固溶体层也能够缓解因不可逆相变带来的容量损失。同年，华为公司与美国阿贡实验室也合作研发了 La、Al 共掺杂的钴酸锂材料，该材料在 4.5V 下能够释放 190mA·h/g 的可逆容量，并具有较好的循环稳定性与倍率性能。2019 年，中科院物理所研究团队成功开发了 Ti-Mg-Al 痕量元素共掺杂的 LiCoO$_2$ 材料，并利用同步辐射 X 射线纳米三维成像、共振非弹性 X 射线散射等实验技术细致地研究了掺杂元素对钴酸锂材料性能提升的作用机制，揭示了不同掺杂元素对材料性能改善的独特作用。结果表明，Mg 和 Al 元素更容易掺杂进入材料的晶体结构中，对材料的相变过程进行调控。而 Ti 元素则倾向于在钴酸锂颗粒表面与界面处富集，调节材料内部应力应变及其周边氧原子活性。基于这些研究结果，未来对于高电压钴酸锂的研究，需要从晶体结构、电子结构和材料亚微米尺度微观结构等不同维度进行综合设计，才能逐步突破技术瓶颈，达到更高的应用电压。

目前世界范围内，$LiCoO_2$ 产量较大的国际企业为日亚化学（Nichia Chemical）和优美科（Umicore），国内产量较大的企业包括厦门钨业、天津巴莫、湖南杉杉、当升科技、中信国安盟固利等。采用 $LiCoO_2$ 材料的锂离子电池的主要应用领域仍然为传统 3C 电子产品。

6.2.2 立方尖晶石结构 $LiMn_2O_4$ 正极材料

在锂离子电池正极材料研究中，另外一个受到重视并且已经商业化的正极材料是 Thackeray 等在 1983 年提出的尖晶石 $LiMn_2O_4$ 正极材料。$LiMn_2O_4$ 具有三维 Li 输运特性。其具有低价、稳定和优良的导电、导锂性能。其分解温度高，且氧化性远低于 $LiCoO_2$，即使出现短路、过充电，也能够避免燃烧、爆炸的危险。

$Li_xMn_2O_4$ 的脱嵌锂过程是一个相变过程，主要有 2 个脱嵌锂电位：4V 和 3V。$0<x\leqslant 1$ 时，锂离子的脱嵌发生在 4V 左右，对应于锂从四面体 $8a$ 位置的脱嵌。在此范围内，锂离子的脱嵌能够保持尖晶石结构的立方对称性，电极循环良好。$Li_xMn_2O_4$ 在过放电（$1\leqslant x\leqslant 2$）的情况下，在 3V 左右出现电压平台，锂离子嵌入到空的 $16c$ 八面体位置，结构扭曲，原来的立方体 $LiMn_2O_4$ 转变为四面体 $Li_2Mn_2O_4$，Mn 从 +3.5 价还原为 +3.0 价。该转变伴随着严重的 Jahn-Teller 畸变，c/a 变化达到 16%，晶胞体积增加 6.5%，导致表面的尖晶石粒子发生破裂。

$LiMn_2O_4$ 的最大缺点是容量衰减较为严重，特别是在较高的温度下。目前认为主要是由以下原因引起的：①Jahn-Teller 效应及钝化层的形成，有文献表明经过循环或者存储后的 $LiMn_2O_4$ 表面 Mn 的价态比内部低，即表面有较多的 Mn^{3+}。在放电过程中，材料表面生成 $Li_2Mn_2O_4$，由于表面畸变的四方晶系与颗粒内部的立方晶系不相容，严重破坏了结构的完整性和颗粒间的有效接触，影响了 Li^+ 扩散和颗粒间的电导性，造成容量损失。②Mn 的溶解，电解液中存在的痕量水分会与电解液中的 $LiPF_6$ 反应生成 HF，导致 $LiMn_2O_4$ 发生歧化反应，Mn^{2+} 溶解到电解液中，尖晶石结构被破坏。③电解液在高电位下分解，在循环过程中电解液会发生分解反应，在材料表面形成 Li_2CO_3 膜，使电池极化增大，从而造成尖晶石 $LiMn_2O_4$ 在循环过程中容量衰减。

为了改善 $LiMn_2O_4$ 的高温循环与储存性能，科研工作者采用了如下方法对其进行改性：使用其他金属离子部分替换 Mn（如 Li，Mg，Al，Ti，Cr，Ni，Co 等）；减小材料尺寸以减少颗粒表面与电解液的接触面积；对材料进行表面改性处理；使用与 $LiMn_2O_4$ 兼容性更好的电解液等。Sun 等在其表面包覆 $LiAlO_2$，经热处理后，发现在尖晶石颗粒表面形成了 $LiMn_{2-x}Al_xO_4$ 的固溶体，对电极表面起到了保护作用，同时提高了晶体结构的稳定性，改善了 $LiMn_2O_4$ 的高温循环性能和储存性能，还提高了倍率性能。纳米单晶颗粒也是提高 $LiMn_2O_4$ 材料性能的手段，因为纳米单晶可以同时满足高电极材料密度和小尺寸的条件，在不降低电极密度的条件下提高其倍率性能。

$LiMn_2O_4$ 材料成本低、无污染、制备容易，适用于大功率低成本动力电池，可用于电动汽车、储能电站以及电动工具等方面。缺点是高温下循环性差，储存时容量衰减快。目前世界范围内，$LiMn_2O_4$ 产量最大的国际企业为日本户田工业（Toda），国内企业包括无锡晶石和湖南桑顿、中信大锰等企业。

6.2.3 正交橄榄石结构 LiFePO$_4$ 材料

1997 年，由 Goodenough 等提出橄榄石结构的磷酸铁锂材料（LiFePO$_4$）可以用作锂离子电池正极材料，其结构如图 6-1（c）所示，参数见表 6-1。与 LiMn$_2$O$_4$ 和 LiCoO$_2$ 等之前的正极材料不同，LiFePO$_4$ 材料反应机理为两相反应（LiFePO$_4$/FePO$_4$），而非固溶体（Li$_{1-x}$CoO$_2$）类型反应。关于其相转变机理，先后提出了核壳（core/shell）机构模型、收缩核（shrinking shell）模型、多米诺骨牌（domino-cascade）模型、相转变波（phase transformation wave）模型、二阶（staging II）模型等模型来解释这个过程。详细讨论。

LiFePO$_4$ 的缺点在于其电子电导率比较差，在 10^{-9} S/cm 量级。锂离子传导被认为是小极化子传导机制，活化能约 0.3~0.5eV，表观扩散系数约 10^{-10}~10^{-15} cm^2/s，导致材料的倍率性能差。为提高其倍率性能，Armand 等提出碳包覆的方法显著提高了 LiFePO$_4$ 的电化学活性，Takahashi 等和 Yamada 等把材料纳米化，缩短扩散路径。随后科研工作者发现掺杂提高电子电导率是优化其电化学性能的重要方法；关于掺杂一直存在争议，主要存在以下几个问题：是否能够掺入到 LiFePO$_4$ 晶格中以及掺杂的位置；掺杂能否提高离子电导率。第一性原理计算从原子水平上研究了 Li 在 LiFePO$_4$ 晶格中的存储与输运问题，证明 Li 的一维输运特性和 3.70V 的禁带宽度。

LiFePO$_4$ 材料主要金属元素是 Fe，因此在成本和环保方面有着很大的优势。LiFePO$_4$ 材料循环寿命可达 2000 次以上，快速充放电寿命也可达到 1000 次以上。与其他正极材料相比，LiFePO$_4$ 具有更长循环寿命、更高稳定性、更安全可靠、更环保且价格低廉、更好的充放电倍率性能。磷酸铁锂电池已被大规模应用于电动汽车、规模储能、备用电源等。

6.3 其他正极材料

在第 6.1 节的讨论中，我们可以根据结构的不同，将常见锂离子电池正极材料分为层状正极、尖晶石结构正极以及聚阴离子型正极（橄榄石型的 LiFePO$_4$ 是其中一种）。在这三类体系中，通过改变过渡金属或聚阴离子的种类，还发展出了一系列的正极材料，它们当中的一部分已经被应用在工业中，如 NiCoMn 三元正极和 Li$_3$V$_2$(PO$_4$)$_3$；一部分目前还没有广泛的应用，但被认为是有希望的下一代锂离子电池正极材料，如 LiNi$_{0.5}$Mn$_{1.5}$O$_4$ 和富锂相等；还有一部分材料仍处于基础研究的起步阶段，如阳离子无序的 Li$_{1+x}$TM$_{1-x}$O$_2$ 材料，虽然距离产业化还很遥远，但能够为未来高能量密度锂离子电池的设计提供新的思路，并对理解正极材料容量来源、电荷补偿机制等基础科学问题提供帮助。

6.3.1 层状结构正极材料

类似于 $LiCoO_2$，在过渡金属层中，处于 $3b$ 位置的 Co 元素可以被 Ni、Mn、Li 以及其他元素取代。当 Co 被 Ni、Mn 部分取代时，称为三元材料，具有多种组合，只要其所占位置的平均电荷为 +3 即可。$LiNi_{1-x-y}Co_yMn_xO_2$ 与 $LiCoO_2$ 一样，具有 α-$NaFeO_2$ 型层状结构（$R-3m$ 空间群），理论容量约为 275mA·h/g。在三元材料中，Mn 始终保持 +4 价，没有电化学活性，Ni 和 Co 有电化学活性。Mn^{4+} 的存在能稳定结构，Co^{3+} 的存在能提高材料的电子电导率，同时抑制 Li、Ni 互占位。且在一定范围，倍率性能随着 Co 的掺杂量提高而变好。Li-Ni-Co-Mn-O 三元材料相比于 $LiCoO_2$ 有成本上的优势，目前已经在商品锂离子电池中大量使用。目前市场上常见的三元材料 Ni、Co、Mn 比例类型为 424、333、523、622、811 等。

随着电动汽车等领域的不断发展，高镍三元材料逐步成为重要的发展方向。高镍三元材料能够在更低的电压区间内释放更多的容量。目前，高镍的 NCM 材料中 Ni 的比例已经能够达到 90% 以上，可逆比容量达到 210mA·h/g，是目前为止能量密度最高的体系之一。但是，随着 Ni 含量的提升，材料也面临着更严峻的挑战。随着材料脱锂量的提升，材料的晶体结构变得不稳定，在高温和长循环等条件下更容易出现失氧、起火、爆炸等极端安全事件。近期，出现了多起电动汽车起火事件，其中三元材料占比较大。此外，随着镍含量的提升，材料表面碱性增加，材料对空气中的水分及 CO_2 更加敏感，因此对合成、存储、匀浆、涂布等工艺条件的要求更加苛刻。不规范的工艺条件会导致材料在应用过程中出现产气、容量下降等问题。

Al 取代 Ni/Co 能够在一定程度上改善材料的结构稳定性，提高材料的循环性能。$LiNi_{0.80}Co_{0.15}Al_{0.05}O_2$（NCA）材料的可逆容量超过 200mA·h/g，Co、Al 的复合掺杂能促进 Ni^{2+} 的氧化，减少 $3a$ 位 Ni^{2+} 含量，抑制充放电过程中从 H2 到 H3 的不可逆相变，从而提高材料本身的循环稳定性。目前，以 NCA 为正极，硅碳复合材料为负极的圆柱电池已经成功产业化。结合 NCM 与 NCA 的设计经验，科研人员近期提出了 $LiNi_{0.89}Co_{0.05}Mn_{0.05}Al_{0.01}O_2$（NCMA）正极。该类正极材料能够综合 NCM 与 NCA 的优势，使材料在发挥更高容量的同时保持更好的稳定性与安全性。NCMA 正极还能够进一步降低材料中 Co 元素的用量，因此成本上更具优势。

2001 年，Ohzuku 等首先合成了 $LiNi_{0.5}Mn_{0.5}O_2$ 正极材料，该材料在 2.5～4.3V 内循环时具有 150mA·h/g 的放电容量。随后 Yang 等通过研究其充放电过程的相变情况，认为该材料可以抑制 $LiNiO_2$ 体系中发生的 H3 相变，提高材料热稳定性。Ohzuku 等指出虽然 $LiNi_{0.5}Mn_{0.5}O_2$ 材料具有很高的可逆容量（200mA·h/g）、低的不可逆容量和极化，但其依然拥有 8%～9% 的 Li^+/Ni^{2+} 混排。Ceder 等提出了该材料的结构模型与 Li^+/Ni^{2+} 混排机理并通过离子交换法制备了 Li^+/Ni^{2+} 混排程度很低的 $LiNi_{0.5}Mn_{0.5}O_2$ 材料，其具有优于其他方法的化学性能，通过计算发现该材料是层状结构材料中具有最大晶胞参数 c 值的材料，这预示着其具有快速充放电的能力。

如果层状结构中 Co 的位置同时还包含 Li，$3a$ 位置的 Li 含量不变，$3b$ 位置含有 Li，一般称为富锂相。富锂相的微观结构根据制备条件的不同可能存在着固溶体和纳米相共存的微观结构上的差异。Dahn 研究了 800℃ O_2 气氛中退火冷却和随炉冷却时 Li-Ni-Mn-O 相图边界的变化，发现缓慢冷却时，相图中尖晶石单相区、层状单相区和岩盐单相区扩大，NiMn 固溶程度增加；而快速冷却时，则相边界更易形成两相分离。

富锂相正极材料的充放电机理涉及许多内容：如材料高容量的来源，首周位于 4.5V 的平台，首周极大的不可逆容量，循环过程中放电平台的逐渐降低等。

对材料的首周脱锂过程，按充电电压 4.4V 分为两个部分。当充电电压低于 4.4V 时，$LiMO_2$ 的 Li 层中的 Li^+ 脱出，同时 Li_2MnO_3 的过渡金属层中位于八面体位置的 Li 扩散到 $LiMO_2$ 中 Li 层的四面体位置（这个现象已被实验和计算证明）。这将起到两个作用：一是稳定材料的结构；二是提供锂源。Yabuuchi 等认为扩散的 Li 也可能存在于八面体位置，且在这个充电范围内，Mn 的价态保持不变，主要是其他过渡金属的价态在升高，提供电荷转移的来源。当充电电压继续升高到达位于 4.5V 的平台时，材料中的 Li 继续脱出，形成 MnO_2 以及 MO_2，在这个过程中会伴随 O_2 析出。Yabuuchi 等认为在这个过程中可能伴随 Mn 向 Li 层的扩散。在随后的嵌锂过程中，脱出的 Li_2O 不能回到材料中（不包括表面反应），且一般认为过渡金属层中的锂空位将无法被再次填充，使得富锂相材料在半电池的测试中首周循环的效率较低。材料的析氧会导致离子的重新排布，Armstrong 等提出了两种模型。第一种模型认为，当 Li 和 O 同时从电极材料表面脱出时，O 从材料内部扩散到表面维持反应进行，同时在材料内部形成氧空位；第二种模型则认为表面过渡金属离子会扩散到材料内部，占据过渡金属层中 Li 脱出留下的八面体空位，当所有的八面体空位都被占据时，析氧过程就结束。第二种模型计算的不可逆容量与实验结果不一致。Wu 等认为，首次充放电过程中，O 离子空位并未完全消失，仍然有一部分 O 空位存在于晶格中。富锂相的母体材料 Li_2MnO_3 在充放电过程中结构变化及电荷转移机制的研究，对于理解高容量富锂相正极材料的电化学性质有着非常重要的意义。Wang 等通过球差校正 ABF-STEM 技术发现，Li_2MnO_3 电化学脱锂过程并不均匀，Li 层和 LiM_2 层的 Li 均能脱出，在一定的脱嵌锂范围内，脱嵌锂导致 Mn 的可逆位移。近年来，随着人们对富锂锰基材料认识的加深，富锂锰基材料的容量及首周库伦效率也不断取得突破。目前容量 280～300mA·h/g 的富锂锰基材料已经进入中试阶段，Xia 等人开发的混合相富锂锰基材料首周可逆容量已经达到 400mA·h/g。除了对首周充放机制进行理解外，科研人员认为，富锂锰基材料中的高容量也可能与材料内部氧阴离子参与氧还原反应过程有关。在高脱锂态下，由于 O 与过渡金属之间存在共用电子对，因此氧离子也会承担部分电荷转移功能，使得富锂材料的容量高于一般的层状材料。也有科研人员认为，在高脱锂态下，材料内部可能存在更高价的过渡金属而不是完全由氧阴离子承担电荷补偿，目前这一观点正等待有效的实验手段对其进行验证。

对于富锂相材料，另一个值得关注的现象是材料在循环过程中放电平均电压持续降低。目前，已有一些文献对其进行研究，科研工作者普遍认为这是由于在循环过程中 Mn、Ni 进入 Li 层，导致材料向尖晶石结构进行转变，使得放电平台下降。富锂相材料在第一周循环中结构将会有一个较明显的改变，在之后的循环过程中结构仍然在持续地变化。另外由于工作电压较高（充电 4.8V），有可能使电解液氧化分解。2018 年，Yu 等人利用同步辐射 X 射线吸收光谱技术并结合特殊设计的模拟电池研究了富锂锰基层状氧化物正极材料在不同充放电周期的氧化还原反应机制。结果表明，抑制富锂材料电压衰减需要提高材料中晶格氧离子在高电压充电时的稳定性，同时，材料中不同元素对电压衰减也存在不同的影响。

因此在实际使用中，需要对材料进行各种改性，以保持材料结构稳定，容量、电压稳定，抑制材料与电解液之间的副反应以及提高材料的电子电导率和离子电导率以提高其倍率性能。目前已经提出了多种改性方法，包括表面包覆改性、表面酸处理、氟掺杂改性、预循环处理等。表面包覆改性后，第一周充电后材料中的氧空位保留率提高了，可以降低其首周不可逆容量，提高材料的循环性能；表面酸处理主要是通过 H^+-Li^+ 离子交换或者溶解刻蚀的方式进行，在酸性环境中，H^+ 和 Li^+ 之间会发生交换现象，表面的 Li_2MnO_3 被活化，可以提高其首周效率；氟掺杂改性即经过在含 F 溶液中钝化的材料，其表面结构破坏程度要小于酸处理的样品，材料的首周效率和循环性能可以得到较大提高；预循环处理有效地减少和延缓了表层材料结构的破坏，提高了其效率和循环性。

在典型的富锂锰基材料研究基础上，为了进一步提升材料容量及稳定性，科研人员设计了一种新型的阳离子无序正极材料。阳离子无序正极材料与传统的富锂锰基正极材料具有相似的岩盐结构，都是以氧阴离子构成材料的基本框架。两者的区别在于，阳离子无序正极材料中的 Li 离子、过渡金属离子完全无序地分布在氧阴离子构成的框架中，而不是与富锂锰基及其他层状结构正极材料一样分层、有序地排布。

在这种结构中，锂离子的传输需要满足两个条件，第一是具有锂离子传输通道，能够使锂离子在局部结构中运动起来；第二就是在整个晶体结构中，锂离子的传输通道能够构成渗流网络，使锂离子在材料中无障碍地迁移。在以岩盐结构为基础的层状材料中，锂离子的迁移通道是通过材料中氧离子构成的八面体-四面体-八面体进行迁移的，在这种迁移模式下，锂离子的迁移活性与其周围过渡金属元素的种类、数量、价态具有密切关系。根据锂离子扩散路径中四面体间隙位周围过渡金属原子（TM）的个数，可以将锂离子传输的四面体分为 0-TM、1-TM、2-TM 三种类型。在传统的层状结构中，锂离子均在 1-TM 类型的四面体间进行传输。而以 2-TM 型四面体为主的阳离子无序正极材料因为四面体周围阳离子间较强的相互作用，锂离子传输活性非常低，因此无法取得实际应用。2014 年，Ceder 课题组通过计算表明，当层状材料中 Li 与过渡金属离子的比例超过 1.1∶1 时，也就是构成富锂状态时，层状材料内部可以形成 0-TM 型连续锂离子传输网络，这一网络会激活以 2-TM 为主的阳离子无序正极材料的离子输运活性，从而使其能够可逆地循环。

目前，阳离子无序正极材料首周可逆比容量可达 300mA·h/g，与常规的富锂锰基材料基本一致。阳离子无序正极材料在循环的过程中具有较小的体积变化以及非常稳定的氧离子晶格框架。与富锂锰基材料类似，其体系内的氧阴离子也能够可逆地进行氧化还原贡献容量，且可逆性要高于富锂锰基材料。但是，目前所报道的阳离子无序正极材料的循环稳定性仍然需要进一步加强，还需要开展更多细致的研究。

6.3.2　高电压尖晶石结构正极材料

高电压正极材料 $LiNi_{0.5}Mn_{1.5}O_4$ 可以看作是 Ni 掺杂的 $LiMn_2O_4$。$LiNi_{0.5}Mn_{1.5}O_4$

为立方尖晶石结构，有两种空间结构，一种空间群为 $Fd\text{-}3m$；另一种空间群为 $P4_332$。在 $Fd\text{-}3m$ 结构中，因为 Ni/Mn 原子随机占位，这种结构也被称为"无序"结构；在 $P4_332$ 结构中，Ni 原子占据 $4a$ 位，Mn 原子占据 $12d$ 位，是一种有序结构。$LiNi_{0.5}Mn_{1.5}O_4$ 具有 4.7V 的平台，其理论放电比容量为 146.7mA·h/g。在 $LiNi_{0.5}Mn_{1.5}O_4$ 中，锰全部为 +4 价，在充放电过程中不发生氧化还原反应，起到稳定晶体结构的作用。同时，由于没有 Mn^{3+} 的存在，就避免了在充放电过程中惰性 Mn^{2+} 的生成。Ni 全部为 +2 价，为材料中的电化学活性金属离子，在充放电过程中 $LiNi_{0.5}Mn_{1.5}O_4$ 对应于 Ni^{2+}/Ni^{3+} 和 Ni^{3+}/Ni^{4+} 的两个平台都处于 4.7V 左右，电压差别很小。但高温合成时，部分氧会从尖晶石结构中脱出，继而使得材料中的 Mn^{3+} 增多，Mn^{3+} 的存在严重影响了材料的结构稳定性。其在高电压下与电解液的反应是阻碍其应用的主要因素，Wang 等通过 LiF 和 Al_2O_3 包覆，可以在物理上隔离材料与电解液的接触，提高循环性能。除了包覆之外，目前对 $LiNi_{0.5}Mn_{1.5}O_4$ 材料改性研究的方法还包括掺杂（如 Cr、Ru 等）和 Ni、Mn 比例微调等方法。

Fe 和 Co 也可以被用作掺杂，形成 $LiFe_{0.5}Mn_{1.5}O_4$ 和 $LiCo_{0.5}Mn_{1.5}O_4$，但是并不像 $LiNi_{0.5}Mn_{1.5}O_4$ 有两种不同的晶体结构，$LiFe_{0.5}Mn_{1.5}O_4$ 和 $LiCo_{0.5}Mn_{1.5}O_4$ 只有无序的 $Fd\text{-}3m$ 结构，原位 XRD 分析 $LiFe_{0.5}Mn_{1.5}O_4$ 和 $LiCo_{0.5}Mn_{1.5}O_4$ 表明，这两种材料在充放电过程中经历了固溶体反应，这两种材料的首周放电容量均不及 $LiNi_{0.5}Mn_{1.5}O_4$ 且随着充放电循环，其容量迅速衰减。

6.3.3 聚阴离子类正极材料

在 $LiMPO_4$ 中，Mn^{2+}/Mn^{3+}、Co^{2+}/Co^{3+}、Ni^{2+}/Ni^{3+} 氧化还原电对的电压较 Fe^{2+}/Fe^{3+} 高，相对应的能量密度也较 $LiFePO_4$ 高，其中 $LiNiPO_4$ 的充放电电压为 5.2V，几乎超出了所有电解质的电化学窗口，所以关于其电化学的报告基本没有，而 $LiMnPO_4$ 和 $LiCoPO_4$ 的充放电电压分别为 4.0V 和 4.8V。$LiMPO_4$ 共同的缺陷是电子电导率和离子电导率都非常低，它们的室温离子扩散系数 $D_{Li}<10^{-14}cm^2/s$。同 $LiFePO_4$ 一样，减小材料颗粒尺寸和碳包覆被认为是提高 $LiMPO_4$ 材料动力学性能的最有效方法。日本 Yamada 小组和 Tohda 小组报道 $LiFe_{0.4}Mn_{0.6}PO_4$ 和 $LiFe_{0.25}Mn_{0.75}PO_4$ 可以在室温小倍率下获得 160mA·h/g 的容量和良好的循环性能，一些多元磷酸盐化合物 $LiFe_{1/4}Mn_{1/4}Co_{1/4}Ni_{1/4}PO_4$ 和 $LiFe_{1/3}Mn_{1/3}Co_{1/3}PO_4$ 也呈现出良好的电化学性能并且具有类似的活性增强效果。$LiFe_{1-x}Mn_xPO_4$ 材料的电化学性能随着 Mn 含量的升高而下降，其倍率特性与 Li 在 $LiMPO_4$ 和 MPO_4 中的扩散相对快慢相关，呈现出路径依赖的特点。

$Li_3V_2(PO_4)_3$ 正极材料是一种在工业中已经获得小批量应用的锂离子电池正极材料，属 NASICON 结构。单斜晶系 $Li_3V_2(PO_4)_3$ 属于 $P21/n$ 空间群，与斜方晶系相比，具备更高的 Li^+ 扩散系数和更高的能量密度，因此受到了广泛的关注。$Li_3V_2(PO_4)_3$ 具有与 $LiCoO_2$ 同样的放电平台和比能量，而 $Li_3V_2(PO_4)_3$ 的热稳定性、安全性远远优于 $LiCoO_2$。但材料的电子电导率只有 10^{-7}S/cm 数量级，高倍率充放电性能较差。研究发现，在 Li 位掺杂 Mg^{2+}、Al^{3+}、Ti^{4+}、Co^{3+} 和 Cr^{3+} 等金属离子，可以同时提高电子电导率和离子电导率；而表面进行碳包覆，可以改善颗粒之间的电接触，改善其导电性，避免 V^{3+} 氧化。随着其研究的不断深入，$Li_3V_2(PO_4)_3$ 作为一种高电势的正极材料，以其毒性较小、成本较低、扩散系数高、比容量高及稳定性能好等显著特点，成为动力锂电池正极材料的发展趋势。

与磷酸盐类似，硅酸盐、硫酸盐、硼酸盐、碳酸盐的研究也引起了广泛关注。

正硅酸盐（Li_2MSiO_4）材料具有与 Li_3PO_4 类似的晶体结构，其中所有阳离子均以四面体与氧离子配位，氧离子以近乎六方密堆方式排列，阳离子占据氧四面体空隙的一半，存在两套可能的占据方式，分别形成与 Li_3PO_4 低温相（β相）和高温相（γ相）同构的结构。但由于不同 Li_2MSiO_4 多形体之间的形成能差异较小，通常情况下合成的 Li_2MSiO_4 为多种多形体的混合物。2005 年，Dominko 等和 Nyten 等报道了 Li_2FeSiO_4 作为锂离子电池正极材料的性能，在 60℃ 以 $C/16$ 电流充放电，其放电比容量约为 130mA·h/g。制约 Li_2MSiO_4 材料性能发挥的主要问题是低的 Li^+ 扩散系数和电子电导率，碳包覆和降低材料颗粒尺寸等方法也被广泛应用于提高 Li_2MSiO_4 的电化学活性。Li_2MnSiO_4 理论上更容易实现每分子两个 Li 的可逆嵌脱，获得高的理论容量（333mA·h/g），但其存在严重的循环容量衰减问题。Li_2MnSiO_4 经充电后处于一个不稳定的状态，脱出一个锂后，$LiMnSiO_4$ 易发生相分离，生成 Li_2MnSiO_4 和 $MnSiO_4$。Li_2CoSiO_4 材料具有相对于锂负极 4.1V 的 Li^+ 脱出/嵌入电位；首次充电过程中约 1.2 个 Li 可以被脱出，在随后的放电过程中只能回嵌约 0.4 个 Li，且伴随着高的首次不可逆容量损失。

硫酸盐也是一种研究较多的聚阴离子型锂离子电池正极材料，主要包括 $Li_2Fe_2(SO_4)_3$ 和 $LiFeSO_4F$，$Li_2Fe_2(SO_4)_3$ 为 NASICON 结构，包括单斜（$P21/n$）和六方（$R3$）两种晶系，理论比容量为 130mA·h/g，脱、嵌锂的理论电压平台为 3.6V。$Li_2Fe_2(SO_4)_3$ 的充放电反应为两相反应，在六方晶系材料中 Li^+ 的扩散通道更为通畅。因此，六方晶系 $Li_2Fe_2(SO_4)_3$ 的电化学性能较单斜晶系的好。$LiFeSO_4F$ 属三斜晶系 $P1$ 空间群。$LiFeSO_4F$ 的理论比容量为 151mA·h/g，放电电压为 3.6V。$LiFeSO_4F$ 的充放电反应也是一种两相反应。由于 $LiFeSO_4F$ 中的 SO_4^{2-} 在高温下易分解，且 $LiFeSO_4F$ 与水反应分解为 FeOOH 和 LiF，所以 $LiFeSO_4F$ 一般不能通过高温固相或水热等方法合成，这限制了其发展和应用。

$LiMBO_3$ 因为拥有最轻的聚阴离子基团，为提供更高的容量提供了可能。Zhao 等首先将 $LiFeBO_3$ 应用于锂离子电池正极材料，并通过碳包覆对材料进行改性。Yamada 等通过实验将其容量提高到 200mA·h/g。相比于 $LiFePO_4$，$LiFeBO_3$ 具有更高的容量、更好的导电性（电导率 3.9×10^7S/cm）和极小的体积变化率（2%）。在此基础上，人们又发展了 $LiMnBO_3$、$Li(Mn_xFe_{1-x})BO_3$ 等正极材料。

6.3.4 基于相转变反应的正极材料

电化学相转变反应（electrochemical conversion reaction）是一种重要的储锂机制，它为多变价正极材料设计提供了一条新的路径，原因是它可以利用高价过渡金属化合物的所有氧化态。最近几年，一系列的过渡金属氟化物、氧化物、硫化物以及氮化物被证实可以实现多电子转移，实现很高的容量。但是这类材料作为正极材料普遍存在难以克服的缺点：①本身并不含有锂源，需要负极提供锂源；②在充放电过程中极化很大，能源利用效率低；③充放电过程中，体积变化大。目前此类材料的研究仅限于基础研究。

6.3.5　有机正极材料

1969 年，Alt 等报道了以羰基化合物二氯异氰尿酸（DCA）为正极的锂-有机物电池，尽管其仍然是一次电池，但为人们寻找合适的锂离子电池正极材料提供了一个方向。与传统的锂离子电池相比，锂电池有机电极材料具有理论比容量高、原料丰富、成本低（不涉及昂贵元素）、容易设计加工和体系安全等优点。按照正极材料分类，锂电池有机电极材料可分为导电聚合物、含硫化合物、氮氧自由基化合物和羰基化合物等。有机正极材料面临的问题是目前开发的有机正极材料在循环性、有效能量密度（特别是体积能量密度）、功率特性方面与现有无机材料相比还有较大差距，而且也不能作为锂源正极材料，这些缺点没有明确的解决办法，目前此类材料的研究也仅限于基础研究。

6.4　本章结语

目前，正极材料的主要发展思路是在层状结构、尖晶石结构、橄榄石结构等材料的基础上，发展相关的各类衍生材料，通过掺杂、包覆、调整微观结构、控制材料形貌、尺寸分布、比表面积、杂质含量等技术手段来综合提高其比容量、倍率、循环性、压实密度、电化学性、化学及热稳定性。最迫切的仍然是在保证安全性与可靠性的前提下，尽可能地提高能量密度，其关键是提高正极材料的容量或者电压。目前，无论是容量还是电压的升高都要求电解质及相关辅助材料能够在宽电位范围工作，因此下一代高能量密度锂离子电池正极材料的发展除了正极材料自身的更新换代外，还取决于高电压电解质技术及相关辅材技术的进步。

本章需要进一步思考的基础科学问题：

1. 正极材料结构与其储锂机制、电化学性能之间的关联是否是一一对应的？是否能寻找出新的具有应用潜力的结构类型？
2. 正极材料中相变的微观过程是怎样的？是否存在完全可逆的相变过程？
3. 正极材料中的电荷补偿机制与正极材料结构、成分之间的关联是什么？阴阳离子之间的电子对分布规律是如何影响电荷补偿的？
4. 在有机液态电解液体系下，正极表界面膜是否稳定？是否具有离子、电子电导？其成分的主要来源及产生机制有哪些？
5. 在有机液态电解液体系下，正极表界面膜与材料结构、成分、电压之间的关联有哪些？

6. 在固态电池中，正极材料与电解质材料的界面又会面临哪些新的挑战？
7. 正极材料内部应力应变的产生机制、裂痕的扩散原理是什么？如何精确表征？
8. 层状结构正极材料脱锂后结构还能保持稳定的脱锂极限是多少？由什么因素决定？
9. 正极材料内部的元素分布是否均匀？元素分布规律对材料性能的影响是什么？
10. 正极材料的最大功率密度或者说倍率特性是多少？如何预测或预判？

参考文献

[1] Armand M. Materials for Advanced Batteries [M]. New York: Plenum Press, 1980.

[2] Whittingham M S. Lithium batteries and cathode materials [J]. Chem Rev, 2004, 104 (10): 4271-4302.

[3] Mizushima K, Jones P C, Wiseman P J, Goodenough J B. Li_xCoO_2 ($0<x<1$): A new cathode material for batteries of high energy density [J]. Solid State Ionics, 1981, 3-4: 171-174.

[4] Thackeray M, David W, Bruce P, et al. Lithium insertion into manganese spinels [J]. Materials Research Bulletin, 1983, 18 (4): 461-472.

[5] Padhi A K, Nanjundaswamy K S, Goodenough J B. Phospho-olivines as positive-electrode materials for rechargeable lithium batteries [J]. Journal of the Electrochemical Society, 1997, 144 (4): 1188-1194.

[6] Thomas M, Bruce P, Goodenough J. AC impedance analysis of polycrystalline insertion electrodes: Application to $Li_{1-x}CoO_2$ [J]. Journal of the Electrochemical Society, 1985, 132 (7): 1521-1528.

[7] Thomas M, Bruce P, Goodenough J. Lithium mobility in the layered oxide $Li_{1-x}CoO_2$ [J]. Solid State Ionics, 1985, 17 (1): 13-19.

[8] Hong J S, Selman J. Relationship between calorimetric and structural characteristics of lithium-ion cells Ⅱ. Determination of Li transport properties [J]. Journal of the Electrochemical Society, 2000, 147 (9): 3190-3194.

[9] Chen L, Huang X, Kelder E, et al. Diffusion enhancement in $Li_xMn_2O_4$ [J]. Solid State Ionics, 1995, 76 (1): 91-96.

[10] Barker J, Pynenburg R, Koksbang R. Determination of thermo-dynamic, kinetic and interfacial properties for the $Li/Li_xMn_2O_4$ system by electrochemical techniques [J]. Journal of Power Sources, 1994, 52 (2): 185-192.

[11] Bach S, Farcy J, Pereira-ramos J. An electrochemical investigation of Li intercalation in the sol-gel $LiMn_2O_4$ spinel oxide [J]. Solid State Ionics, 1998, 110 (3): 193-198.

[12] Prosini P P, Lisi M, Zane D, et al. Determination of the chemical diffusion coefficient of lithium in $LiFePO_4$ [J]. Solid State Ionics, 2002, 148 (1-2): 45-51.

[13] Delmas C, Fouassier C, Hagenmuller P. Structural classification and properties of the layered oxides [J]. Physica B+C, 1980, 99 (1-4): 81-85.

[14] Mendiboure A, Delmas C, Hagenmuller P. New layered structure obtained by electrochemical deintercalation of the metastable $LiCoO_2$ variety [J]. Materials Research Bulletin, 1984, 19 (10): 1383-1392.

[15] Chang K, Hallstedt B, Music D, et al. Thermodynamic description of the layered O_3 and O_2 structural $LiCoO_2$-CoO_2 pseudo-binary systems [J]. Calphad, 2013, 41: 6-15.

[16] Delmas C, Braconnier J J, Hagenmuller P. A new variety of $LiCoO_2$ with an unusual oxygen packing obtained by exchange reaction [J]. Materials Research Bulletin, 1982, 17 (1): 117-123.

[17] Reimers J N, Dahn J R. Electrochemical and insitu X-ray diffraction studies of lithium intercalation in Li_xCoO_2 [J]. Journal of the Electrochemical Society, 1992, 139 (8): 2091-2097.

[18] Goodenough J B M K, Takeda T. Solid-solution oxides for storage battery electrodes [J]. Japanese Journal of Applied Physics, 1980, 19 (3): 305-313.

[19] Ohzyku T, Uuda A. Solid-state redox reactions of $LiCoO_2$ (R-$3m$) for 4V secondary lithium cells [J]. Journal of the Electrochemical Society, 1994, 141 (11): 2972-2977.

[20] Shao-horn Y, Leasseur S, Weill F, et al. Probing lithium and vacancy ordering in O_3 layered Li_xCoO_2 ($x\approx0.5$) an electron diffraction study [J]. Journal of the Electrochemical Society, 2003, 150 (3): A366-A373.

[21] Reimers J N, Dahn J. Electrochemical and in situ X-ray diffraction studies of lithium intercalation in Li_xCoO_2 [J]. Journal of the Electrochemical Society, 1992, 139 (8): 2091-2097.

[22] Yang S H, Croguennec L, Delmas C, et al. Atomic resolution of lithium ions in LiCoO$_2$ [J]. Nat Mater, 2003, 2 (7): 464-467.

[23] Huang R, Hitosugi T, Findlay S D, et al. Real-time direct observation of Li in LiCoO$_2$ cathode material [J]. Applied Physics Letters, 2011, 98 (5): 51913-51915.

[24] Lu X, Sun Y, Jian Z, et al. New insight into the atomic structure of electrochemically delithiated O$_3$-Li$_{1-x}$CoO$_2$ ($0 \leqslant x \leqslant 0.5$) nanoparticles [J]. Nano Lett, 2012, 12 (12): 6192-6197.

[25] Cho J, Kim Y J, Park B. LiCoO$_2$ cathode material that does not show a phase transition from hexagonal to monoclinic phase [J]. Journal of the Electrochemical Society, 2001, 148 (10): A1110-A1115.

[26] Cho J, Kim Y J, Kim T J, et al. Zero-strain intercalation cathode for rechargeable Li-ion cell [J]. Angewandte Chemie International Edition, 2001, 40 (18): 3367-3369.

[27] Chen Z H, Dahn J R. Effect of a ZrO$_2$ coating on the structure and electrochemistry of Li$_x$CoO$_2$ when cycled to 4.5V [J]. Electrochem Solid State Lett, 2002, 5 (10): A213-A216.

[28] Wang Z X, Huang X J, Chen L Q. Performance improvement of surface-modified LiCoO$_2$ cathode materials: An infrared absorption and X-ray photoelectron spectroscopic investigation [J]. Journal of the Electrochemical Society, 2003, 150 (2): A199-A208.

[29] Li H, Wang Z X, Chen L Q, et al. Research on advanced materials for Li-ion batteries [J]. Advanced Materials, 2009, 21 (45): 4593-4607.

[30] Cho J, Lee J G, Kim B, et al. Effect of P$_2$O$_5$ and AlPO$_4$ coating on LiCoO$_2$ cathode material [J]. Chemistry of Materials, 2003, 15 (16): 3190-3193.

[31] Cho J, Kim Y J, Park B. Novel LiCoO$_2$ cathode material with Al$_2$O$_3$ coating for a Li ion cell [J]. Chemistry of Materials, 2000, 12 (12): 3788-3791.

[32] Liu L J, Wang Z X, Li H, et al. Al$_2$O$_3$-coated LiCoO$_2$ as cathode material for lithium ion batteries [J]. Solid State Ionics, 2002, 152-153: 341-346.

[33] Wang Z X, Wu C A, Liu L J, et al. Electrochemical evaluation and structural characterization of commercial LiCoO$_2$ surfaces modified with MgO for lithium-ion batteries [J]. Journal of the Electrochemical Society, 2002, 149 (4): A466-A471.

[34] Stoyanova R, Zhecheva E, Zarkova L. Effect of Mn-substitution for Co on the crystal structure and acid delithiation of LiMn$_y$Co$_{1-y}$O$_2$ solid solutions [J]. Solid State Ionics, 1994, 73 (3): 233-240.

[35] Waki S, Dokko K, Itoh T, et al. High-speed voltammetry of Mn-doped LiCoO$_2$ using a microelectrode technique [J]. Journal of Solid State Electrochemistry, 2000, 4 (4): 205-209.

[36] Ceder G, Chiang Y M, Sadoway D, et al. Identification of cathode materials for lithium batteries guided by first-principles calculations [J]. Nature, 1998, 392 (6677): 694-696.

[37] Goodenough J B. Rechargeable batteries: Challenges old and new [J]. Journal of Solid State Electrochemistry, 2012, 16 (6): 2019-2029.

[38] Gao Y, Yakovleva M V, Ebner W B. Novel LiNi$_{1-x}$Ti$_{x/2}$Mg$_{x/2}$O$_2$ compounds as cathode materials for safer lithium-ion batteries [J]. Electrochemical and Solid-State Letters, 1998, 1 (3): 117-119.

[39] Hosono E, Kudo T, Honma I, et al. Synthesis of single crystalline spinel LiMn$_2$O$_4$ nanowires for a lithium ion battery with high power density [J]. Nano Lett, 2009, 9 (3): 1045-1051.

[40] Amatucci G, Tarascon J M. Optimization of insertion compounds such as LiMn$_2$O$_4$ for Li-ion batteries [J]. Journal of the Electro-chemical Society, 2002, 149 (12): K31-K46.

[41] Arora P, Popov B N, White R E. Electrochemical investigations of cobalt-doped LiMn$_2$O$_4$ as cathode material for lithium-ion batteries [J]. Journal of the Electrochemical Society, 1998, 145 (3): 807-815.

[42] Eriksson T, Gustafsson T, Thomas J O. Surface structure of LiMn$_2$O$_4$ electrodes [J]. Electrochemical and Solid-State Letters, 2002, 5 (2): A35-A38.

[43] Du P A, Blyr A, Cougral P, et al. Mechanism for limited 55 ℃ storage performance of Li$_{1.05}$Mn$_{1.95}$O$_4$ electrodes [J]. Journal of the Electrochemical Society, 1999, 146 (2): 428-436.

[44] Kim J S, Johnson C, Vaughey J, et al. The electrochemical stability of spinel electrodes coated with ZrO$_2$, Al$_2$O$_3$, and SiO$_2$ from colloidal suspensions [J]. Journal of the Electrochemical Society, 2004, 151 (10): A1755-A1761.

[45] Park S B, Shin H C, Lee W G, et al. Improvement of capacity fading resistance of LiMn$_2$O$_4$ by amphoteric oxides [J]. Journal of Power Sources, 2008, 180 (1): 597-601.

[46] Zhan C, Lu J, Jeremy K A, et al. Mn (II) deposition on anodes and its effects on capacity fade in spinel lithium Manganate—Carbon systems [J]. Nat Commun, 2013, 4: 2437.

[47] Xia Y, Zhou Y, Yoshio M. Capacity fading on cycling of 4V Li/LiMn$_2$O$_4$ cells [J]. Journal of the Electrochemical Society, 1997, 144 (8): 2593-2600.

[48] Lee J H, Hong J K, Jang D H, et al. Degradation mechanisms in doped spinels of LiM$_{0.05}$Mn$_{1.95}$O$_4$ (M=Li, B, Al, Co, and Ni) for Li secondary batteries [J]. Journal of Power Sources, 2000, 89 (1): 7-14.

[49] Treuil N, Labrug R E C, Menetrier M, et al. Relationship between chemical bonding nature and electrochemical property of LiMn$_2$O$_4$ spinel oxides with various particle sizes: "Electrochemical grafting" concept [J]. The Journal of Physical Chemistry B, 1999, 103 (12): 2100-2106.

[50] Kosova N, Asanov I, Devyatkina E, et al. State of manganese atoms during the mechanochemical synthesis of LiMn$_2$O$_4$ [J]. Journal of Solid State Chemistry, 1999, 146 (1): 184-188.

[51] Gao Y, Richard M, Dahn J. Photoelectron spectroscopy studies of Li$_{1+x}$Mn$_{2-x}$O$_4$ for Li ion battery applications [J]. Journal of Applied Physics, 1996, 80 (7): 4141-4152.

[52] Sun X, Lee H, Yang X, et al. Improved elevated temperature cycling of LiMn$_2$O$_4$ spinel through the use of a composite LiF-based electrolyte [J]. Electrochemical and Solid-State Letters, 2001, 4 (11): A184-A186.

[53] Sun Y, Wang Z, Chen L, et al. Improved electrochemical performances of surface-modified spinel LiMn$_2$O$_4$ for long cycle life lithium-ion batteries [J]. Journal of the Electrochemical Society, 2003, 150 (10): A1294-A1298.

[54] Lee S, Cho Y, Song H K, et al. Carbon-coated single-crystal LiMn$_2$O$_4$ nanoparticle clusters as cathode material for high-energy and high-power lithium-ion batteries [J]. Angewandte Chemie International Edition, 2012, 51 (35): 8748-8752.

[55] Morgan D, Van D V A, Ceder G. Li conductivity in Li$_x$MPO$_4$ (M=Mn, Fe, Co, Ni) olivine materials [J]. Electrochemical and Solid-State Letters, 2004, 7 (2): A30-A32.

[56] Andersson A S, Thomas J O. The source of first-cycle capacity loss in LiFePO$_4$ [J]. Journal of Power Sources, 2001, 97 (8): 498-502.

[57] Srinivasan V, Newman J. Discharge model for the lithium iron-phosphate electrode [J]. Journal of the Electrochemical Society, 2004, 151 (10): A1517-A1529.

[58] Delmas C, Maccario M, Croguennec L, et al. Lithium deintercalation in LiFePO$_4$ nanoparticles via a domino-cascade model [J]. Nature Materials, 2008, 7 (8): 665-671.

[59] Singh G, Burch D, Ceder G, et al. Phase-transformation wave dynamics in LiFePO$_4$ [J]. Solid State Phenomena, 2008, 139: 95-100.

[60] Gu L, Zhu C, Li H, et al. Direct observation of lithium staging in partially delithiated LiFePO$_4$ at atomic resolution [J]. Journal of the American Chemical Society, 2011, 133 (13): 4661-4663.

[61] Suo L M, Han W Z, Lu X, et al. Highly ordered staging structural interface between LiFePO$_4$ and FePO$_4$ [J]. Phys Chem Chem Phys, 2012, 14 (16): 5363-5367.

[62] Malik R, Zhou F, Ceder G. Kinetics of non-equilibrium lithium incorporation in LiFePO$_4$ [J]. Nature Materials, 2011, 10 (8): 587-590.

[63] Sun Y, Lu X, Xiao R J, et al. Kinetically controlled lithium-staging in delithiated LiFePO$_4$ driven by the Fe center mediated interlayer Li-Li interactions [J]. Chemistry of Materials, 2012, 24 (24): 4693-4703.

[64] Gao Jian, Lv Yingchun, Li Hong. Fundamental scientific aspects of lithium batteries (III) —Phase transition and phase diagram [J]. Energy Storage Sciene and Technology, 2013, 2 (3): 250-266.

[65] Chung S Y, Bloking J T, Chiang Y M. Electronically conductive phospho-olivines as lithium storage electrodes [J]. Nature Materials, 2002, 1 (2): 123-128.

[66] Ravet N, Goodenough J B, Besner S, Simoneau M, Hovington P, Armand M. Abstract127 [C] //The Electrochemical Society and the Electrochemical Society of Japan Meeting Abstracts. Honolulu, 1999, 99 (2): 17-22.

[67] Takahashi M, Tobishima S I, Takei K, et al. Reaction behavior of LiFePO$_4$ as a cathode material for rechargeable lithium batteries [J]. Solid State Ionics, 2002, 148 (3): 283-289.

[68] Yamada A, Chung S C, Hinokuma K. Optimized LiFePO$_4$ for lithium battery cathodes [J]. Journal of the Electrochemical Society, 2001, 148 (3): A224-A229.

[69] Herle P S, Ellis B, Coombs N, et al. Nano-network electronic conduction in iron and nickel olivine phosphates [J]. Nature Materials, 2004, 3 (3): 147-152.

[70] Ouyang C, Shi S, Wang Z, et al. First-principles study of Li ion diffusion in LiFePO$_4$ [J]. Physical Review B, 2004, 69 (10): 104303.

[71] Shaju K M, Rao G V S, Chowdari B V R. Performance of layered Li (Ni$_{1/3}$Co$_{1/3}$Mn$_{1/3}$) O$_2$ as cathode for Li-ion batteries [J]. Electrochimica Acta, 2002, 48 (2): 145-151.

[72] Macneil D D, Lu Z, Dahn J R. Structure and electrochemistry of LiNi$_x$Co$_{1-2x}$Mn$_x$O$_2$ ($0 \leqslant x \leqslant 1/2$) [J]. Journal of the Electrochemical Society, 2002, 149 (10): A1332-A1336.

[73] Ohzuku T, Makimura Y. Layered lithium insertion material of LiCo$_{1/3}$Ni$_{1/3}$Mn$_{1/3}$O$_2$ for lithium-ion batteries [J]. Chemistry Letters, 2001, (7): 642-643.

[74] Li D C, Muta T, Zhang L Q, et al. Effect of synthesis method on the electrochemical performance of LiNi$_{1/3}$Mn$_{1/3}$Co$_{1/3}$O$_2$ [J]. Journal of Power Sources, 2004, 132 (1-2): 150-155.

[75] Sun Y, Ouyang C, Wang Z, et al. Effect of Co content on rate performance of LiMn$_{0.5-x}$Co$_{2x}$Ni$_{0.5-x}$O$_2$ cathode materials for lithium-ion batteries [J]. Journal of the Electrochemical Society, 2004, 151 (4): A504-A508.

[76] Ohzuku T, Makimura Y. Layered lithium insertion material of LiNi$_{1/2}$Mn$_{1/2}$O$_2$: A possible alternative to LiCoO$_2$ for advanced lithium-ion batteries [J]. Chemistry Letters, 2001, 30 (8): 744-745.

[77] Yang X Q, Mcbreen J, Yoon W S, et al. Crystal structure changes of LiMn$_{0.5}$Ni$_{0.5}$O$_2$ cathode materials during charge and discharge studied by synchrotron based in situ XRD [J]. Electrochemistry Communications, 2002, 4 (8): 649-654.

[78] Ohzuku T, Brodd R J. An overview of positive-electrode materials for advanced lithium-ion batteries [J]. Journal of Power Sources, 2007, 174 (2): 449-456.

[79] Reed J, Ceder G. Charge, potential, and phase stability of layered Li (Ni$_{0.5}$Mn$_{0.5}$) O$_2$ [J]. Electrochemical and Solid-State Letters, 2002, 5 (7): A145-A148.

[80] Kang K, Meng Y S, Breger J, et al. Electrodes with high power and high capacity for rechargeable lithium batteries [J]. Science, 2006, 311 (5763): 977-980.

[81] Thackeray M M, Kang S H, Johnson C S, et al. Li$_2$MnO$_3$-stabilized LiMO$_2$ (M=Mn, Ni, Co) electrodes for lithium-ion batteries [J]. Journal of Materials Chemistry, 2007, 17 (30): 3112-3125.

[82] Johnson C S, Kim J S, Lefief C, et al. The significance of the Li$_2$MnO$_3$ component in 'composite' xLi$_2$MnO$_3$·$(1-x)$ LiMn$_{0.5}$Ni$_{0.5}$O$_2$ electrodes [J]. Electrochemistry Communications, 2004, 6 (10): 1085-1091.

[83] Yu X, Lyu Y, Gu L, Wu H, et al. Understanding the rate capability of the high energy density Li-rich layered Li$_{1.2}$Ni$_{0.15}$Co$_{0.1}$Mn$_{0.55}$O$_2$ cathode material [J]. Advanced Energy Materials, 2013. doi: 101002/aenm201300950.

[84] Thackeray M M, Kang S H, Johnson C S, et al. Li$_2$MnO$_3$-stabilized LiMO$_2$ (M=Mn, Ni, Co) electrodes for lithium-ion batteries [J]. Journal of Materials Chemistry, 2007, 17 (30): 3112-3125.

[85] Grey C P, Yoon W S, Reed J, et al. Electrochemical activity of Li in the transition-metal sites of O$_3$ Li [Li$_{(1-2x)/3}$Mn$_{(2-x)/3}$Ni$_x$] O$_2$ [J]. Electrochemical and Solid-Sate Letters, 2004, 7 (9): A290-A293.

[86] Kang K, Ceder G. Factors that affect Li mobility in layered lithium transition metal oxides [J]. Physical Review B, 2006, 74 (9): 094105.

[87] Yabuuchi N, Yoshii K, Myung S T, et al. Detailed studies of a high-capacity electrode material for rechargeable batteries, Li$_2$MnO$_3$-LiCo$_{1/3}$Ni$_{1/3}$Mn$_{1/3}$O$_2$ [J]. Journal of the American Chemical Society, 2011, 133 (12): 404-4419.

[88] Armstrong A R, Holzapfel M, Nov K P, et al. Demonstrating oxygen loss and associated structural reorganization in the lithium battery cathode Li [Ni$_{0.2}$Li$_{0.2}$Mn$_{0.6}$] O$_2$ [J]. Journal of the American Chemical Society, 2006, 128 (26): 8694-8698.

[89] Wu Y, Manthiram A. Effect of surface modifications on the layered solid solution cathodes $(1-z)$ Li [Li$_{1/3}$Mn$_{2/3}$] O$_2$-(z) Li [Mn$_{0.5-y}$Ni$_{0.5-y}$Co$_{2y}$] O$_2$ [J]. Solid State Ionics, 2009, 180 (1): 50-56.

[90] Wang R, He X, He L, et al. Atomic structure of Li$_2$MnO$_3$ after partial delithiation and re-lithiation [J]. Advanced Energy Materials, 2013, 3 (10): 1358-1367.

[91] Croy J R, Kim D, Balasubramanian M, et al. Countering the voltage decay in high capacity xLi$_2$MnO$_3$·$(1-x)$ LiMO$_2$ electrodes (M=Mn, Ni, Co) for Li-ion batteries [J]. Journal of the Electrochemical Society, 2012, 159 (6): A781-A790.

[92] Johnson C S, Li N, Lefief C, et al. Synthesis, characterization and electrochemistry of lithium battery electrodes: xLi$_2$MnO$_3$·$(1-x)$ LiMn$_{0.333}$Ni$_{0.333}$Co$_{0.333}$O$_2$ ($0 \leqslant x \leqslant 0.7$) [J]. Chemistry of Materials, 2008, 20 (19): 6095-6106.

[93] Gu M, Belharouak I, Zheng J, et al. Formation of the spinel phase in the layered composite cathode used in Li-ion batteries [J]. ACS Nano, 2013, 7 (1): 760-767.

[94] Gu M, Geng A, Belharouak I, et al. Nanoscale phase separation, cation ordering, and surface oxygen vacancy formation in pristine $Li_{1.2}Ni_{0.2}Mn_{0.6}O_2$ for Li-ion batteries [J]. Chem Mater, 2013, 25 (11): 2319-2326.

[95] Zheng J, Li J, Zhang Z, et al. The effects of TiO_2 coating on the electrochemical performance of Li $[Li_{0.2}Mn_{0.54}Ni_{0.13}Co_{0.13}]O_2$ cathode material for lithium-ion battery [J]. Solid State Ionics, 2008, 179 (27-32): 1794-1799.

[96] Kim J S, Johnson C, Vaughey J, et al. Pre-conditioned layered electrodes for lithium batteries [J]. Journal of Power Sources, 2006, 153 (2): 258-264.

[97] Kang S H, Thackeray M. Stabilization of $xLi_2MnO_3 \cdot (1-x)LiMO_2$ electrode surfaces (M = Mn, Ni, Co) with mildly acidic, fluorinated solutions [J]. Journal of the Electrochemical Society, 2008, 155 (4): A269-A275.

[98] Kumagai N, Kim J M, Tsuruta S, et al. Structural modification of Li $[Li_{0.27}Co_{0.20}Mn_{0.53}]O_2$ by lithium extraction and its electrochemical property as the positive electrode for Li-ion batteries [J]. Electrochimica Acta, 2008, 53 (16): 5287-5293.

[99] Kim J H, Myung S T, Yoon C S, et al. Comparative study of $LiNi_{0.5}Mn_{1.5}O_{4-\delta}$ and $LiNi_{0.5}Mn_{1.5}O_4$ cathodes having two crystallographic structures: Fd-3m and $P4_332$ [J]. Chemistry of Materials, 2004, 16 (5): 906-914.

[100] Santhanam R, Rambabu B. Research progress in high voltage spinel $LiNi_{0.5}Mn_{1.5}O_4$ material [J]. Journal of Power Sources, 2010, 195 (17): 5442-5451.

[101] Liu J, Manthiram A. Understanding the improvement in the electrochemical properties of surface modified 5 V $LiMn_{1.42}Ni_{0.42}Co_{0.16}O_4$ spinel cathodes in lithium-ion cells [J]. Chemistry of Materials, 2009, 21 (8): 1695-1707.

[102] Shaju K M, Bruce P G. Nano-$LiNi_{0.5}Mn_{1.5}O_4$ spinel: A high power electrode for Li-ion batteries [J]. Dalton Transactions, 2008, 40: 5471-5475.

[103] Arrebola J C, Caballero A, Cruz M, et al. Crystallinity control of a nanostructured $LiNi_{0.5}Mn_{1.5}O_4$ spinel via polymer-assisted synthesis: A method for improving its rate capability and performance in 5 V lithium batteries [J]. Adv Funct Mater, 2006, 16 (14): 1904-1912.

[104] Wang L P, Li H, Huang X J, et al. A comparative study of Fd-3m and $P4_332$ "$LiNi_{0.5}Mn_{1.5}O_4$" [J]. Solid State Ionics, 2011, 193 (1): 32-38.

[105] Aklalouch M, Amarilla J M, Rojas R M, et al. Chromium doping as a new approach to improve the cycling performance at high temperature of 5V $LiNi_{0.5}Mn_{1.5}O_4$-based positive electrode [J]. Journal of Power Sources, 2008, 185 (1): 501-511.

[106] Wang H, Tan T A, Yang P, et al. High-rate performances of the Ru-doped spinel $LiNi_{0.5}Mn_{1.5}O_4$: Effects of doping and particle size [J]. The Journal of Physical Chemistry C, 2011, 115 (13): 6102-6110.

[107] Bhaskar A, Bramnik N N, Senyshyn A, et al. Synthesis, characterization and comparison of electrochemical properties of $LiM_{0.5}Mn_{1.5}O_4$ (M = Fe, Co, Ni) at different temperatures [J]. Journal of the Electrochemical Society, 2010, 157 (6): A689-A695.

[108] Bhasksr A, Bramnik N N, Trots D M, et al. In situ synchrotron diffraction study of charge-discharge mechanism of sol-gel synthesized $LiM_{0.5}Mn_{1.5}O_4$ (M = Fe, Co) [J]. Journal of Power Sources, 2012, 217: 464-469.

[109] Oh S M, Oh S W, Yoon C S, et al. High-performance carbon-$LiMnPO_4$ nanocomposite cathode for lithium batteries [J]. Advanced Functional Materials, 2010, 20 (19): 3260-3265.

[110] Wang F, Yang J, Nuli Y, et al. Highly promoted electrochemical performance of 5V $LiCoPO_4$ cathode material by addition of vanadium [J]. Journal of Power Sources, 2010, 195 (19): 6884-6887.

[111] Hu C, Yi H, Fang H, et al. Improving the electrochemical activity of $LiMnPO_4$ via Mn-site co-substitution with Fe and Mg [J]. Electrochemistry Communications, 2010, 12 (12): 1784-1787.

[112] Li G, Azuma H, Tohda M. $LiMnPO_4$ as the cathode for lithium batteries [J]. Electrochemical and Solid-State Letters, 2002, 5 (6): A135-A137.

[113] Wang X, Yu X, Li H, et al. Li-storage in $LiFe_{1/4}Mn_{1/4}Co_{1/4}Ni_{1/4}PO_4$ solid solution [J]. Electrochemistry Communications, 2008, 10 (9): 1347-1350.

[114] Zhang B, Wang X, Li H, et al. Electrochemical performances of $LiFe_{1-x}Mn_xPO_4$ with high Mn content [J]. Journal of Power Sources, 2011, 196 (16): 6992-6996.

[115] Zhang B, Wang X, Liu Z, et al. Enhanced electrochemical performances of carbon coated mesoporous $LiFe_{0.2}Mn_{0.8}PO_4$ [J]. Journal of the Electrochemical Society, 2010, 157 (3): A285-A288.

[116] Yin S C, Grondey H, Strobel P, et al. Electrochemical property: Structure relationships in monoclinic $Li_{3-y}V_2(PO_4)_3$ [J]. Journal of the American Chemical Society, 2003, 125 (34): 10402-10411.

[117] Huang H, Yin S C, Kerr T, et al. Nanostructured composites: A high capacity, fast rate $Li_3V_2(PO_4)_3$/carbon cathode for rechargeable lithium batteries [J]. Advanced Materials, 2002, 14 (21): 1525-1528.

[118] Yin S C, Strobel P, Grondey H, et al. $Li_{2.5}V_2(PO_4)_3$: A room-temperature analogue to the fast-ion conducting high-temperature γ-phase of $Li_3V_2(PO_4)_3$ [J]. Chemistry of Materials, 2004, 16 (8): 1456-1465.

[119] Dai C S, Chen Z Y, Jin H Z, et al. Synthesis and performance of $Li_3(V_{1-x}Mg_x)_2(PO_4)_3$ cathode materials [J]. Journal of Power Sources, 2010, 195 (17): 5775–5779.

[120] Kuang Q, Zhao Y M, An X N, et al. Synthesis and electrochemical properties of Co-doped $Li_3V_2(PO_4)_3$ cathode materials for lithium-ion batteries [J]. Electrochimica Acta, 2010, 55 (5): 1575-1581.

[121] Chen Y H, Zhao Y M, An X N, et al. Preparation and electrochemical performance studies on Cr-doped $Li_3V_2(PO_4)_3$ as cathode materials for lithium-ion batteries [J]. Electrochimica Acta, 2009, 54 (24): 5844-5850.

[122] Ai D J, Liu K Y, Lu Z G, et al. Aluminothermal synthesis and characterization of $Li_3V_{2-x}Al_x(PO_4)_3$ cathode materials for lithium ion batteries [J]. Electrochimica Acta, 2011, 56 (7): 2823-2827.

[123] Li Y, Zhou Z, Ren M, et al. Electrochemical performance of nanocrystalline $Li_3V_2(PO_4)_3$/carbon composite material synthesized by a novel sol-gel method [J]. Electrochimica Acta, 2006, 51 (28): 6498-6502.

[124] Fu P, Zhao Y M, An X N, et al. Structure and electrochemical properties of nanocarbon-coated $Li_3V_2(PO_4)_3$ prepared by sol-gel method [J]. Electrochimica Acta, 2007, 52 (16): 5281-5285.

[125] Arroyo-de D M E, Armand M, Tarascon J M, et al. On-demand design of polyoxianionic cathode materials based on electronegativity correlations: An exploration of the Li_2MSiO_4 system (M=Fe, Mn, Co, Ni) [J]. Electrochemistry Communications, 2006, 8 (8): 1292-1298.

[126] Dominko R, Bele M, Gaberscek M, et al. Structure and electrochemical performance of Li_2MnSiO_4 and Li_2FeSiO_4 as potential Li-battery cathode materials [J]. Electrochemistry Communications, 2006, 8 (2): 217-222.

[127] Nyten A, Abouimrane A, Armand M, et al. Electrochemical performance of Li_2FeSiO_4 as a new Li-battery cathode material [J]. Electrochemistry Communications, 2005, 7 (2): 156-160.

[128] Belharouak I, Abouimrane A, Amine K. Structural and electro-chemical characterization of Li_2MnSiO_4 cathode material [J]. J Phys Chem C, 2009, 113 (48): 20733-20737.

[129] Lyness C, Delobel B, Armstrong A R, et al. The lithium intercalation compound Li_2CoSiO_4 and its behaviour as a positive electrode for lithium batteries [J]. Chemical Communications, 2007, 46: 4890-4892.

[130] Padhi A K, Manivannan V, Goodenough J B. Tuning the position of the redox couples in materials with NASICON structure by anionic substitution [J]. Journal of the Electrochemical Society, 1998, 145 (5): 1518-1520.

[131] Nanjundaswamy K S, Padhi A K, Goodenough J B, et al. Synthesis, redox potential evaluation and electrochemical characteristics of NASICON-related-3D framework compounds [J]. Solid State Ionics, 1996, 92 (1-2): 1-10.

[132] Ati M, Melot B C, Chotard J N, et al. Synthesis and electro-chemical properties of pure $LiFeSO_4F$ in the triplite structure [J]. Electrochemistry Communications, 2011, 13 (11): 1280-1283.

[133] Liu Z J, Huang X J. Structural, electronic and Li diffusion properties of $LiFeSO_4F$ [J]. Solid State Ionics, 2010, 181 (25-26): 1209-1213.

[134] Ben Y M, Lemoigno F, Rousse G, et al. Origin of the 3.6V to 3.9V voltage increase in the $LiFeSO_4F$ cathodes for Li-ion batteries [J]. Energy & Environmental Science, 2012, 5 (11): 9584-9594.

[135] Dong J, Yu X, Sun Y, et al. Triplite $LiFeSO_4F$ as cathode material for Li-ion batteries [J]. Journal of Power Sources, 2013, 244 (15): 716-720.

[136] Dong Y Z, Zhao Y M, Fu P, et al. Phase relations of Li_2O-FeO-B_2O_3 ternary system and electrochemical properties of $LiFeBO_3$ compound [J]. Journal of Alloys and Compounds, 2008, 461 (1-2): 585-590.

[137] Dong Y Z, Zhao Y M, Shi Z D, et al. The structure and electrochemical performance of $LiFeBO_3$ as a novel Li-battery cathode material [J]. Electrochimica Acta, 2008, 53 (5): 2339-2345.

[138] Yamada A, Iwane N, Harada Y, et al. Lithium iron borates as high-capacity battery electrodes [J]. Advanced Materials, 2010, 22 (32): 3583-3587.

[139] Chen L, Zhao Y M, An X N, et al. Structure and electrochemical properties of $LiMnBO_3$ as a new cathode material for lithium-ion batteries [J]. Journal of Alloys and Compounds, 2010, 494 (1-2): 415-419.

[140] Kim J C, Moore C J, Kang B, et al. Synthesis and electrochemical properties of monoclinic $LiMnBO_3$ as a Li in-

tercalation material [J]. Journal of the Electrochemical Society, 2011, 158 (3): A309-A315.

[141] Zu C X, Li H. Thermodynamic analysis on energy densities of batteries [J]. Energy & Environmental Science, 2011, 4 (8): 2614-2624.

[142] Li H, Richter G, Maier J. Reversible formation and decomposition of LiF clusters using transition metal fluorides as precursors and their application in rechargeable Li batteries [J]. Advanced Materials, 2003, 15 (9): 736-739.

[143] Li T, Li L, Cao Y L, et al. Reversible three-electron redox behaviors of FeF_3 nanocrystals as high-capacity cathode-active materials for Li-ion batteries [J]. The Journal of Physical Chemistry C, 2010, 114 (7): 3190-3195.

[144] Poizot P, Laruelle S, Grugeon S, et al. Nano-sized transition-metal oxides as negative-electrode materials for lithium-ion batteries [J]. Nature, 2000, 407 (6803): 496-499.

[145] Li H, Wang Z, Chen L, et al. Research on advanced materials for Li-ion batteries [J]. Advanced Materials, 2009, 21 (45): 4593-4607.

[146] Alt H, Binder H, Khling A, et al. Investigation into the use of quinone compounds-for battery cathodes [J]. Electrochimica Acta, 1972, 17 (5): 873-887.

[147] Liang Y, Tao Z, Chen J. Organic electrode materials for rechargeable lithium batteries [J]. Advanced Energy Materials, 2012, 2 (7): 742-769.

[148] Wang L L, et al. Reviving lithium cobalt oxide-based lithium secondary batteries-toward a higher energy density [J]. Chem Soc Rev, 2018, 47: 6505-6602.

[149] Lu Y X et al. Research and development of advanced battery materials in China [J]. Energy Storage Mater, 2019, 23: 144-153.

[150] Qian J, et al. Electrochemical surface passivation of $LiCoO_2$ particles at ultrahigh voltage and its applications in lithium-based batteries [J]. Nat Commun, 2018, 9: 4918.

[151] Liu Q, et al. Approaching the capacity limit of lithium cobalt oxide in lithium ion batteries via lanthanum and aluminium doping [J]. Nat Energy, 2018, 3: 936-943.

[152] Zhang J N, et al. Trace doping of multiple elements enables stable battery cycling of $LiCoO_2$ at 4.6 V [J]. Nat Energy, 2019, 4: 594-603.

[153] Hu E, et al. Evolution of redox couples in Li-and Mn-rich cathode materials and mitigation of voltage fade by reducing oxygen release [J]. Nat Energy, 2018, 3: 690-698.

[154] Kim U H, et al. Quaternary Layered Ni-Rich NCMA Cathode for Lithium-Ion Batteries [J]. ACS Energy Lett, 2019, 4: 576-582.

[155] Urban A, et al. The configurational space of rocksalt-type oxides for high capacity lithium battery electrodes [J]. Advanced Energy Materials, 2014, 4: 1400478.

[156] Lee J, et al. Unlocking the potential of cation-disordered oxides for rechargeable lithium batteries [J]. Science, 2014, 343: 519-522.

[157] Yabuuchi N, et al. Organic electrode materials for rechargeable lithium batteries [J]. Nature Communications, 2016, 7: 13814.

[158] Zhao E, et al. Structural and mechanistic revelations on high capacity cation-disordered Li-rich oxides for rechargeable Li-ion batteries [J]. Energy Storage Materials, 2019, 16: 354-363.

[159] Maxwell D R, et al. Manganese oxidation as the origin of the anomalous capacity of Mn-containing Li-excess cathode materials [J]. Nat Energy, 2019, 4: 639-646.

[160] Li M, et al. 30 Years of lithium-ion batteries [J]. Advanced Materials, 2018, 30: 1800561.

[161] Li W, et al. High-voltage positive electrode materials for lithium-ion batteries [J]. Chem Soc Rev, 2017, 46: 3006-3059.

[162] Myung S T, et al. Nickel-rich layered cathode materials for automotive lithium-ion batteries: achievements and perspectives [J]. ACS Energy Lett, 2017, 2: 196-223.

[163] Radin M D, et al. Narrowing the gap between theoretical and practical capacities in Li-ion layered oxide cathode materials [J]. Advanced Energy Materials, 2017, 7: 1602888.

第 7 章

负极材料

1972 年，Armand 等[1]提出了摇椅式电池（rocking chair battery）的概念，正负极材料采用嵌入化合物（intercalation compounds），在充放电过程中，Li^+ 在正负极之间来回穿梭。寻找适合这一概念的正负极材料经历了较长的时间。1981 年，Goodenough 等提出了 $LiMO_2$（M=Co、Ni、Mn）化合物用于正极材料，这些材料均为层状结构化合物，能够可逆地嵌入和脱出 Li。1981 年后大部分有关负极材料的研究主要集中在含 Li 源负极，如 LiAl 合金[2]、LiC 合金[3]、$Li_xMo_6Se_6$[4]、$LiWO_2$[5]、$Li_6Fe_2O_3$[6]等，这些材料价格高、能量密度低、循环性能不稳定、难以实用化。

石墨也具有层状结构，早在 20 世纪 50 年代就已经合成了 Li 的石墨嵌入化合物[7]。1970 年，Dey 等[8]发现 Li 可以通过电化学方法在有机电解质溶液中嵌入石墨，1983 年法国 INPG 实验室[3]第一次在电化学电池中成功地实现了 Li 在石墨中的可逆脱嵌。20 世纪 80 年代世界各地尤其在日本开展了碳负极材料的广泛研究。1989 年，日本 Sony 公司的研究人员[9]终于寻找到了合适的正负极材料、电解质材料的组合，申请了以 $LiCoO_2$ 作 Li 源正极、石油焦作负极、$LiPF_6$ 溶于丙烯碳酸酯（PC）和乙烯碳酸酯（EC）作电解液的二次锂电池体系的专利，并在 1991 年开始商业化生产[10]。1993 年后，商品化的锂离子电池开始采用性能稳定的人造石墨（如中间相碳微球 MCMB、改性天然石墨）为负极材料[11-13]。

由于这一可充放锂电池体系不含金属 Li，日本学者西美绪（Nichi）等就把此类摇椅式电池称为锂离子电池（lithium ion battery），这种方便易懂的提法最终被学术界和产业界接受。自从 Sony 公司商业化锂离子电池以来[14]，锂离子电池产业迅猛发展。目前，主要应用于手机、笔记本电脑、摄像机等便携式设备[15]，同时还涉及太阳能和风力发电储能[16]、航空航天、军事、医疗等方面。目前，锂离子电池正向电动汽车领域以及大规模工业储能系统这两个重要的新兴领域发展[17-20]。

与正极材料一样，负极材料在锂离子电池的发展中也起着关键的作用。近年来，为了使锂离子电池具有较高的能量密度、功率密度，较好的循环性能以及可靠的安全性能，负极材料作为锂离子电池的关键组成部分受到了广泛的关注。对负极材料的选择应满足以下条件[21]：①嵌脱 Li^+ 反应具有低的氧化还原电位，以满足锂离子电池具有较高的输出电压；②Li^+ 嵌入脱出的过程中，电极电位变化较小，这样有利于电池获得稳定的工作电压；③可逆容量大，以满足锂离子电池具有高的能量密度；④脱嵌 Li^+ 过程中结构稳定性好，以使电池具有较高的循环寿命；⑤嵌入 Li^+ 电位如果在 1.2V vs. Li^+/Li 以下，负极表面应能生成致密稳定的固体电解质膜（SEI），从而防止电解质在负极表面持续还原，不断消耗来自正极的 Li^+；⑥具有比较低的 e^- 和 Li^+ 的输运阻抗，以获得较高的充放电倍率和低温充放电性能；⑦充放电后材料的化学稳定性好，提高电池的安全性、循环性，降低自放电率；⑧环境友好，制造过程及电池废弃的过程不对环境造成严重污染和毒害；⑨制备工艺简单，易于规模化，制造和使用成本低；⑩资源丰富。

迄今为止，石墨类碳负极材料是能同时满足以上要求的综合性能最好的负极材料，用途最为广泛。开发新型负极材料面临的最大挑战是需要根据应用需求寻找具有某项或多项突出优点，同时还能兼顾其他综合性能的材料，而材料能否在电池中获得应用取决于该材料最差的某项性能是否满足应用的最低要求，这是典型的如图 7-1 所示的"木桶效应"。由于这些相互制约的要求，过去 20 多年，尽管数千种以上的负极材料获得了研究，但能够最终获得商业应用的负极材料种类实际上非常少。

图 7-1 电极材料开发过程中需要考虑的因素

7.1 典型的锂离子电池负极材料

目前，商业上广泛使用的锂离子电池负极材料主要分为以下两类：①六方或菱形层状结构的人造石墨和天然改性石墨；②立方尖晶石结构的 $Li_4Ti_5O_{12}$。其晶体结构如图 7-2 所示，结构参数、Li 扩散系数及理论容量等见表 7-1。

表 7-1 商业化锂离子电池负极材料及其性能

项目	石墨[22]	钛酸锂[23]
化学式	C	$Li_4Ti_5O_{12}$
结构	层状	立方尖晶石
空间点群	$P6_3/mmc$（或 $R3m$）	$Fd\text{-}3m$

续表

项目	石墨[22]	钛酸锂[23]
晶胞参数	$a=b=0.2461$nm,$c=0.6708$nm;$\alpha=\beta=90°$,$\gamma=120°$(或$a=b=c$,$\alpha=\beta=\gamma\neq 90°$)	$a=b=c=0.8359$nm;$\alpha=\beta=\gamma=90°$
理论密度/(g/cm³)	2.25	3.5
振实密度/(g/cm³)	1.2~1.4	1.1~1.6
压实密度/(g/cm³)	1.5~1.8	1.7~3
理论容量/(mA·h/g)	372	175
实际容量/(mA·h/g)	290~360	约165
电压(vs. Li/Li⁺)/V	0.01~0.2	1.4~1.6
体积变化/%	12	1
表观化学扩散系数/(cm²/s)	10^{-10}~10^{-11}	10^{-8}~10^{-9}
完全嵌锂化合物	LiC_6	$Li_7Ti_5O_{12}$
循环性/次	500~3000	10000(10C,90%)
环保性	无毒	无毒
安全性能	好	很好
适用温度/℃	-20~55	-20~55
价格/(万元/t)	3~14	14~16
主要应用领域	便携式电子产品、动力电池、规模储能	动力电池及大规模储能

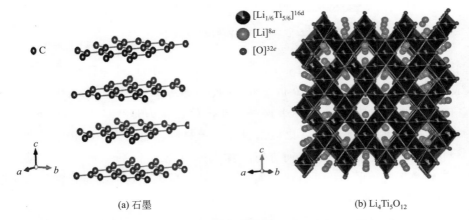

(a) 石墨　　　　(b) $Li_4Ti_5O_{12}$

图 7-2　商业化广泛使用的锂离子电池负极材料的结构

7.1.1　层状石墨类负极材料

石墨具有两种晶体构成，一种是六方石墨，空间点群为 $P6_3/mmc$，$a=b=0.2461$nm，$c=0.6708$nm，$\alpha=\beta=90°$，$\gamma=120°$，碳原子层以 ABAB 方式排列；另一种是菱形石墨，空间群为 $R3m$，$a=b=c$，$\alpha=\beta=\gamma\neq 90°$，碳原子层以 ABCABC 方式排列[22]。石墨中的碳原子是 sp^2 杂化，层与层之间通过范德华力结合，层内原子通过共价键结合，嵌入的 Li 插在石墨层间可以形成不同的"阶"结构。在这里，"阶"的定义为相邻的两个嵌入 Li 原子层之间所间隔的石墨层的个数，如"1 阶"，意味着相邻的两个 Li 嵌入层之间只有一个石墨层也即-Li-C-Li-的顺序，石墨"阶"结构示意图如图 7-3 所示。通过化学合成的方法，Li 与石墨可以形成一系列的插层化合物，如 LiC_{24}、LiC_{18}、LiC_9、LiC_6 等[24]，通常称为石墨层间化合物（graphite intercalated compound，GIC）。1991 年，Dahn[25]研究了有机电解液体系 Li 在石墨中嵌入过程的碳结构变化，形成了一系列插层化合物，随着 Li 的嵌入量增加，Li-GIC 按 4 阶→3 阶→稀释的 2 阶→2 阶→1 阶的顺序发生相变，最后形成了 1 阶的插层化合物 LiC_6，对应石墨的理论容量为 372mA·h/g。"阶结构"之间转换的热力学、动力学及不同"阶结构"之间演化的原子尺度的图像目前仍然不是很清楚，值得深入研究。LiC_6 中的 Li 并不是石墨材料中含 Li 最高的，在高温和高压下 Li 与 HOPG（高定向裂解石墨）反应可以生成 LiC_2，其中 Li 的体积浓度甚至超过了金属 Li[26]。

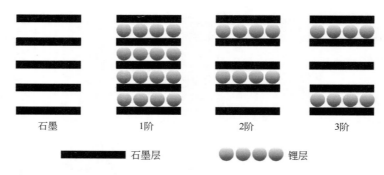

图 7-3 石墨"阶结构"示意图

中间相碳微球（mesophase carbon microbeads，MCMB）是一种重要的人造石墨材料。MCMB 最早出现可以追溯到 20 世纪 60 年代，研究人员在研究煤焦化沥青中发现一些光学各向异性的小球体[27]，实际上这些小球体就被认为是 MCMB 的雏形。1973 年，Yamada 等[28]从中间相沥青中制备出微米级球形碳材料，命名为中间相碳微球，之后引起了碳材料研究者极大的兴趣，进而进行了深入的研究[29]。1993 年，大阪煤气公司将 MCMB 用于锂离子电池的负极并且成功实现产业化。后来，我国上海杉杉和天津铁城等单位相继研发成功并产业化[30]。

MCMB 电化学性能优越的主要原因是颗粒的外表面均为石墨结构的边缘面，反应活性均匀，易于形成稳定的 SEI 膜，有利于 Li^+ 的嵌入脱出。MCMB 的制造成本相对较高，研究人员尝试对天然石墨类材料进行改性以降低负极材料成本。天然石墨颗粒存在的主要问题是外表面反应活性不均匀，晶粒粒度较大，循环过程中表面晶体结构容易被破坏，表面 SEI 膜覆盖不均匀，初始库仑效率低，倍率性能不好。为了解决这一问题，采用了多种方法对天然石墨进行改性，包括颗粒球形化、表面氧化（包括氟化）、表面包覆软碳或硬碳材料以及其他表面修饰策略等[31-36]。改性后天然石墨的电化学性能有了较大的提高，首次效率可以达到 90%～93%，100%DOD 循环寿命达到 500 次，可以基本满足消费电子产品对电池性能的要求。

目前，电动汽车领域对下一代锂离子电池的能量密度、功率密度、寿命提出了更高的要求，人们对纳米孔、微米孔石墨和多面体石墨进一步开展了研究，以期解决锂离子电池高功率的需求[37]。

从储量上看，我国是世界上石墨储量最丰富的国家，晶质石墨储量3068万吨，石墨储量占世界的70%以上[38]。从锂离子电池负极材料产量上看，人造石墨（38%）与天然石墨负极材料（59%）占据了锂离子电池负极材料全球市场的97%。目前世界范围内，石墨负极材料产量最大的企业是日本日立化成有限公司（Hitachi Chemical）与贝特瑞新能源材料股份有限公司（BTR New Energy），较大的企业有上海杉杉科技有限公司、日本吴羽化工（Kureha）、日本炭黑（Nippon Carbon Co. Ltd.）、日本JFE化学、湖南摩根海容新材料股份有限公司等。采用石墨材料的锂离子电池主要应用领域为便携式电子产品，改性石墨已开始在动力电池与储能电池中应用。

7.1.2 立方尖晶石结构 $Li_4Ti_5O_{12}$ 负极材料

在锂离子电池负极材料研究中，另外一个受到重视并且已经进入市场的负极材料是Jonker等[39]于1956年提出的具有尖晶石结构的 $Li_4Ti_5O_{12}$ 负极材料。之后，1983年Murphy等[40]首先对这种材料的嵌Li性能进行了报道，但是当时没有引起足够重视。1994年Ferg等[41]研究了其作为锂离子电池的负极材料，Ohzuku等[23]随后对 $Li_4Ti_5O_{12}$ 在锂离子电池中的应用进行了系统研究，强调其零应变的特点。

高纯 $Li_4Ti_5O_{12}$ 呈白色，密度为 $3.5g/cm^3$，其空间群属于 $Fd\text{-}3m$，为半导体材料（能带宽度为2eV），室温下电子电导率为 $10^{-9}S/m$。$Li_4Ti_5O_{12}$ 的结构式可表示为 $(Li)^{8a}(Li_{1/3}Ti_{5/3})^{16d}(O_4)^{32e}$，每个晶体单胞中含有8个 $(Li)^{8a}(Li_{1/3}Ti_{5/3})^{16d}(O_4)^{32e}$ 分子。$(Li)^{8a}(Li_{1/3}Ti_{5/3})^{16d}(O_4)^{32e}$ 在共面的 $8a$ 四面体位置和 $16c$ 八面体位置的三维间隙空间为 Li^+ 扩散提供了通道[42]。$Li(Li_{1/3}Ti_{5/3})O_4$ 晶体结构中的 $16c$ 位置嵌入一个Li，60%的 Ti^{4+} 还原成 Ti^{3+}，理论嵌Li容量为 $175mA \cdot h/g$；如果全部的 Ti^{4+} 还原成 Ti^{3+}，理论容量可达到 $291.7mA \cdot h/g$[43]。一般认为电解质在 $1.2V$ 以下发生分解，而材料的嵌Li相变电位在 $1.55V$ 附近，因而认为在此电位区间没有SEI膜的生长[44]。其初次循环的库仑效率可达到98.8%。$Li_4Ti_5O_{12}$ 作为锂离子电池负极材料，Li嵌入脱出前后材料的体积变化不到1%，是较为少见的零应变材料[23]，有利于电池以及电极材料结构的稳定，能够实现长的循环寿命。

$Li(Li_{1/3}Ti_{5/3})O_4$ 理论上能嵌入一个 Li^+，普遍认为 Li^+ 嵌入晶格后晶体由尖晶石结构转变成岩盐结构[23]，这个过程可以表示为 $(Li)^{8a}(Li_{1/3}Ti_{5/3})^{16d}(O_4)^{32e} + e^- + xLi^+ \longrightarrow (Li_{1+x})^{8a/16c}(Li_{1/3}Ti_{5/3})^{16d}(O_4)^{32e} \longrightarrow (Li_2)^{16c}(Li_{1/3}Ti_{5/3})^{16d}(O_4)^{32e}$，材料在嵌入一个单位 Li^+ 后八面体位置就没有空余的空间了，这一机理被普遍接受。非原位XRD、高角衍射XRD[45]、Raman光谱[44]也证实了这种机理。理论上完全嵌Li后的结构式为 $Li_2(Li_{1/3}Ti_{5/3})O_4$，该化合物为淡紫色或深蓝色[44]。在 $Li_{1+x}(Li_{1/3}Ti_{5/3})$

O_4 中只要有 Li^+ 嵌入就会有钛离子从 4 价转变成 3 价,形成的 Ti^{4+}/Ti^{3+} 体系有利于电子传递,使材料的电子导电性变好。当 x 达到 1 时,$Li_{1+x}(Li_{1/3}Ti_{5/3})O_4$ 电子导电性最好,可以达到 10^{-2} S/m。

目前,$Li_4Ti_5O_{12}$ 的合成方法主要有固相反应法[45-49]和溶胶-凝胶法[50-52]。溶胶-凝胶法所得到的材料粒径较小,但制备工艺较复杂。此外,还有微波化学法[53]、熔融浸渍法[54]、水热反应法[55,56]等也可以合成制备 $Li_4Ti_5O_{12}$。针对 $Li_4Ti_5O_{12}$ 低的室温电子电导率,大倍率时容量衰减严重,离子掺杂[57-59]、减小颗粒尺寸[60]、金属纳米颗粒包覆[61]、碳包覆[62,63]等方法被用来改善 $Li_4Ti_5O_{12}$ 的倍率性能。

$Li_4Ti_5O_{12}$ 在应用中面临的一个问题是使用时嵌锂态的 $Li_7Ti_5O_{12}$ 与电解液发生化学反应导致胀气[64],特别是在较高温度下。胀气会引起锂离子电池容量衰减、寿命缩短、安全性下降。为了解决 $Li_4Ti_5O_{12}$ 的胀气问题和对材料的电化学性能进行改善,科研者先后提出多种方法对其进行改性:①严格控制材料及电池中的水含量;②控制 $Li_4Ti_5O_{12}$ 中的杂质、杂相含量;③通过掺杂、表面修饰降低表面的反应活性、降低材料的电阻;④优化电池化成工艺;⑤控制 $Li_4Ti_5O_{12}$ 的一次颗粒与二次颗粒大小。

虽然 $Li_4Ti_5O_{12}$ 工作电压较高,但是由于循环性能和倍率性能特别优异,相对于碳材料而言具有安全性方面的优势,因此这种材料在动力型和储能型锂离子电池方面存在着不可替代的应用需求。目前,世界范围内,$Li_4Ti_5O_{12}$ 产量较大的国外企业为日本富士钛工业公司(Fuji Titan)和美国阿尔泰纳米技术公司(Altair Nanomaterials),国内产能较大的企业为贝特瑞新能源材料股份有限公司(BTR New Energy)、珠海银通新能源有限公司以及四川兴能新材料有限公司。

7.2 小批量应用的负极材料

7.2.1 硬碳负极材料

除了石墨以外,碳材料中的硬碳、软碳也是两类很重要的负极材料,不同的是硬碳和软碳的结晶度低,片层结构度没有石墨规整有序,如图 7-4 所示。其中硬碳是难以石墨化的碳,是高分子聚合物的热解碳的无定形结构,即使在 2500℃也不能完全石墨化。Sony 公司在 1991 年首次用聚糠醇(PFA)热解得到的硬碳作为负极材料使用[65]。硬碳表现在 $0\sim1.5V$(vs. Li^+/Li)有比较高的容量(200~600mA·h/g),电压平台曲线由两部分组成:第一部分为斜坡,电压范围位于 0.1~1.0V,容量约为 150~250mA·h/g;另一部分为平台,这个平台表现的容量为 100~400mA·h/g。

常见的硬碳原料有树脂碳(如酚醛树脂)和有机聚合物热解碳(如PVDF)等。其中,以 PFA-C 作原料得到的硬碳容量可达 400mA·h/g,以聚苯酚为原料得到的硬碳容量为 580mA·h/g[66]。Wang 等[67]通过蔗糖溶液的水热脱水成球以及碳化,制备了具有完美的外观、光滑的表面和单分散颗粒分布的含微孔的硬碳球(HCS)材料。在 1000℃下处理得到的 HCS 材料的容量可达 500mA·h/g 以上，首次充放电效率超过 80%，循环性能优异，且倍率特性优异。在 HCS 表面沉积一层石墨化程度较高的 C_2H_2 热解碳后再包覆一层 SiO_x 化合物，材料的充放电效率可达到 83% 以上。将纳米 SnSb 合金颗粒钉扎在 HCS 的表面，可以有效缓解充放电时合金颗粒之间的团聚，循环性得到提高[68]。Hu 等[69]发现，利用微乳液作为媒介，通过晶体生长水热法制备的含微孔的硬碳球具有更小的微孔，显示了很好的循环性能，首周效率为 83.2%，嵌 Li 容量为 566mA·h/g。Fey 等[70]用稻壳热裂解的方法得到的硬碳负极材料具有迄今为止报道的硬碳负极材料中最高的容量，其可逆容量可达 1055mA·h/g。

图 7-4 石墨、软碳和硬碳的结构模型

针对无定形碳材料，提出了多种储锂机理，主要有锂分子机理、多层锂机理、晶格点阵机理、弹性球-弹性网模型、层-边端-表面机理、纳米级石墨储锂机理、碳-锂-氢机理等[65]。虽然硬碳材料具有很多优点，如循环性能好、比容量高等，但是首周效率过低、电位滞后、低电位储锂的倍率性能较差等缺点都影响了硬碳的应用。人们可以通过包覆掺杂的方法改善硬碳的电化学性能。Sony 公司[71]在聚糠醇树脂中掺杂磷化物从而提高了可逆容量。Schonfelder 等[72]发现磷掺杂可以降低硬碳中的微孔数目，增加微孔的大小，降低层间距离，得到的硬碳首周充放电效率为 83%，可逆容量为 550mA·h/g。Buiel 等[73]在硬碳表面沉积裂解的乙烯降低不可逆容量。

从目前的实际应用看，硬碳中低电位储存的锂由于较差的倍率性能和锂析出问题基本不能利用，硬碳材料斜坡段储锂的倍率性能较好，因此硬碳材料更适合高功率的动力锂电池和混合动力汽车。本田（Honda）采用硬碳为负极材料推出 HEV（混合动力汽车）。日产采用硬碳材料的 HEV，输出密度为 3550～4000W/kg，比镍氢电池的 1200W/kg 提高了约 3 倍。我国的杉杉科技等企业也在积极开发用于 HEV 的硬碳负极材料。

7.2.2 软碳负极材料

相比于硬碳的难石墨化，软碳是易石墨化碳，是指在高温 2500℃ 以下可以石墨化的无定形碳。常见的软碳主要有石油焦、碳纤维、针状焦等。1991 年 Sony 推出的第一代锂离子电池负极材料就是石油焦。软碳的晶面间距（d_{002}）比硬碳小，但与石墨相比仍较大。Rosamaria 等[74]认为由于焦炭是无定形碳，结构是乱层排列，首周具有较大的不可逆容量，库仑效率比较低。Stevens 等[75]发现 Li 和 Na 可以嵌在无序的软碳结构中，造成层间距变大，产生一个电压斜坡，没有平台产生。Bonino 等[76]用六苯基苯做前驱体制备软碳，发现随着球磨时间的增长软碳的循环性和容量都衰减。Endo 等[11]发现球磨后的中间相沥青基碳纤维具有良好的充电容量和循环效率。Guo 等[77]发现用 2200℃石墨化处理的纳米碳纤维材料作为负极的循环性能很好，但首周不可逆容量很大。

软碳对电解液的适应性较强，耐过充、过放性能良好，容量比较高并且循环性能比较好。但其充放电电位曲线上无平台，在 0～1.2V 内呈斜坡式，造成平均对锂电位较高，为 1V 左右，因此造成电池端电压较低，限制了电池的能量密度[65]，而且由于插 Li 时，碳质材料会发生体积膨胀，缩减电池寿命。因此需对其进行改性处理，如 Kawai 等[78]发现用中间相沥青碳修饰的针状焦具有较高的容量（400mA·h/g）和很好的循环性，比没有修饰的针状焦的容量提高了 90mA·h/g。由于避免了石墨化处理，软碳负极材料成本较低，循环性能提升至 1500～3000 次后，预计可应用于储能电池、HEV、EV、电动工具电池等。

7.2.3 高容量硅负极材料

在目前的锂离子电池体系中，尽管商业化的石墨类材料容量是现有正极材料容量的两倍，通过模拟，在负极材料容量不超过 1200mA·h/g 的情况下，提高现有负极材料的容量对整个电池的能量密度仍然有较大贡献[79]。在电池的生产和制造过程中，负极材料的成本占到总材料成本的 10%左右。制备成本低廉的同时兼具高容量的负极材料是目前锂离子电池研究的热点。硅材料因其理论容量（4200mA·h/g）高、环境友好、储量丰富等特点而被考虑作为下一代高能量密度锂离子电池的负极材料。

早在 20 世纪 70 年代发展 Li/FeS_2 高温熔融盐锂热电池时，就开始采用 Li-Si 合金来替代金属 Li 作为负极材料以解决金属 Li 熔点较低、在电池中会与隔膜材料和电极材料发生反应的问题，Lai[80]和 Seefurth 等[81-83]以及 Wen 等[84]研究了一系列 Li-Si 化合物（$Li_{22}Si_5$、$Li_{15}Si_4$、$Li_{21}Si_8$、Li_2Si）作为其中的负极材料，在 650～780℃高温区间，这些 Li-Si 合金可以在 50～400mA/cm^2 的电流密度下可逆地充放电，利用率为 60%～90%。1981 年，Huggins 等[85]提出全固态复合微结构电极，即采用活性合金与非活性合金复合材料来提高活性物质的利用率，如将活性的 Li_xSi（$0<x<1.71$）与 $Li_{2.6}Sn$ 组成复合材料。1999 年，Huggins 等[86]和 Gao 等[87]对 Si 负极在室温下的电化学性能进行了报道，发现在首次放电过程中，并没有出现平台，在 0.1V 区间内表现为两相曲线特点，同时发现可逆性非常差。同年，Li 等[88]在国际上首次报道了采用纳米尺寸的 Si 颗粒可以有效地提高循环性能，首次效率可达 76%，第 10 周容量还能保持在 1700mA·h/g，同时利用 Si 纳米线研究了室温下 Li^+ 嵌入 Si 中的反应机理。研究发现随着 Li^+ 的嵌入，Si 的晶体结构逐渐被破坏形成非晶态的 Li-Si 合金[89]。2003 年，Limthongkul 等[90]再次证实了 Li^+ 嵌入导致晶态 Si 形成非晶相，并称为电化学驱动无定形化。2004 年，Dahn 等[91]通过原位和非原位的 XRD 手段对 Si 在充放电过程中的结构变化进行了定量研究，他们也发现了 Li^+ 嵌入导致晶态 Si 非晶化的过程，进一步研究还发现，当放电到 50mV 时原来长程无序的 Li-Si 合金相很快结晶形成了 $Li_{15}Si_4$ 的晶态相，具体反应机理表述如下。

首次放电

$$Si(crystalline) + xLi^+ + xe^- \longrightarrow Li_xSi(amorphous) + (3.75-x)Li^+ +$$
$$(3.75-x)e^-(1) \longrightarrow Li_{15}Si_4(crystalline) \tag{7-1}$$

首次充电

$$Li_{15}Si_4(crystalline) \longrightarrow Si(amorphous) + yLi^+ + ye^- + Li_{15}Si_4(residual) \tag{7-2}$$

该研究结果也被 Grey 等[92,93]和 Huang 等分别用核磁共振（NMR）和原位透射电子显微镜（TEM）的手段所证实。Si 负极材料在储锂过程中存在较大的体积变化，导致活性物质从导电网络中脱落，并导致 Si 颗粒产生裂纹粉化，从而严重地影响了 Si 负极的循环性能。表 7-2 为不同 Li-Si 合金化合物的晶胞参数，当 Li^+ 完全嵌入生成 $Li_{22}Si_5$ 时，其理论体积变化达到 320%[85]。

表 7-2　硅以及硅锂合金化合物的晶胞参数和对应的储锂比容量

硅的不同嵌锂状态	体积/Å³	理论比容量/(mA·h/g)
Si	19.6	0
LiSi	31.4	954
$Li_{12}Si_7$	43.5	1635
Li_2Si	51.5	1900
$Li_{13}Si_4$	67.3	3100
$Li_{15}Si_4$	76.4	3590
$Li_{22}Si_5$	82.4	4200

注：1Å=0.1nm。

上述结果根据晶体结构计算而得，由于 Li^+ 嵌入 Si 会导致无定形化，为了进一步定量了解 Si 在充放电过程中的体积变化，2003 年，Beaulieu 等[94]利用原子力显微镜（AFM）对无定形 Si 微米柱子在充放电过程中的体积变化进行了原位研究。实验结果显示，Si 柱子在 Li^+ 嵌入过程中在水平方向并没有明显的变化，而在垂直方向上发生约 300% 的变化，整体的体积变化也约为 300%，并且其体积变化与嵌锂量呈线性关系。这种线性关系也被 Notten 等[95]利用光学曝光和反应离子刻蚀的手段制备蜂窝状的 Si 薄膜电极所证实。2011 年，Choi 等[96]和 Lee 等[97]利用反应离子刻蚀的方法以 SiO_2 颗粒为掩膜对单晶 Si 衬底进行了刻蚀，得到了具有（１００）、（１１０）、（１１１）三种晶面取向的纳米柱子，这 3 种不同取向纳米柱子在嵌 Li^+ 之后其截面分别呈现出十字、椭圆、六角形的结构，他们认为造成这种不规则体积膨胀的原因主要是由于晶态 Si 中 Li^+ 沿（110）方向的扩散比较快，导致了在该方向上优先膨胀。2011 年，He 等[98-100]进一步定量地研究证实了 Li^+ 嵌入 Si 的线性各向异性体积变化，研究还发现，图形化的薄膜电极可以在一定程度上抑制裂纹的产生，同时通过氧化铝的包覆也有助于改善电极的循环性能。裂纹的出现一般发生在 Li^+ 脱出的情况下，并且这个过程伴随着材料的相变和 Li^+ 的传输，这导致材料在体积发生变化的同时，其本身的力学性质也在发生变化。因此如何改进 Si 负极材料的抗裂强度，抑制或缓解其在循环过程中的裂纹成为了目前 Si 负极材料的研究热点和难点。Wang等[101]通过对比非晶硅薄膜和Ti-Si复合薄膜断裂后的形貌，统计了断裂后裂块的大小、角度和裂纹宽度等信息，认为通过Ti的1~2nm尺度的复合可以有效地改善Si材料的力学性质。

另一个阻碍Si负极材料商业化应用的主要原因是不稳定的SEI膜。由于商用电解液电化学窗口的限制，对于放电电压小于1.2V（vs. Li$^+$/Li）的负极材料，材料表面在放电时能否形成稳定的SEI膜是这种材料能否被广泛应用的关键。对于Si负极材料，由于其放电电压低，且在循环过程中伴随着巨大的体积膨胀而导致裂纹，从而新鲜的Si表面暴露在电解液中将会持续产生SEI，因此如何在Si负极表面形成稳定有效、能适应硅负极充放电过程中体积膨胀收缩的SEI膜是Si负极研究的难点。2007年，Notten等[102,103]观察到Si薄膜电极经过长时间循环后生成了很厚的SEI膜。2009年，Chan等[104]用X射线光电子能谱（XPS）的手段对Si纳米线在1mol/L LiPF$_6$（EC：DMC=1：1）的电解液体系中对表面SEI膜成分进行了仔细的研究。结果显示，Si电极的SEI膜在靠近电解液的表面主要由碳氢化合物等有机物组成，而在靠近电极表面主要是碳酸锂、氟化锂、氧化锂等无机物。2013年，Zheng等[105]利用原子力显微镜力曲线模式研究了Si薄膜在含VC添加剂的电解液体系中首周循环过程不同状态的SEI膜的结构、厚度、杨氏模量以及覆盖度的情况。研究发现，Si负极表面的SEI膜存在单层、双层及多层结构，SEI膜厚度从几纳米至几十纳米不等，在Si薄膜表面呈不均匀分布，覆盖度随放电变大，而充电过程覆盖度减小。SEI膜的厚度、覆盖度与电解质添加剂密切相关。

值得关注的是Si负极材料的体积形变和不稳定的SEI膜是Si的本征问题，因此要进一步提高Si负极材料的循环性能、库仑效率等就必须解决Si负极材料在体积变化过程中产生的与导电网络脱离、裂纹和不能与现有电解液体系形成稳定SEI膜以及SEI膜持续生长等问题。目前，很多研究小组努力尝试从减小颗粒尺寸[88,106-113]、表面修饰[111,114,115]、形貌和结构设计[116-118]、SEI膜调控[119]、电解液和添加剂[120-125]、黏结剂（Alg[126]、Alg-C[127]、CMC[128-131]、CMC-SBR[132]、PVA[133]、PAA[134]、PAA-BP[135]、PAA-PCD[136]、PAA-CMC[137]、PEI[138,139]、PEO[140]、PI[141]、PVDF[142]等）、集流体[143-148]等方面来改善Si负极材料的性能。

研究发现减小Si颗粒尺寸在电极中发挥着重要的作用，相对于微米Si颗粒能够有效地改善循环性[88]；Si颗粒中空隙等微孔的存在[149,150]，可以缓解充放电过程中Si体积变化所带来的负效应；Si-M（M=C、Ni、Fe、Ti、Cu）等复合材料以及SiO$_x$和Si$_3$N$_4$等材料被认为有助于缓解体积膨胀[151,152]；包覆固态电解质作为人工SEI膜可以稳定存在并且有效提高了Si电极的循环性能[102,103]；黏结剂会直接影响Si负极材料的性能，尤其是多功能黏结剂，这种影响更为明显，特别是含有羧基基团和具有导电性能的黏结剂[127,128,134]，如海藻酸钠[126]作为锂离子电池负极材料的黏结剂，纳米Si材料表面有大量羟基，而这类黏结剂聚合物链上的羧基可以和羟基发生酯化作用，进而增加黏结性能，稳定电极结构，同时海藻酸钠在电解液中溶胀小，这就意味着除了直接暴露在电解液中极片表面层的Si之外，只有极少电解液才能到达Si颗粒的表面生成SEI膜，在一定程度上能够调控SEI膜，循环过程中有效地提高了库仑效率；导电的黏结剂（如PFFOMB聚合物[153]）可以缓解Si负极材料体积形变带来的应力，同时可以提高整个电极的电子电导率，提高循环性能。

目前，尽管 Si 负极因其高容量而引起了广泛的关注，研究人员也从多个方面对其进行改性，尤其纳米材料（纳米颗粒[88]、纳米线[154]）对循环的改善有明显的作用，但是纳米材料的性能测试多数是在半电池、大倍率（0.5C 或 1C）、单位活性物质负载量低（<1mg/cm^2）的情况下获得的。在小倍率（0.1C 或 0.2C）充放电下，因为纳米材料比表面积大，容易生成大量 SEI 膜，消耗电池正极中有限的 Li 源，导致实际应用中全电池的能量密度和循环寿命严重衰减，同时纳米材料的振实密度不高，体积能量密度低，因此今后 Si 负极研究还需要继续围绕缓解体积形变和稳定 SEI 膜两个方面展开。

目前，世界范围内，日本大阪钛业科技公司（Osaka Titanium Technologies Co.）生产的 SiO 在工业中能够小批量应用，为了解决 SiO 首周效率不高的问题还需要开发补锂技术。无定形硅合金和纳米硅碳复合材料也开始了小批量的试制和评价。预计在未来，硅负极材料将开始批量进入市场。与锂离子电池中多数材料不同，我国在纳米硅及硅碳负极材料方向上拥有核心材料的专利，具备一定的优势。

7.2.4 SnMC 合金负极材料

1995 年，Fuji Photo Film 公司提出了无定形 Sn 基复合负极材料，并于 1997 年发表在 *Science* 上[155]。该材料显示了较好的循环性，后期的研究发现相对于直接的 Sn 合金，循环性改善的主要原因是纳米尺寸的 Li$_x$Sn 产物分散在无定形的惰性氧化物介质中[156-158]。虽然无定形 Sn 基合金因为循环性、倍率特性离实际应用还有差距而没有获得实际应用[159]，但这一发现实际上对后期高容量合金负极材料的结构设计提供了重要的参考依据，并显著促进了合金负极材料的研究。

直到 2005 年，日本 Sony 公司首次实现了具有高比容量 SnCoC 合金负极材料的产业化，也是合金类负极材料中首个产业化的材料，基于 SnCoC 材料的"Nexelion"电池与石墨负极相比体积容量可以获得 50% 的提升，和锂离子全电池相比可以获得 30% 的提升[160]，SnCoC 是一种将 Sn、Co、C 三种元素复合在一起的无定形合金负极材料，是将 Sn、Co、C 在原子水平上均匀混合并进行非晶化处理的材料[161]，这种负极材料能够有效抑制充放电时电极材料的体积变化，可以提高循环周次，到 2011 年，Sony 公司再次宣布开发了使用 Sn 系非晶材料做电池负极的 3.5A·h 高容量锂离子电池"Nexelion"。这种电池为直径 18mm、长 65mm 的圆柱形电池（18650 型），主要用途为笔记本电脑。利用新开发的负极材料的锂离子电池性能更为优异，充电性能好，可以实现快速充电，同时新材料在低温环境下（−10～0℃）也可以保持高容量，低温性能良好[162]。

有关这类材料的基础研究具代表性的是在 2006 年，Dahn 等[163,164]采用磁控溅射法合成了多种 SnMC（M=Ti、V、Co）三元材料，通过高通量分析法建立了材料组分、结构与性能的关系。相比其他金属而言，Co 与 Sn 能在较宽区域形成非晶态的合金，此外 Co 不容易与 C 形成金属碳化物从而不会诱发 Sn 晶粒的结构重排，找到了 SnCoC 三元材料性能最优的原因，之后 Dahn 等[165-167]采用高能球磨法来规模制备 SnCoC 合金负极材料。通过制备不同配比的 SnCoC 材料，得出了最优化学组成为 $Sn_{30}Co_{30}C_{40}$，该化合物具有优异的电化学性能，容量为 470mA·h/g，能够循环 100 周。

此后，世界各国研究小组对 SnCoC 合金负极材料进行了大量的研究[168-171]，包括在 SnCoC 体系中加入 Fe[172]、Zn[173]、Cu[174]等金属的固溶作用和形成多相的共存，起到了稳定材料结构和提高循环性能的效果。SnCoC 合金负极材料可以通过模板法[175]、溶剂热法[176]、溶胶-凝胶法[177]、固相烧结法[177]、喷雾干燥法[177]、高能球磨法[178]等方法合成制备，在众多合成方法中，高能球磨法制备的 SnCoC 合金材料电化学性能稳定且工艺简单，具有成本上的优势，有利于规模化生产。SnMC 负极最终能否获得商业化应用还要看其综合性能、成本能否在与 Si 负极材料的竞争中取得优势。

7.3 其他负极材料

7.3.1 其他合金类负极材料

除了上文提到的 Si 以及 SnMC 负极材料，还有其他众多的合金负极材料。其中，合金类负极材料包括含锂合金和不含锂合金负极材料。含锂合金的研究出现在锂离子电池早期时候，为了寻找新的取代金属锂的负极材料，在以锂源作为负极时，主要对多种体系的锂合金进行了研究。相对于金属锂，锂合金负极可以在一定程度上避免枝晶的生长，提高了安全性。然而，在反复循环过程中，锂合金将经历较大的体积变化，电极材料逐渐粉化失效，合金结构遭到破坏[179,180]，为了缓解上述问题，一般采用多种复合体系：①Ranher[180]提出将锂合金与相应金属的金属间化合物混合，如将 Li_xAl 合金与 Al_3Ni 混合；②Besenhard 等[181]提出将小颗粒的锂合金嵌入到一个稳定的网络支撑体中；③Maxfield 等[182]提出将锂合金分散在导电聚合物中，如将 Li_xAl、Li_xPb 分散在聚乙炔（polyacetylene）或聚丙苯（polyphenylene）中，其中导电聚合物提供了一个弹性、多孔、具有较高电子和离子电导率的支撑体。这些措施在一定程度上缓解了锂合金体积形变的问题，但离实用化的程度仍然还有一段距离，而且该类材料的其他综合性能没有得到全面的研究。今后随着可充放电金属锂电池的发展可能需要再次系统深入的研究。

20 世纪 90 年代以后，锂离子电池正极提供 Li 源，负极材料可以不含金属 Li。因而在合金类材料的制备上有了更多的选择，比如 Li 能与许多金属 M（M=Mg、Ca、Al、Si、Ge、Sn、Pb、As、Sb、Bi、Pt、Ag、Au、Zn、Cd、Hg）在室温下形成金属间化合物[183]，因此这些能够和 Li 形成合金的金属理论上都能够作为锂电池的负极材料。其中，Sn 基合金是目前得到广泛研究的合金负极材料，单质 Sn 理论容量为 994mA·h/g，对应 $Li_{4.4}Sn$，最大体积形变可达 260%，平均电位 0.6V。Sn 基合金的种类很多，包括单质 Sn、Sn 的氧化物（SnO_2、SnO）[184,185]、非晶态 Sn 基复合氧化物（TCO，分子式为 SnM_xO_y）[155]，Sn 与其他金属形成的金属间化合物（SnSb、SnCu/SnAgCu、SnFe/SnFeC、SnNi、SnCa、SnCo/SnCoC、SnMg、SnAg、SnMn）[186-194]以及 Sn 与其他非金属元素形成的复合物（SiSn[195]

等），其中除了SnCoC被Sony公司商业化之外[160]，其他Sn基合金材料也得到了广泛的研究，如1996年，Yang等[196,197]采用电沉积的方法制备了纳米级的Sn及SnSb、SnAg金属间化合物，发现其循环性得到明显改善。Mukaibo等[198]用电沉积法制备了SnNi合金，其中Sn原子比为62%的材料循环性能最好，70次循环后可逆容量可达650mA·h/g。此外，之前提到的Fuji Photo Film公司的非晶态Sn基复合氧化物也是一类合金负极材料。该材料通过向Sn的氧化物中掺入B、P、Al及金属元素的方法，制备出非晶态（无定形）结构的Sn基复合氧化物，如$SnSi_{0.4}Al_{0.2}P_{0.6}O_{3.6}$，其可逆容量为557mA·h/g，循环寿命为383次[199,200]，1997年在 Science 杂志发表后，引起了人们的广泛关注[155]。

除了SnCoC和SiO小批量应用之外，大多数高比容量合金负极材料并未实现产业化，这是因为电极材料在脱嵌Li过程中其巨大的体积变化导致裂纹、粉化甚至活性物质从集流体上剥离，从而导致电极性能迅速失效，同时不断暴露的新鲜表面与电解液接触将会持续产生不稳定的SEI膜，这是合金负极材料的普遍现象。因此，从体积形变上考虑，纳米合金复合材料在充放电过程中容忍体积变化的能力强，电极结构具有较高的稳定性。但是纳米材料的比表面积很大，需要更多的非活性黏结剂，充放电过程容易发生团聚进而影响电化学性能，振实密度和体积能量密度低，SEI膜会大量产生，在空气中不易存放。实际材料的设计需要兼顾优缺点，在动力学与稳定性之间找到平衡点。可以预见，纳微复合结构材料可能是今后合金类负极材料在结构设计上的重要考虑。

7.3.2 $LiVO_2$层状负极材料

1988年MRS会议，Goodenough等[201]在其综述中提到了层状结构的$LiVO_2$，1997年Ceder等[202]计算了其作为正极的理论电位。$Li_{1+x}V_{1-x}O_2$作为负极材料最早在2002年由日本住友金属工业株式会社提出，韩国三星公司在2005年也申请了专利[203]。$Li_{1.1}V_{0.9}O_2$作为负极材料电位在0.1V（vs. Li^+/Li）左右[204]，理论容量为313.6mA·h/g，可逆容量为200～250mA·h/g。具有与$LiCoO_2$相同的层状结构，空间群为 R-$3m$，立方密堆积的O原子在$6c$位置，Li和V原子分别在八面体位置的交替层上。$LiVO_2$是由中间夹着V原子的两组O原子面组成的一个夹心层，层与层间靠范德华力结合[205]。层与层间嵌入Li后可以转化为Li_2VO_2，V原子占据八面体位，Li占据四面体位，空间群为 P-$3m1$[206]。$LiVO_2$的真实密度（4.3g/cm³）约是石墨密度（2.25g/cm³）的两倍，可以获得高的体积能量密度（理论体积密度为1360mA·h/cm³），因此自三星公司发表文章后引起了一定的关注[207]。目前，$LiVO_2$主要通过固相法合成和制备[204]，$LiVO_2$商业化需要克服：①首周效率低（主要是因为结晶度和纯度不高）；②长循环容量保持不如石墨等缺点[208]。此外V的成本不低，从一定程度上也限制了$LiVO_2$的应用。

7.3.3 过渡金属氧化物负极材料

从20世纪70年代初到80年代中期，二次锂电池均以负极作为Li源。如金属Li或

Li 合金作为负极，TiS_2、MoS_2、V_2O_5 等作为正极材料。在摇椅式体系刚提出时，Li 源负极的观念仍未转变。负极材料曾考虑用 Li_xMoO_2、$LiWO_2$、$Li_6Fe_2O_3$、$LiNb_2O_5$ 等过渡金属氧化物[209-211]。然而这些材料价格昂贵、能量密度低、Li^+ 的扩散速率慢、无法高倍率充放电，没有得到继续发展，这些化合物渐渐被淘汰。后来，正极作为 Li 源，C 作为负极材料逐渐发展起来，成为主流方向。之后出现了不含 Li 源的过渡金属氧化物作为负极材料。

不含 Li 的过渡金属氧化物分为两类，第一类是 Li 的嵌入只有材料结构的改变，没有 Li_2O 的形成，如 MoO_2、WO_2、Fe_2O_3、TiO_2、Nb_2O_5 等材料；第二类是 M_xO_y（M= Co、Ni、Cu、Fe）为代表的过渡金属氧化物，材料嵌 Li 时有 Li_2O 的形成。如 2000 年，Tarascon 等[212]在 Nature 上报道了具有高可逆容量（400～1000mA·h/g）、高倍率性能、优异循环性能的锂离子电池纳米尺寸的 3d 过渡金属氧化物负极材料。该小组系统地研究了 CoO、Co_3O_4、NiO、FeO、Cu_2O 以及 CuO 的电化学性能，他们通过 TEM（SAED）和 XANES 手段研究，提出了新的锂离子电池反应机理，即金属氧化物 M_xO_y（M 为 Co、Ni、Fe、Cu 等过渡金属）与金属 Li 发生反应，生成过渡金属和 Li_2O 的纳米弥散相，可逆的储锂机理如下：$M_xO_y+2yLi \Longleftrightarrow xM+yLi_2O$。由于这些氧化物晶体为岩盐结构，没有额外的空位供给 Li^+ 嵌入和脱出，所以很难发生嵌入反应，此外，这些 3d 过渡金属不可能与 Li 反应形成合金，所以也就不能发生氧化还原和生成合金的二步反应。尽管体相 Li_2O 既不是电子导体，也不是离子导体，不能在室温下参与电化学反应，但是 Li 插入到过渡金属氧化物后，形成了纳米尺度的复合物，过渡金属 M 和 Li_2O 的颗粒尺寸在 5nm 以下，这样微小的尺度从动力学考虑是非常有利的，它是导致 Li_2O 室温显示高电化学活性的主要原因[213-215]。后来发现，这一反应体系也适用于过渡金属氟化物[216,217]、硫化物[218]、磷化物[219]、氮化物[220]等，是一个普遍现象；另外一个有意思的现象是在 M=Sn、Sb、In、Pb、Zn 的氧化物里面类似的可逆反应不是很明显[156]，这些材料中可逆储锂的机制主要是 Li 与其中金属的合金化反应。

从基础研究方面考虑，由于该类材料储锂展示的尺寸效应而引起了广泛的关注。从应用考虑，与碳负极材料相比，尽管过渡金属氧化物 M_xO_y 的可逆容量比较高，但是限制其实际应用的缺点包括：①较高的平均脱 Li 电压（1.0～2.5Vvs.Li^+/Li）；②较大的极化（1V 左右）；③低的首次库仑效率（＜80%）；④充放电过程中体积变化较大；⑤SEI 膜较厚；⑥较差的循环性。目前，大量的研究是通过材料设计，改善其循环性和倍率特性，如：①掺杂包覆，提高过渡金属氧化物的电导率、抑制纳米颗粒的团聚、改善其电化学性能[221-225]；②对过渡金属氧化物负极材料进行形貌和结构设计（空心、核壳、多孔等），发挥微米、纳米多孔等多级结构的优势，做成微/纳米复合材料[226-229]；③将过渡金属氧化物与其他功能材料复合[230-232]。前 5 个问题在大量的研究中并没有解决，特别是极化偏大的问题。目前看这是相变过程中室温下骨架原子扩散较慢导致的，不太可能改善，因此此类材料在室温下没有实际应用价值。

7.4 负极材料的基础科学问题小结

① 结构演化。负极材料的脱嵌锂过程中材料结构反复变化，例如石墨的阶结构，硅的嵌锂非晶化，过渡金属氧化物的转化反应等。为了提高材料的循环性能，一方面，需要从微观尺度上掌握材料体系的结构演变过程以及机理；另一方面，还需要掌握复合材料体系（如硅碳负极）以及内部界面在长循环过程的变化机理和趋势。

② 体积效应。手机和电脑等消费电子应用领域在追求电池质量能量密度的同时也对其充放电过程中的体积膨胀提出了较为严苛的要求。虽然硅等合金类负极材料能够提供较高的容量，但其脱嵌锂过程中体积变化较大，在实际应用中需要与石墨等碳材料复合[233]。因此，需要在电池的全寿命周期，建立电池内部各组元（硅，氧化硅，碳，石墨，黏结剂等）对整个电池体系体积效应贡献的数据模型。

③ 界面反应。在实际电池体系中，有限锂源的消耗是造成循环衰减的主要原因之一，而对嵌锂电位低于 1.2V 的负极而言，更需要掌握负极材料与电解液之间的界面化学与电化学反应的机理和产物。固态电解质界面层（SEI）可以抑制副反应的发生，但是不同体系下，如何尽快形成稳定的界面层也需要深入的研究。

④ 力学特性。高容量负极材料在体积变化的同时伴随着应力的产生，造成颗粒和极片的粉化，甚至会造成电池严重的形变导致安全问题。但体积变化是材料的固有特性，为了缓解以上问题，需要加深对负极体系的力学问题的了解，例如复合材料中裂纹的产生与生长，SEI 膜的弹塑性，以及缓冲材料和黏结剂等应该具备的力学特性等。

⑤ 离子与电子输运特性。电动汽车和动力工具等应用领域要求电池的倍率达到 3C 甚至 10C。电池大倍率充放电需要负极材料具有良好的电子和离子传导性，同时应避免负极出现析锂等情况引发安全问题。因此在负极材料中尤其是复合负极中，锂离子和电子在材料的表面、界面以及体相的输运特性、传输路径和电荷交换等过程需要被关注。

⑥ 锂沉积效应。金属锂负极在实现电池高能量密度方面具备很大的潜力，但锂枝晶导致的安全问题阻碍了其实际应用[234]。锂负极在充放电过程中同时存在金属沉积/溶解和离子氧化/还原过程，而锂表面不均匀的形貌和杂质均会导致电极表面上金属沉积速度分布的不均匀，导致锂枝晶的生成。此外，在实际电池体系中，上百周次的反复沉积/溶解过程更加复杂，因此需要对影响锂沉积的因素进行系统地分析。而如何控制金属锂的沉积位点和解决锂枝晶问题是发展锂负极的挑战。

⑦ 高低温稳定性。实验室电池多在恒温环境下工作，然而真实电池体系需要适应不同的工作环境。一方面，在低温下需要保持电子电导和离子电导，抑制析锂现象的产生；另一方面，在高温下电池界面的稳定性，锂化颗粒的反应活性以及安全性都需要深入研究。

⑧ 黏结特性。在高容量负极体系中，黏结剂对维持电极导电网络的完整性显得尤为重要[235]。例如，硅负极在充放电过程中存在约 300% 的体积膨胀，引起颗粒粉化和电子导电缺失现象，此时需要黏结剂能够适应该体积变化，使电池可持续运行。此外，黏结剂还需要具有化学与电化学稳定性，能承受电解液以及离子和电子的攻击。

7.5 本章结语

目前，负极材料的主要发展思路是朝高功率密度（如含孔石墨、软碳、硬碳、钛酸锂）、高能量密度、高循环性能和低成本的方向发展。高容量的合金类负极（如硅负极材料）将会在下一代锂离子电池中逐渐获得应用，然而合金类负极材料面临的问题是其高容量伴随的体积变化。即便解决了循环性、倍率特性等问题，由于实际应用时电池电芯体积不允许发生较大的变化（一般＜5%，最大允许30%），而合金类材料的容量与体积变化成正比，因此合金类负极材料在实际电池中的容量发挥中受到了限制，虽然其使用能在一定程度上提高现有锂离子电池的能量密度（如20%～30%，与高能量正极材料匹配达到300W·h/kg），但达不到理论预期。特别是体积能量密度相对于石墨负极的优势远不如理论计算结果。相对而言，金属锂负极的体积膨胀限制不突出。目前，金属锂可充放锂电池、全固态锂电池、锂硫电池以及锂空电池等新型电池正在被大量研究。可以预见，尽管已经研究了近50年，金属锂很可能在不久的将来成为高能量密度可充放锂电池负极材料的最终选择方案。

本章需要进一步思考的基础科学问题：

1. 石墨阶结构是如何形成的？高阶结构形成是热力学决定的吗？从3阶到2阶结构，微观上是如何实现的？在任何倍率下嵌入锂都能形成阶结构吗？只有石墨能形成阶结构吗？
2. 硬碳、软碳、钛酸锂、硅的储锂和脱锂的动力学，也有类似于石墨的现象吗？容易脱出不容易嵌入？
3. 负极表面SEI膜生长的均匀性和致密性，与负极表面的电子电导及其均匀性有何依赖关系？
4. 负极表面SEI膜的生长是电化学反应直接成核长大还是先电化学反应，形成产物，后成核累计成大？
5. 负极是否需要避免析锂？锂会析出在SEI膜之外吗？为什么？
6. 负极在不同的SOC下热稳定性是一致的吗？由什么因素决定？
7. 100% SOC充放电过程中，各类负极材料的理论体积膨胀收缩比例和实际体积变化率都是多少？随着循环次数的不同，体积变化率是稳定的吗？
8. 负极如何实现高面容量下较低的体积膨胀率，较高的倍率特性？
9. 从正极扩散而来的过渡金属离子在负极上面都会发生什么反应？对材料的哪些电化学性能有影响？为什么？
10. 影响负极高温储存和自放电行为的因素有哪些？

参考文献

[1] Armand M, Murphy D, Broadhead J, et al. Materials for Advanced Batteries [M]. New York: Plenum Press, 1980: 145.

[2] Garreau M, Thevenin J, Fekir M. On the processes responsible for the degradation of the aluminum lithium electrode used as anode material in lithium aprotic electrolyte batteries [J]. Journal of Power Sources, 1983, 9 (3-4): 235-238.

[3] Yazami R, Touzain P. A reversible graphite-lithium negative electrode for electrochemical generators [J]. Journal of Power Sources, 1983, 9 (3): 365-371.

[4] Tarascon J. Mo_6Se_6: A new solid-state electrode for secondary lithium batteries [J]. Journal of the Electrochemical Society, 1985, 132 (9): 2089-2093.

[5] Scrosati B. Non aqueous lithium cells [J]. Electrochimica Acta, 1981, 26 (11): 1559-1567.

[6] Abraham K. Ambient temperature secondary lithium batteries using LiAl lithium insertion anodes [R]. California: California University Berkeley Lawrence Berkeley Lab, 1987.

[7] Hrold A. Recherches sur les composes d'insertion du graphite [J]. Bull Soc Chim Fr, 1955, 187 (7-8): 999-1012.

[8] Dey A, Sullivan B. The electrochemical decomposition of propylene carbonate on graphite [J]. Journal of the Electrochemical Society, 1970, 117 (2): 222-224.

[9] Sony. Non-aqueous electrolyte secondary cell: EP, EP0391281 [P]. 1989-04-03.

[10] Nagaura T, Tozawa K. Lithium ion rechargeable battery [J]. Prog Batteries Solar Cells, 1990, 9: 209.

[11] Endo M, Kim C, Nishimura K, et al. Recent development of carbon materials for Li ion batteries [J]. Carbon, 2000, 38 (2): 183-197.

[12] Mabuchi A. A survey on the carbon anode materials for rechargeable lithium batteries [J]. Tanso, 1994, 165: 298-306.

[13] Yamaura J, Ozaki Y, Morita A, et al. High voltage, rechargeable lithium batteries using newly-developed carbon for negative electrode material [J]. Journal of Power Sources, 1993, 43 (1): 233-239.

[14] Tarascon J M, Armand M. Issues and challenges facing rechargeable lithium batteries [J]. Nature, 2001, 414 (6861): 359-367.

[15] Van S W, Scrosati B. Advances in Lithium-Ion Batteries [M]. Berlin: Springer-Verlag, 2002.

[16] Kang B, Ceder G. Battery materials for ultrafast charging and discharging [J]. Nature, 2009, 458 (7235): 190-193.

[17] Armand M, Tarascon J M. Building better batteries [J]. Nature, 2008, 451 (7179): 652-657.

[18] Jansen A, Kahaian A, Kepler K, et al. Development of a high-power lithium-ion battery [J]. Journal of Power Sources, 1999, 81: 902-905.

[19] Smith K, Wang C Y. Power and thermal characterization of a lithium-ion battery pack for hybrid-electric vehicles [J]. Journal of Power Sources, 2006, 160 (1): 662-673.

[20] Zhang X, Ross P, Kostecki R, et al. Diagnostic characterization of high power lithium-ion batteries for use in hybrid electric vehicles [J]. Journal of the Electrochemical Society, 2001, 148 (5): A463-A470.

[21] Zhou H H, Ci L C, Liu C Y. Progress in studies of the electrode materials for Li ion batteries [J]. Progress in Chemistry, 1998, 10 (1): 85-92.

[22] Hao R R, Fang X Y, Niu S C. Chemistry of the Elements (Ⅲ) [M]. Beijing: Science Press, 1998, 404-425.

[23] Ohzuku T, Ueda A, Yamamoto N. Zero-strain insertion material of Li ($Li_{1/3}Ti_{5/3}$) O_4 for rechargeable lithium cells [J]. Journal of the Electrochemical Society, 1995, 142 (5): 1431-1435.

[24] Woo K C, Mertwoy H, Fischer J, et al. Experimental phase diagram of lithium-intercalated graphite [J]. Physical Review B, 1983, 27 (12): 7831.

[25] Dahn J. Phase diagram of Li_xC_6 [J]. Physical Review B, 1991, 44 (17): 9170.

[26] Nalamova V, Guerard D, Lelaurain M, et al. X-ray investigation of highly saturated Li-graphite intercalation compound [J]. Carbon, 1995, 33 (2): 177-181.

[27] Feng Z Z, Song S Q. Preparation and application of mesophase pitch [J]. Carbon, 2013, 2: 42-45.

[28] Honda H, Yamada Y. Meso-carbon microbeads [J]. J Japan Petrol Inst, 1973, 16: 392-397.

[29] Xu B, Chen P. Intermediate development phase carbon microbeads (MCMB), properties and applications [J].

New Carbon Materials, 1996, 11 (3): 4-8.

[30] Niu Y J, Zhang H G, Zhou A M, et al. Non-Ferrous Progress: 1996—2005 [M]. Changsha: Central South University Press, 2007.

[31] Choi W C, Byun D, Lee J K. Electrochemical characteristics of silver-and nickel-coated synthetic graphite prepared by a gas suspension spray coating method for the anode of lithium secondary batteries [J]. Electrochimica Acta, 2004, 50 (2): 523-529.

[32] Lee H Y, Baek J K, Lee S M, et al. Effect of carbon coating on elevated temperature performance of graphite as lithium-ion battery anode material [J]. Journal of Power Sources, 2004, 128 (1): 61-66.

[33] Tanaka H, Osawa T, Moriyoshi Y, et al. Improvement of the anode performance of graphite particles through surface modification in RF thermal plasma [J]. Thin Solid Films, 2004, 457 (1): 209-216.

[34] Guoping W, Bolan Z, Min Y, et al. A modified graphite anode with high initial efficiency and excellent cycle life expectation [J]. Solid State Ionics, 2005, 176 (9): 905-909.

[35] Lee J H, Lee S, Paik U, et al. Aqueous processing of natural graphite particulates for lithium-ion battery anodes and their electrochemical performance [J]. Journal of Power Sources, 2005, 147 (1): 249-255.

[36] Yamauchi Y, Hino T, Ohzeki K, et al. Gas desorption behavior of graphite anodes used for lithium ion secondary batteries [J]. Carbon, 2005, 43 (6): 1334-1336.

[37] Zhao X, Hayner C M, Kung M C, et al. In-plane vacancy-enabled high-power Si-graphene composite electrode for lithium-ion batteries [J]. Advanced Energy Materials, 2011, 1 (6): 1079-1084.

[38] 王广驹. 世界石墨生产, 消费及国际贸易 [J]. 中国非金属矿工业导刊, 2006, 27 (1): 61-65.

[39] Jonker G H. Magnetic compounds with perovskite structure Ⅳ conducting and non-conducting compounds [C] // Madrid: Proceedings 3rd Symposium on Reactivity of Solids, 1956: 707-722.

[40] Murphy D, Cava R, Zahurak S, et al. Ternary $Li_x TiO_2$ phases from insertion reactions [J]. Solid State Ionics, 1983, 9: 413-417.

[41] Ferg E, Gummow R, De K A, et al. Spinel anodes for lithium-ion batteries [J]. Journal of the Electrochemical Society, 1994, 141 (11): L147-L150.

[42] Robertson A, Trevino L, Tukamoto H, et al. New inorganic spinel oxides for use as negative electrode materials in future lithium-ion batteries [J]. Journal of Power Sources, 1999, 81: 352-357.

[43] Peramunage D, Abraham K. Preparation of micron-sized $Li_4 Ti_5 O_{12}$ and its electrochemistry in polyacrylonitrile electrolyte-based lithium cells [J]. Journal of the Electrochemical Society, 1998, 145 (8): 2609-2615.

[44] Julien C, Massot M, Zaghib K. Structural studies of $Li_{4/3} Me_{5/3} O_4$ (Me=Ti, Mn) electrode materials: Local structure and electrochemical aspects [J]. Journal of Power Sources, 2004, 136 (1): 72-79.

[45] Scharner S, Weppner W, Schmid B P. Evidence of two-phase formation upon lithium insertion into the $Li_{1.33} Ti_{1.67} O_4$ spinel [J]. Journal of the Electrochemical Society, 1999, 146 (3): 857-861.

[46] Zaghib K, Simoneau M, Armand M, et al. Electrochemical study of $Li_4 Ti_5 O_{12}$ as negative electrode for Li-ion polymer rechargeable batteries [J]. Journal of Power Sources, 1999, 81: 300-305.

[47] Pecharroman C, Amarilla J. Thermal evolution of infrared vibrational properties of $Li_{4/3} Ti_{5/3} O_4$ measured by specular reflectance [J]. Physical Review B, 2000, 62 (18): 12062.

[48] Guerfi A, Charest P, Kinoshita K, et al. Nano electronically conductive titanium-spinel as lithium ion storage negative electrode [J]. Journal of Power Sources, 2004, 126 (1): 163-168.

[49] Gao L, Qiu W, Zhao H L. Lithiated titanium complex oxide as negative electrode [J]. Journal of University of Science and Technology Beijing, 2005, 27 (1): 82-85.

[50] Bach S, Pereira R J, Baffier N. Electrochemical properties of sol-gel $Li_{4/3} Ti_{5/3} O_4$ [J]. Journal of Power Sources, 1999, 81: 273-276.

[51] Kavan L, Grtzel M. Facile synthesis of nanocrystalline $Li_4 Ti_5 O_{12}$ (spinel) exhibiting fast Li insertion [J]. Electrochemical and Solid-State Letters, 2002, 5 (2): A39-A42.

[52] Hao Y, Lai Q Y, Liu D, et al. Synthesis by citric acid sol-gel method and electrochemical properties of $Li_4 Ti_5 O_{12}$ anode material for lithium-ion battery [J]. Materials Chemistry and Physics, 2005, 94 (2-3): 382-387.

[53] 王虹. 微波法制备钛酸锂的方法: 中国, 101333001A [P]. 2008-07-18.

[54] 白莹. 一种用于锂二次电池负极材料尖晶石钛酸锂的制备方法: 中国, 1919736A [P]. 2006-08-17.

[55] Li J, Tang Z, Zhang Z. Controllable formation and electrochemical properties of one-dimensional nanostructured spinel $Li_4 Ti_5 O_{12}$ [J]. Electrochemistry Communications, 2005, 7 (9): 894-899.

[56] 杨立. 一种应用于锂离子电池的钛酸锂负极材料的制备方法：中国，101409341A [P]. 2008-11-20.

[57] Huang S，Wen Z，Zhu X，et al. Effects of dopant on the electrochemical performance of $Li_4Ti_5O_{12}$ as electrode material for lithium ion batteries [J]. Journal of Power Sources，2007，165（1）：408-412.

[58] Tian B，Xiang H，Zhang L，et al. Niobium doped lithium titanate as a high rate anode material for Li-ion batteries [J]. Electrochimica Acta，2010，55（19）：5453-5458.

[59] Huang Y，Qi Y，Jia D，et al. Synthesis and electrochemical properties of spinel $Li_4Ti_5O_{12-x}Cl_x$ anode materials for lithium-ion batteries [J]. Journal of Solid State Electrochemistry，2012，16（5）：2011-2016.

[60] Venkateswarlu M，Chen C，Do J，et al. Electrochemical properties of nano-sized $Li_4Ti_5O_{12}$ powders synthesized by a sol-gel process and characterized by X-ray absorption spectroscopy [J]. Journal of Power Sources，2005，146（1）：204-208.

[61] Cai R，Yu X，Liu X，et al. $Li_4Ti_5O_{12}$/Sn composite anodes for lithium-ion batteries：Synthesis and electrochemical performance [J]. Journal of Power Sources，2010，195（24）：8244-8250.

[62] Yuan T，Yu X，Cai R，et al. Synthesis of pristine and carbon-coated $Li_4Ti_5O_{12}$ and their low-temperature electrochemical performance [J]. Journal of Power Sources，2010，195（15）：4997-5004.

[63] Hu X，Lin Z，Yang K，et al. Effects of carbon source and carbon content on electrochemical performances of $Li_4Ti_5O_{12}$/C prepared by one-step solid-state reaction [J]. Electrochimica Acta，2011，56（14）：5046-5053.

[64] Martha S K，Haik O，Borgel V，et al. $Li_4Ti_5O_{12}$/$LiMnPO_4$ lithium-ion battery systems for load leveling application [J]. Journal of the Electrochemical Society，2011，158（7）：A790-A797.

[65] Huang K L，Wang Z X，Liu S Q. Lithium-Ion Battery Technology and Key Principles [M]. Beijing：Chemical Industry Press，2008.

[66] Xu K，Wang X Y，Xiao L X. Lithium Ion Battery [M]. Changsha：Central South University Press，2002.

[67] Wang Q，Li H，Chen L，et al. Novel spherical microporous carbon as anode material for Li-ion batteries [J]. Solid State Ionics，2002，152：43-50.

[68] Li H，Wang Q，Shi L，et al. Nanosized SnSb alloy pinning on hard non-graphitic carbon spherules as anode materials for a Li ion battery [J]. Chemistry of Materials，2002，14（1）：103-108.

[69] Hu J，Li H，Huang X. Influence of micropore structure on Li-storage capacity in hard carbon spherules [J]. Solid State Ionics，2005，176（11）：1151-1159.

[70] Fey G T K，Chen C L. High-capacity carbons for lithium-ion batteries prepared from rice husk [J]. Journal of Power Sources，2001，97：47-51.

[71] Yin G P，Zhou D R，Xia B J，et al. Preparation of phosphorus-doped carbon and its performance Lithium intercalation [J]. Battery Bimonthly，2000，30（4）：147-149.

[72] Schonfelder H H，Kitoh K，Nemoto H. Nanostructure criteria for lithium intercalation in non-doped and phosphorus-doped hard carbons [J]. Journal of Power Sources，1997，68（2）：258-262.

[73] Buiel E，Dahn J. Li-insertion in hard carbon anode materials for Li-ion batteries [J]. Electrochimica Acta，1999，45（1）：121-130.

[74] Rosamaria F，Ulrich V S，Dahn J R. Studies of lithium intercalation into carbons using nonaqueous electrochemical-cells [J]. Journal of the Electrochemical Society，1990，137（7）：2009-2013.

[75] Stevens D，Dahn J. The mechanisms of lithium and sodium insertion in carbon materials [J]. Journal of the Electrochemical Society，2001，148（8）：A803-A811.

[76] Bonino F，Brutti S，Piana M，et al. Structural and electrochemical studies of a hexaphenylbenzene pyrolysed soft carbon as anode material in lithium batteries [J]. Electrochimica Acta，2006，51（17）：3407-3412.

[77] Guo M，Wang J C，Wu L B，et al. Study of carbon nanofibers as negative materials for Li-ion batteries [J]. Battery Bimonthly，2004，34（5）：384-385.

[78] Sato Y，Kikuchi Y，Kawai T，et al. Characteristics of coke carbon modified with mesophase-pitch as a negative electrode for lithium ion batteries [J]. Journal of Power Sources，1999，81：182-186.

[79] Yoshio M，Tsumura T，Dimov N. Electrochemical behaviors of silicon based anode material [J]. Journal of Power Sources，2005，146（1）：10-14.

[80] Lai S C. Solid lithium-silicon electrode [J]. Journal of the Electrochemical Society，1976，123：1196.

[81] Sharma R A，Seefurth R N. Thermodynamic properties of the lithium-silicon system [J]. Journal of the Electrochemical Society，1976，123（12）：1763-1768.

[82] Seefurth R N，Sharma R A. Investigation of lithium utilization from a lithium-silicon electrode [J]. Journal of the

Electrochemical Society, 1977, 124 (8): 1207-1214.

[83] Seefurth R N, Sharma R A. Dependence of lithium-silicon electrode potential and lithium utilization on reference electrode location [J]. Journal of the Electrochemical Society, 1980, 127 (5): 1101-1104.

[84] Wen C J, Huggins R A. Chemical diffusion in intermediate phases in the lithium-silicon system [J]. Journal of Solid State Chemistry, 1981, 37 (3): 271-278.

[85] Boukamp B A, Lesh G C, Huggins R A. All-solid lithium electrodes with mixed-conductor matrix [J]. Journal of the Electrochemical Society, 1981, 128 (4): 725-729.

[86] Weydanz W J, Wohlfahrt M M, Huggins R A. A room temperature study of the binary lithium-silicon and the ternary lithium-chromium-silicon system for use in rechargeable lithium batteries [J]. Journal of Power Sources, 1999, 81: 237-242.

[87] Gao B, Sinha S, Fleming L, et al. Alloy formation in nanostructured silicon [J]. Advanced Materials, 2001, 13 (11): 816-819.

[88] Li H, Huang X J, Chen L Q, et al. A high capacity nano-Si composite anode material for lithium rechargeable batteries [J]. Electrochem Solid State, 1999, 2 (11): 547-549.

[89] Li H, Huang X J, Chen L Q, et al. The crystal structural evolution of nano-Si anode caused by lithium insertion and extraction at room temperature [J]. Solid State Ionics, 2000, 135 (1-4): 181-191.

[90] Limthongkul P, Jang Y I, Dudney N J, et al. Electrochemically-driven solid-state amorphization in lithium-silicon alloys and implications for lithium storage [J]. Acta Materialia, 2003, 51 (4): 1103-1113.

[91] Hatchard T D, Dahn J R. In situ XRD and electrochemical study of the reaction of lithium with amorphous silicon [J]. Journal of the Electrochemical Society, 2004, 151 (6): A838-A842.

[92] Key B, Bhattacharyya R, Grey C P, et al. Real-time NMR investigations of structural changes in silicon electrodes for lithium-ion batteries [J]. Journal of the American Chmical Society, 2009, 131 (26): 9239-9249.

[93] Key B, Morcrette M, Grey C P, et al. Pair distribution function analysis and solid State NMR studies of silicon electrodes for lithium ion batteries: Understanding the (De) lithiation mechanisms [J]. Journal of the American Chemical Society, 2011, 133 (3): 503-512.

[94] Beaulieu L Y, Hatchard T D, Bonakdarpour A, et al. Reaction of Li with alloy thin films studied by in situ AFM [J]. Journal of the Electrochemical Society, 2003, 150 (11): A1457-A1464.

[95] Baggetto L, Danilov D, Notten P H L. Honeycomb-structured silicon: Remarkable morphological changes induced by electrochemical (De) lithiation [J]. Advanced Materials, 2011, 23 (13): 1563-1566.

[96] Lee S W, Mcdowell M T, Choi J W, et al. Anomalous shape changes of silicon nanopillars by electrochemical lithiation [J]. Nano Letters, 2011, 11 (7): 3034-3039.

[97] Lee S W, Mcdowell M T, Berla L A, et al. Fracture of crystalline silicon nanopillars during electrochemical lithium insertion [J]. Proceedings of the National Academy Sciences of the USA, 2012, 109 (11): 4080-4085.

[98] He Y, Yu X Q, Wang Y H, et al. Alumina-coated patterned amorphous silicon as the anode for a lithium-ion battery with high coulombic efficiency [J]. Advanced Materials, 2011, 23 (42): 4938-4941.

[99] He Y, Wang Y H, Yu X Q, et al. Si-Cu thin film electrode with kirkendall voids structure for lithium-ion batteries [J]. Journal of the Electrochemical Society, 2012, 159 (12): A2076-A2081.

[100] He Y, Yu X Q, Li G, et al. Shape evolution of patterned amorphous and polycrystalline silicon microarray thin film electrodes caused by lithium insertion and extraction [J]. Journal of Power Sources, 2012, 216: 131-138.

[101] Wang Y, He Y, Xiao R, et al. Investigation of crack patterns and cyclic performance of Ti-Si nanocomposite thin film anodes for lithium ion batteries [J]. Journal of Power Sources, 2012, 202: 236-245.

[102] Notten P H L, Roozeboom F, Niessen R A H, et al. 3-D integrated all-solid-state rechargeable batteries [J]. Advanced Materials, 2007, 19 (24): 4564-4567.

[103] Baggetto L, Oudenhoven J F M, Van D T, et al. On the electrochemistry of an anode stack for all-solid-state 3D-integrated batteries [J]. Journal of Power Sources, 2009, 189 (1): 402-410.

[104] Chan C K, Ruffo R, Hong S S, et al. Surface chemistry and morphology of the solid electrolyte interphase on silicon nanowire lithium-ion battery anodes [J]. Journal of Power Sources, 2009, 189 (2): 1132-1140.

[105] Zheng J Y, Zheng H, Wang R, et al. An investigation on the sold electrolyte interphase of silicon anode for Li-ion batteries through force curve method [J]. Journal of Electrochemistry, 2013, 19 (6): 530-536.

[106] Zhang X W, Patil P K, Wang C S, et al. Electrochemical performance of lithium ion battery, nano-silicon-based, disordered carbon composite anodes with different microstructures [J]. Journal of Power Sources, 2004,

[107] Chan C K, Ruffo R, Hong S S, et al. Structural and electrochemical study of the reaction of lithium with silicon nanowires [J]. Journal of Power Sources, 2009, 189 (1): 34-39.

[108] Cui L F, Ruffo R, Chan C K, et al. Crystalline-amorphous core-shell silicon nanowires for high capacity and high current battery electrodes [J]. Nano Letters, 2009, 9 (1): 491-495.

[109] Mcdowell M T, Lee S W, Ryu I, et al. Novel size and surface oxide effects in silicon nanowires as lithium battery anodes [J]. Nano Letters, 2011, 11 (9): 4018-4025.

[110] Ryu I, Choi J W, Cui Y, et al. Size-dependent fracture of Si nanowire battery anodes [J]. Journal of the Mechanics and Physics of Solid, 2011, 59 (9): 1717-1730.

[111] Xu W L, Vegunta S S S, Flake J C. Surface-modified silicon nanowire anodes for lithium-ion batteries [J]. Journal of Power Sources, 2011, 196 (20): 8583-8539.

[112] Yue L, Wang S Q, Zhao X Y, et al. Nano-silicon composites using poly (3,4-ethylenedioxythiophene): Poly (styrenesulfonate) as elastic polymer matrix and carbon source for lithium-ion battery anode [J]. Journal of Materials Chemistry, 2012, 22 (3): 1094-1099.

[113] Zang J L, Zhao Y P. Silicon nanowire reinforced by single-walled carbon nanotube and its applications to anti-pulverization electrode in lithium ion battery [J]. Composites Part B: Engineering, 2012, 43 (1): 76-82.

[114] Yoshio M, Wang H Y, Fukuda K, et al. Carbon-coated Si as a lithium-ion battery anode material [J]. Journal of the Electrochemical Society, 2002, 149 (12): A1598-A1603.

[115] Qu J, Li H Q, Henry J J, et al. Self-aligned Cu-Si core-shell nanowire array as a high-performance anode for Li-ion batteries [J]. Journal of Power Sources, 2012, 198: 312-317.

[116] Jia H P, Gao P F, Yang J, et al. Novel three-dimensional mesoporous silicon for high power lithium-ion battery anode material [J]. Advanced Energy Materials, 2011, 1 (6): 1036-1039.

[117] Yao Y, Mcdowell M T, Ryu I, et al. Interconnected silicon hollow nanospheres for lithium-ion battery anodes with long cycle life [J]. Nano Letters, 2011, 11 (7): 2949-2954.

[118] Fu K, Yildiz O, Bhanushali H, et al. Aligned carbon nanotube-silicon sheets: A novel nano-architecture for flexible lithium ion battery electrodes [J]. Advanced Materials, 2013, 25 (36): 5109-5114.

[119] Min J H, Bae Y S, Kim J Y, et al. Self-organized artificial SEI for improving the cycling ability of silicon-based battery anode materials [J]. B Korean Chem Soc, 2013, 34 (4): 1296-1299.

[120] Choi N S, Yew K H, Lww K Y, et al. Effect of fluoroethylene carbonate additive on interfacial properties of silicon thin-film electrode [J]. Journal of Power Sources, 2006, 161 (2): 1254-1259.

[121] Chakrapani V, Rusli F, Filler M A, et al. Quaternary ammonium ionic liquid electrolyte for a silicon nanowire-based lithium ion battery [J]. J Phys Chem C, 2011, 115 (44): 22048-22053.

[122] Etacheri V, Haik O, Goffer Y, et al. Effect of fluoroethylene carbonate (FEC) on the performance and surface chemistry of Si-nanowire Li-ion battery anodes [J]. Langmuir, 2011, 28 (1): 965-976.

[123] Buddie M C. High performance silicon nanoparticle anode in fluoroethylene carbonate-based electrolyte for Li-ion batteries [J]. Chemical Communications, 2012, 48 (58): 7268-7270.

[124] Profatilova I A, Stock C, Schmitz A, et al. Enhanced thermal stability of a lithiated nano-silicon electrode by fluoroethylene carbonate and vinylene carbonate [J]. Journal of Power Sources, 2013, 222: 140-149.

[125] Leung K, Rempe S B, Foster M E, et al. Modeling electrochemical decomposition of fluoroethylene carbonate on silicon anode surfaces in lithium ion batteries [J]. Journal of the Electrochemical Society, 2014, 161 (3): A213-A221.

[126] Kovalenko I, Zdyrko B, Magasinski A, et al. A major constituent of brown algae for use in high-capacity Li-ion batteries [J]. Science, 2011, 334 (6052): 75-79.

[127] Ryou M H, Kim J, Lee I, et al. Mussel-inspired adhesive binders for high-performance silicon nanoparticle anodes in lithium-ion batteries [J]. Advanced Materials, 2012, 25 (11): 1571-1576.

[128] Li J, Lewis R, Dahn J. Sodium carboxymethyl cellulose a potential binder for Si negative electrodes for Li-ion batteries [J]. Electrochemical and Solid-State Letters, 2007, 10 (2): A17-A20.

[129] Bridel J S, Azais T, Morcrette M, et al. Key parameters governing the reversibility of Si/carbon/CMC electrodes for Li-ion batteries [J]. Chem. Mater, 2009, 22 (3): 1229-1241.

[130] Mazouzi D, Lestriez B, Roue L, et al. Silicon composite electrode with high capacity and long cycle life [J]. Electrochemical and Solid-State Letters, 2009, 12 (11): A215-A218.

[131] Guo J C, Wang C S. A polymer scaffold binder structure for high capacity silicon anode of lithium-ion battery [J]. Chemical Communications, 2010, 46 (9): 1428-1430.

[132] Liu W R, Yang M H, Wu H C, et al. Enhanced cycle life of Si anode for Li-ion batteries by using modified elastomeric binder [J]. Electrochemical and Solid-State Letters, 2005, 8 (2): A100-A103.

[133] Park H K, Kong B S, Oh E S. Effect of high adhesive polyvinyl alcohol binder on the anodes of lithium ion batteries [J]. Electrochem Commun, 2011, 13 (10): 1051-1053.

[134] Magasinski A, Zdyrko B, Kovalenko I, et al. Toward efficient binders for Li-ion battery Si-based anodes: Polyacrylic acid [J]. ACS Applied Materials & Interfaces, 2010, 2 (11): 3004-3010.

[135] Yun J B, Soo K J, Tae L K. A photo-cross-linkable polymeric binder for silicon anodes in lithium ion batteries [J]. RSC Advances, 2013, 3 (31): 12625-12630.

[136] Han Z J, Yabuuchi N, Hashimoto S, et al. Cross-linked poly (acrylic acid) with polycarbodiimide as advanced binder for Si/graphite composite negative electrodes in Li-ion batteries [J]. ECS Electrochemistry Letters, 2013, 2 (2): A17-A20.

[137] Koo B, Kim H, Cho Y, et al. A highly cross-linked polymeric binder for high-performance silicon negative electrodes in lithium ion batteries [J]. Angewandte Chemie International Edition, 2012, 51 (35): 8762-8767.

[138] Bae J, Cha S H, Park J. A new polymeric binder for silicon-carbon nanotube composites in lithium ion battery [J]. Macromol Res, 2013, 21 (7): 826-831.

[139] Yim C H, Abu L Y, Courtel F M. High capacity silicon/graphite composite as anode for lithium-ion batteries using low content amorphous silicon and compatible binders [J]. Journal of Materials Chemistry A, 2013, 1 (28): 8234-8243.

[140] Erk C, Brezesinski T, Sommer H, et al. Toward silicon anodes for next-generation lithium ion batteries: A comparative performance study of various polymer binders and silicon nanopowders [J]. ACS Applied Materials & Interfaces, 2013, 5 (15): 7299-7307.

[141] Kim J S, Choi W, Cho K Y, et al. Effect of polyimide binder on electrochemical characteristics of surface-modified silicon anode for lithium ion batteries [J]. Journal of Power Sources, 2013, 144: 521-526.

[142] Li J, Christensen L, Obrovac M, et al. Effect of heat treatment on Si electrodes using polyvinylidene fluoride binder [J]. Journal of the Electrochemical Society, 2008, 155 (3): A234-A238.

[143] Kim Y L, Sun Y K, Lee S M. Enhanced electrochemical performance of silicon-based anode material by using current collector with modified surface morphology [J]. Electrochimca Acta, 2008, 53 (13): 4500-4504.

[144] Guo J C, Sun A, Wang C S. A porous silicon-carbon anode with high overall capacity on carbon fiber current collector [J]. Electrochem Commun, 2010, 12 (7): 981-984.

[145] Choi J Y, Lee D J, Lee Y M, et al. Silicon nanofibrils on a flexible current collector for bendable lithium-ion battery anodes [J]. Adv Funct Mater, 2013, 23 (17): 2108-2114.

[146] Hang T, Nara H, Yokoshima T, et al. Silicon composite thick film electrodeposited on a nickel micro-nanocones hierarchical structured current collector for lithium batteries [J]. Journal of Power Sources, 2013, 222: 503-509.

[147] Luais E, Sakai J, Desploban S, et al. Thin and flexible silicon anode based on integrated macroporous silicon film onto electrodeposited copper current collector [J]. Journal of Power Sources, 2013, 242: 166-170.

[148] Tang X X, Liu W, Ye B Y, et al. Preparation of current collector with blind holes and enhanced cycle performance of silicon-based anode [J]. T Nonferr Metal Soc, 2013, 23 (6): 1723-1727.

[149] Kim H, Han B, Choo J, et al. Three-dimensional porous silicon particles for use in high-performance lithium secondary batteries [J]. Angewandte Chemie, 2008, 120 (52): 10305-10308.

[150] Bang B M, Kim H, Song H K, et al. Scalable approach to multi-dimensional bulk Si anodes via metal-assisted chemical etching [J]. Energy & Environmental Science, 2011, 4 (12): 5013-5019.

[151] Kasavajjula U, Wang C, Appleby A J. Nano-and bulk-silicon-based insertion anodes for lithium-ion secondary cells [J]. Journal of Power Sources, 2007, 163 (2): 1003-1039.

[152] Magasinski A, Dixon P, Hertzberg B, et al. High-performance lithium-ion anodes using a hierarchical bottom-up approach [J]. Nature Materials, 2010, 9 (4): 353-358.

[153] Liu G, Xun S, Vukmirovic N, et al. Polymers with tailored electronic structure for high capacity lithium battery electrodes [J]. Advanced Materials, 2011, 23 (40): 4679-4683.

[154] Chan C K, Peng H, Liu G, et al. High-performance lithium battery anodes using silicon nanowires [J]. Nature

Nanotechnology, 2007, 3 (1): 31-35.

[155] Idota Y, Kubota T, Matsufiti A, et al. Tin-based amorphous oxide: A high-capacity lithium-ion-storage material [J]. Science, 1997, 276 (5317): 1395-1397.

[156] Courtney I A, Dahn J. Key factors controlling the reversibility of the reaction of lithium with SnO_2 and Sn_2BPO_6 glass [J]. Journal of the Electrochemical Society, 1997, 144 (9): 2943-2948.

[157] Li H, Huang X J, Chen L Q. Direct imaging of the passivating film and microstructure of nanometer-scale SnO anodes in lithium rechargeable batteries [J]. Electrochemical and Solid-State Letters, 1998, 1 (6): 241-243.

[158] Liu W, Huang X J, Wang Z, et al. Studies of stannic oxide as an anode material for lithium-ion batteries [J]. Journal of the Electrochemical Society, 1998, 145 (1): 59-62.

[159] Li H, Wang Z, Chen L, et al. Research on advanced materials for Li-ion batteries [J]. Advanced Materials, 2009, 21 (45): 4593-4607.

[160] David M. New materials extend Li-ion performance [J]. Power Electronics Technology, 2006, 1 (5): 50.

[161] Ogisu K. R&D activities & results for sony batteries [C] //San Francisco: Portable Power 2005 Conference, 2005.

[162] 索尼公司. 索尼成功开发3.5 A·h高容量锂离子电池"Nexelion" [EB/OL]. [2011-07-15]. http//www.sony.com.cn/news_center/press_release/technology/1955_3787.htm.

[163] Dahn J, Mar R, Abouzeid A. Combinatorial study of $Sn_{1-x}Co_x$ ($0 < x < 0.6$) and $(Sn_{0.55}Co_{0.45})_{1-y}C_y$ ($0 < y < 0.5$) alloy negative electrode materials for Li-ion batteries [J]. Journal of the Electrochemical Society, 2006, 153 (2): A361-A365.

[164] Todd A, Mar R, Dahn J. Tin-transition metal-carbon systems for lithium-ion battery negative electrodes [J]. Journal of the Electrochemical Society, 2007, 154 (6): A597-A604.

[165] Ferguson P, Martine M, Dunlap R, et al. Structural and electrochemical studies of $(Sn_xCo_{1-x})_{60}C_{40}$ alloys prepared by mechanical attriting [J]. Electrochimica Acta, 2009, 54 (19): 4534-4539.

[166] Ferguson P, Rajora M, Dunlap R, et al. $(Sn_{0.5}Co_{0.5})_{1-y}C_y$ alloy negative electrode materials prepared by mechanical attriting [J]. Journal of the Electrochemical Society, 2009, 156 (3): A204-A208.

[167] Ferguson P, Todd A, Dahn J. Comparison of mechanically alloyed and sputtered tin-cobalt-carbon as an anode material for lithium-ion batteries [J]. Electrochemistry Communications, 2008, 10 (1): 25-31.

[168] Hassoun J, Mulas G, Panero S, et al. Ternary Sn-Co-C Li-ion battery electrode material prepared by high energy ball milling [J]. Electrochemistry Communications, 2007, 9 (8): 2075-2081.

[169] Lavela P, Nacimiento F, Ortiz G F, et al. Sn-Co-C composites obtained from resorcinol-formaldehyde gel as anodes in lithium-ion batteries [J]. Journal of Solid State Electrochemistry, 2010, 14 (1): 139-148.

[170] Liu B, Abouimrane A, Ren Y, et al. New anode material based on $SiO-Sn_xCo_yC_z$ for lithium batteries [J]. Chemistry of Materials, 2012, 24 (24): 4653-4661.

[171] Zhong X C, Jiang F Q, Xin P A, et al. Preparation and electrochemical performance of Sn-Co-C composite as anode material for Li-ion batteries [J]. Journal of Power Sources, 2009, 189 (1): 730-732.

[172] Yang S, Li Q, Shen D. Influence of Fe on electrochemical performance of Sn_xCo_y/C anode materials [J]. Chinese Journal of Power Sources, 2011, 35 (2): 148-152.

[173] Shaobin Y, Ding S, Qiang L. Synthesis and electrochemical properties of $Sn_{0.35-0.5x}Co_{0.35-0.5x}Zn_xC_{0.30}$ composite [J]. Acta Metallurgica Sinica, 2010, 46 (1): 6-12.

[174] Yang S B, Shen D, Wu X G, et al. Effects of Cu on structures and electrochemical properties of Sn-Co/C composite [J]. Chinese Journal of Nonferrous Metals, 2012, 22 (4): 1163-1168.

[175] Cui W, Wang F, Wang J, et al. Nanostructural CoSnC anode prepared by $CoSnO_3$ with improved cyclability for high-performance Li-ion batteries [J]. Electrochimica Acta, 2011, 56 (13): 4812-4818.

[176] Li M Y, Liu C L, Shi M R, et al. Nanostructure Sn-Co-C composite lithium ion battery electrode with unique stability and high electrochemical performance [J]. Electrochimica Acta, 2011, 56 (8): 3023-3028.

[177] Xin L, Jing Y X, Hai L Z, et al. Synthesis and properties of $Sn_{30}Co_{30}C_{40}$ ternary alloy anode material for lithium ion battery [J]. Acta Chimica Sinica, 2013, 71 (7): 1011-1016.

[178] Lee S I, Yoon S, Park C M, et al. Reaction mechanism and electrochemical characterization of a Sn-Co-C composite anode for Li-ion batteries [J]. Electrochimica Acta, 2008, 54 (2): 364-369.

[179] Fauteux D, Koksbang R. Rechargeable lithium battery anodes: Alternatives to metallic lithium [J]. Journal of Applied Electrochemistry, 1993, 23 (1): 1-10.

[180] Rahner D, Machill S, Schlorb H, et al. Intercalation materials for lithium rechargeable batteries [J]. Solid State Ionics, 1996, 86: 891-896.

[181] Besenhard J, Hess M, Komenda P. Dimensionally stable Li-alloy electrodes for secondary batteries [J]. Solid State Ionics, 1990, 40: 525-529.

[182] Maxfield M, Jow T, Gould S, et al. Composite electrodes containing conducting polymers and Li alloys [J]. Journal of the Electrochemical Society, 1988, 135 (2): 299-305.

[183] Winter M, Besenhard J O. Electrochemical lithiation of tin and tin-based intermetallics and composites [J]. Electrochimica Acta, 1999, 45 (1): 31-50.

[184] Du C W, Chen Y B, Wu M S, et al. Advances in lithium-ion battery anode materials for non-carbon [C] // Tianjin: Proceedings of the 24th Chinese Chemistry and Physical Power Source Academic Conferences, 2000.

[185] Wu Y P, Wan C R. Study on materials for lithium-ion batteries tin-based negative [J]. Chinese Journal of Power Source, 1999, 23 (3): 191-193.

[186] Kepler K D, Vaughey J T, Thackeray M M. $Li_xCu_6Sn_5$ ($0 < x < 13$): An intermetallic insertion electrode for rechargeable lithium batteries [J]. Electrochemical and Solid-State Letters, 1999, 2 (7): 307-309.

[187] Mao O, Dunlap R, Dahn J. Mechanically alloyed Sn-Fe (-C) powders as anode materials for Li-ion batteries: I. The Sn_2Fe-C system [J]. Journal of the Electrochemical Society, 1999, 146 (2): 405-413.

[188] Larcher D, Beaulieu L, Macneil D, et al. In situ X-ray study of the electrochemical reaction of Li with η'-Cu_6Sn_5 [J]. Journal of the Electrochemical Society, 2000, 147 (5): 1658-1662.

[189] Li H, Zhu G, Huang X, et al. Synthesis and electrochemical performance of dendrite-like nanosized SnSb alloyprepared by co-precipitation in alcohol solution at low temperature [J]. Journal of Materials Chemistry, 2000, 10 (3): 693-696.

[190] Kim H, Kim Y J, Kim D, et al. Mechanochemical synthesis and electrochemical characteristics of Mg_2Sn as an anode material for Li-ion batteries [J]. Solid State Ionics, 2001, 144 (1): 41-49.

[191] Wang L, Kitamura S, Sonoda T, et al. Electroplated Sn-Zn alloy electrode for Li secondary batteries [J]. Journal of the Electrochemical Society, 2003, 150 (10): A1346-A1350.

[192] Yin J, Wada M, Yoshida S, et al. New Ag-Sn alloy anode materials for lithium-ion batteries [J]. Journal of the Electrochemical Society, 2003, 150 (8): A1129-A1135.

[193] Tamura N, Fujimoto M, Kamino M, et al. Mechanical stability of Sn-Co alloy anodes for lithium secondary batteries [J]. Electrochimica Acta, 2004, 49 (12): 1949-1956.

[194] Wang L, Kitamura S, Obata K, et al. Multilayered Sn-Zn-Cu alloy thin-film as negative electrodes for advanced lithium-ion batteries [J]. Journal of Power Sources, 2005, 141 (2): 286-292.

[195] Beauleiu L, Hewitt K, Turner R, et al. The electrochemical reaction of Li with amorphous Si-Sn alloys [J]. Journal of the Electrochemical Society, 2003, 150 (2): A149-A156.

[196] Besenhard J, Yang J, Winter M. Will advanced lithium-alloy anodes have a chance in lithium-ion batteries? [J]. Journal of Power Sources, 1997, 68 (1): 87-90.

[197] Yang J, Winter M, Besenhard J. Small particle size multiphase Li-alloy anodes for lithium-ionbatteries [J]. Solid State Ionics, 1996, 90 (1): 281-287.

[198] Mukaibo H, Sumi T, Yokoshima T, et al. Electrodeposited Sn-Ni alloy film as a high capacity anode material for lithium-ion secondary batteries [J]. Electrochemical and Solid-State Letters, 2003, 6 (10): A218-A220.

[199] Photo F. Nonaqueous secondary battery: EP, 651450A1 [P]. 1995-03-03.

[200] Photo F. Nonaqueous secondary battery: US, 5478671 [P]. 1995-12-26.

[201] Goodenough J, Manthiram A, James A, et al. Lithium insertion compounds [C] //Proceedings of the MRS, 1988.

[202] Aydinol M, Kohan A, Ceder G. Abinitio calculation of the intercalation voltage of lithium-transition-metal oxide electrodes for rechargeable batteries [J]. Journal of Power Sources, 1997, 68 (2): 664-668.

[203] 三星SDI株式会社. 用于非水电解液电池的负极活性材料, 其制备方法和非水电解液电池: 中国, 1783551A [P]. 2005-10-27.

[204] Song J H, Park H J, Kim K J, et al. Electrochemical characteristics of lithium vanadate, $Li_{1+x}VO_2$, new anode materials for lithium ion batteries [J]. Journal of Power Sources, 2010, 195 (18): 6157-6161.

[205] Chang J J. Synthesis and electrochemical: Properties of lithium-ion battery anode material $Li_{1+x}VO_2$ [D]. Dalian: Dalian University of Technology, 2012.

[206] Armstrong A R, Lyness C, Panchmatia P M, et al. The lithium intercalation process in the low-voltage lithium

battery anode $Li_{1+x}V_{1-x}O_2$ [J]. Nature Materials, 2011, 10 (3): 223-229.

[207] Chen H, Xiang K X, Hu Z L, et al. Synthesis and electrochemical performance of new anode materials $Li_{1.1}V_{0.9}O_2$ for lithium ion batteries [J]. Transactions of Materials and Heat Treatment, 2012, 33 (5): 34-38.

[208] Choi N S, Kim J S, Yin R Z, et al. Electrochemical properties of lithium vanadium oxide as an anode material for lithium-ion battery [J]. Materials Chemistry and Physics, 2009, 116 (2): 603-606.

[209] Lazzari M, Scrosati B. A cyclable lithium organic electrolyte cell based on two intercalation electrodes [J]. Journal of the Electrochemical Society, 1980, 127 (3): 773-774.

[210] Dipietro B, Patriarco M, Scrosati B. On the use of rocking chair configurations for cyclable lithium organic electrolyte batteries [J]. Journal of Power Sources, 1982, 8 (2): 289-299.

[211] Ktakata H O, Mori T, Koshita N. Procedures of the symposium on primary and secondary lithium batteries [J]. Electrochem Soc Inc, 1988, 347: 91-93.

[212] Poizot P, Laurelle S, Grugeon S, et al. Nano-sized transition-metal oxides as negative-electrode materials for lithium-ion batteries [J]. Nature, 2000, 407 (6803): 496-499.

[213] Debart A, Dupont L, Poizot P, et al. A transmission electron microscopy study of the reactivity mechanism of tailor-made CuO particles toward lithium [J]. Journal of the Electrochemical Society, 2001, 148 (11): A1266-A1274.

[214] Dedryvere R, Laruelle S, Grugeon S, et al. Contribution of X-ray photoelectron spectroscopy to the study of the electrochemical reactivity of CoO toward lithium [J]. Chemistry of Materials, 2004, 16 (6): 1056-1061.

[215] Xin C, Naiqing Z, Kening S. 3d transition-metal oxides as anode micro/nano-materials for lithium ion batteries [J]. Progress in Chemistry, 2011, 23 (10): 2045-2054.

[216] Li H, Richter G, Maier J. Reversible formation and decomposition of LiF clusters using transition metal fluorides as precursors and their application in rechargeable Li batteries [J]. Advanced Materials, 2003, 15 (9): 736-739.

[217] Badway F, Mansour A, Pereira N, et al. Structure and electrochemistry of copper fluoride nanocomposites utilizing mixed conducting matrices [J]. Chemistry of Materials, 2007, 19 (17): 4129-4141.

[218] Dbart A, Dupont L, Patrice R, et al. Reactivity of transition metal (Co, Ni, Cu) sulphides versus lithium: The intriguing case of the copper sulphide [J]. Solid State Sciences, 2006, 8 (6): 640-651.

[219] Gillot F, Boyanov S, Dupont L, et al. Electrochemical reactivity and design of NiP_2 negative electrodes for secondary Li-ion batteries [J]. Chemistry of Materials, 2005, 17 (25): 6327-6337.

[220] Pereira N, Dupont L, Tarascon J, et al. Electrochemistry of Cu_3N with lithium a complex system with parallel processes [J]. Journal of the Electrochemical Society, 2003, 150 (9): A1273-A1280.

[221] Zhang W M, Wu X L, Hu J S, et al. Carbon coated Fe_3O_4 nanospindles as a superior anode material for lithium-ion batteries [J]. Advanced Functional Materials, 2008, 18 (24): 3941-3946.

[222] Rahman M, Chou S L, Zhong C, et al. Spray pyrolyzed NiO-C nanocomposite as an anode material for the lithium-ion battery with enhanced capacity retention [J]. Solid State Ionics, 2010, 180 (40): 1646-1651.

[223] Wang Y, Zhang H J, Lu L, et al. Designed functional systems from peapod-like Co@carbon to Co_3O_4@carbon nanocomposites [J]. ACS Nano, 2010, 4 (8): 4753-4761.

[224] Zhou G, Wang D W, Li F, et al. Graphene-wrapped Fe_3O_4 anode material with improved reversible capacity and cyclic stability for lithium ion batteries [J]. Chemistry of Materials, 2010, 22 (18): 5306-5313.

[225] Wang Y, Zhang L. Simple synthesis of CoO-NiO-C anode materials for lithium-ion batteries and investigation on its electrochemical performance [J]. Journal of Power Sources, 2012, 209: 20-29.

[226] Zhang P, Guo Z, Kang S, et al. Three-dimensional Li_2O-NiO-CoO composite thin-film anode with network structure for lithium-ion batteries [J]. Journal of Power Sources, 2009, 189 (1): 566-570.

[227] Zhu X J, Guo Z P, Zhang P, et al. Highly porous reticular tin-cobalt oxide composite thin film anodes for lithium ion batteries [J]. Journal of Materials Chemistry, 2009, 19 (44): 8360-8365.

[228] Wang C, Wang D, Wang Q, et al. Fabrication and lithium storage performance of three-dimensional porous NiO as anode for lithium-ion battery [J]. Journal of Power Sources, 2010, 195 (21): 7432-7437.

[229] Xia Y, Zhang W, Xiao Z, et al. Biotemplated fabrication of hierarchically porous NiO/C composite from lotus pollen grains for lithium-ion batteries [J]. Journal of Materials Chemistry, 2012, 22 (18): 9209-9215.

[230] Yu Y, Chen C H, Shi Y. A tin-based amorphous oxide composite with a porous, spherical, multideck-cage

morphology as a highly reversible anode material for lithium-ion batteries [J]. Advanced Materials, 2007, 19 (7): 993-997.

[231] Li F, Zou Q Q, Xia Y Y. Co-loaded graphitable carbon hollow spheres as anode materials for lithium-ion battery [J]. Journal of Power Sources, 2008, 177 (2): 546-552.

[232] Wu Z S, Ren W, Wen L, et al. Graphene anchored with Co_3O_4 nanoparticles as anode of lithium ion batteries with enhanced reversible capacity and cyclic performance [J]. ACS Nano, 2010, 4 (6): 3187-3194.

[233] Luo F, Liu B N, Zheng J Y, et al. Review—nano-silicon/carbon composite anode materials towards practical application for next generation Li-ion batteries [J]. Journal of the Electrochemical Society, 2015, 162 (14): A2509-A2528.

[234] Jang W C, Doron A. Promise and reality of post-lithium-ion batteries with high energy densities [J]. Nature Review Materials, 2016, 1 (2): 16013.

[235] Sujong C, Minseong K, Kyungho K. Confronting issues of the practical implementation of si anode in high-energy lithium-ion batteries [J]. Joule, 2017, 1 (3): 47-60.

第 8 章

非水液体电解质材料

非水有机液体电解质（通常称为"电解液"）是锂离子电池的关键成分之一，通常被称为锂离子电池的"血液"。电解液在正负极之间起到离子传导的作用，其组成包括非水有机溶剂、锂盐和添加剂。本章将对非水液体电解质材料进行介绍。

8.1 液态电解质的性质

液体电解质材料一般应当具备如下特性：①电导率高，要求电解液黏度低，锂盐溶解度和解离度高；②离子迁移数高；③稳定性高，要求电解液具备高的闪点、高的分解温度、低的电极反应活性，搁置无副反应等；④界面稳定，具备较好的正负极材料表面成膜特性，能在前几周充放电过程中形成稳定的低阻抗固体电解质中间相（solid electrolyte interphase，SEI）；⑤宽的电化学窗口，能够使电极表面钝化，从而在较宽的电压范围内工作；⑥工作温度范围宽；⑦浸润性好；⑧不易燃烧；⑨环境友好，无毒或毒性小；⑩较低的成本。

8.1.1 离子电导率

离子电导率（σ）反映的是电解液传输离子的能力，是衡量电解液性能的重要指标之一。离子电导率满足式（8-1）

$$\sigma = \sum n_i u_i z_i e \tag{8-1}$$

式中，n_i 为参与输运的离子的浓度；u_i 为参与输运的离子的迁移率；z_i 为第 i 种离子的电荷量。实验中测量离子电导率主要根据式（8-2），电解质电阻可以采用交流阻抗方法测量，通过等效电路对数据进行拟合，获得等效电路中元件的数值，得到电阻值，计算得到电导率

$$\sigma = \frac{d}{R_b S} \tag{8-2}$$

式中，d 为电导池两电极间距离；R_b 为电解质电阻；S 为电极面积。液态电解质的

离子电导率一般符合 Arrhenius 方程

$$\sigma = A\exp\left(-\frac{E_a}{RT}\right) \tag{8-3}$$

式中，E_a 为离子导电活化能；T 为绝对温度；A 为指前因子也称频率因子；R 为气体常数。实验上将得到的电导率的自然对数与温度的倒数作图，得到 $\ln\sigma$-$1/T$ 关系曲线，拟合得到的直线斜率为 E_a/R，截距为 $\ln A$，进而可以得到体系的活化能 E_a 和频率因子 A。

液体电解质在温度较低时，黏度增大，满足 Vogel-Tamman-Fulcher（VTF）方程，此类行为在聚合物电解质中也较为常见，可以通过自由体积模型解释

$$\sigma = AT^{1/2}\exp\frac{-B}{T-T_0} \tag{8-4}$$

式中，A 为指前因子；B 为活化能 E_a。通过对电导率与温度关系的非线性拟合，可以得到电解液体系活化能的数值、指前因子 A 和 T_0。

8.1.2 离子迁移数

实验测定的电解质溶液的离子电导率包括了电解质中各种离子的贡献。离子迁移数是对某一种离子迁移能力的反映，每一种离子所传输的电荷量在通过溶液的总电荷量中所占的分数，称为该种离子的迁移数，用符号 t（transference number 的第一个字母）表示。对于锂离子电池和可充放电金属锂电池而言，充放电过程中需要传输的是 Li^+，Li^+ 的迁移数越高，参与储能反应的有效输运的离子也就越多。Li^+ 迁移数较低将导致有效传导的离子电阻较高，同时阴离子更容易富集在正负极表面，导致电极极化增大，并增大了阴离子分解的概率，不利于获得较好的循环性和倍率特性。

Li^+ 迁移数可以通过直流极化和交流阻抗相结合的办法获得。采用此法测量 Li^+ 迁移数时考虑三个假设：①直流极化后，所有的电流值只由 Li^+ 的运输造成；②测量结果忽略 SEI 膜对电阻的贡献；③在直流极化过程中电解液保持稳定，不发生分解。记录初始电流值（I_0），经过一定时间电流趋于稳态，记录此时电流值（I_s），可将 Li^+ 的迁移数 t_{Li^+} 表示为式（8-5）

$$t_{Li^+} = \frac{I_s}{I_0} \tag{8-5}$$

考虑到极化前后电解质电阻变化对 Li^+ 迁移数的影响，采用 Bruce 和 Vincent 的修正公式进行计算

$$t_{Li^+} = \frac{I_s(\Delta V - I_0 R_0)}{I_0(\Delta V - I_s R_s)} \tag{8-6}$$

式中，ΔV 为极化电压；R_0、R_s 分别为电极极化前、后的电阻，数值由电池极化前后的 Nyquist 曲线得出，电极电阻是电池传输电阻（R_{ct}）和电极钝化层电阻（R_{SEI}）之和，可由等效电路得出，通过式（8-6）计算可以得出 Li^+ 的迁移数 t_{Li^+}。

还可以按照如下方法测量：首先组成两电极的 Li/电解液/Li 电池体系，然后用交流阻抗谱仪测得总电阻 R_{total}，接着用此电池进行直流极化测试。设置直流电压为 10～

100mV，经过一段时间平衡后，在阻塞了其他离子迁移的情况下，只剩下 Li^+ 迁移的电流。由公式 $R_{DC}=V/I_{DC}$，可以得到 Li^+ 的迁移数 $t_{Li^+}=R_{total}/R_{DC}$。

8.1.3 电化学窗口

在充放电过程中，要求电解质在正负极材料发生的氧化还原反应电位之间保持稳定，超出这个电位范围，电解质就会发生电化学反应而分解。电化学窗口是指电解质能够稳定存在的电压范围，是选择锂离子电池电解质的重要参数之一。电化学稳定窗口可以由循环伏安（CV）方法测定。在较宽的电位扫描范围内，没有明显的电流，意味着电解质的电化学稳定性较好。

循环伏安测量电化学窗口，需要注意以下问题。①过高估计电化学窗口。大量的文献报道结果中，往往出现 CV 测量的电解质电化学窗口很宽，在实际的锂离子电池中，存在 CV 在测量电位范围内并不稳定的现象。主要有 6 个原因：循环伏安采用惰性电极；扫速较快；仪器电流测量精度较低；研究者制图时将电流轴设定较大的范围，导致一些弱电流信号在发表的图中看不出来；采用过电位较高的两电极电化学池测量 CV；测量温度范围不够宽。②过低估计电解质电化学窗口。许多电解质虽然在第一次扫描时出现了显著的氧化还原峰，但在后续扫描中，氧化还原峰电流大幅度下降，这意味着正负极表面已经在第一次反应之后形成 SEI 膜钝化，提高了后续反应的稳定性。

溶剂和锂盐的电化学窗口，可以通过第一性原理计算出材料的最高占据轨道（HOMO）和最低未占据轨道（LUMO）的相对差值来大致判断。但对于通过表面钝化而拓宽了电化学窗口范围的电解质体系，目前还不能准确预测，因此电化学窗口的判断主要以实验为主，理论预测可以在开发新电解质体系时提供一定的参考。

8.1.4 黏度

黏度是考察锂离子电池电解液的一个重要参数，它的数值直接影响离子在电解质体系中的扩散性能[28]，通常使用的有机电解液的溶剂分子是靠分子间较弱的范德华力相互作用，黏度相对较低。离子液体中阴阳离子通过较强的静电库仑力相互作用，导致离子液体室温黏度较大。如 PP_{13}TFSI 型离子液体室温黏度达到 117cP（$1cP=10^{-3}$ Pa·s），而 DMC 的黏度仅为 0.59cP。

Stokes-Einstein 方程给出了离子迁移率与液体黏度的关系式

$$\mu_i = \frac{1}{6\pi\eta r_i} \tag{8-7}$$

式中，η 为黏度；μ_i 为溶液中离子的迁移速率；r_i 为溶剂化半径。液体电解质的电导率与离子迁移速率成正比，所以它随着黏度的升高而降低。

电解液的各项性能与溶剂的许多其他性能参数密切相关。如溶剂的熔点、沸点、闪点等因素对电池的使用温度、电解质盐的溶解度、电极电化学性能和电池的安全性能有重要的影响。表 8-1 给出了典型有机溶剂的相对介电常数、黏度、熔点、沸点等参数。

8.2 液态电解质在锂离子电池中的反应

液态电解质在锂离子电池充放电过程中会发生一系列的反应，以下分类简述。

8.2.1 负极表面形成 SEI 膜的反应

一般负极材料（如石墨、硅碳等），嵌锂电位低于 1.2V(vs. Li^+/Li)，在锂离子电池首次充电过程中不可避免地要与电解液发生反应，溶剂分子、锂盐或添加剂在一定电位下被还原，从而在负极表面形成一层 SEI 膜。

SEI 膜的组成非常复杂，一般认为其由 Li_2CO_3、LiF、Li_2O、LiOH、烷基酯锂（$ROCO_2Li$）、烷氧基锂（ROLi）、聚合物锂（如 PEO-Li）等多种无机、有机、聚合物组成，具体组分与所用的电解液、电极材料、充放电条件、反应温度等因素有关。它是一层电子绝缘离子导电的膜，能够阻止电解液与负极材料的进一步反应，不影响 Li^+ 的通过。SEI 膜的形成对改善电极的性质和提高电极的寿命具有不可忽视的作用。SEI 膜的存在会降低电极材料颗粒之间的电子接触，降低首周效率，所以一般不希望 SEI 膜厚度很大。关于 SEI 膜的具体性质，郑杰允等在此系列文章的界面问题部分给出了详细介绍，包括 SEI 膜的结构及生长机理，影响 SEI 膜的因素，表征 SEI 膜形貌、组成、电化学稳定性、力学特性及覆盖度、输运特性和生长过程的一些基本手段以及如何对 SEI 膜进行改性。

在研究过程中，研究者发现很多添加剂具有改善 SEI 膜性能的作用，这些添加剂在锂离子电池中起着非常重要的作用，将在 8.3.3 节进一步讨论。

表 8-1 一些锂离子电池用有机溶剂的基本物理性质

种类	状态	溶剂	熔点 T_m/℃	沸点 T_b/℃	相对介电常数 ε(25℃)	黏度 η(25℃)/cP
碳酸酯	环状	乙烯碳酸酯(EC)	36.4	248	89.78	1.90(40℃)
		丙烯碳酸酯(PC)	−48.8	242	64.92	2.53
		丁烯碳酸酯(BC)	−53	240	53	3.2
	链状	碳酸二甲酯(DMC)	4.6	91	3.107	0.59(20℃)
		碳酸二乙酯(DEC)	−74.3	126	2.805	0.75
		碳酸甲乙酯(EMC)	−53	110	2.958	0.65
羧酸酯	环状	γ-丁内酯(γBL)	−43.5	204	39	1.73
	链状	乙酸乙酯(EA)	−84	77	6.02	0.45
		甲酸甲酯(MF)	−99	32	8.5	0.33
醚类	环状	四氢呋喃(THF)	−109	66	7.4	0.46
		2-甲基-四氢呋喃(2-Me-THF)	−137	80	6.2	0.47
	链状	二甲氧基甲烷(DMM)	−105	41	2.7	0.33
		1,2-二甲氧基乙烷(DME)	−58	84	7.2	0.46
腈类	链状	乙腈(AN)	−48.8	81.6	35.95	0.341

8.2.2 与正极之间的反应

正极材料脱锂时具有较高的氧化电位，有的充电电位高达 4.9V(vs. Li^+/Li)。电极材料在充电态处于高氧化态，因此电解液容易在正极表面发生电化学氧化和化学氧化反应，溶剂或阴离子失去电子，产生一些副产物。

1985 年，Thomas 等经过实验发现 $LiCoO_2$ 正极在 PC 基电解液中会有表面钝化膜的形成。后来，Tarascon 等研究尖晶石 $LiMn_2O_4$ 电化学性能时发现该材料表面也存在界面膜，从而阻止电解液与正极材料的进一步反应。随后，Aurbach 等也发现了 $LiCoO_2$、$LiMn_2O_4$ 和 $LiNiO_2$ 表面的正极钝化膜。研究者们通过 FTIR、SEM、XPS 和交流阻抗谱等技术认为，正极表面 SEI 膜的成分主要是烷氧基碳酸锂、聚醚和 Li_2CO_3 等。

尽管已有上述报道，但在很长一段时间内，正极表面的 SEI 膜是否存在及其形成机制仍然有一定的争议。正极表面钝化膜的主要成分类似于负极，包括 Li_2CO_3、$ROCO_2Li$；负极表面 SEI 膜的形成机理是溶剂和锂盐的电化学还原反应。而氧化反应无法直接产生 Li_2CO_3 和 $ROCO_2Li$。一种推测认为 Li_2CO_3 以及 $ROCO_2Li$ 是来自正极材料在空气中储存时与 CO_2 和原材料中 CO_3^{2-} 反应形成的产物；也有人认为正极表面的 SEI 膜是从负极漂移过来的物质沉积在正极表面形成。此外，即便在正极材料表面形成 SEI 膜，由于正极表面的高氧化态，SEI 膜也有可能被分解，难以在正极材料上稳定生长。已经发表的大量 TEM 照片显示，并不是在所有的充放电之后正极材料表面都能发现 SEI 膜。

柳娜等[40]的研究发现，纳米 $LiCoO_2$ 与电解液及 DMC 之间在浸泡过程中均能发生自发的化学反应而生成表面钝化膜。将纳米 $LiCoO_2$ 浸入电解液或 DMC 一周后，其表面会有一层 2~5nm 的表面膜生成，其组成主要是 $ROCO_2Li$、ROLi 和 Li_2CO_3，$LiCoO_2$ 被部分还原为 Co_3O_4。这说明 $LiCoO_2$ 与电解液之间会发生自发的化学反应，而进一步研究发现电化学反应能够促进纳米 $LiCoO_2$ 表面 SEI 膜的进一步生长。

正极表面膜具有与负极 SEI 膜相似的性质，是一层电子绝缘离子导电的膜，能阻止电解液与电极材料之间的进一步反应。虽然正极表面 SEI 膜的形成机理还存在争议，但对影响正极表面膜的因素已有一定的了解，包括正极表面稳定性、正极材料的结构、电解液组分、正极与电解液的相容性、电解液对正极集流体的腐蚀性等因素。改善正极界面特性也应该从这些方面着手进行，主要手段包括：①正极材料比表面积优化，如通过提高结晶度、表面包覆来降低比表面积、减少表面缺陷；②体相掺杂，掺杂一方面可以增加电极材料的稳定性，另一方面可以改变电极材料的表面催化活性；③表面包覆，避免电解液与正极材料直接接触，从而改善电解液与正极材料的相容性；④优化电解质锂盐，如可以使用混合盐或者开发新型锂盐来降低与电极材料的反应活性；⑤优化溶剂组成，电解液的性质一般由溶剂性质决定，所以溶剂的选择非常重要；⑥开发能够改善正极界面膜性质的电解液添加剂。

8.2.3 过充反应

在锂离子电池过度充电时，会发生一系列的反应。首先是过多的 Li^+ 从正极材料中脱出，嵌入负极材料中，可能会导致正极材料结构的坍塌和负极锂枝晶的生成，这主要发生在 Li 不能完全脱出的正极材料，如 $LiCoO_2$、$Li_{1+x}(NiCoMn)O_2$ 等；其次是电解液组分（主要是溶剂）在正极表面发生不可逆的氧化分解反应，产生气体并释放大量热量，从而导致电池内压增加和温度升高，给电池的安全性带来严重影响。对于电解液来说，过充反应主要是碳酸酯分子失去电子分解成大量 CO_2、少量烯烃、CO 以及含氟含磷化合物的过程。

避免过充反应的发生，除了电极材料与电解质的优化，还包括优化外电路保护，使用过充保护添加剂。关于防过充添加剂的内容参见"8.3.3.4 过充保护添加剂"。

8.2.4 受热反应

电解液的热稳定性关系到锂离子电池的安全性。目前，研究电解液受热反应的方法主要包括 DSC、TG-MS 和绝热加速量热（ARC）实验等，以 $LiPF_6$ 的碳酸酯类电解液为例来说明电解液受热反应的过程。电解液受热会引起 $LiPF_6$ 的分解，其分解产物 PF_5 会攻击碳酸酯类溶剂中氧原子上的孤对电子，从而导致溶剂的分解，在分解产物中会有大量的 CO_2 等气体生成，这些反应会造成电解液温度越来越高，继而引起电解液的燃烧和爆炸。除此之外，溶剂中两种不同组分之间也会发生开环聚合反应，生成 DMDOHC，其具体过程如下式所示

$$LiPF_6(s) \longrightarrow LiF(s) + PF_5(g) \qquad (8\text{-}8)$$

$$H_3C\text{—}O\text{—}CO\text{—}O\text{—}CH_3 + PF_5 \longrightarrow$$
$$H_3C\text{—}F, H_3C\text{—}O\text{—}CH_3, 烃类, CO_2 \qquad (8\text{-}9)$$

$$n \text{(EC)} \xrightarrow{PF_5/HF} \ce{+CH_2-CH_2+}_n + \ce{+O-CO-(OCH_2CH_2)_n+} + nCO_2 \qquad (8\text{-}10)$$

$$n \text{(EC)} + H_3C\text{—}O\text{—}CO\text{—}O\text{—}CH_3 \longrightarrow \text{DMDOHC} \qquad (8\text{-}11)$$

商品电解液储存时，在 85℃ 会发生明显的电解液分解，从 ARC 的表征可以看出，一般商品电池的热失控温度在 123～167℃。针对电解液受热分解的问题，一方面应该注意电池的使用温度，防止受热反应的发生；另一方面应该开发热稳定性比较高的电解液。

8.3 溶剂、锂盐和添加剂

锂离子电池液体电解质一般由非水有机溶剂和电解质锂盐两部分组成。由于单一的溶剂很难满足电解质的各项性能要求，所以溶剂主要是几种性质不同的有机溶剂的混合。在常规锂离子电池体系对于混合盐的使用则相对较少，但在以金属锂为负极的电池体系中，双盐或者三盐的混合体系使用较多。

8.3.1 溶剂

锂离子电池电解质的性质与溶剂的性质密切相关，一般来说溶剂的选择应该满足如下一些基本要求：①一种有机溶剂应该具有较高的介电常数 ε，从而使其有足够高的溶解锂盐的能力；②有机溶剂应该具有较低的黏度 η，从而使电解液中 Li^+ 更容易迁移；③有机溶剂对电池中的各个组分必须是惰性的，尤其是在电池工作电压范围内必须与正极和负极有良好的兼容性；④有机溶剂或者其混合物必须有较低的熔点和较高的沸点，换言之有比较宽的液程，使电池有比较宽的工作温度范围；⑤有机溶剂必须具有较高的安全性（高的闪点）、无毒无害、成本较低。

醇类、胺类和羧酸类等质子性溶剂虽然具有较高的解离盐的能力，但是它们在 2.0~4.0V（vs. Li^+/Li）会发生质子的还原和阴离子的氧化，所以它们一般不用来作为锂离子电池电解质的溶剂。从溶剂需要具有较高的介电常数出发，可以应用于锂离子电池的有机溶剂应该含有羰基（C═O）、氰基（C≡N）、磺酰基（S═O）和醚链（─O─）等极性基团。锂离子电池溶剂的研究主要包括有机醚和有机酯，这些溶剂分为环状的和链状的，一些主要有机溶剂的物理性质参见表 8-1。

对于有机酯来说，其中大部分环状有机酯具有较宽的液程、较高的介电常数和较高的黏度，而链状的溶剂一般具有较窄的液程、较低的介电常数和较低的黏度。其原因主要是环状的结构具有比较有序的偶极子阵列，而链状结构比较开放和灵活，导致偶极子会相互抵消，所以一般在电解液中会使用链状和环状的有机酯混合物来作为锂离子电池电解液的溶剂。对于有机醚来说，不管是链状的还是环状的化合物，都具有比较适中的介电常数和比较低的黏度。

8.3.1.1 碳酸丙烯酯

1958 年，Tobias 等发现 $LiClO_4$ 可以通过电沉积方法从碳酸丙烯酯（PC）溶剂中沉淀下来，从此 PC 溶剂成为受关注的研究对象。PC 熔点（-49.2℃）低，沸点（241.7℃）和闪点（132℃）高，因此含有 PC 的电解液具有好的高低温性能和安全性能。碳酸丙烯酯（PC）具有宽的液程、高的介电常数和对锂的稳定性，所以它是最早被研究的，也是最早被 Sony 公司商业化的锂离子电池溶剂材料。在第一代商业化的锂离子电池中，Sony 公司采用 PC 作为溶剂的主要成分，针状焦作为负极，$LiCoO_2$ 作为正极。

但是 PC 作为电解质溶剂也有很多不足，在开发碳负极材料时，研究发现，溶剂化的 Li^+ 容易向结晶度较高的石墨系碳负极材料共嵌入，使石墨层发生剥离，导致锂离子电池循环性能下降。而结晶度低的碳不容易发生共嵌入，与 PC 有一定的兼容性。

在早期锂电池中，其负极材料是金属 Li，循环过程中会有新的 Li 单质生成，这种 Li 单质具有比较高的比表面积和反应活性，PC 与金属 Li 的反应是不可避免的；其次是在锂离子电池中石墨负极的溶剂共嵌入导致的剥落分解和首周不可逆容量问题，这主要是由 PC 在充电过程中的共嵌入造成的；除此之外，使用 PC 的早期可充放锂电池存在非常严重的安全问题。在循环过程中，Li^+ 的不均匀沉积会导致锂枝晶的形成，随着枝晶的长大，隔膜被刺穿，造成电池短路。

综上所述，PC 很难作为单一的溶剂应用于锂电池和锂离子电池中。

8.3.1.2 碳酸乙烯酯

相比于 PC，碳酸乙烯酯（EC）具有比较高的分子对称性，具有比较高的熔点，关于 EC 最早的研究是由 Elliot 等在 1964 年开始，他们使用 EC 作为一种共溶剂加入电解液中，从而提高了电解液的离子电导率。20 世纪 70 年代，Scrosati 等发现加入电解质锂盐和少量 PC 之后，电解液的熔点将会大幅降低。从此之后，EC 开始作为一种共溶剂应用于锂电池和锂离子电池中，并且最终取得了商业化的大规模应用。EC 基的电解质相对于 PC 基的来说，具有较高的离子电导率、较好的界面性质、能够形成稳定的 SEI 膜，解决了石墨负极的溶剂共嵌入问题。

EC 的高熔点限制了电解质在低温的应用，EC 也很难作为单一溶剂使用，通常和低沸点的链状碳酸酯混合使用。

8.3.1.3 二甲基碳酸酯

通过一系列的研究发现 EC 是电解液中必不可少的部分，为了使 EC 基的电解液能够应用于低温，科研工作者试图在电解液中加入其他的共溶剂来实现。这些共溶剂主要包括 PC 和一系列的醚基溶剂，但是 PC 的加入会导致很大的首周不可逆容量，醚的加入会降低电解液的电化学窗口，所以大家开始考虑线型碳酸酯。关于线型碳酸酯的研究主要是从 1994 年 Tarascon 和 Guyomard 对二甲基碳酸酯（DMC）的研究开始的。

DMC 具有低黏度、低沸点、低介电常数，它能与 EC 以任意比例互溶，得到的混合溶剂以一种协同效应的方式集合了两种溶剂的优势：具有高的锂盐解离能力、高的抗氧化性、低的黏度。这种性质与有机醚类是不同的，该协同效应的机理目前还不是很清楚。

除了 DMC 以外，还有很多其他的线型碳酸酯（如 DEC、EMC 等）也渐渐地被应用于锂离子电池中，其性能与 DMC 相似。目前，常用的锂离子电池电解质溶剂主要是由 EC 和一种或几种线型碳酸酯混合而成。

8.3.1.4 醚类溶剂

在 20 世纪 80 年代，醚类溶剂曾经引起广泛的关注，因为它们具有低的黏度、高的离子电导率和相对于 PC 改善的 Li 负极表面形貌。其中主要研究集中于 THF、2-Me-THF、DME 和聚醚等，发现它们虽然循环效率有所提高，但是也存在很多问题，限制了它们的实际应用。首先是容量保持率比较差，随着循环进行，容量衰减较快；其次是在长循环过程中仍然会有锂枝晶的产生，导致安全问题；此外，醚类溶剂抗氧化性比较差，在低电位下很容易被氧化分解。如在 Pt 表面，THF 的氧化电位仅为 4.0V(vs. Li^+/Li)，而环状碳酸酯能够达到 5.0V。很多高电压的正极材料需要在 4.0V 或以上工作，这就限制了醚类电解质的应用。在目前研究的锂硫电池和锂空气电池中（其充电电压低于 4.0V），醚类电解质有希望得到应用。

8.3.1.5 砜类溶剂

砜类作为高电压电解质溶剂的研究是从 1998 年徐康等对乙基甲基砜（EMS）的研究开始的，后来 Dahn 等对 EMS 进行了深入研究，发现基于 EMS 的电解液具有很高的抗氧化电位且能与正极材料有很好的相容性。该溶剂熔点比较高（36.5℃），低温性能不是很好，Angell 等对该溶剂分子结构进行改性，得到了熔点比较低的甲氧基乙基甲基砜（MEMS，熔点接近 0℃）。基于 MEMS 的电解液使用 LiTFSI 作为锂盐，具有比较宽的电化学窗口，但是和石墨负极的相容性不好。2009 年，Amine 等研究了不同体系的砜基电解质，发现乙基甲基砜和四甲基砜（TMS）具有比较高的离子电导率和比较宽的电化学窗口，能与高电压正极材料 $LiMn_2O_4$ 和 $LiNi_{0.5}Mn_{1.5}O_4$ 有很好的相容性。2011 年，Jiang 等通过计算在氧化电位方面给出了同样的结论。

砜类电解质也有其局限性，限制了它的大规模应用。首先，砜类电解液与石墨负极的相容性比较差，不能在石墨负极表面形成稳定的 SEI 膜，负极只能选择还原电位较高的 $Li_4Ti_5O_{12}$；其次，砜类电解液与 Celgard 隔膜的浸润性较差，需要使用价格较高的玻璃纤维隔膜来组装电池，该性能在加入线型碳酸酯或乙酸乙酯之后有所改善。

8.3.1.6 腈类溶剂

腈类溶剂具有较宽的液程、较高的介电常数和较低的黏度，因此有希望作为锂离子电池电解质溶剂使用。腈类电解液首先被 Ue 等应用于双电层电容器，戊二腈和己二腈的氧化电位高达 $8.3V(vs. Li^+/Li)$，是已知的氧化电位最高的溶剂。腈类（如戊二腈、己二腈和癸二腈等），作为共溶剂也被用于锂离子电池中，组成的电解液具有较高的电导率、较低的黏度和较强的抗氧化性，能够与高电压正极材料 Li_2NiPO_4F 等有较好的兼容性。除此之外，腈类电解液具有非常优越的低温性能。但是腈类溶剂与负极的兼容性差，需要使用碳酸酯作为共溶剂或者使用 VC 和 LiBOB 等成膜添加剂改善其性能。

8.3.1.7 氟取代碳酸酯溶剂

氟取代碳酸酯具有低熔点、高氧化稳定性和不易燃等优势，从而适合用于锂离子电池电解质，很多研究者把它作为锂离子电池电解质添加剂或共溶剂使用。Frisch 等通过计算发现，相比于没有取代的碳酸酯，氟代碳酸酯确实具有比较高的抗氧化电位。2013 年，Zhang 等把 F-AEC、F-EMC 和 F-EPE 等氟代碳酸酯用作锂离子电池电解质溶剂，发现它们具有比较高的氧化电位和比较好的高电压电极材料（$LiNi_{0.5}Mn_{1.5}O_4$）相容性。

除此之外，氟类化合物非常稳定，在高温下循环性和储存性能都明显提高。氟取代碳酸酯有望应用于高电压体系的电解质溶剂，但是它对 $LiPF_6$ 的溶解性较差，需要进一步的改善。

8.3.2 锂盐

尽管锂盐的种类非常多，但是能应用于锂离子电池电解质的锂盐却非常少，目前文献报道的溶剂有 150 多种，而锂盐只有几种。如果要应用于锂离子电池，它需要满足如下一些基本要求：①在有机溶剂中具有比较高的溶解度，易于解离，从而保证电解液具有比较高的电导率；②具有比较高的抗氧化还原稳定性，与有机溶剂、电极材料和电池部件不发生电化学和热力学反应；③锂盐阴离子必须无毒无害，环境友好；④生产成本较低，易于制备和提纯。实验室和工业生产中一般选择阴离子半径较大、氧化和还原稳定性较好的锂盐，以尽量满足以上特性。

常见的阴离子半径较小的锂盐（如 LiF、LiCl 和 Li_2O 等）虽然成本较低，但是其在有机溶剂中溶解度较低，很难满足实际需求。虽然硼基阴离子受体化合物的使用大大提高了它们的溶解度，但是会带来电解液黏度增加等问题。如果使用 Br^-、I^-、S^{2-} 和羧酸根等弱路易斯碱离子取代这些阴离子，锂盐的溶解度会得到提高，但是电解液的抗氧化性将会降低。

$LiAlX_4$（X 代表卤素）是在一次锂电池中经常使用的锂盐，但是 AlX_3 是比较强的路易斯酸，导致这一系列的锂盐容易与有机溶剂反应。除此之外，AlX_4^- 容易与电池部件发生反应，这些不足限制了它们在锂离子电池中的应用。

目前，经常研究的锂盐主要是基于温和路易斯酸的一些化合物，这些化合物主要包括高氯酸锂（$LiClO_4$）、硼酸锂、砷酸锂、磷酸锂和锑酸锂等（简称 $LiMF_n$，其中 M 代表 B、As、P、Sb 等，n 等于 4 或者 6）。除此之外，有机锂盐〔如 $LiCF_3SO_3$、$LiN(SO_2CF_3)_2$ 及其衍生物〕也被广泛研究和使用。一些常用锂盐的物理化学性质参见表 8-2，下文将介绍一些代表性的锂盐。

8.3.2.1 六氟磷酸锂

六氟磷酸锂（$LiPF_6$）是目前商品锂离子电池中广泛使用的电解质锂盐，虽然它单一的性质并不是最优的，但是其综合性能是最有优势的（表 8-2）。$LiPF_6$ 在常用有机溶剂中具有比较适中的离子迁移数、适中的解离常数、良好的铝箔钝化能力，尽管通常认为在 4.4V 以上 $LiPF_6$ 会发生分解，但是配合成膜添加剂后，$LiPF_6$ 能够与各种正负极材料匹配。

表 8-2 一些锂离子电池常用锂盐的物理化学性质

锂盐	摩尔质量 /(g/mol)	是否对铝箔腐蚀	是否对水敏感	电导率 σ(1mol/L,EC/DMC,20℃)/(mS/cm)
六氟磷酸锂（$LiPF_6$）	151.91	否	是	10.00
四氟硼酸锂（$LiBF_4$）	93.74	否	是	4.50
高氯酸锂（$LiClO_4$）	106.40	否	否	9.00
六氟砷酸锂（$LiAsF_6$）	195.85	否	是	11.10(25℃)
三氟甲基磺酸锂（$LiCF_3SO_3$）	156.01	是	是	1.70(PC,25℃)
双（三氟甲基磺酰）亚胺锂（LiTFSI）	287.08	是	是	6.18
双（全氟乙基磺酰）亚胺锂（LiBETI）	387.11	是	是	5.45
双氟磺酰亚胺锂（LiFSI）	187.07	是	是	10.40(25℃)
（三氟甲基磺酰）（正全氟丁基磺酰）亚胺锂（LiTNFSI）	437.11	否	是	1.55
（氟磺酰）（正全氟丁基磺酰）亚胺锂（LiFNFSI）	387.11	否	是	4.70
双草酸硼酸锂（LiBOB）	193.79	否	是	7.50(25℃)

但是 $LiPF_6$ 也有其缺点，限制了它在很多体系中的应用。首先，$LiPF_6$ 是化学和热力学不稳定的，即使在室温下也会发生如下反应：$LiPF_6(s) \longrightarrow LiF(s) + PF_5(g)$，该反应的气相产物 PF_5 会使反应向右移动，在高温下分解尤其严重。PF_5 是很强的路易斯酸，容易进攻有机溶剂中氧原子上的孤对电子，导致溶剂的开环聚合和醚键裂解。

其次，$LiPF_6$ 对水比较敏感，痕量水的存在就会导致 $LiPF_6$ 的分解，这也是 $LiPF_6$ 难以制备和提纯的主要原因。其分解产物主要是 HF 和 LiF，其中 LiF 的存在会导致界面电阻的增大，影响锂离子电池的循环寿命。

在 2011 年以前，$LiPF_6$ 的主要生产者是日本和韩国的一些公司，近年来我国也取得了重要进展，实现了高纯 $LiPF_6$ 的产业化。目前，国内 $LiPF_6$ 的主要生产厂家包括多氟多、九九久、天津金牛等，其中多氟多化工股份有限公司能够批量生产晶体 $LiPF_6$。全球产量较大的 $LiPF_6$ 生产公司包括日本森田、关东电化、SUTERAKEMIFA 和韩国蔚山等公司。

寻找能够替代 $LiPF_6$ 的新型锂盐的研究工作一直在持续，主要包括以下三类化合物：①以 C 为中心原子的锂盐，如 $LiC(CF_3SO_2)_3$ 和 $LiCH(CF_3SO_2)_2$ 等，$LiC(CF_3SO_2)_3$ 的热稳定性比较好，$LiCH(CF_3SO_2)_2$ 的电化学性能比较稳定；②以 N 为中心原子的锂盐，如 $LiN(CF_3SO_2)_2$，由于阴离子电荷的高度离域分散，该盐在有机电解液中极易解离，其电导率与 $LiPF_6$ 相当，也能在负极表面形成均匀的钝化膜，但是其从 3.6V 左右开始就对正极集流体铝箔有很强的腐蚀作用；③以 B 为中心原子的锂盐，如双硼酸酯锂（LiBOB），其分解温度为 320℃，同时其具有电化学稳定性高、分解电压＞4.5V 等优点，但其还原电位较高［约 $1.8V(vs. Li^+/Li)$］，作为主盐无法在负极表面形成有效的钝化膜抑制其持续分解，因此不能作为锂离子电池电解质的导电盐。

2011 年，Zhou 等合成了新的锂盐 LiFNFSI，该盐在 220℃下不分解，具有较高的电导率，高温 60℃条件下，在石墨/$LiCoO_2$ 电池中表现出较好的循环性能。

8.3.2.2 四氟硼酸锂

由于 $LiPF_6$ 存在易分解和对水分敏感的问题，关于 $LiPF_6$ 的替代锂盐的研究工作一直在进行，四氟硼酸锂（$LiBF_4$）便是其中的一种。相对于 $LiPF_6$ 来说，$LiBF_4$ 的高温性能和低温性能均比较好。除此之外，相对于 $LiClO_4$ 来说它具有比较高的安全性。但是它的解离常数相对于其他锂盐要小很多，导致 $LiBF_4$ 基电解质电导率不高，$LiBF_4$ 容易与金属 Li 发生反应，与石墨负极的兼容性较差，这些因素限制了它的大规模应用。

8.3.2.3 高氯酸锂

高氯酸锂（$LiClO_4$）由于其价格低廉、对水分不敏感、高稳定性、高溶解性、高离子电导率和正极表面高氧化稳定性［约 $5.1V(vs. Li^+/Li)$］一直受到广泛关注。研究发现，相比于 $LiPF_6$ 和 $LiBF_4$ 来说，$LiClO_4$ 基的电解质在负极表面形成的 SEI 膜具有更低的电阻，这与前者容易形成 HF 和 LiF 有关。

$LiClO_4$ 是一种强氧化剂，它在高温和大电流充电的情况下很容易与溶剂发生剧烈反应；其次在运输过程中不安全，因此 $LiClO_4$ 一般在实验室应用而几乎不应用于工业生产。

8.3.2.4 六氟砷酸锂

六氟砷酸锂（$LiAsF_6$）的各项性能均比较好，与 $LiPF_6$ 接近，它作为锂盐的电解液具有比较高的离子电导率，比较好的负极成膜性能[约 1.15V(vs. Li^+/Li)]，并且 SEI 膜中不含 LiF，原因是 As—F 键比较稳定，不容易水解，该类电解液还具有比较宽的电化学窗口，$LiAsF_6$ 曾经广泛应用于一次锂电池中。

但是 $LiAsF_6$ 有毒，成膜过程中会有剧毒的 As（Ⅲ）生成，其反应为：$AsF_6^- + 2e^- \longrightarrow AsF_3 + 3F^-$，并且在一次锂电池中还存在锂枝晶的生长，导致了 $LiAsF_6$ 主要用于研究。

8.3.2.5 三氟甲基磺酸锂

磺酸盐是一类重要的锂离子电池电解质锂盐，这类有机锂盐存在强的全氟烷基吸电子基团，强的吸电子基团和共轭结构的存在导致负电荷被离域，所以其阴离子比较稳定，酸性明显提高。因此，这些锂盐即使在低介电常数的溶剂中解离常数也非常高，由于全氟烷基的存在导致这些锂盐在有机溶剂中溶解度也很大。相比于羧酸盐、$LiPF_6$ 和 $LiBF_4$ 来说，磺酸盐的抗氧化性好、热稳定性高、无毒、对水分不敏感。综上所述，有机磺酸锂盐比较适合作为锂离子电池电解质锂盐。

其中三氟甲基磺酸锂（$LiCF_3SO_3$）是一种组成和结构最简单的磺酸盐，它是最早工业化的锂盐之一，它具有比较好的电化学稳定性，与 $LiPF_6$ 接近。但是它存在的一些缺点限制了它的大规模应用：首先是一次电池中锂枝晶的生长问题；其次是这种锂盐所组成的电解液电导率较低；最后是这种盐存在严重的铝箔腐蚀问题。

8.3.2.6 双（三氟甲基磺酰）亚胺锂

双（三氟甲基磺酰）亚胺锂（LiTFSI）是一种酰胺基的锂盐，它的结构式为

$$F_3C-\underset{\underset{O}{\overset{O}{\|}}}{S}-\underset{Li}{N}-\underset{\underset{O}{\overset{O}{\|}}}{S}-CF_3$$

从结构式可以看出该盐的阴离子由两个三氟甲基磺酸基团稳定，同样存在较强的吸电子基团和共轭结构，所以它也是一种酸性很强的化合物，与硫酸相近。Armand 等将此盐应用于聚合物锂离子电池，3M 公司在 20 世纪 90 年代将此盐进行了商业化，作为动力电池的添加剂使用，具有改善正负极 SEI 膜、稳定正负极界面、抑制气体产生、改善高温性能和循环性等多种功能。

LiTFSI 具有高的离子电导率，宽的电化学窗口[玻璃碳作为工作电极，5.0V(vs. Li^+/Li)]，能够抑制锂枝晶的生长，所以引起了广泛的关注。但是 LiTFSI 也有其不足之处，它对正极集流体铝箔存在严重的腐蚀，需要加入能够钝化铝箔的添加剂，例如 $LiPF_6$ 或含氰基的化合物，才能在一定程度上抑制该反应。

8.3.2.7 双氟磺酰亚胺锂

双氟磺酰亚胺锂（LiFSI）具有与 LiTFSI 相似的物理化学性质。该盐是由 Armand 等于 1999 年合成并报道的，它具有比较高的电导率，接下来 Zaghib 等对此盐及其在锂离子电池中的应用进行了初步的研究。该盐各项性能都比较好：具有高的热稳定性，在碳酸酯体系中具有高的溶解度，相比于 $LiPF_6$ 体系具有较高的电导率和锂离子迁移数。但是存在腐蚀铝箔的问题，这主要是由合成过程中引入的 Cl^- 杂质和电解液中痕量水分造成的。该盐的铝箔腐蚀问题可以通过加入 $LiClO_4$ 等添加剂来解决。

除此之外，华中科技大学的周志斌教授合成了一系列的氟磺酰亚胺锂盐，它们的结构式为

$$R^1-\underset{\underset{O}{\overset{O}{\|}}}{S}-\underset{Li}{N}-\underset{\underset{O}{\overset{O}{\|}}}{S}-R^2$$

其中，$R^1 = C_nF_{2n+1}$，$R^2 = C_mF_{2m+1}$，n 和 m 是 ≥ 0 的整数。在这一系列的锂盐中，随着链长的增加，铝箔腐蚀会得到一定程度的抑制，其中 LiFNFSI（$n=0$，$m=4$）是一种热稳定性非常高的锂盐，而且对水分不敏感，具有较高的溶解度和较高的电导率，经过研究发现它具有比较好的室温和高温性能。目前苏州氟特公司正在准备量产该材料。

8.3.2.8 双草酸硼酸锂

双草酸硼酸锂（LiBOB）首先由 Lischka 等在 1999 年合成，它是一种配位螯合物，正交晶系，属于 $Pnma$ 空间点群。它的结构式为

从图中可以看出 BOB 以硼原子为中心，呈四面体结构，这种五重配位的形式使得 Li^+ 很容易再结合其他分子形成正八面体配位结构，所以 LiBOB 具有很强的吸湿性。这种结构电荷分布比较分散，阴阳离子相互作用较弱，在有机溶剂中具有较高的溶解度。此外，徐康等发现 LiBOB 还原电位较高［约 1.6V（vs. Li^+/Li）］，可以防止石墨电极的 PC 共嵌入问题，也有助于在负极表面成膜。但由于实际溶解度较小、电导率较低，可以作为添加剂在锂离子电池中使用。

8.3.2.9 二氟草酸硼酸锂

作为结合了 LiBOB 和 $LiBF_4$ 优点的锂盐二氟草酸硼酸锂（LiDFOB）具有良好的热稳定性和溶解度，不易水解，能够钝化铝箔，是常用的新型锂盐。

LiDFOB 也可以作为添加剂使用，在正负极表面均具有良好的成膜性，由于其溶解度较高，LiDFOB 可以作为辅助锂盐配合 $LiPF_6$ 使用。但是 LiDFOB 离子电导率较低，低温性能较差，作为单一锂盐使用倍率性能不佳。

经过多年的努力，锂离子电池非水液体电解质的基本组分已经确定：主要是 EC 加一种或几种线型碳酸酯作为溶剂，$LiPF_6$ 作为电解质锂盐。但是这种体系的电解质也存在一

些难以解决的问题：①EC 导致的熔点偏高问题，致使这种体系的电解质无法在低温下应用；②$LiPF_6$ 的高温分解导致该电解质无法在高温下使用。该电解质体系的工作温度范围为 -20～50℃，低于 -20℃ 时性能下降是暂时的，高温可以恢复，但是高于 60℃ 的性能变化则是永久性的。

鉴于该体系的种种问题，很多关于其他有机溶剂和电解质锂盐的研究工作一直在继续，目前也取得了一些研究进展，但是离商业化应用还有些距离。

8.3.3 添加剂

商品的锂离子电池可能包含 10 种以上的添加剂，这些添加剂的特点是用量少但是能显著改善电解液某一方面的性能。它们的作用一般分为提高电解液的电导率，提高电池的循环效率，增大电池的可逆容量，改善电极的成膜性能等，以下将简单介绍。

8.3.3.1 成膜添加剂

成膜添加剂的作用是改善电极与电解质之间的 SEI 膜成膜性能，改善负极与电解液之间的界面化学。最早的成膜添加剂是由美国 Covalent 公司在 1997 年提出的 SO_2 添加剂，它能够有效防止 PC 共嵌入，防止电极腐蚀，提高电池的安全性。接下来又有很多成膜添加剂被发现，其中最重要的有碳酸亚乙烯酯（VC），它能够有效防止 PC 共嵌入，提高 SEI 膜高温稳定性。除此之外经常使用的成膜添加剂还有氟代碳酸乙烯酯（FEC）、亚硫酸丙烯酯（PS）和亚硫酸乙烯酯（ES）等。

8.3.3.2 离子导电添加剂

离子导电添加剂的作用是提高电解液的电导率，它主要是通过阴阳离子配体或中性配体来提高锂盐的解离度从而达到该目的的。最早使用的导电添加剂是由法国科学院在 1996 年提出的 NH_3 和一些低分子量胺类。目前常用的导电添加剂有 12-冠-4-醚、阴离子受体化合物和无机纳米氧化物等，它们均能有效提高电解液的离子电导率。

8.3.3.3 阻燃添加剂

安全性是锂离子电池一直以来最为关注的问题，阻燃添加剂的加入能够在一定程度上提高电解液的安全性。它的主要作用机理是：锂离子电池电解液在受热的情况下会发生自由基引发的链式加速反应，而阻燃添加剂能够捕获自由基，阻断链式反应。目前常用的阻燃添加剂有磷酸三甲酯（TMP）、磷酸三乙酯（TEP）等磷酸酯，二氟乙酸甲酯（MFA）、二氟乙酸乙酯（EFA）等氟代碳酸酯和离子液体等。其中磷酸酯在加热过程中生成的磷自由基能够将氢自由基有效捕获，氟代碳酸酯能够将电解液的放热峰提高 100℃ 以上。

8.3.3.4 过充保护添加剂

锂离子电池在过充时会产生安全问题，通过外电路的控制和保护可以解决这个问题，除此之外，使用过充保护添加剂也是一种有效的方法。它的基本原理是：添加剂的氧化电位略高于正极脱锂电位,当电池超过工作电压之后,添加剂优先发生反应,造成电池的断路或微短路,从而使电池停止工作并缓慢放热,这个过程不破坏电极材料和电解液。但是在正常的工作电压范围内,添加剂不参与电池反应。过充保护添加剂主要包括氧化还原电对、电聚合和气体发生三种类型的添加剂。其中氧化还原电对添加剂最为常用,其原理是:当电压超过电池截止电压时,添加剂在正极表面被氧化,氧化产物扩散到负极表面被还原,还原产物再扩散到正极表面继续被氧化,按照此过程循环进行,直到充电结束。比较常见的过充保护添加剂有邻位和对位的二甲氧基取代苯、丁基二茂铁和联苯等。

8.3.3.5 控制电解液中酸和水含量的添加剂

电解液中微量的 HF 和水分会造成 $LiPF_6$ 的分解和电极材料表面的破坏,所以要控制电解液中水分和酸的含量。这类添加剂的作用机理主要是靠与电解液中酸和水结合来降低它们的含量。目前常用的控制 HF 含量的添加剂主要有锂或钙的碳酸盐、氧化铝、氧化镁和氧化钡等,它们能够与电解液中微量的 HF 发生反应,阻止它的影响。控制水含量的添加剂主要是六甲基二硅烷(HMDS)等吸水性较强的化合物。

8.3.3.6 高电压添加剂

添加剂的使用是改善碳酸酯体系电解液抗氧化性的一种有效手段,这类添加剂主要是正极成膜添加剂,它的 HOMO 能量略高于溶剂分子,能够提前在正极表面成膜,从而改善正极材料的表面性能,减少正极材料与电解液的接触以达到抑制电解液分解的目的。

目前,常用的高电压添加剂主要有苯的衍生物(如联苯、三联苯)、杂环化合物(如呋喃、噻吩及其衍生物)、1,4-二氧环乙烯醚和三磷酸六氟异丙基酯等。它们均能有效改善电解液在高电压下的氧化稳定性,在高电压锂离子电池中起着非常重要的作用。研究发现,溶剂的纯度对电解液的抗氧化性也有重要影响,溶剂纯度的提高能够大幅度提高电解液的抗氧化性。关于溶剂纯度的影响,日本佐贺大学的 Yoshio 在第十届中国国际电池技术交流会上给出了详细介绍。如当 EC 的纯度从 99.91% 提高到 99.979% 时,它的氧化电位从 4.87V 提高至 $5.5V(vs. Li^+/Li)$。

添加剂的种类和作用非常多,除了上文详细描述的这些之外还有很多其他的多功能添加剂,例如甲基乙烯碳酸酯(MEC)和氟代碳酸乙烯酯(FEC)等改善高低温性能的添加剂,LiBOB 和 LiODFB 等抑制铝箔腐蚀的添加剂,联苯和邻三联苯等改善正极成膜性能的添加剂,三(2,2,2-三氟乙基)磷酸(TTFP)等提高 $LiPF_6$ 稳定性的添加剂。开发能够提高电解液各项性能指标的添加剂是提高锂离子电池性能的重要手段。

8.4 离子液体电解质

离子液体由阴阳离子组成,由于阴离子或者阳离子体积较大,阴阳离子之间的相互作用力较弱,电子分布不均匀,阴阳离子在室温下能够自由移动,使之呈液体状态。相比于锂离子电池，离子液体有如下几个优势：①离子液体的饱和蒸气压非常低，几乎可以忽略，不易挥发；②离子液体具有较宽的液程，大约300℃；③离子液体不易燃，具有较高安全性；④具有较高的离子电导率；⑤化学或电化学稳定性好，通常具有5V以上的电化学窗口；⑥对水和空气不敏感；⑦无污染且易回收，比较环保。离子液体被认为是一种绿色溶剂，以上这些优势可以消除传统液态电解质带来的安全隐患，有希望应用于特殊条件下和安全系数要求高的系统中。

离子液体的种类较多，根据有机阳离子的不同可以分为如下几类：含氮杂环类、季铵盐类和季鏻盐类等，其中含氮杂环类又包括咪唑盐类、哌啶盐类和吡啶盐类等。组成离子液体的阴离子种类也比较多，按照阴阳离子随机组合，理论上能够组成千万种离子液体。

离子液体于1914年由Walden首次合成，20世纪70年代开始应用于锂离子电池电解质中。在国内，中国科学院过程工程研究所的离子液体清洁过程与节能创新团队已经实现了离子液体的产业化。由于其存在很多独特的性质，在锂离子电池中的应用引起了众多研究者的兴趣。但是离子液体电解质也存在诸多问题：首先是大部分离子液体黏度较高，比普通有机溶剂高至少1~2个数量级，导致电池的浸润性较差，虽然离子电导率高，但是离子迁移数较低，因此倍率性能不佳；其次是与正负极材料的兼容性差；除此之外，离子液体目前价格较高，限制了其在锂离子电池中的大规模应用。

离子液体是一种室温熔融盐，高温熔融盐也具有较高的应用前景。中国科学院物理研究所李泓团队与华中科技大学周志斌研究团队合作研究的LiFSI-KFSI（4:6，摩尔比）高温熔融盐电解质取得了一定的进展，它在85℃具有较高的电导率和锂离子迁移数，具有宽的电化学窗口，并且能够钝化铝箔，与正极有很好的兼容性，该体系有希望应用于高温锂离子电池中。

离子液体电解质是有望彻底解决锂离子电池安全性的一类电解质。降低离子液体的黏度，提高锂离子在其中的迁移速率，是离子液体电解质能够在锂离子电池中商业化的前提。研究发现，使用含有不对称结构的离子液体有助于降低其黏度。

将离子液体与碳酸酯电解质溶液、聚合物或无机陶瓷混合形成的三类电解质也获得了大量研究，并取得了较好的结果，这些新体系的发展有望促进离子液体在电池中获得实际应用。

8.5 凝胶聚合物电解质

Feuillade等在1975年首先提出了凝胶型电解质，后来由Abraham等作了深入的研究。凝胶型聚合物电解质通常被定义为，聚合物作为整个电解质的骨架部分，起到力学支撑的作用，其网络结构使得碱金属盐和有机增塑剂形成的电解液能够均匀地分布在网络中，而离子的输运主要发生在包含液体电解质的部分。这类电解质的电导率与碱金属盐的有机溶液相当，在室温下一般可以达到10^{-3}S/cm以上。1995年美国Bellcore公司开发了新型凝胶聚合物电解质，发展聚合物锂离子电池产业。1996年Bellcore公司将凝胶聚合物电解质应用在锂离子电池上,并率先申请了有关塑料锂离子电池的专利,很多公司随后购买了专利的使用权,将聚合物锂离子电池推广向市场。在第三代锂离子电池中,Sony 公司将凝胶聚合物电解质实现了商业化,并自此开始对凝胶型聚合物锂离子电池展开研究。从 1975 年凝胶型聚合电解质首次被提出以来,多种体系的凝胶聚合物电解质得到了不断的研究和发展。其中常用的凝胶型聚合物电解质基体有:聚丙烯腈(PAN)、聚氧化乙烯(PEO)、聚甲基丙烯酸甲酯(PMMA)、聚偏氟乙烯(PVDF)等。凝胶型聚合物电解质的电导率还与有机溶剂的种类有关,常用的增塑剂有 EC、PC、γ-BL 等,也可以将几种增塑剂混合使用。ATL 采用 Bellcore 技术,利用 PVDF-HFP 共聚物经过一定工艺自成膜,然后卷绕,注液后 PVDF 吸水膨胀形成胶状电解质。新一代的聚合物锂离子电池在形状上可做到薄形化,ATL 电池最薄可达 0.5mm,相当于一张卡片的厚度,任意面积化和任意形状化,大大提高了电池造型设计的灵活性,从而可以配合产品需求。

8.5.1 聚丙烯腈

聚丙烯腈(PAN)是较早被研究的一种聚合物,它与碳酸酯类增塑剂相容性好,有良好的阻燃性,具有很宽的电化学窗口,并且具有较高的结晶度。Feuillade 等对 PAN 基的凝胶型聚合物电解质进行了深入的研究,通过配比不同比例的增塑剂、锂盐,制备出凝胶型聚合物电解质,发现 PAN 基的聚合物电解质具有良好的电导率,可以达到 10^{-3}S/cm,满足实用的要求,而且 Li^+ 迁移数可以达到 0.5 以上。由于 PAN 上的氰基与 Li 电极相容性差,通过掺入无机填料可以对其进行改性。将沸石粉掺在 PAN 凝胶聚合物电解质中形成复合电解质材料,抑制了低温时结晶过程的发生,增加了无定形区面积,从而降低了玻璃化转变温度,提高低温的离子电导率。5%的沸石可以有效降低金属 Li 表面阻挡层的生长,改善了电解质/电极的界面稳定性。

8.5.2 聚氧化乙烯

1978 年 Armand 提出聚氧化乙烯(PEO)与碱金属配位的可充电离子导体,从此开始了对 PEO 为基的聚合物电解质的研究工作。PEO 是结晶度较高的线型化合物,它的导电机理是:在其非晶相中,通过锂盐与聚合物链段不断发生的络合解络合,来完成离子的传导过程。PEO 能和锂盐形成络合物,在没有任何增塑剂的情况下,离子在 PEO 聚合物的链段上进行跳跃传导,而且具有较高的电导率。由于增塑剂可以降低 PEO 的结晶度,提高链段的运动能力和锂盐的解离度,从而进一步提高离子电导率。1990 年,Morita 等制备的凝胶 PEO 电解质,室温下电导率可以达到 10^{-3}S/cm。加了增塑剂的 PEO 力学性能较差,研究表明加入纳米无机陶瓷填料可以增强其力学性能。Appetecch 等报道了一种新型材料,他们将 PEO 作为基体,混入 PEGDME,增塑剂 DMC、PC、DEC,陶瓷 γ-$LiAlO_2$,共混制备得到凝胶型聚合物电解质,其电导率可以达到 1.9×10^{-3}S/cm,改善了凝胶聚合物电解质的力学性能,而且具有较宽的电化学窗口。

8.5.3 聚甲基丙烯酸甲酯

1985 年 Iijima 等首次提出了用 PMMA 作为溶胶电解质,当 PMMA 含量超过 15%

（质量分数）时，25℃的电导率可以达到 10^{-3} S/cm。PMMA 含有羧基官能团，与碳酸酯相容性好，对 Li 电极的界面稳定性高，价格便宜，制备简单。由于 PMMA 的力学强度较差，导致力学性能降低。徐康等通过对聚合物的改性，将 PMMA 进行化学交联，得到了力学性能较高、性质均一的凝胶型聚合物电解质，由于含液率较高，具有较好的离子电导率。在 $-50 \sim 200$℃ 内，没有出现相转变，提高了聚合物锂离子电池的使用温度范围。

8.5.4 聚偏氟乙烯

20 世纪 80 年代已开始了对聚偏氟乙烯（PVDF）的研究。PVDF 具有高的电化学稳定性、高的介电常数以及高的结晶度。Bellcore 公司最早实现的聚合物锂电池，就是采用了 P（VDF-HFP）为基体的微孔型聚合物电解质膜作为核心组件。但是由于与金属 Li 的界面稳定性差，而限制了在金属 Li 作负极材料的锂电池中的应用。Jiang 等以 PVDF 作为聚合物基体，加入增塑剂 EC、PC 制备了凝胶聚合物电解质。发现该电解质的电导率与电解质的黏度和载流子的浓度有关系，而且聚合物的力学强度受 PVDF 含量的影响。当锂盐选择 $LiN(SO_2CF_3)_2$ 时，室温电导率可以达到 2.2×10^{-3} S/cm。而且在不锈钢作为电极集流体时，电化学窗口满足应用要求。唯一一种在文献中公开商业化的凝胶聚合物电解质是采用 Bellcore 技术成膜的 PVDF 基电解质。

除了上述介绍的聚丙烯腈（PAN）、聚氧化乙烯（PEO）、聚甲基丙烯酸甲酯（PMMA）、聚偏氟乙烯（PVDF）等聚合物，其他凝胶聚合物电解质种类也很多。聚合物共混体系、无机陶瓷复合体系已经成为目前聚合物电解质研究的热点。凝胶型聚合物电解质对环境的污染小，使用的安全性能更好，在电池市场中倍受青睐。

8.6 本章结语

锂离子电池电解质未来发展的方向需要重点解决以下问题。

① 电解液和电池的安全性。通过离子液体，氟代碳酸酯，加入过充添加剂、阻燃剂，采用高稳定性锂盐来解决，最终可能需要通过固体电解质来彻底解决安全性，将在后续章节讨论。

② 提高电解质的工作电压。可以通过提纯溶剂，采用离子液体、氟代碳酸酯，添加正极表面膜添加剂等来解决，同样发展固体电解质也能显著提高电压范围。

③ 拓宽工作温度范围。低温电解质体系需要采用熔点较低的醚、腈类体系，高温需要采用离子液体（熔融盐）、新锂盐、氟代酯醚来提高。固体电解质可以在很高的温度工作，但低温性能可能较差。

④ 延长电池寿命。需要精确调控 SEI 膜的组成与结构，主要通过加入 SEI 膜成膜添加剂、游离过渡金属离子捕获剂等来实现。固体电解质应该在界面稳定性方面具有优势。

⑤ 降低成本。需要降低锂盐和溶剂的成本，解决锂盐和溶剂纯度较低时如何提高电池性能的技术问题，这方面目前仍需要深入研究。

本章需要进一步思考的基础科学问题：

1. 液态电解质的设计如何平衡对离子电导率、迁移数、黏度、活化能的要求？
2. 液态电解质如何提高阳离子迁移数？
3. 如何判断液态电解质在多孔粉末电极中充分浸润电极？如何实现？
4. 如何通过控制液态电解质的溶剂、盐和添加剂以及化成过程控制 SEI 膜的组成、厚度、结构、热稳定性？
5. 高盐浓度电解质如何在多孔粉末电极中实现均匀分布？
6. 如何提高液态电解质的高温稳定性？如何防止高温储存或过充胀气？
7. 液态电解质能够在最低多低的温度下保持较高的离子电导率？
8. 相转变类电解质在电池中应用的主要问题是什么？
9. 在浸润了液态电解质的复杂多孔粉末电极中，离子在不同尺度上是如何输运的？如何判断局部是否出现贫锂或富锂？
10. 液态电解质能否在任意工作状态下防止与锂的持续副反应？

参考文献

[1] Tobias C W. Electrochemical studies in cyclic carbonate esters [J]. Journal of the Electrochemical Society, 1957, 104 (8): C171.

[2] Eineli Y, Thomas S R, KochV R. New electrolyte system for Li-ion battery [J]. Journal of the Electrochemical Society, 1996, 143 (9): L195-L197.

[3] Ma S H, Li J, Jing X B, et al. A study of cokes used as anodic materials in lithium ion rechargeable batteries [J]. Solid State Ionics, 1996, 86 (8): 911-917.

[4] Nishi Y, Azuma H, Omaru A. Non-aq. electrolyte cell having improved cycling characteristics | by close control of the interlayer spacings of the carbonaceous anode, the true density and the temp. at which exothermic peaks appear: EP, 357001-A1; JP, 2066856-A; US, 4959281-A; EP, 357001-B1; DE, 68919943-E; JP, 2674793-B2; JP, 10003948-A; JP, 2812324-B2; KR, 9711198-B1 [P/OL]. 1990-03-07. http://www.google.com/patents/US4959281.

[5] Fong R, Vonsacken U, Dahn J R. Studies of lithium intercalation into carbons using nonaqueous electrochemicalcells [J]. Journal of the Electrochemical Society, 1990, 137 (7): 2009-2013.

[6] Guyomard D, Tarascon J M. Rechargeable $Li_{1+x}Mn_2O_4$/carbon cells with a new electrolyte-composition-potentiostatic studies and application to practical cells [J]. Journal of the Electrochemical Society, 1993, 140 (11): 3071-3081.

[7] Tarascon J M, Guyomard D. New electrolyte compositions stable over the 0V to 5V voltage range and compatible with the $Li_{1+x}Mn_2O_4$ carbon Li-ion cells [J]. Solid State Ionics, 1994, 69 (3-4): 293-305.

[8] Dahn J R, Vonsacken U, Juzkow M W, et al. Rechargeable $LiNiO_2$ carbon cells [J]. Journal of the Electrochemical Society, 1991, 138 (8): 2207-2211.

[9] Ohzuku T, Iwakoshi Y, Sawai K. Formation of lithium-graphite intercalation compounds in nonaqueous electrolytes and their application as a negative electrode for a lithium ion (shuttlecock) cell [J]. Journal of the Electrochemical Society, 1993, 140 (9): 2490-2498.

[10] KochV R, Young J H. Stability of secondary lithium electrode in tetrahydrofuran-based electrolytes [J]. Journal of the Electrochemical Society, 1978, 125 (9): 1371-1377.

[11] Desjardins C D, Cadger T G, Salter R S, et al. Lithium cycling performance in improved lithium hexafluoroarsenate 2-methyl tetrahydrofuran electrolytes [J]. Journal of the Electrochemical Society, 1985, 132 (3): 529-533.

[12] Abraham K M, Goldman J L, Natwig D L. Characterization of ether electrolytes for rechargeable lithium cells

[J]. Journal of the Electrochemical Society, 1982, 129 (11): 2404-2409.

[13] Yoshimatsu I, Hirai T, Yamaki J. Lithium electrode morphology during cycling in lithium cells [J]. Journal of the Electrochemical Society, 1988, 135 (10): 2422-2427.

[14] Ue M, Mori S. Mobility and ionic association of lithium-salts in a propylene carbonate-ethyl methyl carbonate mixed-solvent [J]. Journal of the Electrochemical Society, 1995, 142 (8): 2577-2581.

[15] Methlie I G J. Electric current producing cell: US, 3415687A [P/OL]. 1968.

[16] Hu Y S, Li H, Huang X J, et al. Novel room temperature molten salt electrolyte based on LiTFSI and acetamide for lithium batteries [J]. Electrochemistry Communications, 2004, 6 (1): 28-32.

[17] Ozawa K. Lithium-ion rechargeable batteries with $LiCoO_2$ and carbon electrodes: The $LiCoO_2$/C system [J]. Solid State Ionics, 1994, 69 (3-4): 212-221.

[18] Xu K, Zhang S S, Jow T R, et al. LiBOB as salt for lithium-ion batteries: A possible solution for high temperature operation [J]. Electrochemical and Solid State Letters, 2002, 5 (1): A26-A29.

[19] Kelley B, Northrup M, Hurley P D. Effect of riboflavin and pteroylglutamic acid on growth and white cell production of rats [J]. Proceedings of the Society for Experimental Biology and Medicine, 1951, 76 (4): 804-806.

[20] Hongbo H, Jun G, Zhou Z B, et al. Lithium (fluorosulfonyl) (nonafluorobutanesulfonyl) imide (LiFNFSI) as conducting salt to improve the high-temperature resilience of lithium-ion cells [J]. Electrochemistry Communications, 2011, 13 (3): 265-268.

[21] Murugavel S. Origin of non-Arrhenius conductivity in fast ion conducting glasses [J]. Physical Review B, 2005, 72 (13): 134204-134213.

[22] Gu G Y, Bouvier S, Wu C, et al. 2-Methoxyethyl (methyl) carbonate-based electrolytes for Li-ion batteries [J]. Electrochimica Acta, 2000, 45 (19): 3127-3139.

[23] Doyle M, Fuller T F, Newman J. The importance of the lithium ion transference number in lithium polymer cells [J]. Electrochimica Acta, 1994, 39 (13): 2073-2081.

[24] Fujinami T, Buzoujima Y. Novel lithium salts exhibiting high lithium ion transference numbers in polymer electrolytes [J]. Journal of Power Sources, 2003, 119: 438-441.

[25] Liaw B Y, Roth E P, Jungst R G, et al. Correlation of arrhenius behaviors in power and capacity fades with cell impedance and heat generation in cylindrical lithium-ion cells [J]. Journal of Power Sources, 2003, 119: 874-886.

[26] Tsunashima K, Sugiya M. Physical and electrochemical properties of low-viscosity phosphonium ionic liquids as potential electrolytes [J]. Electrochemistry Communications, 2007, 9 (9): 2353-2358.

[27] Matsumoto H, Sakaebe H, Tatsumi K. Preparation of room temperature ionic liquids based on aliphatic onium cations and. asymmetric amide anions and their electrochemical properties as a lithium battery electrolyte [J]. Journal of Power Sources, 2005, 146 (1-2): 45-50.

[28] Ding M S, Jow T R. Conductivity and viscosity of PC-DEC and PC-EC solutions of $LiPF_6$ [J]. Journal of the Electrochemical Society, 2003, 150 (5): A620-A628.

[29] Webber A. Conductivity and viscosity of solutions of $LiCF_3SO_3$, Li $(CF_3SO_2)_2$N, and their mixtures [J]. Journal of the Electrochemical Society, 1991, 138 (9): 2586-2590.

[30] Shu Z X, Mcmillan R S, Murray J J. Electrochemical intercalation of lithium into graphite [J]. Journal of the Electrochemical Society, 1993, 140 (4): 922-927.

[31] Jean M, Chausse A, Messina R. Analysis of the passivating layer and the electrolyte in the system: Petroleum coke/solution of $LiCF_3SO_3$ in mixed organic carbonates [J]. Electrochimica Acta, 1998, 43 (12-13): 1795-1802.

[32] Zheng Jieyun, Li Hong. Fundamental scientific aspects of lithium batteries (V) —Interfaces [J]. Energy Storage Science and Technology, 2013, 2 (5): 503-513.

[33] 郑洪河. 锂离子电池电解质 [M]. 北京: 化学工业出版社, 2006.

[34] Thomas M, Bruce P G, Goodenough J B. AC impedance analysis of polycrystalline insertion electrodes: Application to $Li_{1-x}CoO_2$ [J]. Journal of the Electrochemical Society, 1985, 132 (7): 1521-1528.

[35] Guyomard D, Tarascon J M. The carbon $Li_{1+x}Mn_2O_4$ system [J]. Solid State Ionics, 1994, 69 (3-4): 222-237.

[36] Tarascon J M, Guyomard D. The $Li_{1+x}Mn_2O_4$/C rocking-chair system: A review [J]. Electrochimica Acta, 1993, 38 (9): 1221-1231.

[37] Aurbach D, Markovsky B, Levi M D, et al. New insights into the interactions between electrode materials and

electrolyte solutions for advanced nonaqueous batteries [J]. Journal of Power Sources, 1999, 81: 95-111.

[38] Ostrovskii D, Ronci F, Scrosati B, et al. Reactivity of lithium battery electrode materials toward non-aqueous electrolytes: Spontaneous reactions at the electrode-electrolyte interface investigated by FTIR [J]. Journal of Power Sources, 2001, 103 (1): 10-17.

[39] Edstrom K, Gustafsson T, Thomas J O. The cathode-electrolyte interface in the Li-ion battery [J]. Electrochimica Acta, 2004, 50 (2-3): 397-403.

[40] Liu N, Li H, Wang Z X, et al. Origin of solid electrolyte interphase on nanosized $LiCoO_2$ [J]. Electrochemical and Solid State Letters, 2006, 9 (7): A328-A331.

[41] Xu K. Nonaqueous liquid electrolytes for lithium-based rechargeable batteries [J]. Chem Rev, 2004, 104 (10): 4303-4317.

[42] Arora P, White R E, Doyle M. Capacity fade mechanisms and side reactions in lithium-ion batteries [J]. Journal of the Electrochemical Society, 1998, 145 (10): 3647-3667.

[43] Campion C L, Li W T, Lucht B L. Thermal decomposition of $LiPF_6$-based electrolytes for lithium-ion batteries [J]. Journal of the Electrochemical Society, 2005, 152 (12): A2327-A2334.

[44] Maleki H, Deng G P, Anani A, et al. Thermal stability studies of Li-ion cells and components [J]. Journal of the Electrochemical Society, 1999, 146 (9): 3224-3229.

[45] Ravdel B, Abraham K M, Gitzendanner R, et al. Thermal stability of lithium-ion battery electrolytes [J]. Journal of Power Sources, 2003, 119: 805-810.

[46] Sloop S E, Pugh J K, Wang S, et al. Chemical reactivity of PF_5 and $LiPF_6$ in ethylene carbonate/dimethyl carbonate solutions [J]. Electrochemical and Solid State Letters, 2001, 4 (4): A42-A44.

[47] Yang H, Zhuang G V, Ross P N. Thermal stability of $LiPF_6$ salt and Li-ion battery electrolytes containing $LiPF_6$ [J]. Journal of Power Sources, 2006, 161 (1): 573-579.

[48] Blomgren G E. Electrolytes for advanced batteries [J]. Journal of Power Sources, 1999, 81: 112-118.

[49] Zhang X, Ross P N, Kostecki R, et al. Diagnostic characterization of high power lithium-ion batteries for use in hybrid electric vehicles [J]. Journal of the Electrochemical Society, 2001, 148 (5): A463-A470.

[50] Zhuang Quanchao, Wu Shan, Liu Wenyuan, et al. The research of organic electrolyte solutions for Li-ion batteries [J]. Electrochemistry, 2001, 4 (7): 403-412.

[51] Fry A J. Synthetic Organic Electrochemistry [M]. London: John Wiley Press, 1989.

[52] Aurbach D. Nonaqueous Electrochemistry [M]. New York: Marcel-Dekker Press, 1999.

[53] Harris W S. Electrochemical studies in cyclic esters [D] California: University of California, 1958.

[54] Sugeno N, Anzai M, Nagaura T. Non-aq. electrolyte secondary battery-has carbon@ material negative electrode, lithium complex oxide, positive electrode and mixed solvent electrolyte: EP, 486950-A; EP, 486950-A1; CA, 2055305-A; JP, 4184872-A; JP, 4280082-A; US, 5292601-A; EP, 486950-B1; DE, 69103384-E; JP, 3079613-B2; JP, 3089662-B2; JP, 2000268864-A; CA, 2055305-C; JP, 3356157-B2 [P/OL]. 1992-05-27. http://www.thomsonpatentstore.net/portal/servlet/DIIDirect? CC = EP&PN = 486950&DT = A&SrcAuth = Wila&Token = uxTOe.GiY2bHu08Rjd6xHXjcM_VTwlq3vOi381uNAPAwH.cAlHY8oLoVC1_0DHp7-WmziB3PJWDsx_TyLKY5NC3EJugKKbcHx4Ohk6W_mhI0.

[55] Selim R, Bro P. Some observations on rechargeable lithium electrodes in a propylene carbonate electrolyte [J]. Journal of the Electrochemical Society, 1974, 121 (11): 1457-1459.

[56] Rauh R D, Brummer S B. Effect of additives on lithium cycling in propylene carbonate [J]. Electrochimica Acta, 1977, 22 (1): 75-83.

[57] Rauh R D, Reise T F, Brummer S B. Efficiencies of cycling lithium on a lithium substrate in propylene carbonate [J]. Journal of the Electrochemical Society, 1978, 125 (2): 186-190.

[58] Aurbach D, Daroux M L, Faguy P W, et al. Identification of surface-films formed on lithium in propylene carbonate solutions [J]. Journal of the Electrochemical Society, 1987, 134 (7): 1611-1620.

[59] Chung G C, Kim H J, Yu S I, et al. Origin of graphite exfoliation: An investigation of the important role of solvent cointercalation [J]. Journal of the Electrochemical Society, 2000, 147 (12): 4391-4398.

[60] Newman G H, Francis R W, Gaines L H, et al. Hazard investigations of $LiClO_4$-dioxolane electrolyte [J]. Journal of the Electrochemical Society, 1980, 127 (9): 2025-2027.

[61] Xu K, Ding M S, Jow T R. Quaternary onium salts as nonaqueous electrolytes for electrochemical capacitors [J]. Journal of the Electrochemical Society, 2001, 148 (3): A267-A274.

[62] Elliott W. Contract NAS 3-6015 (N 65-11518) [R]. 1964.

[63] Pistoia G, Derossi M, Scrosati B. Study of behavior of ethylene carbonate as a nonaqueous battery solvent [J]. Journal of the Electrochemical Society, 1970, 117 (4): 500-502.

[64] Abraham K M, Foos J S, Goldman J L. Long cycle-life secondary lithium cells utilizing tetrahydrofuran [J]. Journal of the Electrochemical Society, 1984, 131 (9): 2197-2199.

[65] Geronov Y, Puresheva B, Moshtev RV, et al. Rechargeable compact Li cells with $Li_xCr_{0.9}V_{0.1}S_2$ and $Li_{1+x}V_3O_8$ cathodes and ether-based electrolytes [J]. Journal of the Electrochemical Society, 1990, 137 (11): 3338-3344.

[66] Takami N, Ohsaki T, Inada K. The impedance of lithium electrodes in $LiPF_6$-based electrolytes [J]. Journal of the Electrochemical Society, 1992, 139 (7): 1849-1854.

[67] Yamaura J, Ozaki Y, Morita A, et al. High-voltage, rechargeable lithium batteries using newly-developed carbon for negative electrode material [J]. Journal of Power Sources, 1993, 43 (1-3): 233-239.

[68] Zhang S S, Liu Q G, Yang L L. Polyacene as an anode in lithium ion batteries [J]. Journal of the Electrochemical Society, 1993, 140 (7): L107-L108.

[69] Guyomard D, Tarascon J M. Li metal-free rechargeable $LiMn_2O_4$/carbon cells: Their understanding and optimization [J]. Journal of the Electrochemical Society, 1992, 139 (4): 937-948.

[70] Aurbach D, Eineli Y, Markovsky B, et al. The study of electrolyte-solutions based on ethylene and diethyl carbonates for rechargeable Li batteries II. Graphite-electrodes [J]. Journal of the Electrochemical Society, 1995, 142 (9): 2882-2890.

[71] Eineli Y, Thomas S R, KochV, et al. Ethylmethylcarbonate, a promising solvent for Li-ion rechargeable batteries [J]. Journal of the Electrochemical Society, 1996, 143 (12): L273-L277.

[72] Campbell S A, Bowes C, Mcmillan R S. The electrochemical-behavior of tetrahydrofuran and propylene carbonate without added electrolyte [J]. Journal of Electroanalytical Chemistry, 1990, 284 (1): 195-204.

[73] Laoire C O, Mukerjee S, Plichta E J, et al. Rechargeable lithium/tegdme-$LiPF_6$/O_2 battery [J]. Journal of the Electrochemical Society, 2011, 158 (3): A302-A308.

[74] Jung H G, Hassoun J, Park J B, et al. An improved high-performance lithium-air battery [J]. Nature Chemistry, 2012, 4 (7): 579-585.

[75] Li L F, Lee H S, Li H, et al. New electrolytes for lithium ion batteries using LiF salt and boron based anion receptors [J]. Journal of Power Sources, 2008, 184 (2): 517-521.

[76] Xie B, Lee H S, Li H, et al. New electrolytes using Li_2O or Li_2O_2 oxides and tris (pentafluorophenyl) borane as boron based anion receptor for lithium batteries [J]. Electrochemistry Communications, 2008, 10 (8): 1195-1197.

[77] Linden D. Handbook of Batteries [M]. New York: Mcgraw-Hill Press, 1995.

[78] Ue M. Mobility and ionic association of lithium and quaternary ammonium-salts in propylene carbonate and gamma-butyrolactone [J]. Journal of the Electrochemical Society, 1994, 141 (12): 3336-3342.

[79] Hayashi K, Nemoto Y, Tobishima S, et al. Mixed solvent electrolyte for high voltage lithium metal secondary cells [J]. Electrochimica Acta, 1999, 44 (14): 2337-2344.

[80] Krause L J, Lamanna W, Summerfield J, et al. Corrosion of aluminum at high voltages in non-aqueous electrolytes containing perfluoroalkylsulfonyl imides: new lithium salts for lithium-ion cells [J]. Journal of Power Sources, 1997, 68 (2): 320-325.

[81] Behl W K, Plichta E J. Stability of aluminum substrates in lithium-ion battery electrolytes [J]. Journal of Power Sources, 1998, 72 (2): 132-135.

[82] Plichta e J, Behl W K. A low-temperature electrolyte for lithium and lithium-ion batteries [J]. Journal of Power Sources, 2000, 88 (2): 192-196.

[83] Aurbach D. Nonaqueous Electrochemistry [M]. New York: Marcel Dekker, 1999.

[84] Zhang S S, Xu K, Jow T R. Study of $LiBF_4$ as an electrolyte salt for a Li-ion battery [J]. Journal of the Electrochemical Society, 2002, 149 (5): A586-A590.

[85] Takami N, Ohsaki T, Hasebe H, et al. Laminated thin Li-ion batteries using a liquid electrolyte [J]. Journal of the Electrochemical Society, 2002, 149 (1): A9-A12.

[86] Zhang S S, Xu K, Jow T R. A new approach toward improved low temperature performance of Li-ion battery [J]. Electrochemistry Communications, 2002, 4 (11): 928-932.

[87] Ue M, Murakami A, Nakamura S. Anodic stability of several anions examined by abinitio molecular orbital and

[88] Ue M, Takeda M, Takehara M, et al. Electrochemical properties of quaternary ammonium salts for electrochemical capacitors [J]. Journal of the Electrochemical Society, 1997, 144 (8): 2684-2688.

[89] Takata K, Morita M, Matsuda Y, et al. Cycling characteristics of secondary Li electrode in $LiBF_4$ mixed ether electrolytes [J]. Journal of the Electrochemical Society, 1985, 132 (1): 126-128.

[90] Aurbach D, Zaban A, Schechter A, et al. the study of electrolyte-solutions based on ethylene and diethyl carbonates for rechargeable Li batteries I. Li metal anodes [J]. Journal of the Electrochemical Society, 1995, 142 (9): 2873-2882.

[91] Nanjundiah C, Goldman J L, Dominey L A, et al. Electrochemical stability of $LiMF_6$ (M=P, As, Sb) in tetrahydrofuran and sulfolane [J]. Journal of the Electrochemical Society, 1988, 135 (12): 2914-2917.

[92] Plichta E, Slane S, Uchiyama M, et al. An improved Li/Li_xCoO_2 rechargeable cell [J]. Journal of the Electrochemical Society, 1989, 136 (7): 1865-1869.

[93] Naoi K, Mori M, Naruoka Y, et al. The surface film formed on a lithium metal electrode in a new imide electrolyte, lithium bis (perfluoroethylsulfonylimide) [LiN $(C_2F_5SO_2)_2$] [J]. Journal of the Electrochemical Society, 1999, 146 (2): 462-469.

[94] Foropoulos J, Desmarteau D D. Synthesis, properties, and reactions of bis [(trifluoromethyl) sulfonyl] imide, $(CF_3SO_2)_2NH$ [J]. Inorganic Chemistry, 1984, 23 (23): 3720-3723.

[95] Sylla S, Sanchez J Y, Armand M. Electrochemical study of linear and cross-linked poe-based polymer electrolytes [J]. Electrochimica Acta, 1992, 37 (9): 1699-1701.

[96] Yang H, Kwon K, Devine T M, et al. Aluminum corrosion in lithium batteries: An investigation using the electrochemical quartz crystal microbalance [J]. Journal of the Electrochemical Society, 2000, 147 (12): 4399-4407.

[97] Nakajima T, Mori M, Gupta V, et al. Effect of fluoride additives on the corrosion of aluminum for lithium ion batteries [J]. Solid State Sciences, 2002, 4 (11-12): 1385-1394.

[98] Dicenso D, Exnar I, Graetzel M. Non-corrosive electrolyte compositions containing perfluoroalkylsulfonyl imides for high power Li-ion batteries [J]. Electrochemistry Communications, 2005, 7 (10): 1000-1006.

[99] Michot C, Armand M, Sanchez J, et al. New ionically conductive material-contains ionic cpd. with fluoro: Sulphonyl substituent, useful as electrolyte for lithium battery, etc: WO, 9526056-A; EP, 699349-A; FR, 2717612-A; FR, 2717612-A1; WO, 9526056-A1; EP, 699349-A1; JP, 8511274-W; US, 5916475-A; US, 6254797-B1; US, 2001025943-A1; US, 6682855-B2; CA, 216336-C; JP, 2006210331-A; JP, 3878206-B2; EP, 699349-B1; DE, 69535612-E; DE, 69535612-T2 [P/OL]. 1995-09-22. http: //www. thomsonpatentstore. net/portal/servlet/DIIDirect? CC = WO&PN = 9526056&DT = A&SrcAuth = Wila&Token = Io6 _ blYNv8J9zDfT62whrf3t8N3eZb3waLjBKuZlK3b03XxIu3Iwd1 _ aZcLct9dvPJxc6K8fkTiGlt-41Jqt2k3B _ MbkAy-WE _ w1nDpCw9LO1.

[100] Zaghib K, Charest P, Guerfi A, et al. Safe Li-ion polymer batteries for HEV applications [J]. Journal of Power Sources, 2004, 134 (1): 124-129.

[101] Zaghib K, Charest P, Guerfi A, et al. $LiFePO_4$ safe Li-ion polymer batteries for clean environment [J]. Journal of Power Sources, 2005, 146 (1-2): 380-385.

[102] Li L, Zhou S, Han H, et al. Transport and electrochemical properties and spectral features of non-aqueous electrolytes containing LiFSI in linear carbonate solvents [J]. Journal of the Electrochemical Society, 2011, 158 (2): A74-A82.

[103] Zhou S, Han H, Nie J, et al. Improving the high-temperature resilience of $LiMn_2O_4$ based batteries: LiFNFSI an effective salt [J]. Journal of the Electrochemical Society, 2012, 159 (8): A1158-A1164.

[104] Lischka U, Wietelmann U, Wegner M. Easily prepared, environmentally compatible, stable lithium borate complex salts: WO, 200000495-A; DE, 19829030-C; EP, 1091963-A; DE, 19829030-C1; WO, 200000495-A1; EP, 1091963-A1; KR, 2001072657-A; JP, 2002519352-W; EP, 1091963-B1; DE, 59902958-G; US, 6506516-B1; ES, 2185354-T3; CA, 2336323-C; JP, 3913474-B2; KR, 716373-B1 [P/OL]. http: //www. thomsonpatentstore. net/portal/servlet/DIIDirect? CC = WO&PN = 9807729&DT = A&SrcAuth = Wila&Token = xqe _ w0UirTZEPeDTdK9CexJ7MaNr3lTbxB3FfyrnG2XFY1eMRGFaJNN _ TTO8lQY5 _ 1 _ 0etmwZiy _ GokZWUmD2VX. rNh8BURFkScvjBpk4JE.

[105] Zavalij P Y, Yang S, Whittingham M S. Structural chemistry of new lithium bis (oxalato) borate solvates [J]. Acta

Crystallographica Section B: Structural Science, 2004, 60: 716-724.

[106] Xu K, Lee U, Zhang S S, et al. Chemical analysis of graphite/electrolyte interface formed in LiBOB-based electrolytes [J]. Electrochemical and Solid State Letters, 2003, 6 (7): A144-A148.

[107] Xu K, Zhang S S, Jow T R. Formation of the graphite electrolyte interface by lithium bis (oxalato) borate [J]. Electrochemical and Solid State Letters, 2003, 6 (6): A117-A120.

[108] Xu K, Zhang S S, Poese B A, et al. Lithium bis (oxalato) borate stabilizes graphite anode in propylene carbonate [J]. Electrochemical and Solid State Letters, 2002, 5 (11): A259-A262.

[109] Eineli Y, Thomas S R, KochV R. The role of SO_2 as an additive to organic Li-ion battery electrolytes [J]. Journal of the Electrochemical Society, 1997, 144 (4): 1159-1165.

[110] Aurbach D, Gamolsky K, Markovsky B, et al. On the use of vinylene carbonate (VC) electrolyte solutions for Li-ion as an additive to batteries [J]. Electrochimica Acta, 2002, 47 (9): 1423-1439.

[111] Wrodnigg G H, Wrodnigg T M, Besenhard J O, et al. Propylene sulfite as film-forming electrolyte additive in lithium ion batteries [J]. Electrochemistry Communications, 1999, 1 (3-4): 148-150.

[112] Wrodnigg G H, Besenhard J O, Winter M. Ethylene sulfite as electrolyte additive for lithium-ion cells with graphitic anodes [J]. Journal of the Electrochemical Society, 1999, 146 (2): 470-472.

[113] Girard H, Simon N, Ballutaud D, et al. Effect of anodic and cathodic treatments on the charge transfer of boron doped diamond electrodes [J]. Diamond and Related Materials, 2007, 16 (2): 316-325.

[114] Herlem G, Fahys B, Szekely M, et al. *n*-Butylamine as solvent for lithium salt electrolytes, structure and properties of concentrated solutions [J]. Electrochimica Acta, 1996, 41 (17): 2753-2760.

[115] Aurbach D, Eineli Y, Chusid O, et al. The correlation between the surface-chemistry and the performance of Li-carbon intercalation anodes for rechargeable rocking-chair type batteries [J]. Journal of the Electrochemical Society, 1994, 141 (3): 603-611.

[116] Shu Z X, mcmillan R S, Murray J J. Effect of 12 crown-4 on the electrochemical intercalation of lithium into graphite [J]. Journal of the Electrochemical Society, 1993, 140 (6): L101-L103.

[117] Lee H S, Sun X, Yang X Q, et al. Synthesis of cyclic aza-ether compounds and studies of their use as anion receptors in nonaqueous lithium halide salts solution [J]. Journal of the Electrochemical Society, 2000, 147 (1): 9-14.

[118] Lee H S, Yang X Q, Mcbreen J, et al. The synthesis of a new family of anion receptors and the studies of their effect on ion pair dissociation and conductivity of lithium salts in nonaqueous solutions [J]. Journal of the Electrochemical Society, 1996, 143 (12): 3825-3829.

[119] Richard M N, Dahn J R. Accelerating rate calorimetry study on the thermal stability of lithium intercalated graphite in electrolyte I. Experimental [J]. Journal of the Electrochemical Society, 1999, 146 (6): 2068-2077.

[120] Richard M N, Dahn J R. Accelerating rate calorimetry study on the thermal stability of lithium intercalated graphite in electrolyte II. Modeling the results and predicting differential scanning calorimeter curves [J]. Journal of the Electrochemical Society, 1999, 146 (6): 2078-2084.

[121] Richard M N, Dahn J R. Predicting electrical and thermal abuse behaviours of practical lithium-ion cells from accelerating rate calorimeter studies on small samples in electrolyte [J]. Journal of Power Sources, 1999, 79 (2): 135-142.

[122] Richard M N, Dahn J R. Accelerating rate calorimetry studies of the effect of binder type on the thermal stability of a lithiated mesocarbon microbead material in electrolyte [J]. Journal of Power Sources, 1999, 83 (1-2): 71-74.

[123] Xianming W, Yasukawa E, Kasuya S. Nonflammable trimethyl phosphate solvent-containing electrolytes for lithium-ion batteries: I. Fundamental properties [J]. Journal of the Electrochemical Society, 2001, 148 (10): A1058-A1065.

[124] Xianming W, Yasukawa E, Kasuya S. Nonflammable trimethyl phosphate solvent-containing electrolytes for lithium-ion batteries: II. The use of an amorphous carbon anode [J]. Journal of the Electrochemical Society, 2001, 148 (10): A1066-A1071.

[125] Yamaki J, Tanaka T, Ihara M, et al. Thermal stability of methyl difluoroacetate as a novel electrolyte solvent for lithium batteries electrolytes [J]. Electrochemistry, 2003, 71 (12): 1154-1156.

[126] Yamaki J I, Yamazaki I, Egashira M, et al. Thermal studies of fluorinated ester as a novel candidate for electrolyte solvent of lithium metal anode rechargeable cells [J]. Journal of Power Sources, 2001, 102 (1-2): 288-293.

[127] Buhrmester C, Chen J, Moshurchak L, et al. Studies of aromatic redox shuttle additives for LiFePO$_4$-based Li-ion cells [J]. Journal of the Electrochemical Society, 2005, 152 (12): A2390-A2399.

[128] Chen J, Buhrmester C, Dahn J R. Chemical overcharge and overdischarge protection for lithium-ion batteries [J]. Electrochemical and Solid State Letters, 2005, 8 (1): A59-A62.

[129] Shima K, Ue M, Yamaki J. Redox mediator as an overcharge protection agent for 4V class lithium-ion rechargeable cells [J]. Electrochemistry, 2003, 71 (12): 1231-1235.

[130] Tobishima S, Ogino Y, Watanabe Y. Influence of electrolyte additives on safety and cycle life of rechargeable lithium cells [J]. Journal of Applied Electrochemistry, 2003, 33 (2): 143-150.

[131] Guoying C, Richardson T J. Overcharge protection for rechargeable lithium batteries using electroactive polymers [J]. Electrochemical and Solid-State Letters, 2004, 7 (2): A23-A26.

[132] Adachi M, Tanaka K, Sekai K. Aromatic compounds as redox shuttle additives for 4V class secondary lithium batteries [J]. Journal of the Electrochemical Society, 1999, 146 (4): 1256-1261.

[133] Abraham K M, Pasquariello D M, Willstaedt E B. Normal-butylferrocene for overcharge protection of secondary lithium batteries [J]. Journal of the Electrochemical Society, 1990, 137 (6): 1856-1857.

[134] Xiao L F, Ai X P, Cao Y L, et al. Electrochemical behavior of biphenyl as polymerizable additive for overcharge protection of lithium ion batteries [J]. Electrochimica Acta, 2004, 49 (24): 4189-4196.

[135] Barker J, Stux A M. Electrolytic cells containing additives for inhibiting the decomposition of lithium salts: Comprising anode, cathode and electrolyte containing a lithium salt, a solvent and a carbonate additive, and useful for non-aqueous batteries: US, 5707760-A [P/OL]. 1998-01-13. http://www.thomsonp-atentstore.net/portal/servlet/DIIDirect? CC = US&PN = 5707760&DT = A&SrcAuth = Wila&Token = toT09zn _. SGVVv-H8NEzjcm9b. HOX24bOxhaZbBuQhcmzf9SDtC-iZv4wSJufl2naVPpV9BN1vmaxSQ1GyP3hs199AMK0vc. Gjt739-pV-La38A.

[136] Inventor U. Electrolyte used for lithium ion battery, comprises specified amount of lithium salt, organic solvent and additive: CN, 102324563-A [P/OL]. 2012-01-18. http://www.google.com/patents/CN102324563A? cl=zh.

[137] Shang Z, Yang H, Hou H, et al. Electrolyte comprises lithium hexafluorophosphate, mixed solvent comprising diethyl carbonate, ethyl methyl carbonate, dimethyl carbonate, ethylene carbonate and propylene carbonate, and additive: CN, 103107358-A [P/OL]. 2013-05-15. http://www.google.com/patents/CN103107358A? cl=zh.

[138] Chalasani D, Li J, Jackson N M, et al. Methylene ethylene carbonate: Novel additive to improve the high temperature performance of lithium ion batteries [J]. Journal of Power Sources, 2012, 208: 67-73.

[139] Liu B X, Li B, Guan S Y. Effect of fluoroethylene carbonate additive on low temperature performance of Li-ion batteries [J]. Electrochemical and Solid-State Letters, 2012, 15 (6): A77-A79.

[140] Liao L, Zuo P, Ma Y, et al. Effects of fluoroethylene carbonate on low temperature performance of mesocarbon microbeads anode [J]. Electrochimica Acta, 2012, 74: 260-266.

[141] Bian F, Zhang Z, Yang Y. Effects of fluoroethylene carbonate additive on low temperature performance of Li-ion batteries [J]. Journal of Electrochemistry, 2013, 19 (4): 355-360.

[142] Liao L, Cheng X, Ma Y, et al. Fluoroethylene carbonate as electrolyte additive to improve low temperature performance of LiFePO$_4$ electrode [J]. Electrochimica Acta, 2013, 87: 466-472.

[143] Xiang R, Li F, Jia G, et al. Effects of FEC additive on the low temperature performance of LiODFB-based lithium-ion batteries [J]. Applied Energy Technology, 2013, 724: 1025-1028.

[144] Tsujioka S, Takase H, Takahashi M, et al. Electrolyte for electrochemical device such as lithium cell and electrical double-layer capacitor, comprises ionic metal complex: EP, 1195834-A2; US, 2002081496-A1; JP, 2002184460-A; JP, 2002184465-A; JP, 2002110235-A; JP, 2002373703-A; US, 6783896-B2; JP, 3722685-B2; JP, 3730860-B2; JP, 3730861-B2; JP, 4076738-B2; EP, 1195834-B1; DE, 60143070-E [P/OL]. 2002-04-10. http://www.google.com/patents/EP1195834A2? cl=en.

[145] Tsujioka S, Takase H, Takahashi M, et al. Electrolyte for electrochemical devices, comprises ionic metal complex, and specific ionic compound (s): US, 2002061450-A1; JP, 2002164082-A; JP, 2002164083-A; JP, 2003068359-A; US, 6787267-B2; JP, 3730855-B2; JP, 3730856-B2; JP, 4076748-B2 [P/OL]. 2002-05-23. http://www.thomsonpatentstore.net/portal/servlet/DIIDirect? CC = US&PN = 2002061450&DT = A1&SrcAuth=Wila&Token = toT09zn _. SGtruRdL2 _ HB1ySfIKpulpMOuDg _ fQd. 8DT _ V8owJlKu3 _ Ru-

vqdqda2lLM2THVjbrCTxrFhSyhKUPP_jSq2gZv3EFfL2jT4f0.

[146] Chen X, Wang N, Xia H, et al. Non-aqueous electrolyte battery, has non-aqueous electrolyte provided with solvent, additive and electrolyte that is made of lithium hexafluorophosphate, where additive is biphenyl, vinyl acetate, divinyl adipate or vinylene carbonate: CN, 102280662-A [P/OL]. 2011-12-14. http://www.google.com/patents/CN102280662A? cl=zh.

[147] Choi S J, Lee M S, Jeong C S, et al. Additive, useful in non-aqueous electrolyte for overcharge prevention of secondary battery, comprises terphenyl derivative e. g. o-terphenyl, and xylene derivative e. g. bromoxylene: EP, 2523247-A1; US, 2012288752-A1; KR, 2012126305-A; JP, 2012238595-A; CN, 102780036-A [P/OL]. 2012-11-14. http://www.google.com/patents/CN102780036A? cl=en.

[148] Zhang S S. A review on electrolyte additives for lithium-ion batteries [J]. Journal of Power Sources, 2006, 162 (2): 1379-1394.

[149] Blanchard L A, Hancu D, Beckman E J, et al. Green processing using ionic liquids and CO_2 [J]. Nature, 1999, 399 (6731): 28-29.

[150] Huddleston J G, Willauer H D, Swatloski R P, et al. Room temperature ionic liquids as novel media for "clean" liquid-liquid extraction [J]. Chemical Communications, 1998, 16: 1765-1766.

[151] Welton T. Room-temperature ionic liquids, solvents for synthesis and catalysis [J]. Chem Rev, 1999, 99 (8): 2071-2083.

[152] Tait S, Osteryoung R A. Infrared study of ambient-temperature chloroaluminates as a function of melt acidity [J]. Inorganic Chemistry, 1984, 23 (25): 4352-4360.

[153] Liu Jianlian. Study on synthesis and characterization of typical N,N'-dialkylimidazolium-based ionic liquids [D]. Xi'an: Northwest University, 2006.

[154] Galinski M, Lewandowski A, Stepniak I. Ionic liquids as electrolytes [J]. Electrochimica Acta, 2006, 51 (26): 5567-5580.

[155] Garcia B, Lavallee S, Perron G, et al. Room temperature molten salts as lithium battery electrolyte [J]. Electrochimica Acta, 2004, 49 (26): 4583-4588.

[156] Matsumoto H, Sakaebe H, Tatsumi K, et al. Fast cycling of $Li/LiCoO_2$ cell with low-viscosity ionic liquids based on bis (fluorosulfonyl) imide FSI [J]. Journal of Power Sources, 2006, 160 (2): 1308-1313.

[157] Sakaebe H, Matsumoto H. N-Methyl-N-propylpiperidinium bis (trifluoromethanesulfonyl) imide (PP 13-TFSI)-novel electrolyte base for Li battery [J]. Electrochemistry Communications, 2003, 5 (7): 594-598.

[158] Armand M, Endres F, Macfarlane D R, et al. Ionic-liquid materials for the electrochemical challenges of the future [J]. Nature Materials, 2009, 8 (8): 621-629.

[159] Lewandowski A, Swiderska M A. Ionic liquids as electrolytes for Li-ion batteries: An overview of electrochemical studies [J]. Journal of Power Sources, 2009, 194 (2): 601-609.

[160] Nakagawa H, Izuchi S, Kuwana K, et al. Liquid and polymer gel electrolytes for lithium batteries composed of room-temperature molten salt doped by lithium salt [J]. Journal of the Electrochemical Society, 2003, 150 (6): A695-A700.

[161] Innis P C, Mazurkiewicz J, Nguyen T, et al. Enhanced electrochemical stability of polyaniline in ionic liquids [J]. Current Applied Physics, 2004, 4 (2-4): 389-393.

[162] Liu Y, Zhou S, Han H, et al. Molten salt electrolyte based on alkali bis (fluorosulfonyl) imides for lithium batteries [J]. Electrochimica Acta, 2013, 105: 524-529.

[163] Chagnes A, Carre B, Willmann P, et al. Modeling viscosity and conductivity of lithium salts in gamma-butyrolactone [J]. Journal of Power Sources, 2002, 109 (1): 203-213.

[164] Feuillade G, Perche P. Ion-conductive macromolecular gels and membranes for solid lithium cells [J]. Journal of Applied Electrochemistry, 1975, 5 (1): 63-69.

[165] Han K N, Seo H M, Kim J K, et al. Development of a plastic Li-ion battery cell for EV applications [J]. Journal of Power Sources, 2001, 101 (2): 196-200.

[166] Huang K Q, Wan J H, Goodenough J B. Increasing power density of LSGM-based solid oxide fuel cells using new anode materials [J]. Journal of the Electrochemical Society, 2001, 148 (7): A788-A794.

[167] Tarascon J M, Gozdz A S, Schmutz C, et al. Performance of Bellcore's plastic rechargeable Li-ion batteries [J]. Solid State Ionics, 1996, 86 (8): 49-54.

[168] Slane S, Salomon M. Composite gel electrolyte for rechargeable lithium batteries [J]. Journal of Power Sources,

1995, 55 (1): 7-10.

[169] Armand M, Dalard F, Deroo D, et al. Modeling the voltammetric study of intercalation in a host structure: Application to lithium intercalation in RuO_2 [J]. Solid State Ionics, 1985, 15 (3): 205-210.

[170] Morita M, Fukumasa T, Motoda M, et al. Polarization behavior of lithium electrode in solid electrolytes consisting of a poly (ethylene oxide)-grafted polymer [J]. Journal of the Electrochemical Society, 1990, 137 (11): 3401-3404.

[171] Appetecchi G B, Croce F, Dautzenberg G, et al. Composite polymer electrolytes with improved lithium metal electrode interfacial properties I. Electrochemical properties of dry PEO-LiX systems [J]. Journal of the Electrochemical Society, 1998, 145 (12): 4126-4132.

[172] Iijima T, Toyoguchi Y, Eda N. Quasi-solid organic electrolytes gelatinized with polymethyl-methacrylate and their applications for lithium batteries [J]. Denki Kagaku, 1985, 53 (8): 619-623.

[173] Xu J J, Ye H. Polymer gel electrolytes based on oligomeric polyether/cross-linked PMMA blends prepared via in situ polymerization [J]. Electrochemistry Communications, 2005, 7 (8): 829-835.

[174] Jiang Z, Carroll B, Abraham K M. Studies of some poly (vinylidene fluoride) electrolytes [J]. Electrochimica Acta, 1997, 42 (17): 2667-2677.

[175] Jang H P, Ju H C, Eun H L, et al. Thickness-tunable polyimide nanoencapsulating layers and their influence on cell performance/thermal stability of high-voltage $LiCoO_2$ cathode materials for lithium-ion batteries [J]. Journal of Power Sources, 2013, 244: 442-449.

[176] Lee E H, Park J H, Kim J M, et al. Direct surface modification of high-voltage $LiCoO_2$ cathodes by UV-cured nanothickness poly (ethylene glycol diacrylate) gel polymer electrolytes [J]. Electrochimica Acta, 2013, 104: 249-254.

[177] Xu K, Angell C A. High anodic stability of a new electrolyte solvent: Unsymmetric noncyclic aliphatic sulfone [J]. Journal of the Electrochemical Society, 1998, 145 (4): L70-L72.

[178] Lu Z H, Dahn J R. Can all the lithium be removed from $T_2Li_{2/3}[Ni_{1/3}Mn_{2/3}]O_2$? [J]. Journal of the Electrochemical Society, 2001, 148 (7): A710-A715.

[179] Seel J A, Dahn J R. Electrochemical intercalation of PF_6 into graphite [J]. Journal of the Electrochemical Society, 2000, 147 (3): 892-898.

[180] Sun X G, Angell C A. New sulfone electrolytes for rechargeable lithium batteries. Part I. Oligoether-containing sulfones [J]. Electrochemistry Communications, 2005, 7 (3): 261-266.

[181] Sun X, Angell C A. Doped sulfone electrolytes for high voltage Li-ion cell applications [J]. Electrochemistry Communications, 2009, 11 (7): 1418-1421.

[182] Abouimrane A, Belharouak I, Amine K. Sulfone-based electrolytes for high-voltage Li-ion batteries [J]. Electrochemistry Communications, 2009, 11 (5): 1073-1076.

[183] Shao N, Sun X G, Dai S, et al. Electrochemical windows of sulfone-based electrolytes for high-voltage Li-ion batteries [J]. Journal of Physical Chemistry B, 2011, 115 (42): 12120-12125.

[184] Watanabe Y, Kinoshita S I, Wada S, et al. Electrochemical properties and lithium ion solvation behavior of sulfone-ester mixed electrolytes for high-voltage rechargeable lithium cells [J]. Journal of Power Sources, 2008, 179 (2): 770-779.

[185] Nanbu N, Takehara M, Watanabe S, et al. Polar effect of successive fluorination of dimethyl carbonate on physical properties [J]. Bulletin of the Chemical Society of Japan, 2007, 80 (7): 1302-1306.

[186] Achiha T, Nakajima T, Ohzawa Y, et al. Electrochemical behavior of nonflammable organo-fluorine compounds for lithium ion batteries [J]. Journal of the Electrochemical Society, 2009, 156 (6): A483-A488.

[187] Zhang Z, Hu L, Wu H, et al. Fluorinated electrolytes for 5V lithium-ion battery chemistry [J]. Energy & Environmental Science, 2013, 6 (6): 1806-1810.

[188] Le M L P, Allion N, Strobel P, et al. Fluorinated solvents for high voltage lithium batteries physicochemical and electrochemical investigations [C] //Montreal: 219th ECS Meeting Abstracts 382, 2011.

[189] Ue M, Ida K, Mori S. Electrochemical properties of organic liquid electrolytes based on quaternary onium salts for electrical double-layer capacitors [J]. Journal of the Electrochemical Society, 1994, 141 (11): 2989-2996.

[190] Abu L Y, Davidson I. High-voltage electrolytes based on adiponitrile for Li-ion batteries [J]. Journal of the Electrochemical Society, 2009, 156 (1): A60-A65.

[191] Abu L Y, Davidson I. New electrolytes based on glutaronitrile for high energy/power Li-ion batteries [J]. Journal of

Power Sources, 2009, 189 (1): 576-579.

[192] Nagahama M, Hasegawa N, Okada S. High voltage performances of Li_2NiPO_4F cathode with dinitrile-based electrolytes [J]. Journal of the Electrochemical Society, 2010, 157 (6): A748-A752.

[193] Gmitter A J, Plitz I, Amatucci G G. High concentration dinitrile, 3-Alkoxypropionitrile, and linear carbonate electrolytes enabled by vinylene and monofluoroethylene carbonate additives [J]. Journal of the Electrochemical Society, 2012, 159 (4): A370-A379.

[194] Abouimrane A D I, Abu L Y. Dinitrilebased liquid electrolytes: WO, 2008/138132A1 [P/OL]. 2008.

[195] Abe K, Ushigoe Y, Yoshitake H, et al. Functional electrolytes: Novel type additives for cathode materials, providing high cycleability performance [J]. Journal of Power Sources, 2006, 153 (2): 328-335.

[196] Lee Y S, Lee K S, Sun Y K, et al. Effect of an organic additive on the cycling performance and thermal stability of lithium-ion cells assembled with carbon anode and $LiNi_{1/3}Co_{1/3}Mn_{1/3}O_2$ cathode [J]. Journal of Power Sources, 2011, 196 (16): 6997-7001.

[197] Lee K S, Sun Y K, Noh J, et al. Improvement of high voltage cycling performance and thermal stability of lithium-ion cells by use of a thiophene additive [J]. Electrochemistry Communications, 2009, 11 (10): 1900-1903.

[198] 许梦清, 邢丽丹, 李伟善. 用于高电压锂离子电池的非水电解液及其制备方法与应用: 中国, 101702447A [P/OL]. 2010.

[199] Von C A, Xu K. Electrolyte additive in support of 5V Li ion chemistry [J]. Journal of the Electrochemical Society, 2011, 158 (3): A337-A342.

第 9 章
全固态锂离子电池

商用锂离子电池由于采用含有易燃有机溶剂的液体电解质，存在着安全隐患。发展全固态锂离子电池是提升电池安全性的可行技术途径之一。目前全固态锂离子电池的应用还需要解决一些科学与技术问题，包括：开发能在宽温度范围使用，兼顾高电导率与高电化学稳定性的固体电解质材料；减小电解质相与电极相界面间离子输运电阻的技术；适合全固态电池使用的正负极材料；相关材料与电池的设计与规模化制造技术。本章从固体电解质材料的研究开发进展，高通量计算用于固体电解质材料的筛选以及电极材料与固体电解质界面问题等方面进行了总结。

随着电动汽车的发展以及电网储能及小型储能需求的发展，开发能够在宽的温度范围使用，具有高安全性、高能量密度及功率密度的电池十分必要。在各种商业化可充放电化学储能装置中，锂离子电池拥有最高的能量密度。现有的商用锂离子电池主要包含两种类型：一种是采用液态电解质的锂离子电池；另外一种是采用凝胶电解质的锂离子电池。液态电解质的锂盐溶于有机溶液中，并包含多种功能添加剂。锂盐为 $LiPF_6$、LiFSI 等；有机溶剂为环状碳酸酯（EC、PC）、链状碳酸酯（DEC、DMC、EDC）、羧酸酯类（MF、MA、EA、MP 等）。凝胶电解质是在多孔的聚合物基体中吸附电解液形成的电解质。与液体电解液相同，凝胶电解质中的电解液起到离子传导及在负极表面形成稳定的固体电解质中间层（SEI）膜的作用。

液态电解质与凝胶电解质拥有较高的室温离子电导率，电解液能够有效地浸润电极颗粒，并能够在正负极活性材料的表面形成稳定的固体电解质膜，因此现有商用锂离子电池在室温附近具有低的电池内阻及较好的循环稳定性。但有机液体电解质在低温下发生液固转化，离子电导率显著下降，导致温度降低时（−20℃以下）电池内阻显著增大，无法满足低温应用要求。当电池外部温度升高或大电流充放电或短路导致电池内部温度升高时，电解液与电极之间的化学反应速度加剧，进一步产生热量，可以导致热失控。这一过程产生气体，最终导致电池密封失效，可燃的气体与有机溶剂在高温下遇到氧气起火燃烧爆炸。凝胶型电解质中电解液的含量相对较少，安全性能有所提高，但无法从根本上解决安全性问题。此外，液体电解液在低电位会被还原，在负极表面形成固体电解质膜；在高电位会发生氧化分解，造成电池充放电库仑效率降低。另外，目前，商用锂离子电池电解液体系中一般采用 $LiPF_6$ 作为电解质盐，$LiPF_6$ 热稳定性差，与水反应生成 HF，进攻正负极材料表面，造成电池性能恶化。目前常用有机电解液体系电化学窗口一般小于 4.5V，限制了高电压正极材料的使用，影响了高能量密度锂离子电池的发展。

电动汽车期望电池寿命达到 15 年，大规模工业储能需要储能器件服役寿命能满足 25～30 年的使用要求，从而显著降低系统的全寿命周期成本。含有液态有机溶剂的锂离子电池，由于液体电解质与电极材料、封装材料缓慢地相互作用和反应，长期服役时溶剂容易干涸、挥发、泄漏，电极材料容易被腐蚀，影响电池寿命。电池的循环寿命与材料中的杂质含量密切相关。杂质的存在，可以催化液体电解质在电极表面发生副反应，导致表面膜不断生长、活性物质不可逆消耗、电解液逐渐耗尽、电池内阻不断增大。目前，为了应对动力电池、储能电池对循环性、一致性的要求，高水平的材料制造企业尽可能做到电池材料杂质含量低于 $10\mu g/g$。多数企业生产的电池材料杂质含量在 $100\sim200\mu g/g$。杂质含量的减少，可以显著延长电池的循环寿命，但同时带来的问题是在所有的制造环节必须考虑防止杂质的引入，导致制造成本显著提高，电池的可靠性无法从根本上保障。如果采用固体电解质，则可以避开液体电解液带来的副反应、泄漏、腐蚀问题，从而有望显著延长服役寿命、降低电池整体制造成本、降低电池制造技术门槛，有利于大规模推广使用。

近年来，大容量锂离子电池在电动汽车、飞机辅助电源方面出现了严重的安全事故，这些问题的起因与锂离子电池中采用可燃的有机溶剂有关。虽然通过添加阻燃剂、采用耐高温陶瓷隔膜、正负极材料表面修饰、优化电池结构设计、优化 BMS、在电芯外表面涂覆相变阻燃材料、改善冷却系统等措施，能在相当程度上提高现有锂离子电池的安全性，但这些措施无法从根本上保证大容量电池系统的安全性，特别是在电池极端使用条件下、在局部电池单元出现安全性问题时。而采用完全不燃的无机固体电解质，则能从根本上保证锂离子电池的安全性。

9.1 全固态锂离子电池概述

为了克服现有商业液态锂离子电池所面临的问题，科研人员也正在大力发展基于固体电解质的锂离子电池[1-10]，它具有显著的优点：①相对于液体电解质，固体电解质不挥发，一般不可燃，因此采用固体电解质的固态电池会具有优异的安全性；②由于固体电解质能在宽的温度范围内保持稳定，因此全固态电池能够在宽的温度范围内工作，特别是高温下；③一些固体电解质对水分不敏感，能够在空气中长时间保持良好的化学稳定性，因此固态电池的制造全流程不一定需要惰性气氛的保护，会在一定程度上降低电池的制造成本；④有些固体电解质材料具有很宽的电化学窗口，这使得高电压电极材料有望应用，从而提高电池能量密度；⑤相对于多孔的凝胶电解质及浸润液体电解液的多孔隔膜，固体电解质致密，并具有较高的强度及硬度，能够有效地阻止锂枝晶的刺穿，因此提高了电池的安全性，同时也使得金属 Li 作为负极的使用成为可能。综上所述，从基本的特性分析考虑，如果寻找到合适的材料体系，采用固体电解质的全固态锂电池，可以具有优异的安全特性、循环特性以及高的能量密度和低的成本。

固体电解质包括聚合物固体电解质、无机固体电解质以及复合电解质。不含液态有机溶剂的全固态聚合物电解质是采用锂盐与聚合物复合形成的电解质材料，也称为干聚合物电解质（dry polymer electrolyte）。其在玻璃化转变温度以上具有较高的电导率，并具有良好的柔韧性及拉伸剪切性能，易于制备成柔性可弯折电池。固态聚合物电解质中，锂盐通过与高分子相互作用，能够在高分子介质中发生一定程度的正负离子解离并与高分子的极性基团络合形成络合物。高分子链段蠕动过程中，正负离子不断地与原有基团解离，并与邻近的基团络合，在外加电场的作用下，可以实现离子的定向移动，从而实现正负离子的传导。

聚合物电解质的发现始于 20 世纪 70 年代。1973 年，Fenton 等[11]发现 PEO 能够溶解碱金属盐形成络合物。1975 年，Wright[12]测量了 PEO-碱金属盐络合物电导率，发现其具有较高的离子电导率。1979 年，Armand 等[13]报道了 PEO 的碱金属盐在 40～60℃时离子电导率达 10^{-5} S/cm，且具有良好的成膜性能，可用作锂离子电池的电解质。之后人们采用不同的方法来提高聚合物电解质的电导率，包括两个方面：抑制聚合物结晶，提高聚合物链段的蠕动性；增加载流子的浓度。抑制聚合物结晶性以提高聚合物链段蠕动性的方法包括：交联、共聚、共混、聚合物合金化、加入无机添加剂。增加载流子浓度的方法包括：使用低解离能的锂盐、增加锂盐的解离度。聚合物电解质采用的常见聚合物基体包括聚氧化乙烯（PEO）、聚丙烯腈（PAN）、聚甲基丙烯酸甲酯（PMMA）、聚偏氟乙烯（PVDF）等。目前采用 PEO 作为电解质，工作温度在 80℃的全固态锂电池已被开发出来，法国 Bollore 及美国 SEEO 公司已尝试制造 Li/PEO electrolyte/$LiFePO_4$ 电芯用于电动汽车、分布式储能设备。目前此类电池需要配备热管理系统，电池从低温到工作温度需要一定的启动时间。此外，常用的与金属 Li 稳定的 PEO 聚合物电解质，电化学窗口小于 4V，因此 PEO 聚合物的全固态电池不能采用高电压电极材料。目前，正在开发复合型多层聚合物固体电解质，能够在高电压下工作，从而显著提高电池的能量密度。开发在室温工作的干聚合物电解质，也是研究的热点和重要目标。

无机固体电解质是一类具有较高离子传输特性的无机快离子导体材料，其具有较高的机械强度，能够阻止锂枝晶穿透电解质造成内短路。可以采用原子层沉积（ALD）、热蒸发、电子束蒸发、磁控溅射、气相沉积、等离子喷涂、流延成型、挤塑成型、喷墨打印、冷冻干燥、陶瓷烧结等方法制备成不同厚度、不同形状的电解质层或薄膜。

相对于聚合物固体电解质，无机固体电解质能够在宽的温度范围内保持化学稳定性，因此基于无机固体电解质的电池具有更高的安全特性。无机固体电解质主要包括氧化物无机固体电解质与硫化物无机固体电解质。氧化物无机固体电解质稳定性较好，但兼具高的离子电导率、宽的电化学窗口、成本较低、易于制造的材料尚未开发成功。硫化物无机电解质的晶界电阻较低，总的电导率高于一般氧化物电解质，最新开发的硫化物电解质 $Li_{10}GeP_2S_{12}$ 室温离子电导率已达到液体电解质的水平（图 9-1）[14]，因此相对于氧化物电解质，基于硫化物的全固态电池具有更加优异的电化学性能。由于目前的正极材料多为氧化物材料，研究发现氧化物正极/硫化物固体电解质的界面电阻较高，对电池容量利用率和高倍率性能有显著影响。改善氧化物正极/硫化物电解质的界面对提高硫基全固态锂离子电池电化学性能具有很重要的作用。日本东京工业大学的研究证明了这一结论。以 $Li_2S-SiS_2-Li_3PO_4$ 玻璃作为固体电解质，$LiCoO_2$ 为正极，In 箔片为负极装配固态电池。

电池的平均充电电压为 3.6V，平均放电电压为 3.1V，电池在第 1、2 次循环中库仑效率由于不可逆的合金化反应快速降低，第 3 次循环后电池库仑效率变为 100%，这表明电极反应是完全可逆的。但这些电池只能在很低的电流下进行充放电，如果提高充放电电流，其容量会明显衰退，其原因在于固态电池中硫化物电解质/氧化物电极界面有非常高的阻抗。为了显著降低界面阻抗，日本物质材料研究机构 Takada 所领导的团队通过在电极（$LiCoO_2$）/电解质（thio-LISICON）界面引入纳米尺度的纯离子导电缓冲层，如 $Li_4Ti_5O_{12}$、$LiNbO_3$ 和 $LiTaO_3$，从而显著降低了界面电阻，使全固态锂离子电池的高倍率容量和循环性能明显改善[15]。这也进一步说明，在全固态锂二次电池中电极/电解质界面的修饰和改性研究对提高整体电池电性能的重要作用。

综上所述，采用固态电解质的电池相对于液体电解质的电池，具有更高的安全性能，但目前固态电池内阻较高，为了发挥全固态电池的优势，发展高电导率的电解质材料、降低固固界面电阻是关键。各种无机固态电解质与液体电解液、离子液体、聚合物电解质等材料电导率的对比如图 9-1 所示。

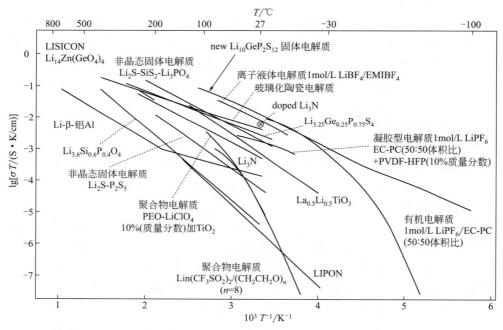

图 9-1　无机固态电解质、液态电解液、聚合物电解质、
离子液体等的电导率随温度变化的曲线[14]

9.2　固体电解质材料

固体电解质按化学组成可分为无机固体电解质、聚合物固体电解质以及聚合物复合电解质。

9.2.1 无机固体电解质

无机固体电解质按结晶形态可分为晶态固体电解质、非晶态固体电解质和无机复合固体电解质。

9.2.1.1 晶态固体电解质

晶态电解质主要有 Perovskite 型、反钙钛矿型、NASICON 型、LISICON 型、Garnet 型、Li_3N 型等。

(1) Perovskite 型

一般将碱土金属的钛酸盐称为钙钛矿 $ATiO_3$（A=Ca、Sr、Ba），通式可写为 ABO_3。结构示意图如图 9-2 所示。Brous 等[16]首次通过三价稀土 La^{3+} 和一价碱土阳离子（Li^+、Na^+、K^+）共同取代 A 位的碱土离子合成了钙钛矿结构的 $Li_{0.5}La_{0.5}TiO_3$。Inaguma 等[17]报道了 $Li_{3x}La_{2/3-x}TiO_3$（$x=0.11$）的室温体相离子电导率高达 10^{-3} S/cm。这引起了大家对 $Li_{3x}La_{2/3-x}TiO_3$ 的广泛关注。研究发现随产物组成和合成条件的变化，$Li_{3x}La_{2/3-x}TiO_3$ 存在着立方、四方、正交等多种结晶形式。Fourquet 等[18]对其结构进行研究，发现在 $0.06<x<0.14$ 时为固溶体，晶体结构为四方结构，空间群为 $P4/mmm$。这一类固溶体是纯的离子导体，电子电导几乎可以忽略。在 $Li_{3x}La_{2/3-x}TiO_3$ 的晶体结构中，高价的 La 占据 A 位，导致 A 位空位的产生，Li^+ 在其中以空位机制传导。钙钛矿材料中 Li^+ 输运遵循渗流模型，$x=0.11$ 时具有最高的电导率。$Li_{3x}La_{2/3-x}TiO_3$（$x=0.11$）的离子电导率在 400K 以下符合经典的 Arrhenius 定律[19]，但以 200K 为界其活化能发生了变化，如图 9-3 所示。

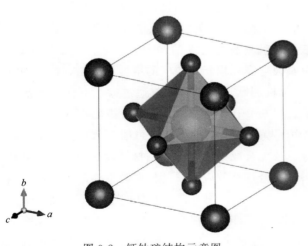

图 9-2 钙钛矿结构示意图

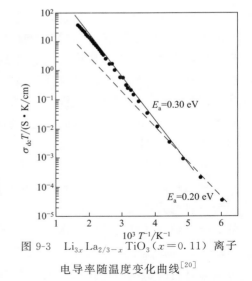

图 9-3 $Li_{3x}La_{2/3-x}TiO_3$（$x=0.11$）离子电导率随温度变化曲线[20]

虽然 $Li_{3x}La_{2/3-x}TiO_3$ 具有较高的离子电导率，但是仍然存在着很多问题。在合成过程中，由于高温导致 Li_2O 的损失，所以较难控制最后得到产物的组分，不易得到较高的电导率，而且由于其具有较高的界面电阻，材料总的电导率较低[17]。此外，$Li_{3x}La_{2/3-x}TiO_3$ 与金属 Li 接触导致 Ti^{4+} 被还原而产生较高的电子电导[21]，因此不能与负极材料直接接触使用，从而影响了其应用。

（2）反钙钛矿型

与目前已知的大多数固体电解质源于已有材料掺杂取代，或者作为其他反应的副产物偶然发现相比，赵予生等[22]提出的富锂反钙钛矿结构固体电解质（lithium-rich anti-perovskites，LiRAP）集中体现了材料设计的思想。以下从实验和计算两方面，介绍富锂反钙钛矿材料 LiRAP 的设计、性能、输运机制，以及通过化学、材料结构、电子结构等手段有目的地提高离子输运，改善材料性能。

钙钛矿材料简式可写作 ABX_3，最常见的钙钛矿型固体电解质 $Li_{3x}La_{2/3-x}TiO_3$（$0.04<x<0.14$）中导电的 Li^+ 处于 A 位。对于另一种高温快离子导体 $NaMgF_3$ 和（K、Na）MgF_3，导电离子为 F^-，处于 X 位。事实上，对于燃料电池来说，以 X 位 O^{2-} 为可输运离子的钙钛矿结构快离子导体，无论对于阳极材料、阴极材料或固体电解质，均有广泛应用。早期的钙钛矿结构离子电导率的研究还包括 $MgSiO_3$ 类型的氧离子导电，主要用来研究下地幔的电导率。受到此类材料的启发，为了得到锂离子导体，赵予生等用一价阳离子 Li^+ 占据 X 位得到富锂材料，并用 −1 价的卤族元素（F、Cl、Br、I）和 −2 价的 O 分别来替代 A 位和 B 位。此时，ABX_3 中各个元素的电性与传统钙钛矿材料相反，且结构单元中有 3 个 Li^+，在现有固体电解质中具有最高的 Li^+ 浓度，因此称为"富锂的反钙钛矿材料"（lithium-rich anti-perovskites，以下简记为 LiRAP）。一个有趣的巧合是，几乎在同一时间，反钙钛矿结构的 Li_3OCl 作为合成 Li_5OCl_3 的副产物被发现[23]。次年，Emly 等[24]从计算和相图的角度同样确认了亚稳态结构 Li_3OCl 的存在。不同发现途径以及不同合成方法，均确认了这一新型材料及其结构。

之所以选择钙钛矿型材料为原型进行材料设计，是因为钙钛矿材料具有比较高的结构容忍度[25]，因此具有较大的结构调控空间，以保持设计材料的稳定性和提高电导率。结构调控可以通过控制合成中的化学取代获得，且具有宽范围的调控空间来保持结构稳定。图 9-4 为多种卤素混合、Li 位高价掺杂以及卤化锂耗损三种效应的结构示意图，这三种方法可以提高缺陷浓度并促进锂-空位的晶格位机制跃迁；同时可以通过掺杂取代扩大晶格参数，使得 Li^+ 经过瓶颈的输运更加容易。图 9-4 中扁平的椭球暗示了浅而平的势垒，使 Li^+ 的跃迁概率被加强。推测其原因，软声子模驱动的结构相变有可能有效地提高 Li^+ 跃迁。基于此推测，赵予生等通过调节组分、改变合成条件等，达到材料设计的目的，以期获得室温超离子相变的高电导率目标材料。整合三种效应，反钙钛矿结构锂离子导体可以记为更普遍的通式——$Li_{3-x-\delta}M_{x/2}O(A_{1-z}A'_z)_{1-\delta}$。图 9-5 为 Li_3OCl 和 $Li_3OCl_{0.5}Br_{0.5}$ 的电导率随温度升高的变化图。注意到，$Li_3OCl_{0.5}Br_{0.5}$ 在 247℃ 发生相转变，电导率突然增大，其室温电导率为 6.05×10^{-3} S/cm。在实验的制备和测量过程中，研究发现制备方法和材料的织构（texture）对电导率有非常大的影响，并猜想虽然该材料富锂，但是 Li^+ 被紧紧束缚在八面体顶角处，导致低的电导率。材料制备的热处理历史导致微观结构的变化，如空位增多、结构扭曲（如八面体的倾斜）、局域无序等，可能是高电导率的原

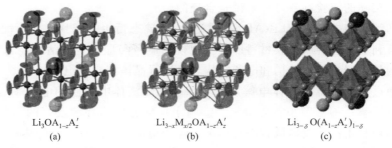

图 9-4 反钙钛矿结构中混合（a）、掺杂（b）、耗损（c）效应示意图[22]

因。然而，制备过程中很多因素不可控，利用第一性原理计算来研究 Li^+ 输运机制就成为必要，在解释以上现象的同时，为下一步结构调控方向提供理论基础[22]。

Zhang 等[26] 首先利用第一性原理的分子动力学方法（*ab initio* molecular dynamics，AIMD）模拟了 Li^+ 输运，证实了 Li^+ 输运伴随着亚晶格熔化，Li^+ 亚晶格熔化温度略低于整个晶体熔融温度，结论与实验符合。计算表明，完美晶体中，升高温度至 Li^+ 均方根位移（MSD）显著增强时，阴离子的 MSD 也开始增加，意味着整个晶体的熔融而非 Li^+ 亚晶格的熔化。当引入锂空位

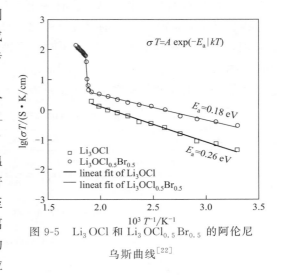

图 9-5 Li_3OCl 和 $Li_3OCl_{0.5}Br_{0.5}$ 的阿伦尼乌斯曲线[22]

和阴离子无序时，Li^+ 的 MSD 显著增强的起始温度大大降低，此时阴离子不动，意味着亚晶格熔化。简言之，锂空位和阴离子无序是超离子导体导电的主要驱动力。这种方法得到的活化能高于实验值，尤其对于 $Li_3OCl_{0.5}Br_{0.5}$ 类混合卤素材料，不同构型的活化能不小于 0.288eV，远大于实验值的 0.18eV。为了解释实验中低的活化能，Emly 等[24] 提出了哑铃形 Li^+ 间隙位模型，并提出三原子协同跃迁机制，如图 9-6 所示。然而，此种 Fren-kel 缺陷对生成能为 1.94eV，意味着此种缺陷浓度很小，虽然解释了低的活化能，但是很难解释高的电导率。一种可能是在制备过程中偏离了化学计量并引入过量的缺陷，

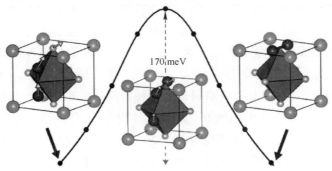

图 9-6 低活化能扩散路径[24]

这种猜测对于实际的实验制备过程来说是合理的。

从实验和计算结果来看，LiRAP 及其掺杂、玻璃态衍生物，具有如下优点[22]：①材料熔点低，有利于直接合成薄膜材料以及大规模的制备与应用；②活化能低且电导率高，尤其是此结构易于调整晶体结构和电子结构，且进行晶体结构、组分、晶型调控后，材料电导率可以超过其他现有的固体电解质，并超过部分液体电解质的电导率（图 9-5）；③极低的电子电导率；④密度小、质量轻；⑤与水反应分解为环境友好的卤化锂和氢氧化锂，且易于循环利用；⑥成本低；⑦与锂金属电极稳定；⑧具有一定的热稳定性，虽然该材料 0 K 时相对于 Li_2O 和 LiA（A＝Cl、Br）为亚稳态，但是当施加一定温度时相会稳定。关于该材料的电化学窗口，尚有一定争议。Zhang 等[26] 通过计算得到电化学窗口接近 5eV；然而，Emly 等[24] 通过计算锂化学势-组分相图发现，虽然 Li_3OCl 具有超过 5eV 的带隙，但是当对其施加 2.5V 偏压时，反钙钛矿材料倾向于分解为 Li_2O_2、LiCl 和 $LiClO_4$。Schroeder 等[27] 研究了 Li_3OBr 相对于常见溶剂的稳定性，此项研究考虑到不限于锂离子全固态电池的其他应用，小结见表 9-1。

表 9-1 Li_3OBr 与溶剂反应产物的相分析[27]

溶剂	现象	产物分析
去离子水	30min 内溶解样品，在空气中 75℃ 加热 48h	$LiBr \cdot xH_2O$ $LiOH \cdot H_2O$ $NaHCO_3$
DEC	倒掉溶剂，将样品在室温下真空干燥 样品在空气中 75℃ 加热，过夜	无定形态/纳晶 $(Li,H)_3OBr$ $LiOH \cdot xH_2O$
DMC	倒掉溶剂，将样品在室温下真空干燥 样品在空气中 75℃ 加热，过夜	无定形态/纳晶 $(Li,H)_3OBr$ $LiOH \cdot xH_2O$
PC	倒掉溶剂，将样品在室温下真空干燥 样品在空气中 75℃ 加热，过夜	$(Li,H)_3OBr$（晶化程度低）
DME	倒掉溶剂，将样品在室温下真空干燥 样品在空气中 75℃ 加热，过夜	$(Li,H)_3OBr$ $LiOH \cdot xH_2O$
Gen2（EC/EMC/$LiPF_6$）	几周之后溶解	

与 2012 年开始发展的 LiRAP 材料相较，卤化锂水合物型反钙钛矿材料的 Li^+ 无机固体电解质材料发现于 1969 年[28]，其对于电导率的研究可以追溯到 1980 年[29,30]，在随后的三十几年中[31-38]，由于其分子动力学无序性、相变和离子电导等特性被多次研究，并研究其与 Al_2O_3 复合后的电导率[33]。其通式可以写做：$Li_{3-n}(OH_n)Cl$（$0.83 \leqslant n \leqslant 2$）和 $Li_{3-n}(OH_n)Br$（$1 \leqslant n \leqslant 2$）[38]。材料的高温相为反钙钛矿型立方相结构，不同组分的相变温度不同。高温立方相结构如图 9-7 所示[37]，不同组分的活化能见表 9-2[38]。可以看到，在非立方相（低温相）时离子电导被抑制，这是由于静态的 OH^- 和 OH_2 阻塞了 Li 的位置，导致 Li^+ 可占据的位置减少。对于立方相，OH^- 和 OH_2 可以"自由"转动，Li^+ 的可占据位增多，电导率提高。

对于晶体中的离子输运，空位和间隙位都是必要的条件；"超离子导体"的先决条件之一，就是具有足够大的、可移动的缺陷浓度。此材料天然存在空位，如对于 $Li_2(OH)X$，只有 2/3 的 Li^+ 位置被占据，而对于 $LiX \cdot H_2O$，只有 1/3 的 Li^+ 位置被占据[38]。同时，中子衍射和 NMR 的研究发现，此材料具有无序的结构[37]；这种固有的无序结构，使得其对于 Li^+ 输运具有过剩的等价位置，有利于 Li^+ 传输。由于其本身的特性，使得通过结构来调控锂离子电导率成为可能，也因此受到部分关注。但是其稳定温度范围过窄且高温下可能脱水（如对于 $LiCl \cdot H_2O$，稳定温度从 95℃ 到 98～100℃ [37]，而对于 LiRAP，即使熔融仍然可以工作，40℃ 仍不会分解[22]）、制备难度较大（相对而言，LiRAP 的熔融制备方法容易大规模生产）、相对较低的电导率，因而没有受到很高的重视。

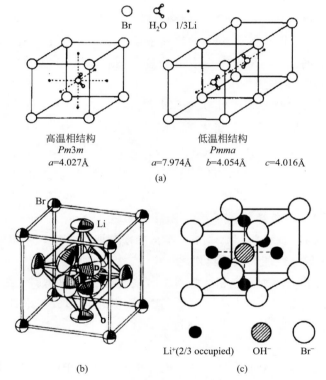

图 9-7 $LiBr \cdot H_2O$ 的高温相与低温相结构（a）[32]；$T=330$ K 时，$LiBr \cdot D_2O$ 的立方结构单元（b）[37]；反钙钛矿相 $Li_2(OH)Br$ 示意图（c）[30]

表 9-2 锂离子运动的活化能[38]

样 品	E_A/(kJ/mol)			
	阻抗谱		$^7Li\ T_1$ NMR	
	升温	降温	线宽	BPP 拟合
$Li_{2.17}(OH_{0.83})Cl$			33.8±0.1	37.2±0.5
$Li_2(OH)Cl$	53.7±1.5	55.3±1.5	34.9±0.1	34.4±0.5
$Li_{1.16}(OH_{1.84})Cl$	59.4±0.9	57.7±0.5	28.8±0.2	

续表

样 品	E_A/(kJ/mol)			
	阻抗谱		^{7}Li T_1 NMR	
	升温	降温	线宽	BPP 拟合
LiCl·H$_2$O		74±4		
Li$_2$(OH)Br	81±1	85±2	35.7±0.1	44.7±1.8
Li$_{1.04}$(OH$_{1.96}$)Br	63±1	61±1	38.6±0.7	
LiBr·H$_2$O	67±2	68±3	25.1±0.2	

(3) NASICON 型

NASICON，是 sodium super ionic conductor 的简称，具有 NASICON 结构的 NaA$_2^{IV}$(PO$_4$)$_3$（A^{IV} = Ge、Ti、Zr）在 1968 年被确认[39]。1976 年 Goodenough 等[40]发现用 Si 取代 NaZr$_2$(PO$_4$)$_3$ 中部分 P 得到 Na$^+$ 的快离子导体 Na$_{1+x}$Zr$_2$P$_{3-x}$Si$_x$O$_{12}$（0≤x≤3），并将类似结构的化合物称为 NASICON 结构化合物。随后，用 Li 取代 Na，同样得到具有较高离子电导率的锂离子导体。

NASICON 结构化合物分子式可写为 AM$_1$M$_2$P$_3$O$_{12}$。其中 A 可为碱金属离子（Li$^+$、Na$^+$、K$^+$、Rb$^+$、Cs$^+$）、碱土金属离子（Mg^{2+}、Ca^{2+}、Sr^{2+}、Ba^{2+}）、H$^+$、H$_3$O$^+$、NH$_4^+$、Cu$^+$、Cu^{2+}、Ag$^+$、Pb^{2+}、Cd^{2+}、Mn^{2+}、Co^{2+}、Ni^{2+}、Zn^{2+}、Al^{3+}、Ln^{3+}（Ln 为稀土元素）、Ge^{4+}、Zr^{4+}、Hf^{4+} 或是空位。M 可为二价离子（Zn^{2+}、Cd^{2+}、Ni^{2+}、Mn^{2+}、Co^{2+}），三价离子（Fe^{3+}、Sc^{3+}、Ti^{3+}、V^{3+}、Cr^{3+}、Al^{3+}、In^{3+}、Ga^{3+}、Y^{3+}、Ln^{3+}），四价离子（Ti^{4+}、Zr^{4+}、Hf^{4+}、Sn^{4+}、Si^{4+}、Ge^{4+}）和五价离子（V^{5+}、Nb^{5+}、Ta^{5+}、Sb^{5+}、As^{5+}）。丰富的掺杂取代，使 NASICON 结构成为研究化学掺杂-晶体结构-导电性能关系的典例之一。

NASICON 化合物一般为六方相，随着组分变化还会出现单斜、三斜、四方等结构。NASICON 六方相结构如图 9-8 所示。XO$_4$ 与 MO$_6$ 共顶点形成三维骨架，A 离子在三维骨架的 A（1）位与 A（2）位，A 离子通过在 A（1）位与 A（2）位的跳跃实现了离子的传输[41]。

与钙钛矿结构类似，NASICON 材料的电导率同样受限于体相与晶界电阻两方面，研究人员从这两方面努力，以期提高其电导率。Subramanian 等[42]发现 LiZr$_2$(PO$_4$)$_3$ 的电导率非常低，用 Ti 部分或完全取代 Zr，LiZr$_{2-x}$Ti$_x$(PO$_4$)$_3$ 电导率有了明显提高，这是因为 Li$^+$ 的输运要通过晶体骨架的"瓶颈"，瓶颈过小或者过大，都不利于 Li$^+$ 的跃迁，而 Ti 离子半径提供了最为适合的瓶颈口径。而用三价的 M^{3+}（Sc^{3+}、In^{3+}）取代 M^{4+}（Zr^{4+}、Ti^{4+}、Hf^{4+}）得到 Li$_{1+x}$M^{3+}M^{4+}(PO$_4$)$_3$ 一系列化合物，也会使电导率显著增加，这主要是由于引入间隙位 Li$^+$，载流子浓度得到了提高。

除了上述对体相电导率的调控以外，提高晶界电导是另一个提高总体电导率的重要手段。Aono 等[43,44]对 Li$_{1+x}$M$_x$Ti$_{2-x}$(PO$_4$)$_3$（M = Al、Cr、Ga、Fe、Sc、In、Lu、Y、La）的电导率随 M 掺杂量的变化研究发现，电导率的增加与材料的致密度有关。如图 9-9

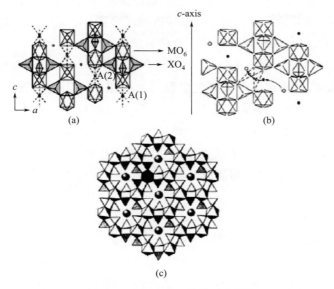

图 9-8 NASICON 结构示意图

(a) A（1）、A（2）占位；(b) 传输路径；(c) $A_2(XO_4)$（0 0 1）面上的投影[41]

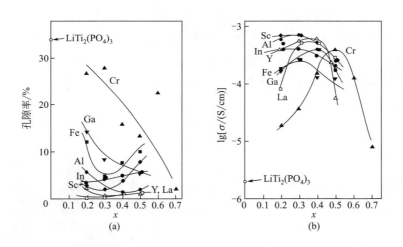

图 9-9 $Li_{1+x}M_xTi_{2-x}(PO_4)_3$（M＝Al、Cr、Ga、Fe、Sc、In、Y、La）中陶瓷孔隙率随掺杂量的变化 (a)；$Li_{1+x}M_xTi_{2-x}(PO_4)_3$（M＝Al、Cr、Ga、Fe、Sc、In、Y、La）中离子电导率随掺杂量的变化 (b)[44]

所示，对同一元素，孔隙率随掺杂量先下降后增加，相应电导率随掺量先增加后下降。在孔隙率最低时，获得最高的电导率。不同掺杂元素间比较，孔隙率越低，相应材料的电导率越高。可以推测是通过掺杂提高材料致密度，并达到提高材料总电导的目的。为了提高致密度，他们引入助烧剂，并研究了助烧剂 Li_3PO_4 和 Li_3BO_3 对 $LiTi_2(PO_4)_3$ 电导率的影响，结果表明，助烧剂的加入同时提高了体电导和晶界电导[45]。

玻璃态材料与晶体材料相比，除了彻底解决了晶界电阻的问题外，对于晶体结构无序度的提高同样提供了改善电导率的可能性。且熔融法制备材料比较简单，容易批量规模应用。Fu[46]采用熔融淬火的方法制备了高致密度的玻璃陶瓷，被 OHARA 公司采用，作为批量制备 NASICON 结构固体电解质陶瓷片的方法。

目前，$Li_{1+x}Al_xTi_{2-x}(PO_4)_3$（LATP）和 $Li_{1+x}Al_xGe_{2-x}(PO_4)_3$（LAGP）研究比较广泛，两者具有高的电导率和宽的电化学窗口，应用于全固态锂离子电池[47]、锂空电池[48]和锂液流电池[41]中。但是由于 $Li_{1+x}Al_xTi_{2-x}(PO_4)_3$ 在低电位时 Ti^{4+} 被还原嵌 Li，所以不能与金属 Li 或低电位嵌 Li 负极材料直接接触使用，可采用 $Li_4Ti_5O_{12}$[47,49]等负极材料，或者通过包覆或者引入缓冲层等方式加以保护利用。LATP 的原材料资源丰富，成本较低，材料稳定性高，相信进一步改进后具有较好的应用前景。

（4）LISICON 型

LISICON，是 lithium super ionic conductor 的简称。1978 年 Hong[50]首次报道了 $Li_{14}Zn(GeO_4)_4$，并把它称作 LISICON，室温离子电导率仅 10^{-7}S/cm，300℃时为 0.125S/cm。随后 Bruce 等[51]和 West 等[52]对 $Li_{2+2x}Zn_{1-x}GeO_4$ 进行了研究。$Li_{14}Zn(GeO_4)_4$ 可认为是 Li_4GeO_4 和 Zn_2GeO_4 的固溶体，然而仅在比较窄的温压条件下可以得到此相，Li_2ZnGeO_4 和 Li_4GeO_4 的二元相图如图 9-10 所示。LISICON 型材料晶体结构与 γ-Li_3PO_4 有关。

$Li_{14}Zn(GeO_4)_4$ 的结构如图 9-11 所示。在 $Li_{14}Zn(GeO_4)_4$ 的结构中有 1 个三维阴离子骨架 $[Li_{11}Zn(GeO_4)_4]^{3-}$，11 个 Li^+ 分布在两个不同的位置上，其中 4 个 Li^+ 占据 $4c$ 位置，7 个 Li^+ 与 Zn^{2+} 共同占据 $8d$ 位置，其余的 3 个 Li^+ 分别占据在 $4c$ 和 $4a$ 位置上（$4c$ 位置占有率约为 55％，$4a$ 位置占有率约为 16％）。这 3 个 Li^+ 构成了离子输运的三维通道[55]。

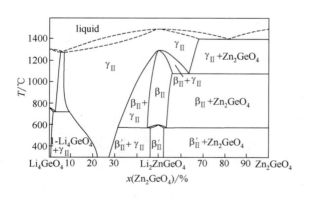

图 9-10　Li_2ZnGeO_4-Li_4GeO_4 的二元相图[53]

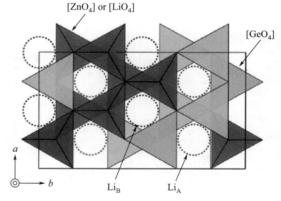

图 9-11　$Li_{14}Zn(GeO_4)_4$ 的结构示意图[54]

另一类具有 γ-Li_3PO_4 结构的固溶体 $Li_{3+x}X_xY_{1-x}O_4$（X＝Si、Sc、Ge、Ti；Y＝P、As、V、Cr），以 $Li_{3.5}Ge_{0.5}V_{0.5}O_4$ 的离子电导率最高，室温可达 4×10^{-5}S/cm。该材料与 LiI 混合的复合型固体电解质，已被美国应用于高温固态电池中[56]。

$Li_{14}Zn(GeO_4)_4$ 虽然在高温下具有较高的电导率，但室温电导率太低。另外，其较高的 Li_2O 含量导致其对空气中的水分及二氧化碳比较敏感，该材料对金属锂也不稳定，这些因素限制了其应用。

Kanno 等[57,58]提出将 LISICON 中的 O 由 S 来替代得到 Thio-LISICON，同样是 γ-Li_3PO_4 结构，S^{2-} 的半径较大，扩大了离子传输通道，且 S^{2-} 极化也大，减小了骨架对 Li^+ 的束缚作用，所以 Thio-LISICON 可以提高离子电导率。但是得到的 $Li_{4-2x}Zn_xGeS_4$ 室温下电导率仍然不是很高，如图 9-12 所示。

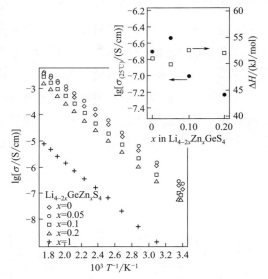

图 9-12　$Li_{4-2x}Zn_xGeS_4$ 离子电导率随温度变化的曲线[57]

随后 Kanno 等[59]又对 $Li_{4-x}Ge_{1-x}P_xS_4$ 进行了研究，当 $x=0.75$ 时电导率最高，其室温离子电导率达 2.2×10^{-3} S/cm。对于含 Si 的一系列固溶体 $Li_{4-x}Ge_{1-x}P_xS_4$，当 $x=0.6$ 时电导率最高，室温离子电导率为 6.4×10^{-4} S/cm[58]。

2011 年 Kamaya 等[14]报道了 $Li_{10}GeP_2S_{12}$，室温下离子电导率为 1.2×10^{-2} S/cm，这也是目前发现的室温离子电导率最高的无机固体电解质。$Li_{10}GeP_2S_{12}$ 是由 $(Ge_{0.5}P_{0.5})S_4$ 四面体、PS_4 四面体、LiS_4 四面体以及 LiS_6 八面体构成的三维网状结构。其中，$(Ge_{0.5}P_{0.5})S_4$ 四面体和 LiS_6 八面体共棱且形成沿 c 轴的一维长链，构成了沿 c 轴方向的一维 Li^+ 长程迁移通道。虽然 Kamaya 认为其电化学窗口约 5V（vs. Li/Li^+），但第一性原理计算认为 $Li_{10}GeP_2S_{12}$ 的电化学稳定电压范围仅 1.7~2.1V（vs. Li^+/Li）[60]，其与金属锂反应生成 Li_3P、Li_2S 以及 $Li_{15}Ge_4$，在 2.14V 时会被氧化成 Li_3PS_4、S 以及 GeS_2。实验中观察到的较宽的电化学窗口可能与反应产物的钝化有关。

（5）Garnet 型

Garnet 结构是一种硅酸盐矿物结构，分子通式可写为 $A_3B_2(SiO_4)_3$。其中 A 为八配位阳离子，B 为六配位阳离子，A 占据八面体的 $24c$ 位。2004 年 Thangadurai 等[61]发现了具有 Garnet 结构的新型固体电解质 $Li_5La_3M_2O_{12}$（M=Nb、Ta），具有较高的离子电导率和较宽的电化学窗口。Cussen[62]通过中子衍射确定了其结构，Li^+ 分别占据在四面体的 $24d$ 位（占据率 80%）和八面体的 $48g$ 位（占据率 40%）。Wilmer 等[63]指出材料合成过程中的温度对 Li^+ 在四面体和八面体位的分布有很大的影响，这也影响着 Li^+ 在 Garnet 型材料结构中的迁移。

Weppner 等[64-66]用低价的 Ca、Sr、Ba、K 等元素替代 $Li_5La_3M_2O_{12}$（M=Nb、Ta）中的 La 得到一系列的材料，其中电导率最高的是 $Li_6BaLa_2Ta_2O_{12}$，室温电导率为 4×10^{-5} S/cm[65]。用 In^{3+} 替代 Ta^{5+} 或 Nb^{5+}，得到的一系列产物中，$Li_{5.5}La_3Nb_{1.75}In_{0.25}O_{12}$ 的电导率最高，50℃时电导率可达 1.8×10^{-4} S/cm[66]。2007 年，Weppner 等[67]用四价

Zr 取代五价位置，得到立方结构的 $Li_7La_3Zr_2O_{12}$，室温电导率最高可达 $3×10^{-4}$ S/cm，活化能约 0.3eV，其电导率与其他固体电解质电导率比较如图 9-13 所示。Awaka 等[68]合成了四方相的 $Li_7La_3Zr_2O_{12}$，室温下电导率约为 $4.2×10^{-7}$ S/cm，活化能为 0.54eV。Geiger 等[69]通过变温 XRD 实验发现，在 100~150℃，四方相的 $Li_7La_3Zr_2O_{12}$ 向立方相转变。另外，固相合成中使用氧化铝坩埚导致 Al 掺杂进入 $Li_7La_3Zr_2O_{12}$，可使高温立方相得以在室温下仍然存在。立方相的电导率高于四方相（图 9-14），这是因为立方结构各向同性的三维离子传输通道，更利于 Li^+ 的迁移，以及少量 Al^{3+} 的高价掺杂提高了空位浓度。

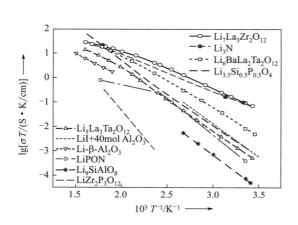

图 9-13 几种常见的无机固体电解质材料的电导率随温度变化曲线[67]

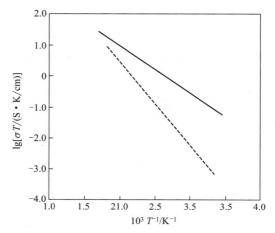

图 9-14 四方相（虚线）和立方相（实线）的 $Li_7La_3Zr_2O_{12}$ 的电导率随温度变化的曲线[69]

Garnet 类型的固体电解质材料具有较高的电导率和宽的电化学窗口，而且在空气中稳定，与金属 Li 接触也很稳定，不考虑 La、Zr 的成本，是较为理想的氧化物 Li^+ 导体。

（6）Li_3N 型

Li_3N 是最早被研究的无机固态电解质，室温下单晶电导率可达 10^{-3} S/cm，但其电导率存在着各向异性，而且稳定性很差，分解电压只有 0.45V[70]。为了提高其稳定性，在 Li_3N 中加入 LiX（X=Cl、Br、I）形成 Li_3N-LiX 固溶体，分解电压有所提高，然而其电导率大大下降。其中，$Li_{1.8}N_{0.4}Cl_{0.6}$ 分解电压高于 2.5V，室温电导率为 $2.5×10^{-6}$ S/cm[71]。Jia 等[72,73]用 Na^+、K^+ 等取代 $Li_{1.8}N_{0.4}Cl_{0.6}$ 中的 Li^+ 得到 $Li_{9-nx}M_xN_2Cl_3$（M=Na、K、Rb、Cs、Mg、Al），提高了材料的离子电导率。Hatake 等[74]在 Li_3N 中加入 MI（M=Li、Na、K）形成 $3Li_3N$-MI 一系列材料，室温电导率为 $7×10^{-5}$~$1.1×10^{-4}$ S/cm，分解电压为 2.5~2.8V。Li_3N 及其一系列衍生物由于离子电导率和分解电压都较低，限制了其在全固态电池中作为主体固态电解质的应用。在金属 Li 表面生长 Li_3N 保护层防止锂枝晶,获得了一些关注[75]。

9.2.1.2 非晶态固体电解质

相对于晶态的固体电解质,非晶态的固体电解质具有各向同性的性质,因此离子通道在宏观尺度看也是各向同性,离子通道连接也更容易些,而且界面电阻小,非晶态结构可能获得更高的总离子电导率。这一结论并不适用于所有的材料,一般而言,晶态材料如果具有大量的互相连通的离子空位,且活化能较低,可迁移离子所占据的格点位具备称为亚晶格无序的特征,也有可能具有较高的离子电导率,这类材料也是典型的快离子导体具备的结构特征。

(1) LiPON

1993 年,Bates 等[76]采用 Li_3PO_4 为靶材,在 N_2 气氛中,采用射频磁控溅射的方法生长 LiPON,室温离子电导率达 10^{-6} S/cm,电化学窗口为 5.5V。采用这种方法制备的 LiPON 膜厚度只有几百纳米,因此电解质膜的电阻较小,适用于薄膜锂离子电池中。使用 LiPON 薄膜做成的全固态薄膜电池结构如图 9-15 所示。LiPON 是目前研究最为广泛、并在微型电池中有实际应用的锂离子固体电解质。如果能够解决低成本大面积 LiPON 薄膜的制备技术以及开发相应成熟的大容量电池技术,则有望进一步拓展其应用空间。

(2) 非晶态硫化物

Kulkarni 等[78]发现在非晶态的固体电解质中,氧化物的非晶态电解质总电导率较低,活化能较高。硫化物材料由于 S 相对 O 极化更大,与 Li 的相互作用更弱,因此同系列材料中,硫化物电解质材料具有更高的电导率。非晶态硫化物固体电解质以 Li_2S 为主要成分,

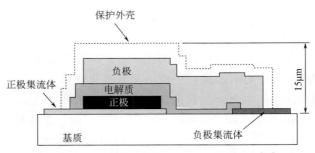

图 9-15 LiPON 固态薄膜电池结构示意图[77]

早期的研究体系主要为 Li_2S 与其他硫化物形成的多元非晶体系,如 Li_2S-P_2S_5[79]、Li_2S-SiS_2[80],图 9-16 给出了 Li_2S-M_xS_y(M=Al、Si、P)室温电导率随 Li_2S 含量的变化[81]。Hayashi 等[82]又发现在二元硫化物体系中加入少量含锂氧化物材料(Li_4SiO_4)能够进一步提高材料的电导率,室温电导率最高可达 10^{-3} S/cm。NMR 及 XPS 结果表明,当加入少量氧化物时,玻璃电解质材料中绝大部分 O 以桥接氧的形式存在,而绝大部分 S 以非桥接硫的形式存在,O 与两个 Si 形成桥接基团,导致与 Si 结合的 S 对 Li 的吸引变弱,因此电导率会增大。当玻璃材料中复合氧化物的含量增加时,其中的桥接氧的比例会下降,非桥接氧比例上升,而非桥接氧对 Li 有较强的吸引力,因此材料的电导率会下降。Hayashi 等[83]对 Li_2S-SiS_2-Li_xMO_y(M=Si、P、B、Al、Ga、In)研究发现玻璃的晶化温度 T_c 和玻璃的转化温度 T_g 差值越大,材料的稳定性越好,离子的电导率也越高,T_c-T_g 的值随 Li_xMO_y 含量的变化如图 9-17 所示。

玻璃态的硫化物无机固体电解质晶化后一般导致电导率下降,但是 Hayashi 等[84]发现球磨得到 $Li_2S-P_2S_5$ 非晶材料室温下电导率为 10^{-4} S/cm,再将这个非晶材料经过热处理后得到玻璃陶瓷,室温下电导率达 10^{-3} S/cm。经过热处理的过程,其中的导电相 Li_7PS_6、Li_3PS_4 和未知相析出,同时软化了玻璃相粉末,减少了晶界,从而显著提高了离子电导率。

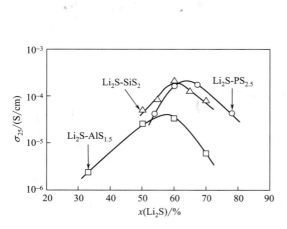

图 9-16　$Li_2S-M_xS_y$(M=Al、Si、P) 室温电导率随 Li_2S 含量的变化[81]

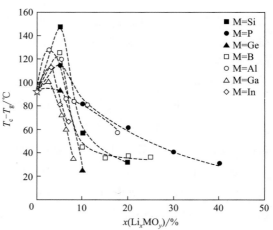

图 9-17　T_c-T_g 的值随 Li_xMO_y(M=Si、P、Ge、B、Al、Ga、In) 含量的变化[83]

非晶态硫化物固态电解质具有较高的电导率,而且容易按照要求尺寸加工,所以在全固态电池中已有小批量生产。但是硫化物在空气中不稳定,容易与空气中的水发生反应。相应的全固态电池的制造需要开发专门的生产线。由于目前全固态硫基锂电池的能量密度、循环性优势尚未体现出来,倍率特性、低温特性还不是很满意,实际应用仍然需要较长的开发周期。

9.2.1.3　无机复合固体电解质

1973 年 Liang[85]发现将 Al_2O_3 加入 LiI 得到的复合物电导率比 LiI 有了很大提高,如图 9-18 所示。Maier[86,87]认为产生这种现象的原因是由于 Al_2O_3 加入 LiI 后 Li^+ 被吸附在亲核的 Al_2O_3 表面,从而形成了空间电荷层,空间电荷层的空穴浓度增大具有高导电性,所以复合物的电导率有了很大提高。Knauth 等[88,89]也做了类似的研究。空间电荷层作用在低温下的低体相载流子浓度的固体材料(未掺杂体系)更为显著。

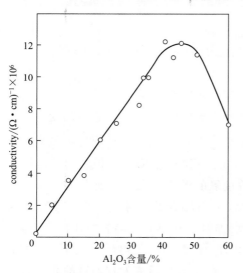

图 9-18　$LiI-Al_2O_3$ 的电导率随 Al_2O_3 含量的变化[85]

9.2.2 聚合物固体电解质

聚合物电解质以聚合物为基体，由强极性聚合物和金属盐通过 Lewis 酸-碱反应模式[90]，不断地发生络合解络合反应，形成了具有离子导电性功能的高分子材料。聚合物电解质一般由聚合物（如聚醚、聚酯和聚胺等）和碱金属盐（如 $LiClO_4$、$LiPF_6$、$LiBF_4$、$LiAsF_6$ 等）组成[91]。与无机固体电解质材料相比，聚合物电解质具有很好的柔顺性，良好的成膜性、黏弹性，质量轻的优点。与传统的液态锂离子电池相比，固态聚合物电解质避免了电解质的泄漏，具有安全性能高、重量轻（比同等规格的液态锂离子电池轻 20%～40%）、容量大（比同等规格的液态锂离子电池高 5%～15%）[92,93]，被公认为最具应用市场的电池产品。一般将聚合物电解质分为凝胶聚合物电解质和固态聚合物电解质。

发展较为成熟的纯固态聚合物电解质为聚醚碱金属盐复合物，不含有增塑剂，离子的传导依赖极性聚合物网络中的离子，PEO 类聚合物主体与盐类简单混合而得到的聚合物电解质是这类材料的典型代表，也被称为"第一代聚合物电解质"[94]。PEO 基固态聚合物电解质（SPE）是研究最早、最广泛的一类。高分子量的 PEO 在 60℃ 以下开始发生结晶，而固体聚合物电解质主要是在无定形区域通过链段蠕动进行离子的传导，所以 PEO 基的聚合物电解质想要达到 10^{-4} S/cm 可应用的电导率，需要在 60～80℃ 进行工作[95]。由于室温离子电导率太低，因此这类纯固态聚合物电解质材料电池的应用需要配备加热与温度控制系统。

由于室温下 PEO 基聚合物电解质的电导率很低，只有温度升高时，无定形区域增大，电导率才会得到提高，然而大部分锂离子电池都是在室温下工作的，所以需要对基体进行处理，以达到降低结晶度、提高聚合物链段的运动能力、提高锂盐解离度、进而提高电导率的目的。通过共混、共聚、接枝、梳化、超支化、交联网络等方法[96-98]，降低聚合物的结晶性或者降低其玻璃化转变温度。交联是一种有效减小晶区比例的方法，它包括无规共聚物、嵌段共聚物或梳状共聚物。在聚合物电解质中，电解质有交联和非交联两种[99]。一般而言，非交联的凝胶聚合物电解质具有较差的力学性能。因而，交联对于聚合物电解质的制备是至关重要的。采用 EO 和 PO 的交联嵌段共聚物，可以将聚合物电解质的室温电导率提高到 $5×10^{-5}$ S/cm[100]。通过将 PEO 链接到聚硅氧烷主链上形成梳状共聚物，可以将聚合物电解质的室温电导率提高到 $2×10^{-4}$ S/cm[101]。如果将 PEO 和 PMMA 共混，再与 $LiClO_4$ 形成络合物，可以将聚合物电解质的室温电导率提高到 10^{-4} S/cm 以上[102]。

要形成较高电导率的聚合物电解质，首先，需要主体聚合物具备极性基团，应含有 O、S、N、P 等能够提供孤对电子的原子与阳离子形成配位键以抵消盐的晶格能[103]；其次，配位中心的间距要适当，这样可以与每个阳离子形成多重键，达到良好的溶解度。此外，聚合物分子链段要足够柔顺，聚合物上官能团的旋转阻力要尽量低，有利于阳离子的移动[104]。除了 PEO 外，常见的聚合物基体还有聚环氧丙烷（PPO）、聚丙烯腈（PAN）、聚甲基丙烯酸甲酯（PMMA）、聚偏氟乙烯（PVDF）、聚碳酸丙烯酯（PPC）等[93,105]。

崔光磊等在中国传统太极图中"刚柔并济"思想的启发下[106]，首次开发出了以纤维素为支撑结构的 PPC 基 ASPEs 薄膜（CPPC-ASPEs），该 ASPEs 具有高离子电导率（20℃时为 3×10^{-4} S/cm，120℃时为 1.4×10^{-3} S/cm）、宽电化学稳定窗口（4.6V）等优点，可用于室温条件下工作的高安全性、高电压全固态锂聚合物电池。CPPC-ASPEs 是目前报道的室温离子电导率最高的固态聚合物电解质。

9.2.3 聚合物复合电解质

20 世纪 80 年代后期，为增强聚合物电解质的机械稳定性，Wieczorek 等[107]首先加入陶瓷填料。之后的实验表明，通过添加无机颗粒形成的复合聚合物电解质，其力学性能、电导率以及离子迁移数都得到了改善[108]。无机陶瓷在聚合物电解质中，具有以下作用：无机填料的加入降低聚合物体系的晶相含量，增大离子输运的无定形相，利于锂离子快速传输[90]；填料颗粒附近也可以形成快速锂离子通道[109]；填料可以作为 Lewis 酸与聚合物中的 N、F、O 等 Lewis 碱发生反应，增加自由载离子数目[109]；填料可以增加聚合物的力学性、成膜性，改善与金属 Li 的界面状况[110]。现在研究较多的无机填料包括惰性陶瓷材料[111-117]，如 Al_2O_3、TiO_2、SiO_2、ZrO_2；锂离子导体材料[118-120]，如 LAGP，$LiAlO_2$，$Li_7La_3Zr_2O_{12}$ 等。无机锂离子导体材料的加入不仅可以通过提高聚合物相的无序度来增强锂离子传导，其本身也可提供锂离子传导通道，此外有研究表明聚合物相与无机相复合界面处可能存在锂离子快速传输通道[121]。目前基于锂离子导体的有机-无机复合电解质由于其较高的离子电导率和良好的电极-电解质界面受到了越来越多的关注。崔光磊等将 PPC 与 $Li_{6.75}La_3Zr_{1.75}Ta_{0.25}O_{12}$ 复合后得到的固态聚合物复合电解质室温离子电导率为 5.2×10^{-4} S/cm，电化学窗口为 4.6V[120]。$LiFePO_4$/Li 电池采用该固态电解质，室温下以 1C 倍率循环 200 周后放电容量仍然可以保持在 95%。前期研究的有机-无机复合固态聚合物电解质中，锂离子导体材料常常是以纳米颗粒的形式存在，只有当锂离子导体材料的含量足够高时才能在无机相中形成连续的锂离子通道。与纳米颗粒相比，锂离子导体材料以一维或三维结构存在更容易提供长程连续的锂离子通道。Yang 等制备了 LATP-PEO 复合固态电解质，其中 LATP 在 PEO 基体中以垂直纳米线的形式存在，垂直取向的 LATP 纳米线提供了一维连续锂离子通道[122]。Yu 等制备了 3D 多孔的 LLTO 骨架并向其中灌注 PEO 聚合物形成有机-无机复合固态电解质[123]。从图 9-19 可以看出，含有 LLTO 纳米颗粒的复合 PEO 固态电解质在低体积分数（<2.7%）时符合渗流模型，而当 LLTO 纳米颗粒体积继续增加时由于纳米颗粒的聚集导致了电导率的降低。然而，含有三维 LLTO 骨架的复合 PEO 固态电解质由于其三维连续的离子通道具有更高的离子电导率。它们的离子传输机制如图 9-19（c）和（d）所示。有机-无机复合固态电解质既克服了固态聚合物电解质电导率低的问题又缓解了无机电解质与电极界面的接触问题，从电解质对多种技术指标的综合需求看，有机-无机复合电解质可能是最能满足实际应用的固体电解质，但是目前关于聚合物复合电解质中有机-无机界面离子传输机制尚不明确，对其中科学问题缺乏系统性认识，未来需要更多的关注。

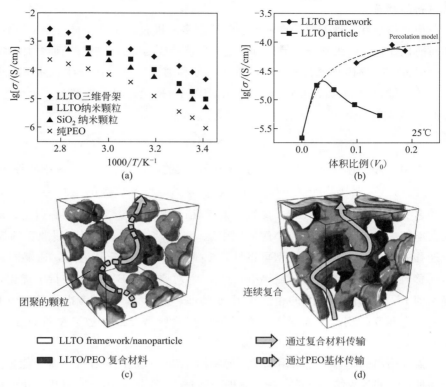

图 9-19 纯 PEO，加入 SiO_2 颗粒、LLTO 纳米颗粒及 LLTO 三维骨架的复合电解质的离子电导率随温度变化曲线（a）；渗流模型（黑色虚线）及 25℃时 LLTO 纳米颗粒及 LLTO 三维骨架复合电解质电导率（b）。含有团聚的纳米颗粒（c）以及三维连续骨架结构（d）的复合电解质中可能的离子传输机制[123]

9.3 高通量计算在固体电解质材料筛选中的应用

以往开发 Li^+ 导体，主要是基于固体离子学、固体化学、固体物理的知识，通过大量的实验以及研究者的思想和经验来实现，虽然已经开发了多种 Li^+ 导体，但目前还没有特别满意的材料，固体电解质材料的研发一般耗时数年甚至数十年，成为电池领域最难研究的一个方向，需要长期的积累和仔细的研究。而且迄今为止，对于离子在固体中输运性质的研究，理论还不成熟，过去的实验手段多为宏观尺度，无法提供广泛的离子电导、扩散系数与材料结构的关系。为了加速对固态电解质材料的研究开发步伐，系统地研究与 Li^+ 导体相关的具有普适性的规律，可以通过高通量计算方法进行尝试，此类方法被 MIT 的 Ceder 称为"材料基因组"方法。通过高通量、多尺度的大范围计算和搜索，借助数据挖掘技术和方法，有希望探索并总结出离子传导的普适性规律，计算出具有离子通道、离子迁移势垒较低的材料[5, 124-127]。这样再通过高通量制备、测试，有可能显著加速对电解质新材料的筛选。

但是第一性原理一般计算量大,消耗较多的机时,能算的材料原子数较少、单胞较小,如何简化计算、逐级筛选,但计算误差又能接受成为能否开展好高通量计算材料的关键。

研究 Li^+ 导体材料首先涉及计算 Li^+ 迁移通道,具有代表性的算法汇总见表 9-3。

关于键价总和(bond valence sum,BVS)理论,可以理解为:氧化态为 $|V_A|$ 的中心原子 A 与配位原子 X 成键的键价之和应满足

$$\sum_X s_{A-X} = \exp[(r_0 - r_{A-X})/b] = V_A$$

式中,A 为阳离子;X 表示与阳离子配位的阴离子;S_{A-X} 为中心原子 A 与成键原子 X 的键价;V_A 是阳离子 A 的键价;r_{A-X} 表示 A-X 成键的实际键长;r_0 是单位键价对应的名义键长;b 是 Softness 参数,对于 Li^+ 指定为 0.37Å(1Å=0.1nm)。通常,Li^+ 仅在空间中键价和失配度较小的区域中运动,如 $|\Delta V| \leqslant 0.1$。以通道的连通程度表示离子迁移的难易。应用上述模型可以计算 Li^+ 在晶体中的迁移通道,遗憾的是没有定量的评估迁移难易程度的参数。随后 Admas 等[128]对原有的 BV 理论进行了优化,引入了 Morse 势,即 Li^+ 与阴离子之间的相互作用以 Morse 势表示,同时考虑了晶体中 Li^+ 与阳离子之间的库仑相互作用,从而对 Li^+ 在晶体中传输所需要的扩散势垒进行了定量计算,即活化能。

需要指出的是,无论采用基于几何结构、基于成键配位亦或是基于能量变化的算法,对于同种材料的计算结果均能得到很好的一致性[129],区别在于采用基于几何结构的算法只能定性分析离子输运通道的连通性;而基于成键配位的方法通过引入 Morse 势的 BV 算法,除了可以计算离子输运通道之外,还可以定量计算离子在晶格中输运时形成连通路径所需要越过的最小势垒。需要指出的是这种算法对能量的计算并不精确,通常高估了离子输运的活化能,但由于其计算速度快,同时存在定量评估参数,对于初步筛选离子导体材料依然具有重要的参考价值。以氧化物 Li^+ 导体为例设计的高通量计算流程如图 9-20 所示。

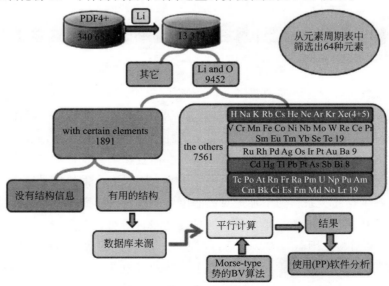

图 9-20 氧化物锂离子导体的虚拟筛选

表 9-3　各种计算锂离子迁移通道的算法比较

类别	基于几何结构	基于成键配位	基于能量变化
原理	Li$^+$ 在晶格结构里足够大的几何空间中运动	Li$^+$ 在晶格结构里键价和接近理想键价值的区域运动	Li$^+$ 在晶格结构中能量势垒低的路径上运动
方法	① Voroni-Dirichlet 分割[130] ② colony 表面[131] ③ procrystal 分析[129]	键价(BV)理论[132]	① 经验势函数＋过渡态 ② DFT＋过渡态 ③ 分子动力学
计算精度	粗	半定量	定量
时间尺度	min	min	h/d
程序包	TOPOS[133]/crystal explorer[129]	3D BVSMAPPER[129]	VASP/materials studio

基于 BV 算法进行的高通量计算，其结果的可靠性在 Gao 等[134]的文章中有详细的论述，此处只是展示部分计算结果，如图 9-21 所示。BV 算法计算出的活化能与实验值的比较见表 9-4。

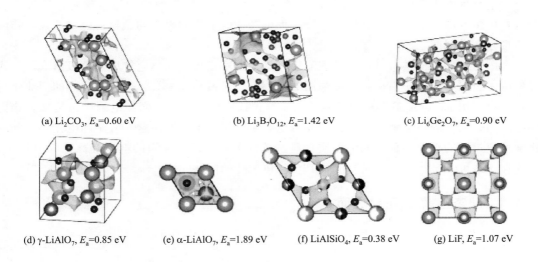

(a) Li$_2$CO$_3$, E_a=0.60 eV　　(b) Li$_3$B$_7$O$_{12}$, E_a=1.42 eV　　(c) Li$_6$Ge$_2$O$_7$, E_a=0.90 eV

(d) γ-LiAlO$_7$, E_a=0.85 eV　(e) α-LiAlO$_7$, E_a=1.89 eV　(f) LiAlSiO$_4$, E_a=0.38 eV　(g) LiF, E_a=1.07 eV

图 9-21　七大晶系的离子通道

表 9-4　BV 算法计算出的活化能与实验值比较

组分	BV 法 E_a/eV	实验值 E_a/eV
(a) Li$_2$CO$_3$	0.60	1.06[135], 0.77[136]
(b) γ-LiAlO$_2$	0.85	0.93[137]
(c) Li$_6$Zr$_2$O$_7$	1.01	0.95[138]
(d) Li$_3$BO$_3$	1.01	0.76[139]
(e) Li$_2$ZrO$_3$	1.04	0.91[140]

9.4 全固态锂电池的界面问题

经过多年的发展，固态电解质本征电导率已经逐渐接近甚至超越液态电解液，但界面阻抗限制了锂离子在全电池中的有效输运，成为制约其性能的瓶颈之一。造成界面阻抗大的主要原因包括：①固-固接触面积较小，固态电解质无法像液态电解液那样具有良好的浸润性。②在全固态电池制备或者充放电过程中，电解质和电极界面化学势与电化学势差异驱动的界面元素互扩散形成的界面相可能不利于离子的传输。此外，固-固界面还存在空间电荷层，也有可能抑制离子垂直界面的扩散和传导。③固体电解质与电极的稳定性问题，包括化学稳定性，如某些电解质与电极之间存在界面反应；电化学稳定性，一些电解质有可能在接触正极或者负极的界面发生氧化或者还原反应。④界面应力问题。在充放电过程中，多数正负极材料在嵌脱锂过程中会出现体积变化，而电解质不发生变化，这使得在充放电过程中固态电极/固态电解质界面应力增大，可能导致界面结构破坏，物理接触变差，内阻升高，活性物质利用率下降。无机陶瓷电解质的以上问题较为突出，聚合物电解质也存在类似问题。

全固态原型电池常被设计为"三明治"结构，由正极、固体电解质和负极组成，并通过集流体引出导线。正极材料一般采用 $LiCoO_2$、$LiFePO_4$、NCM 等较为成熟的商业化材料，尤以使用 $LiCoO_2$ 最多；负极则选取金属锂或锂铟合金等；固体电解质包括聚合物固体电解质、无机固体电解质和聚合物复合电解质。固态电池既包含具有不同电化学势和电化学反应过程的正极、负极和电解质材料，又包含一个相互交织的电子和离子导电网络，因此存在电极/电解质多级界面反应和电子/离子传输问题。

9.4.1 固态电解质/正极界面

9.4.1.1 物理接触

由于无机电解质材料尤其是氧化物材料质地坚硬，与电极材料的物理接触不致密，同时在电池循环过程中电极材料由于脱嵌锂发生较大体积变化，使得界面物理接触恶化，阻抗迅速增加。因此在制备正极时常常在正极活性物质中混入电解质，如果是硫化物则用高压的方法，氧化物则需要高温烧结[141-143]。Han 等引入助溶剂 $Li_{2.3}C_{0.7}B_{0.3}O_3$，在高温烧结下与 $LiCoO_2$ 和 LLZO（无机固体电解质）表面的 Li_2CO_3 反应在界面处形成 $Li_{2.3-x}C_{0.7+x}B_{0.3-x}O_3$，使 $LiCoO_2$ 与 LLZO 紧密接触，降低了界面阻抗[144]。为了解决界面物理接触的问题，还可以制备三维电解质增加电极电解质接触面积，或在正极与无机电解质之间加入质地柔软的聚合物缓冲层，以及采用有机-无机复合电解质材料。

9.4.1.2 空间电荷层

Takada 等在以氧化物为正极、硫化物为电解质的固态电池中发现首周充电曲线出现一个斜坡以及低的容量,他们将此现象归结为高的界面阻抗[145]。此界面阻抗源自界面处的空间电荷层。由于氧化物正极与硫化物电解质之间的化学势的不同,锂离子从硫化物电解质向氧化物正极移动,在两者界面处形成空间电荷层。氧化物正极如 $LiCoO_2$ 是离子-电子混合导体,正极处的电子会消除锂离子浓度梯度,为了保持平衡将会有更多的锂离子从硫化物电解质向正极移动,最终导致界面处形成阻抗非常大的空间电荷层。常用的解决方法是在两者之间引入离子导电电子绝缘的氧化物中间层[145,146]。Ohta 等在 $LiCoO_2$ 表面包覆 $LiNbO_3$ 和 $Li_4Ti_5O_{12}$ 有效抑制了空间电荷层,降低了界面阻抗。

9.4.1.3 界面反应

固态电解质与正极材料在界面处的元素互扩散和界面反应同样是导致界面阻抗大的原因。固态电解质与正极材料简单混合后在室温下并没有明显的反应发生,说明室温下反应动力学缓慢。为了解决氧化物固态电解质与正极材料的接触问题常采用预烧结的方法,高温下两者反应显著,在界面处形成阻抗大的界面层。虽然硫化物电解质不需要与正极材料预烧结,但是制备电池时常用的高压仍然会促使界面元素互扩散。抑制界面反应有效且常用的方法是正极材料的包覆及电解质表面修饰[141-143, 147-149]。常见的包覆材料除 $LiNbO_3$ 和 $Li_4Ti_5O_{12}$ 外还有 Li_3PO_4、$LiAlO_2$、Al_2O_3、$LiTaO_3$、Li_2SiO_3、Li_3BO_3 以及 Li_2ZrO_3。Ceder 等指出理想的正极包覆材料应具有以下特点:①具有宽的电化学窗口,在正极工作电压范围内电化学稳定;②与正极材料和电解质材料接触化学稳定;③较高的离子电导和低的电子电导。他们采用高通量计算筛选出了一系列性能优良的正极包覆材料,如 LiH_2PO_4,$LiTi_2(PO_4)_3$ 及 $LiPO_3$[150]。

9.4.2 固态电解质/金属锂负极界面

9.4.2.1 物理接触

金属锂质地较软,固态电解质与金属锂物理接触似乎比较紧密,但是一些固态电解质与金属锂的界面浸润性较差,固固接触仍然形成较高的界面阻抗。Sakamoto 等通过对 LLZO 固体电解质表面杂质的深入研究分析,LLZO 与金属锂界面阻抗高的原因是其"疏锂性",而 LLZO 表面呈现出疏锂态的原因是杂质 Li_2CO_3 和 LiOH 的存在[151]。纯净的 LLZO 具有亲锂性,但 LLZO 保持表面清洁度难度较大,可以通过加入中间层的方法将疏锂性转变为亲锂性。郭向欣等采用磁控溅射对 Ta 掺杂的 LLZO 陶瓷片表面进行 Au、Nb 和 Si 薄膜的修饰,证明采用 Si 薄膜修饰的固态电解质与金属锂的界面阻抗降至 $5\Omega \cdot cm^2$,锂对称电池可循环 120h 仍然保持良好的界面接触[152]。Han 等使用原子层沉积的方法将 Al_2O_3 层沉积在固态电解质表面,成功地改善了固态电解质 $Li_7La_{2.75}Ca_{0.25}Zr_{1.75}Nb_{0.25}O_{12}$ 和锂金属的浸润性和化学稳定性,使得界面阻抗从 $1710\Omega \cdot cm^2$ 降到 $1\Omega \cdot cm^2$[153]。

9.4.2.2 界面反应

由于金属锂的强还原性,一些固态电解质与金属锂接触后会发生反应,Wenzel[154]根据界面的稳定性将固态锂电池中的界面分为以下三类:①热力学稳定的界面,界面处未发生反应；②界面处发生反应，生成离子-电子导电的混合导体界面层；③界面处发生反应，但生成离子导电电子绝缘的稳定界面层。在固态电池中，我们期望第一种和第三种类型的界面存在，第三种类型界面层的离子电导对电池性能有很大影响。Mo 等计算了一些常见的无机固态电解质的电化学窗口如图 9-22 所示[155]。从图中可以看出，除 LLZO 外常见的无机固态电解质均会与金属锂发生反应。LiPON 与金属锂反应后界面生成 Li_3PO_4、Li_3P、Li_3N 及 Li_2O，它们均是离子导电电子绝缘的，即第三种类型的界面。由于钝化层的存在反应将会很快停止，形成稳定的界面。含有金属元素的固态电解质，如 LGPS、LATP 和 LAGP 与金属锂接触后还原产物中含有金属 Ge、Ti，形成第二种界面，导致界面层不断增厚、界面阻抗持续增加。抑制这类界面产生常用的方法是加入缓冲层避免固态电解质与金属锂的直接接触。Hao 等采用磁控溅射的方法在 LATP 与金属锂中间加入 ZnO 层，抑制了 LATP 与金属锂的反应，降低了界面阻抗[156]。

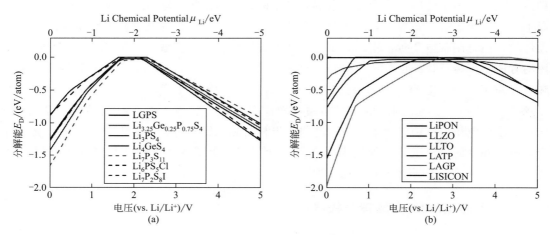

图 9-22 硫化物（a）和氧化物（b）固态电解质在不同电压下（vs. Li/Li^+）的分解能

9.4.2.3 锂枝晶

依据液态锂电池中锂枝晶的生长机理，我们认为固态电解质具有高的机械强度可以有效抑制锂枝晶的生长，但在实验中依然可以观察到锂枝晶在 LLZO、Li_2S-P_2S_5 和 β-Li_3PS_4 中的生长。Cheng 等认为枝晶是沿着固态电解质的晶界或孔洞生长[157]，那么枝晶的生长与电解质的致密度有关，但是在致密度＞97％的 LLZO 陶瓷片中仍然可以观察到锂枝晶。Chiang 等认为当电流密度达到临界值，枝晶就会在电解质表面预先存在的缺陷如裂缝中开始生长[158]，锂枝晶在缺陷中生长带来的应力使裂缝延伸从而更有利于锂枝晶的生长。Han 等发现在 LLZO 和 Li_3PS_4 中存在锂枝晶，但 LiPON 中却没有观察到明显的锂枝晶存在，他们认为形成锂枝晶的原因是固态电解质不可忽略的电子电导[159]。抑制锂枝晶生长最常用的方法有减小晶粒尺寸，减少表面缺陷，引入离子导电电子绝缘的界面缓冲层等。目前固态电解质中锂枝晶的生长机理仍有一定争议，只有深刻理解其中的科学问题才能更好地发展应对策略。

通常界面阻抗和界面稳定性问题总是同时存在，对电极或者电解质的修饰往往能同时解决两者，遗憾的是现有文献报道的固态电池体系的电流密度及循环次数远低于液态电池体系。推测原因是电池在长循环后容量衰减过快，电极以及电解质表面修饰不再起作用，界面性能恶化，没有解决好电极体积膨胀问题。因而，深入理解界面问题对于发展高性能的全固态锂电池至关重要。

9.5 全固态锂电池性能参数

Richter 等综述了目前文献中报道的各种固态电池的性能，提出了固态锂电池中需要关注的性能参数，并提出了未来需要达到的目标，如表 9-5 所示[160]。从全固态锂电池电芯角度考虑，固态锂电池的发展不仅在于电池材料，如正负极材料和电解质材料的筛选和优化，同时与电芯设计也密切相关[161]。例如集流体的种类和尺寸，铝塑膜的厚度，电芯的结构设计及尺寸等对固态电池的性能有着不可忽视的影响。

表 9-5 高能量高功率固态锂电池的研发目标[160]

性能参数	当前水平	目标
电压 V	NCM 约 3.8V LNMO 约 4.7V Li_2S 约 2.2V	保证其他技术指标前提下，继续提高插层型及转换型正极材料的工作电压以提高能量密度和比能量
比容量 q	NMC 约 150mA·h/g Li_2S 约 1600mA·h/g	实际比容量要接近理论值，例如 NCM 约 200mA·h/g，Li_2S 约 1600mA·h/g，LNMO 约 147mA·h/g
活性材料在复合正极中的体积分数 $\phi_{CAM/Ca}$	40%~60%	>70%
正极假想比能量(只考虑正极)$E_{m(Ca)}$ 和面容量 Q_A	大多数可以做到 400W·h/kg，2mA·h/m²	500W·h/kg，5mA·h/m²(只考虑复合正极的质量，负极假设为金属锂)
电流密度 j	使用金属锂负极时枝晶生成的临界电流为 1mA/cm²	锂枝晶形成的临界电流应提高至 5mA/cm²
能量效率 Φ_E	插层型正极材料从循环第二周开始可以做到 90%以上	长循环可以一直保持在 90%以上
内阻 R	14Ω·cm²	只有内阻小于 40Ω·cm² 才能保证在 1C 倍率下能量效率大于 90%
锂箔厚度 l_{An}	20μm	理想情况为不需要锂负极，锂只来源于正极
固态电解质厚度 l_{SE}	30μm	仍需进一步降低以保证足够低的内阻
复合正极厚度 l_{Ca}	<600μm	复合正极厚度的增加会增加比能量但同时同样倍率下电流密度也会增加，需要平衡两者
电池面积	无机固态电解质目前可以做到约 200cm²	提升无机电解质和有机-无机复合电解质膜的制备面积

续表

性能参数	当前水平	目标
外界压力 p	目前固态电池循环时会施加外压	实际应用时不施加外压
添加剂	聚合物黏结剂和导电添加剂	添加剂的作用需要更深入的研究
能量保持率	70℃时，$LiFePO_4$/Li 固态聚合物电池循环 1400 周面能量 $8.45 W·h/cm^2$，能量保持率约 80%	在不低于 $5 W·h/cm^2$ 面能量的条件下，循环 1000 周仍有 80% 以上的能量保持率

9.6 本章结语

经过多年的潜心研究，科研工作者们对固态锂电池中的基础科学问题有了更深刻的认识，在发展固态锂电池方面也取得了长足的进步，但是固态锂电池仍然存在着未解决的科学问题：①固态电解质中锂离子传输机制仍然没有明确和统一，尤其是在有机-无机复合固态电解质中锂离子在有机-无机界面的传输机制；②固态电解质/电极多相界面处离子/电子传输及电荷转移问题；③固态电解质/电极的多相界面由于固态电池的装配方式导致难以表征，尤其是实时追踪界面反应；④固态电解质中锂枝晶的生长机理；⑤固态电池在工作时的行为涉及多时间和多空间尺度，这对实验表征和理论模拟手段都是很大的挑战。从基础研究考虑，固体电解质的核心问题是离子在体相、表面、界面的输运，这正是固体离子学研究的核心内容，过去的理论大多数是宏观唯象的阐述。今后，随着第一性原理分子动力学、格林函数、Monte Carlo 等方法的发展，结合原子尺度、三维空间、时间分辨等研究离子输运的实验技术的发展，相信固体电解质中离子输运构效关系及输运机制的研究最终会超越唯象理论的水平，列席在凝聚态物理的科学殿堂中，并对固体电解质材料的开发起到指导作用。从应用考虑，研发全固态锂离子电池的主要目的是从根本上解决目前所使用的锂离子电池安全性问题，提高能量密度、循环性、服役寿命、降低电池成本[162]。目前性能最好的无机全固态锂离子电池已经有了 10~15 A·h 级原型电池的展示，循环性可以达到 500 次，但其综合性能指标离实际应用还有相当的距离。目前各类无机电解质材料的研发十分活跃，材料体系尚未定型，也未形成全固态电池的综合技术解决方案。而金属锂 PEO 基聚合物全固态锂电池的能量密度已经达到 220~350 W·h/kg，循环次数可达 3000 次，已获得批量生产。相对于无机陶瓷电解质，采用金属锂作为负极的聚合物锂电池更接近应用。虽然全固态锂电池的商业化还需要时间，从长远考虑，全固态电池在规模储能、电动汽车、地质勘探、石油钻井、航空航天、国防安全中具有不可替代的应用前景。正如许晓雄博士[162]针对全固态电池世界范围内的研究现状分析所言："与国际先进水平相比，我国在这一方面需要加大研发力度，进一步提升储能电池的安全性、寿命、能量密度及系统集成技术，并提升相关领域的知识产权。发展大容量全固态锂电池前沿技术刻不容缓。"

第9章 全固态锂离子电池

本章需要进一步思考的基础科学问题：

1. 在两种以上的固体电解质相存在时，离子电导率和离子扩散通道如何测量？
2. 在全固态电池中，固固界面问题包括哪些问题？
3. 影响聚合物电解质氧化电位高低的因素是什么？如何解决聚合物电解质与高电压正极材料的匹配？
4. 全固态锂离子电池和全固态金属锂电池，从理论上分析，有何优势和劣势？劣势能否克服？如果可以克服，这是否是颠覆性技术？
5. 金属锂电极或者含锂的负极如果用在全固态电池中，哪些基础科学问题必须解决？
6. 硫化物电解质氧化反应是放热反应吗？会导致热失控吗？
7. 固态电解质的离子电导率符合 Arrenius 规律，硫化物的离子电导率低温下显著高于液态电解质，硫化物电解质的全固态电池是否低温性能一定好于液态电解质？为什么？
8. 原位聚合的固态电解质与聚合物固态电解质以及液态电解质相比，热稳定性与电化学稳定性会有优势吗？为什么？
9. 混合固液电解质，高盐浓度电解质，凝胶电解质，原位固态化电解质在离子导电机制上有什么本质区别？
10. 全固态电池循环过程中，由于电极体积膨胀收缩导致的应力对接触的固态电解质有何影响？

参考文献

[1] Chen R, Qu W, Guo X, et al. The pursuit of solid-state electrolytes for lithium batteries: From comprehensive insight to emerging horizons [J]. Materials Horizons, 2016, 3 (6), 487-516.

[2] Gao Z, Sun H, Fu L, et al. Promises, challenges, and recent progress of inorganic solid-state electrolytes for all-solid-state lithium batteries [J]. Adv Mater, 2018, 30 (17), e1705702.

[3] Zhang Z, Shao Y, Lotsch B, et al. New horizons for inorganic solid state ion conductors [J]. Energy Environ Sci, 2018, 11: 1945-1976.

[4] Kerman K, Luntz A, Viswanathan V, et al. Review—practical challenges hindering the development of solid state Li ion batteries [J]. J Electrochem Soc, 2017, 164 (7): A1731-A1744.

[5] Nolan A M, Zhu Y, He X, et al. Computation-accelerated design of materials and interfaces for all-solid-state lithium-ion batteries [J]. Joule, 2018, 2 (10): 2016-2046.

[6] Chen R, Li Q, Yu X, et al. Approaching practically accessible solid-state batteries: Stability issues related to solid electrolytes and interfaces [J]. Chem Rev, 2019.

[7] Gurung A, Pokharel J, Baniya A, et al. A review on strategies addressing interface incompatibilities in inorganic all-solid-state lithium batteries [J]. Sustainable Energy & Fuels, 2019, 3 (12): 3279-3309.

[8] He Y, Lu C, Liu S, et al. Interfacial incompatibility and internal stresses in all-solid-state lithium ion batteries [J]. Adv Energy Mater, 2019, 9 (36).

[9] Pervez S A, Cambaz M A, Thangadurai V, et al. Interface in solid-state lithium battery: Challenges, progress, and outlook [J]. ACS Appl Mater Interfaces, 2019, 11 (25): 22029-22050.

[10] Liu X, Li X, Li H, et al. Recent progress of hybrid solid-state electrolytes for lithium batteries [J]. Chemistry, 2018, 24 (69): 18293-18306.

[11] Fenton D E, Parker J M, Wright P V. Complexes of alkali-metal ions with poly (ethylene oxide) [J]. Polymer, 1973, 14 (11): 589-589.

[12] Wright P V. Electrical conductivity in ionic complexes of poly (ethylene oxide) [J]. British Polymer Journal,

1975, 7 (5): 319-327.

[13] Armand M, Chabagno. J M, Duclot M. Fast ion transport in solids [M]. Holland: Eds North Lolland Publishing Co, 1979.

[14] Kamaya N, Homma K, Yamakawa Y, et al. A lithium superionic conductor [J]. Nat Mater, 2011, 10 (9): 682-686.

[15] Takada K, Ohta N, Zhang L, et al. Interfacial modification for high-power solid-state lithium batteries [J]. Solid State Ionics, 2008, 179 (27-32): 1333-1337.

[16] Brous J, Fankuchen I, Banks E. Rare earth titanates with a Perovskite structure [J]. Acta Cryst, 1953, 6 (1): 67-70.

[17] Inaguma Y, Liquan C, Mitsuru, High ionic conductivity in lithium lanthanum titanate [J]. Solid State Commun, 1993, 86 (10): 689-693.

[18] Fourquet J L, Duroy H, Crosnier-Lopez M P. Structural and microstructural studies of the series $La_{2/3-x}Li_{3x}\square_{1/3-2x}TiO_3$ [J]. J Solid State Chem, 1996, 127 (2): 283-294.

[19] Emery J, Buzare J Y, Bohnke O, et al. Lithium-7 NMR and ionic conductivity studies of lanthanum [J]. Solid State Ionics, 1997, 99 (1-2): 41-51.

[20] Bohnke O. The fast lithium-ion conducting oxides $Li_{3x}La_{2/3-x}TiO_3$ from fundamentals to application [J]. Solid State Ionics, 2008, 179 (1-6): 9-15.

[21] Shan Y, Chen L, Inaguma Y, et al. Oxide cathode with perovskite structure for rechargeable lithium batteries [J]. J Power Sources, 1995, 54 (2): 397-402.

[22] Zhao Y, Daemen L L, Superionic conductivity in lithium-rich anti-perovskites [J]. J Am Chem Soc, 2012, 134 (36): 15042-15047.

[23] Reckeweg O, Blaschkowski B, Schleid T. Li_5OCl_3 and Li_3OCl: Two remarkably different lithium oxide chlorides [J]. Z Anorg Allg Chem, 2012, 638 (12-13): 2081-2086.

[24] Emly A, Kioupakis E, Van derVen A, Phase stability and transport mechanisms in antiperovskite Li_3OCl and Li_3OBr Superionic Conductors [J]. Chem Mater, 2013, 25 (23): 4663-4670.

[25] Mouta R, Silva R X, Paschoal C W A. Tolerance factor for pyrochlores and related structures [J]. Acta Crystallographica Section B-Structural Science Crystal Engineering and Materials, 2013, 69: 439-445.

[26] Zhang Y, Zhao Y, Chen C, Ab initio study of the stabilities of and mechanism of superionic transport in lithium-rich antiperovskites [J]. Phys Rev B, 2013, 87 (13).

[27] Schroeder D J, Hubaud A A, Vaughey J T. Stability of the solid electrolyte Li_3OBr to common battery solvents [J]. Mater Res Bull, 2014, 49: 614-617.

[28] Weiss E, Hensel H, Kuhr H. Radiological and nuclear magnetic broad line resonance of lithium halide monohydrate [J]. Chem Ber Recl, 1969, 102 (2): 632.

[29] Rudo K, Hartwig P, Weppner W. Ionic conductivities and phase-equilibria of the lithium iodide hydrates [J]. Revue De Chimie Minerale, 1980, 17 (4): 420-429.

[30] Hartwig P, Rabenau A, Weppner W. Lithium hydroxide halides -phase-equilibria and ionic conductivities [J]. Journal of the Less-Common Metals, 1981, 78 (2): 227-233.

[31] Andersen N H, Kjems J K, Poulsen F W. Neutron-scattering studies of the ionic conductor $LiI \cdot D_2O$ [J]. Phys Scr, 1982, 25 (6): 780-784.

[32] Nakamura O, Goodenough J B. Conductivity enhancement of lithium bromide monohydrate by Al_2O_3 particles [J]. Solid State Ionics, 1982, 7 (2): 119-123.

[33] Nakamura O, Goodenough J B. Fast lithium-ion transport in composites containing lithium bromide dihydrate [J]. Solid State Ionics, 1982, 7 (2): 125-128.

[34] Barlage H, Jacobs H. $Li_2I(OH)$ -A compound with one-dimensional infinite edge-sharing $[Li_{4/2}(OH)^+]$ pyramids [J]. Z Anorg Allg Chem, 1994, 620 (3): 475-478.

[35] Barlage H, Jacobs H. Unusual coordination polyhedra around oxygen in $Li_4Cl(OH)_3$ [J]. Z Anorg Allg Chem, 1994, 620 (3): 471-474.

[36] Barlage H, Jacobs H. $Li_2Br(NH_2)$ -The 1st ternary alkali-metal amide halide [J]. Z Anorg Allg Chem, 1994, 620 (3): 479-482.

[37] Eilbracht C, Kockelmann W, Hohlwein D, et al. Orientational disorder in perovskite like structures of $Li_2X(OD)$ (X=Cl, Br) and $LiBr$ center dot D_2O [J]. Physica B, 1997, 234: 48-50.

[38] Schwering G, Honnerscheid A, van Wullen L, et al. High lithium ionic conductivity in the lithium halide hydrates Li_{3-n} (OH_n) Cl ($0.83 \leqslant n \leqslant 2$) and Li_{3-n} (OH_n) Br ($1 \leqslant n \leqslant 2$) at ambient temperatures [J]. Chemphyschem, 2003, 4 (4): 343-348.

[39] Hagman L-O, Kierkegaard P. The crystal structure of $NaMe_2^{IV}$ (PO_4)$_3$; Me^{IV} = Ge, Ti, Zr [J]. Acta Chem Scand, 1968, 22 (6): 1822-1832.

[40] Goodenough J B, Hong H Y-P, Kafalas J A. Fast Na^+-ion transport in skeleton structures [J]. Mater Res Bull, 1976, 11 (2): 203-220.

[41] Anantharamulu N, Koteswara Rao K, Rambabu G, et al. A wide-ranging review on Nasicon type materials [J]. Journal of Materials Science, 2011, 46 (9): 2821-2837.

[42] Subramanian M A, Subramanian R, Clearfield A. Lithium ion conductors in the system AB (IV)$_2$ (PO_4)$_3$ (B = Ti, Zr and Hf) [J]. Solid State Ionics, 1986, 18-19 (1): 562-569.

[43] Aono H, Sugimoto E. Ionic conductivity of the lithium titanium phosphate ($Li_{1+x}M_xTi_{2-x}$(PO_4), M=Al, Sc, Y, and La) systems [J]. journal of Electrochemical Society, 1989, 136 (2): 590-591.

[44] Aono H, Sugimoto E. Ionic conductivity and sinterablity of lithium titanium phosphate system [J]. Solid State Ionics, 1990, 40-41 (1): 38-42.

[45] Aono H, Sugimoto E. Electrical property and sinterability of $LiTi_2$(PO_4)$_3$ mixed with lithium salt (Li_3PO_4 or Li_3BO_3) [J]. Solid State Ionics, 1991, 47 (3-4): 257-264.

[46] Fu J. Fast Li^+ ion conduction in Li_2O-Al_2O_3-TiO_2-SiO_2-P_2O_2 glass-ceramics [J]. J. Am. Ceram. Soc. 1997, 80 (7), 1901-1903.

[47] Birke P, Salam F, Doring S, et al. A first approach to a monolithic all solid state inorganic lithium [J]. Solid State Ionics, 1999, 118 (1-2): 149-157.

[48] Kitaura H, Zhou H. Electrochemical performance of solid-state lithium-air batteries using carbon nanotube catalyst in the air electrode [J]. Adv Energy Mater, 2012, 2 (7): 889-894.

[49] Ferg E, Gummow R J, Kock A D. Spinel anodes for lithium-ion batteries [J]. Journal of Electrochemical Society, 1994, 141 (11): 147-150.

[50] Hong H Y-P. Crystal structure and ionic conductivity of $Li_{14}Zn$ (GeO_4)$_4$ and other new Li^+ superionic conductors [J]. Mater Res Bull, 1978, 13 (8): 117-124.

[51] Bruce P G, Wes, A R, The AC conductivity of polycrystalline LISICON, $Li_{2+x}ZnGeO_4$, and a model for intergranular constriction resistances [J]. Journal of Electrochemical Society, 1983, 130 (3): 662-669.

[52] Robertson A D, West A R, Ritchie A G. Solid state ionics-review of crystalline lithium-ion conductors suitable for hightemperature battery applications [J]. Solid State Ionics, 1997, 104 (1-2): 1-11.

[53] Bruce P G, West A R. Phase diagram of the LISICON, solid electrolyte system, Li_4GeO_4-Zn_2GeO_4 [J]. Mater Res Bull, 1980, 15 (3): 379-385.

[54] Fujimura K, Seko A, Koyama Y, et al. Accelerated materials design of lithium superionic conductors based on first-principles calculations and machine learning algorithms [J]. Adv Energy Mater, 2013, 3 (8): 980-985.

[55] Bruce P G, West A R, David W I F. Structure determination of LISICON solid solutions by powder neutron diffraction [J]. J Solid State Chem, 1988, 75: 390-396.

[56] Sumathipala H H, Dissanayake M A K L, West A R. Novel Li-ion conductors and mixed conductors, Li_{3+x}-$Si_xCr_{1-x}O_4$ and a simple method for estimating Li^+/e^- transport numbers [J]. Journal of Electrochemical Society, 1995, 142 (7): 2138-2143.

[57] Kanno R, Hata T, Kawamoto Y. et al. Synthesis of a new lithium ionic conductor, thio-LISICON-lithium germanium sulfide system [J]. Solid State Ionics, 2000, 130 (1-2): 97-104.

[58] Murayama M, Kanno R, Irie M. Synthesis of new lithium ionic conductor thio-LISICON-lithium silicon sulfides system [J]. J Solid State Chem, 2002, 168 (1): 140-148.

[59] Kanno R, Murayama M. Lithium ionic conductor thio-LISICON: The Li_2S-GeS_2-P_2S_5 System [J]. J Electrochem Soc, 2001, 148 (7): A742.

[60] Mo Y, Ong S P, Ceder G. First principles study of the $Li_{10}GeP_2S_{12}$ lithium super ionic conductor material [J]. Chem Mater, 2012, 24 (1): 15-17.

[61] Thangadurai V, Kaack H, Weppner W J F. Novel fast lithium ion conduction in Garnet-type $Li_5La_3M_2O_{12}$ (M= Nb, Ta) [J]. J Am Ceram. Soc, 2004, 86 (3): 437-440.

[62] Cussen E J. The structure of lithium garnets: Cation disorder and clustering in a new family of fast Li^+ conductors

[J]. Chem Commun, 2006 (4): 412.

[63] van Wullen L, Echelmeyer T, Meyer H W, Wilmer D. The mechanism of Li-ion transport in the garnet $Li_5La_3Nb_2O_{12}$ [J]. Phys Chem Chem Phys, 2007, 9 (25): 3298-3303.

[64] Thangadurai V, Weppner W. $Li_6ALa_2Nb_2O_{12}$ (A = Ca, Sr, Ba): A new class of fast lithium ion conductors with garnet-like structure [J]. J Am Ceram Soc, 2005, 88 (2): 411-418.

[65] Thangadurai V, Weppner W. $Li_6ALa_2Ta_2O_{12}$ (A=Sr, Ba): Novel garnet-like oxides for fast lithium ion conduction [J]. Adv Funct Mater, 2005, 15 (1): 107-112.

[66] Thangadurai V, Weppner W. Effect of sintering on the ionic conductivity of garnet-related structure $Li_5La_3Nb_2O_{12}$ and In-and K-doped $Li_5La_3Nb_2O_{12}$ [J]. J Solid State Chem, 2006, 179 (4): 974-984.

[67] Murugan R, Thangadurai V, Weppner W. Fast lithium ion conduction in garnet-type $Li_7La_3Zr_2O_{12}$ [J]. Angew Chem Int Ed Engl, 2007, 46 (41): 7778-7781.

[68] Awaka J, Kijima N, Hayakawa H, Akimoto J. Synthesis and structure analysis of tetragonal $Li_7La_3Zr_2O_{12}$ with the garnet-related type structure [J]. J Solid State Chem, 2009, 182 (8): 2046-2052.

[69] Geiger C A, Alekseev E, Lazic B, et al. Crystal chemistry and stability of "$Li_7La_3Zr_2O_{12}$" garnet: A fast lithium-ion conductor [J]. Inorg Chem, 2011, 50 (3): 1089-1097.

[70] Boukamp B A, Huggins R A. Lithium ion conductivity in lithium nitride [J]. Phys Lett A, 1976, 58 (4): 231-233.

[71] Hartwing P, Weppner W, Wichelhaus W. Fast ionic lithium conduction in solid lithium nitride chloride [J]. Mater Res Bull, 1979, 14 (4): 493-498.

[72] Jia Y Z, Yang J X. Study of the lithium solid electrolytes based on lithium nitride chloride ($Li_9N_2Cl_3$) [J]. Solid State Ionics, 1997, 96 (1-2): 113-117.

[73] Jing Yan, Jia Yongzhong, Ma Peihua. Synthesis and characterization of the solid state electrolyte $Li_{9-nx}M_xN_2Cl_3$ (M=Na, Mg, Al) [J]. Chinese Journal of Inorganic Chemistry, 2000, 16 (6): 921-927.

[74] Hatake S, Kuwano J, Miyamori M, Saito Y. New lithium-ion conducting compounds $3Li_3N-MI$ (M = Li, Na, K, Rb) andtheir application to solid-state lithium-ion cells [J]. J Power Sources, 1997, 68 (2): 416-420.

[75] Wu M, Wen Z, Liu Y, et al. Electrochemical behaviors of a Li_3N modified Li metal electrode in secondary lithium batteries [J]. J Power Sources, 2011, 196 (19): 8091-8097.

[76] Bates J B, Dudney N J, Gruzalski G R, et al. Fabrication and characterization of amorphous lithium electrolyte thin films and rechargeable thin-film batteries [J]. Journal of Power Sources, 1993, 43 (1-3): 103-110.

[77] Dudney N J, Neudecker B J. Solid state thin-film lithium battery systems [J]. Curr Opin Solid State Mater Sci, 1999, 4 (5): 479-482.

[78] Kulkarni A R, Maiti H S, Paul A. Fast ion conducting lithium glasses-Review [J]. Bull Mater Sci, 1984, 6 (2): 201-221.

[79] Mercier R, Malugani J P, Fahys B, et al. Superionic conduction in $Li_2S-P_2S_5$-LiI-glasses [J]. Solid State Ionics, 1981, 5 (10): 663-666.

[80] Pardel A, Ribes M. Electrical properties of lithium conductive silicon sulfide glasses prepared by twin roller quenching [J]. Solid State Ionics, 1986, 18-19 (1): 351-355.

[81] Hayashi A. Preparation and characterization of glass materials for all-solid-state lithium secondary batteries [J]. J Ceram Soc Jpn, 2007, 115 (2): 110-117.

[82] Hayashi A, Tatsumisago M, Minami T. Structural investigation of 95 ($0.6Li_2S-0.4SiS_2$) $5Li_4SiO_4$ oxysulfide glass by using X-ray photoelectron spectroscopy [J]. Journal of American Ceramic Society, 1998, 81 (5): 1305-1309.

[83] Minami T, Hayashi A, Tatsumisago M. Preparation and characterization of lithium ion-conducting oxysulfide glasses [J]. Solid State Ionics, 2000, 136: 1015-1023.

[84] Tatsumisago M, Hama S, Hayashi, A, et al. New lithium ion conducting glass-ceramics prepared from mechanochemical $Li_2S-P_2S_5$ glasses [J]. Solid State Ionics, 2002, 154-155: 635-640.

[85] Liang C C. Conduction characteristics of the lithium iodide-aluminum oxide solid electrolytes [J]. J Electrochem Soc, 1973, 120 (10): 1289-1292.

[86] Maier J. Space-charge regions in solid 2-phase systems and their conduction contribution. 1. Conductance enhancement in the system ionic conductor-inert phase and application on $AgCl-Al_2O_3$ and $AgCl-SiO_2$ [J]. J Phys Chem Solids, 1985, 46 (3): 309-320.

[87] Maier J. Progress in solid state chemistry -ionic conduction in space charge regions [J]. Progress in Solid State Chemistry, 1995, 23 (3): 171-263.

[88] Debierre J-M, Knauth P, Albinet G. Enhanced conductivity in ionic conductor-insulator composites: Experiments and numerical model [J]. Appl Phys. Lett, 1997, 71 (10): 1335.

[89] Knauth P. Ionic conductor composites: Theory and materials [J]. J Electroceram, 2000, 5 (2): 111-125.

[90] Croce F, Appetecchi G B, Persi L, et al. Nanocomposite polymer electrolytes for lithium batteries [J]. Nature, 1998, 394 (6692): 456-458.

[91] Aobing D, Jingchao C, Jianjun Z, et al. All-solid-state lithium-ion batteries based on polymer electrolytes: State of the art, challenges and future trends [J]. Energy Storage Science and Technology, 2016, 5 (5): 627-648.

[92] Scrosati B, Garche J. Lithium batteries: Status, prospects and future [J]. J Power Sources, 2010, 195 (9): 2419-2430.

[93] Zhang Q, Liu K, Ding F. et al. Recent advances in solid polymer electrolytes for lithium batteries [J]. Nano Res, 2017, 10 (12): 4139-4174.

[94] Meyer W H. Polymer electrolytes for lithium-ion batteries [J]. Adv Mater, 1998, 10 (6): 439-448.

[95] Lightfoot P, Mehta M A, Bruce P G. Crystal-structure of the polymer electrolyte poly (ethylene oxide)$_3$LiCF$_3$SO$_3$ [J]. Science, 1993, 262 (5135): 883-885.

[96] Armand M. The history of polymer electrolytes [J]. Solid State Ionics, 1994, 69 (3-4): 309-319.

[97] Zhang J, Yue L, Hu P, et al. Taichi-inspired rigid-flexible coupling cellulose-supported solid polymer electrolyte for high-performance lithium batteries [J]. Scientific Reports, 2014, 4: 6272.

[98] Han P, Zhu Y, Liu J. An all-solid-state lithium ion battery electrolyte membrane fabricated by hot-pressing method [J]. J Power Sources, 2015, 284: 459-465.

[99] Murata K, Izuchi S, Yoshihisa Y. An overview of the research and development of solid polymer electrolyte batteries [J]. Electrochim Acta, 2000, 45 (8-9): 1501-1508.

[100] Cui Z Y, Xu Y Y, Zhu L P, et al. Preparation of PVDF/PEO-PPO-PEO blend microporous membranes for lithium ion batteries via thermally induced phase separation process [J]. J Membr Sci, 2008, 325 (2): 957-963.

[101] Pan C-y, Zhang Q, Feng Q, et al. Effect of catalyst on structure of (PEO)$_8$LiClO$_4$-SiO$_2$ composite polymer electrolyte films [J]. Journal of Central South University of Technology, 2008, 15 (4): 438-442.

[102] Ghelichi M, Qazvini N T, Jafari S H, et al. Conformational, thermal, and ionic conductivity behavior of PEO in PEO/PMMA miscible blend: Investigating the effect of lithium salt [J]. J Appl Polym Sci, 2013, 129 (4): 1868-1874.

[103] Abraham K M, Alamgir M. Li$^+$-conductive solid polymer electrolytes with liquid-like conductivity [J]. J Electrochem Soc, 1990, 137 (5): 1657-1657.

[104] Gadjourova Z, Andreev Y G, Tunstall D P, et al. Ionic conductivity in crystalline polymer electrolytes [J]. Nature, 2001, 412 (6846): 520-523.

[105] Hu P, Chai J, Duan Y, et al. Progress in nitrile-based polymer electrolytes for high performance lithium batteries [J]. J Mater Chem A, 2016, 4 (26): 10070-10083.

[106] Zhang J, Zhao J, Yue L, et al. Safety-reinforced poly (propylene carbonate)-based all-solid-state polymer electrolyte for ambient-temperature solid polymer lithium batteries [J]. Adv Energy Mater, 2015, 5 (24): 1501082 (1-10).

[107] Wieczorek W, Such K, Wycislik H, et al. Modifications of crystalline-structure of PEO polymer electrolytes with ceramic additives [J]. Solid State Ionics, 1989, 36 (3-4): 255-257.

[108] Croce F, Curini R, Martinelli A, et al. Physical and chemical properties of nanocomposite polymer electrolytes [J]. J Phys Chem B, 1999, 103 (48): 10632-10638.

[109] Quartarone E, Mustarelli P, Magistris A. PEO-based composite polymer electrolytes [J]. Solid State Ionics, 1998, 110 (1-2): 1-14.

[110] Li Q, Sun H Y, Takeda Y, et al. Interface properties between a lithium metal electrode and a poly (ethylene oxide) based composite polymer electrolyte [J]. J Power Sources, 2001, 94 (2): 201-205.

[111] Weston J E, Steele B C H. Effects of inert fillers on the mechanical and electrochemical properties of lithium salt poly (ethylene-oxide) polymer electrolytes [J]. Solid State Ionics, 1982, 7 (1): 75-79.

[112] Croce F, Persi L, Scrosati B, et al. Role of the ceramic fillers in enhancing the transport properties of composite

polymer electrolytes [J]. Electrochim Acta, 2001, 46 (16): 2457-2461.

[113] Capiglia C, Yang J, Imanishi N, et al. Composite polymer electrolyte: the role of filler grain size [J]. Solid State Ionics, 2002, 154: 7-14.

[114] Itoh T, Miyamura Y, Ichikawa Y, et al. Composite polymer electrolytes of poly (ethylene oxide) /$BaTiO_3$/Li salt with hyperbranched polymer [J]. J Power Sources, 2003, 119: 403-408.

[115] Croce F, Settimi L, Scrosati B. Superacid ZrO_2-added, composite polymer electrolytes with improved transport properties [J]. Electrochem Commun, 2006, 8 (2): 364-368.

[116] Tu Z, Kambe Y, Lu Y, et al. Nanoporous polymer-ceramic composite electrolytes for lithium metal batteries [J]. Adv Energy Mater, 2014, 4 (2): 1300654.

[117] Lin D, Liu W, Liu Y, et al. High ionic conductivity of composite solid polymer electrolyte via in situ synthesis of monodispersed SiO_2 nanospheres in poly (ethylene oxide) [J]. Nano Lett, 2016, 16 (1): 459-465.

[118] Jung Y-C, Lee S-M, Choi J-H, et al. All solid-state lithium batteries assembled with hybrid solid electrolytes [J]. J Electrochem Soc, 2015, 162 (4): A704-A710.

[119] Hu L, Tang Z, Zhang Z. New composite polymer electrolyte comprising mesoporous lithium aluminate nanosheets and PEO/$LiClO_4$ [J]. J Power Sources, 2007, 166 (1): 226-232.

[120] Zhang J, Zang X, Wen H, et al. High-voltage and free-standing poly (propylene carbonate) /$Li_{6.75}La_3Zr_{1.75}Ta_{0.25}O_{12}$ composite solid electrolyte for wide temperature range and flexible solid lithium ion battery [J]. J Mater Chem A, 2017, 5 (10): 4940-4948.

[121] Zhang X, Liu T, Zhang S, et al. Synergistic coupling between $Li_{6.75}La_3Zr_{1.75}Ta_{0.25}O_{12}$ and poly (vinylidene fluoride) induces high ionic conductivity, mechanical strength, and thermal stability of solid composite electrolytes [J]. J Am Chem Soc, 2017, 139 (39): 13779-13785.

[122] Zhai H, Xu P, Ning M, et al. A flexible solid composite electrolyte with vertically aligned and connected ion-conducting nanoparticles for lithium batteries [J]. Nano Lett, 2017, 17 (5): 3182-3187.

[123] Bae J, Li Y, Zhang J, et al. A 3D nanostructured hydrogel-framework-derived high-performance composite polymer lithium-ion electrolyte [J]. Angew Chem Int Ed Engl, 2018, 57 (8): 2096-2100.

[124] Urban A, Seo D-H, Ceder G. Computational understanding of Li-ion batteries [J]. npj Computational Materials, 2016, 2 (1).

[125] Xu H, Yu Y, Wang Z, et al. First principle material genome approach for all solid-state batteries [J]. Energy & Environmental Materials, 2019, 2 (4): 234-250.

[126] He X, Bai Q, Liu Y, et al. Crystal structural framework of lithium super-ionic conductors [J]. Adv Energy Mater, 2019.

[127] Zhang Y, He X, Chen Z, et al. Unsupervised discovery of solid-state lithium ion conductors [J]. Nat Commun, 2019, 10 (1).

[128] Adams S, Rao R P. High power lithium ion battery materials by computational design [J]. Physica Status Solidi a-Applications and Materials Science, 2011, 208 (8): 1746-1753.

[129] Filso M O, Turner M J, Gibbs G V, et al. Visualizing lithium-ion migration pathways in battery materials [J]. Chemistry-a European Journal, 2013, 19 (46): 15535-15544.

[130] Anurova N A, Blatov V A. Analysis of ion-migration paths in inorganic frameworks by means of tilings and Voronoi-Dirichlet partition: a comparison [J]. Acta Crystallographica Section B-Structural Science Crystal Engineering and Materials, 2009, 65: 426-434.

[131] Nuspl G, Takeuchi T, Weiss A, et al. Lithium ion migration pathways in $LiTi_2(PO_4)_3$ and related materials [J]. J Appl Phys, 1999, 86 (10): 5484-5491.

[132] Brown I D. Recent developments in the methods and applications of the BondValence model [J]. Chem Rev, 2009, 109 (12): 6858-6919.

[133] Blatov V A, Shevchenko A P. Analysis of voids in crystal structures: the methods of 'dual' crystal chemistry [J]. Acta Crystallographica Section A, 2003, 59: 34-44.

[134] Gao J, Chu G, He M, et al. Screening possible solid electrolytes by calculating the conduction pathways using BondValence method [J]. Science China Physics, Mechanics & Astronomy, 2014, 57 (8): 1526-1536.

[135] Dissanayake M, Bandara L, Karaliyadda L H, et al. Thermal and electrical properties of solid polymer electrolyte $PEO_9Mg(ClO_4)_2$ incorporating nano-porous Al_2O_3 filler [J]. Solid State Ionics, 2006, 177 (3-4): 343-346.

[136] Shannon R D, Taylor B E, English A D, et al. New Li solid electrolytes [J]. Electrochim Acta, 1977, 22 (7): 783-796.

[137] Shimura T, Tokiwa Y, Iwahara H. Protonic conduction in lanthanum strontium aluminate and lanthanum mobate-based oxides at elevated temperatures [J]. Solid State Ionics, 2002, 154: 653-658.

[138] Rao R P, Reddy M V, Adams S, et al. Preparation and mobile ion transport studies of Ta and Nb doped $Li_6Zr_2O_7$ Li-fast ion conductors [J]. Materials Science and Engineering B-Advanced Functional Solid-State Materials, 2012, 177 (1): 100-105.

[139] Pantyukhina M I, Zelyutin G V, Batalov N N, et al. Effect of substituting Li-6 for Li-7 on ionic conductivity of alpha-Li_3BO_3 [J]. Russ J Electrochem, 2000, 36 (7): 792-795.

[140] Pantyukhina M I, Obrosov V P, Stepanov A P, et al. Study of ion transport in Li_2ZrO_3 solid electrolytes with different lithium isotope ratios [J]. Crystallography Reports, 2004, 49 (4): 676-679.

[141] Woo J H, Trevey J E, Cavanagh A S, et al. Nanoscale interface modification of $LiCoO_2$ by Al_2O_3 atomic layer deposition for solid-state Li batteries [J]. J Electrochem Soc, 2012, 159 (7): A1120-A1124.

[142] Sakuda A, Nakamoto N, Kitaura H, et al. All-solid-state lithium secondary batteries with metal-sulfide-coated $LiCoO_2$ prepared by thermal decomposition of dithiocarbamato complexes [J]. J Mater Chem, 2012, 22 (30): 15247.

[143] Ohtomo T, Hayashi A, Tatsumisago M, et al. All-solid-state lithium secondary batteries using the $75Li_2S \cdot 25P_2S_5$ glass and the $70Li_2S \cdot 30P_2S_5$ glass-ceramic as solid electrolytes [J]. J Power Sources, 2013, 233: 231-235.

[144] Han F, Yue J, Chen C, et al. Interphase engineering enabled all-ceramic lithium battery [J]. Joule, 2018, 2 (3): 497-508.

[145] Ohta N, Takada K, Zhang L, et al. Enhancement of the high-rate capability of solid-state lithium batteries by nanoscale interfacial modification [J]. Adv Mater, 2006, 18 (17): 2226-2229.

[146] Ohta N, Takada K, Sakaguchi I, et al. $LiNbO_3$-coated $LiCoO_2$ as cathode material for all solid-state lithium secondary batteries [J]. Electrochem Commun, 2007, 9 (7): 1486-1490.

[147] Ohtomo T, Hayashi A, Tatsumisago M, et al. All-solid-state batteries with Li_2O-Li_2S-P_2S_5 glass electrolytes synthesized by two-step mechanical milling [J]. J Solid State Electrochem, 2013, 17 (10): 2551-2557.

[148] Ohta S, Komagata S, Seki J, et al. All-solid-state lithium ion battery using garnet-type oxide and Li_3BO_3 solid electrolytes fabricated by screen-printing [J]. J Power Sources, 2013, 238: 53-56.

[149] Noh S, Kim J, Eom M, et al. Surface modification of $LiCoO_2$ with $Li_{3x}La_{2/3-x}TiO_3$ for all-solid-state lithium ion batteries using Li_2S-P_2S_5 glass-ceramic [J]. Ceram Int, 2013, 39 (7): 8453-8458.

[150] Xiao Y, Miara L J, Wang Y, et al. Computational screening of cathode coatings for solid-state batteries [J]. Joule, 2019, 3 (5): 1252-1275.

[151] Sharafi A, Yu S, Naguib M, et al. Impact of air exposure and surface chemistry on Li-$Li_7La_3Zr_2O_{12}$ interfacial resistance [J]. J Mater Chem A, 2017, 5 (26): 13475-13487.

[152] Zhao N, Fang R, He M-H, et al. Cycle stability of lithium/garnet/lithium cells with different intermediate layers [J]. Rare Met, 2018, 37 (6): 473-479.

[153] Han X, Gong Y, Fu K K, et al. Negating interfacial impedance in garnet-based solid-state Li metal batteries [J]. Nat Mater, 2017, 16 (5): 572-579.

[154] Wenzel S, Leichtweiss T, Krüger D, et al. Interphase formation on lithium solid electrolytes—An in situ approach to study interfacial reactions by photoelectron spectroscopy [J]. Solid State Ionics, 2015, 278: 98-105.

[155] Zhu Y, He X, Mo Y. Origin of outstanding stability in the lithium solid electrolyte materials: Insights from thermodynamic analyses based on first-principles calculations [J]. ACS Appl Mat Interfaces, 2015, 7 (42): 23685-23693.

[156] Hao X, Zhao Q, Su S, et al. Constructing multifunctional interphase between $Li_{1.4}Al_{0.4}Ti_{1.6}(PO_4)_3$ and Li metal by magnetron sputtering for highly stable solid-state lithium metal batteries [J]. Adv Energy Mater, 2019, 9 (34).

[157] Cheng E J, Sharafi A, Sakamoto J. Intergranular Li metal propagation through polycrystalline $Li_{6.25}Al_{0.25}La_3Zr_2O_{12}$ ceramic electrolyte [J]. Electrochim Acta, 2017, 223: 85-91.

[158] Porz L, Swamy T, Sheldon B W, et al. Mechanism of lithium metal penetration through inorganic solid electrolytes [J]. Adv Energy Mater, 2017, 7 (20).

[159] Han F, Westover A S, Yue J, et al. High electronic conductivity as the origin of lithium dendrite formation within solid electrolytes [J]. Nature Energy, 2019, 4 (3): 187-196.

[160] Randau S, Weber D A, Kötz O, et al. Benchmarking the performance of all-solid-state lithium batteries [J]. Nature Energy, 2020.

[161] 李文俊, 徐航宇, 杨琪, 等. 高能量密度锂电池开发策略 [J]. 储能科学与技术, 2020, 9 (2): 448-478.

[162] Xu Xiaoxiong, Qiu Zhijun, Guan Yibiao, et al. All-solid-state lithium-ion batteries: State-of-the-artdevelopment and perspective [J]. Energy Storage Science and Technology, 2013, 2 (4): 331-341.

第 10 章

锂空气电池与锂硫电池

随着市场对能量密度的要求越来越高,以金属锂取代石墨作为负极材料逐渐成为发展趋势,由于金属 Li 具有较小的密度和最低的氧化还原电位,金属锂电池相比锂离子电池具有更高的能量密度和容量。其实早在 20 世纪 60 年代,金属锂电池就开始作为一次电池被报道,但是可充放电金属锂电池却存在着很大的安全隐患,循环过程中由于不均匀的电流密度,金属 Li 表面会生成锂枝晶,继而刺穿隔膜造成电池短路,粉化后的锂也更容易燃烧,导致可充放电金属锂电池存在极大的安全隐患。

随后 1990 年 Sony 公司成功以石墨替代金属 Li 成为负极,实现了锂离子电池商业化,宣告了锂离子电池新时代的到来。然而随着对日常消费电子产品、电动汽车和储备电源等领域的更高需求,迫切需要寻找能量密度更高的电池体系,特别是能超越锂离子电池的体系,可充放电的金属锂电池再一次回到人们的视线中。

在所有的锂电池体系中,锂空气电池和锂硫电池由于具有较高的理论能量密度而成为研究的热点,本章将讨论这两种可充放电的金属锂电池体系。

10.1 锂空气电池

10.1.1 锂空气电池基本工作原理

锂空气电池由金属 Li 或 Li 合金作为负极,含可溶性锂盐的导电介质作为电解质,空气(主要是氧气)作为正极,其工作原理如图 10-1 所示。在放电过程中,负极金属锂失去电子变成 Li^+ 溶解在电解质中,Li^+ 经过电解质从锂负极表面迁移至空气正极,电子从外电路迁移至空气正极,O_2 得到电子后与锂离子反应生成 Li_2O_2 或 LiOH,同时向外电路提供电能;在充电过程中,正极的 Li_2O_2 或 LiOH 分解,产生的 Li^+ 回到负极被还原成单质 Li,同时向空气中释放出氧气。

锂空气电池电解质可以是非水有机体系、水系、固体电解质体系,还可以是这 3 种电解质体系的混合。根据电解质体系的不同,目前已经研制的可充放电锂空气电池可以分为

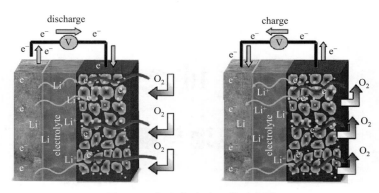

图 10-1 锂空气电池工作示意图

如下 4 种：①非水有机锂空气电池，1996 年由 Abraham 等首次提出，电解质主要是非水有机液体电解质，在该体系中，放电产物主要是 Li_2O_2，产物不溶于电解质，在空气正极表面进行堆积，随着放电产物填满空气正极的孔洞，也限制了该体系锂空气电池的能量密度。②水系锂空气电池，2007 年由 Visco 等提出，电解质以水系作为溶剂，放电产物主要是 LiOH，产物可溶于电解质，不存在所谓的正极限制，但是 LiOH 的增多会影响电池的酸碱平衡，影响电解质的性能，同时，负极锂的保护是个难题，需要研发出离子导电性好且不与金属 Li 反应、与水接触保持稳定的固体电解质材料是该体系的技术关键。③全固态锂空气电池，首先由 Kumar 等在 2010 年开始研究，采用固体电解质，安全性较高、高温性能好，但是电解质与正负极材料的接触电阻较大，需要加入缓冲层减小接触电阻，但一定程度上影响电池的能量密度。作为全固态电池的空气电极还需要解决电极结构的设计，既能具有较多孔隙容纳放电产物，同时也需要保持较好的电子接触。④混合体系锂空气电池，2007 年由 Polyplus 公司提出，并由 Zhou 等进一步发展，电解质由非水有机液体电解质和水系电解质共同组成，正极表面是水系电解质，增大产物 LiOH 的溶解度，负极表面是非水有机液体电解质，避免水分等与金属 Li 的反应，中间由 Li^+ 导电的疏水膜或固体电解质隔开，这种体系能够在一定程度上避免单独电解质体系的问题，但是能够长期稳定工作的电解质尚未研发成功；此外其结构复杂，界面电阻较大，因此尚有许多科学问题需要进一步研究。不同电池体系的结构示意图如图 10-2 所示。每种体系都有其优势和局限性，其发展取决于基础科学研究、关键材料、电池设计的进步，究竟哪种体系能够获得最终应用目前仍不能确定。

以非水有机体系锂空气电池为例，其在正极的反应过程如下所示

放电过程：$O_2 + e^- \longrightarrow O_2^-$；$O_2^- + Li^+ \longrightarrow LiO_2$；

$$2LiO_2 \longrightarrow Li_2O_2 + O_2 \tag{10-1}$$

充电过程：$Li_2O_2 \longrightarrow 2Li^+ + 2e^- + O_2$ (10-2)

该过程是 Bruce 等和 McCloskey 等经过表面增强拉曼光谱（SERS）、微分电化学质谱（DEMS）和同位素标记等测试手段推断得到的结论。虽然更细致的反应机理尚不是很清楚，但是用上述反应过程可以解释锂空气电池中的一些问题。部分文献报道当放电电压低于 2.0V 时，放电产物会有 Li_2O 的生成，但是 Li_2O 非常稳定，在充电过程中分解非常困难。一般锂空气电池的放电电位高于 2.0V，放电产物主要是 Li_2O_2。

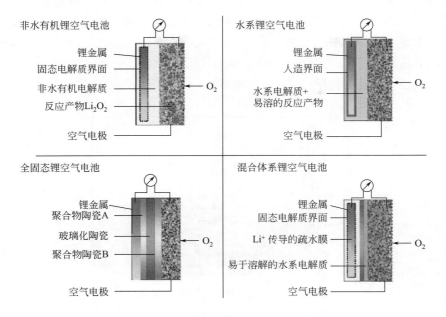

图 10-2 不同锂空气电池体系的结构示意图

锂空气（Li/O_2）电池的典型充放电曲线如图 10-3 所示，可以看出放电过程基本是一个 2.7V 左右的平台，而充电阶段表现为 2 段或者 3 段斜坡。Shao-Horn 等提出 3 段斜坡对应着 3 种不同的反应路径：在充电初期过电势很小的阶段，表面的 Li_2O_2 自发发生 Li^+ 的脱出，变为 $Li_{2-x}O_2$，这个阶段不受电流密度大小的影响；在第二阶段随着电压的推升，体相的 Li_2O_2 发生分解产生氧气；在第三阶段也就是充电末期，电压依然继续升高，在这个阶段，一些报道认为

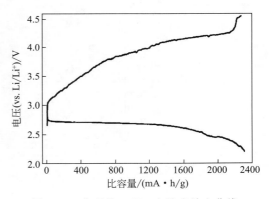

图 10-3 典型的 Li/O_2 电池充放电曲线

是碳材料表面参与了反应或者是电解液发生了分解，而这个阶段的主产物为 CO_2。

相比于其他电池体系，锂空气电池具有最高的理论能量密度（仅次于 Li/F 电池体系，而 F 有毒并且实验难度大，因此 Li/F 电池体系很难实现）。从图 10-4 和图 10-5 可以看出，如果产物按照 Li_2O 计算，锂空气电池是目前电池体系中质量能量密度最高的体系。相对于 Mg/O_2 和 Al/O_2 电池，Li/O_2 电池更容易实现可充放，极化较低。因此，从发展高能量密度可充放电池角度来说，Li/O_2 电池可能具有最高的能量密度。对于非水有机体系锂空气电池来说，根据金属 Li 的质量计算，它的能量密度能够达到 11.238kW·h/kg，如果把氧气的质量也算进去，它的能量密度能达到 3.608kW·h/kg。如果将它应用于电动汽车中，考虑到实际能量密度与理论能量密度的比例，其能量密度有望达到 800~1000W·h/kg，理论上可以使电动汽车的续航里程（一次充电）超过 800km，而现在的锂离子电池电芯能量密度为100~265W·h/kg。采用高容量的正极和负极材料,能量密度预计还可以进一步提高到250~350W·h/kg,但可能还是明显低于锂空气电池。因此,二次锂空气电池的研究引起了大家的广泛关注。根据 *web of science* 的统计,最近20年,以"lithium air battery or Li-air battery or lithium oxygen battery or Li/O$_2$ battery"为关键词的文章数量呈指数上升趋势,呈现高速增长的趋势,如图10-6所示。

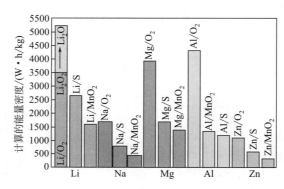

图10-4 不同电池体系的质量能量密度图

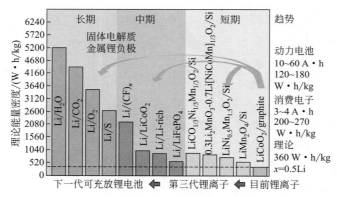

图10-5 不同锂电池体系质量能量密度图及发展趋势预测

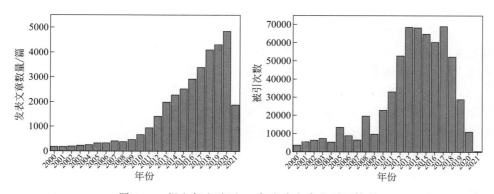

图10-6 锂空气电池近20年发表文章和引用情况

(注:检查日期为2021年6月8日)

10.1.2 锂空气电池组成

锂空气电池电芯主体部分主要由金属 Li 或锂合金负极、空气正极和电解质 3 部分组成。其中空气正极可以包含活性材料、集流体、黏结剂和催化剂等，电解质主要包括液体电解质和隔膜，也可以是聚合物电解质或者是固体电解质。除此之外，锂空气电池系统如果需要直接使用空气，可能还需要防水透气膜、负极保护层、封装材料、气泵和过滤膜等。本节以非水有机体系锂空气电池电芯为例，简要介绍锂空气电池各个组成部分。

10.1.2.1 电解液

与锂离子电池一样，锂空气电池电解液的主要作用也是锂离子传输的媒介。因此，它除了满足锂离子电池电解质的基本性质（高的电导率和锂离子迁移数，低的黏度，高的化学和电化学稳定性）外，还需要满足如下要求：①对锂空气电池中间产物 O_2^- 比较稳定；②蒸气压较低，在使用过程中不易挥发；③氧气在其中具有一定的溶解度和扩散速度。非水有机液体电解质首先采用的是锂离子电池中常用的碳酸酯体系，然后发展出了更稳定的醚类电解质体系。此外，一些新型的电解液体系也被发现在锂空气电池中比较稳定。

(1) 碳酸酯类

锂空气电池早期的电解液都是基于锂离子电池中常用的碳酸酯体系，其中丙烯碳酸酯（PC）由于具有宽的电化学窗口、低的挥发性和宽的液程，被研究得最为广泛。尽管 Aurbach 等很早就发现 PC 体系不稳定，但是并没有引起人们足够的重视。直到 2010 年，Mizuno 等用直接的证据指出 PC 体系中放电产物是 Li_2CO_3 和其他的烷基碳酸盐而不是人们希望看到的 Li_2O_2，碳酸酯体系电解质在氧化还原过程中的不稳定性才被人们重视。Bruce 等通过对放电产物进行红外光谱、质谱、表面增强拉曼光谱和核磁共振谱等研究发现，在烷基碳酸酯电解液中，锂空气电池的主要放电产物是甲酸锂、乙酸锂和碳酸锂，红外光谱观察不到 Li_2O_2 的存在。在充电过程中，这些产物能够被分解，释放出 CO_2 气体。因此在碳酸酯体系锂空气电池中，其反应主要是电解液不断地不可逆氧化分解。关于碳酸酯电解液分解的原因，Bryantsev 等通过第一性原理计算认为是活泼的中间产物 O_2^- 导致的。这些研究使锂空气电池的反应机制更加清晰，为寻找稳定的电解液指出了方向。

(2) 醚类

认识到了碳酸酯体系的不稳定性，众多研究者开始通过实验和计算寻找在 O_2 存在下更稳定的电解质体系。其中醚类电解质引起了大家的关注，乙二醇二甲醚（DME）和四乙二醇二甲醚（TEGDME）是两种比较稳定的体系，并且具有较高的氧化稳定性、不可燃性和高的热稳定性。研究发现，使用醚类电解质时，锂空气电池的放电产物主要是 Li_2O_2。但是经过进一步的研究，McCloskey 等发现 DME 也不稳定，它会与放电产物 Li_2O_2 发生反应生成羧化锂和 LiOH 等产物，并且 DME 具有很高的挥发性，不能长时间循环。而长链的 TEGDME 不仅跟 DME 一样在 O_2^- 存在下具有较高的稳定性，而且不易挥发，放电产物主要是 Li_2O_2，在锂空气电池中取得了广泛的应用。但是 Bruce 等发现随着循环的进行，放电产物中 Li_2O_2 的比例越来越低，证明醚类电解液并不适合作为锂空气电池的电解液。

在研究 TEGDME 的过程中，Jung 等发现锂盐的选择也是非常重要的，三氟甲基磺酸锂（$LiCF_3SO_3$）比 $LiPF_6$ 具有更好的性能。

（3）其他体系电解液

除了上述两种电解液体系之外，锂空气电池电解液的研究还包括乙腈（ACN）、二甲基亚砜（DMSO）、二甲基甲酰胺（DMA）、苯甲醚和离子液体等体系，其中效果最好的是 DMSO 体系。彭章泉等报道了一个非常稳定的锂空气电池，使用 $LiClO_4$/DMSO 作为电解液，多孔金作为正极。该电池具有非常好的循环性能和容量保持率，通过 FTIR、Raman、NMR 和 DEMS 等测试手段确定放电产物是 Li_2O_2，该产物在充电过程中能够可逆完全分解。但是 DMSO 与负极锂片兼容性较差，需要对锂片进行保护。乙腈虽然具有非常好的抗氧化稳定性，但是其挥发性较高且毒性较大，限制了它的实际应用。

在锂离子电池中混合溶剂乙烯碳酸酯（EC）和二甲基碳酸酯（DMC）表现出较好的性能，所以在锂空气电池中研究者也希望通过混合溶剂的使用取得理想的效果。关于混合溶剂的报道首先是由 Scrosati 等提出的，他们使用 TEGDME 和 N-甲基-（N-丁基）吡咯烷双三氟甲基磺酰亚胺锂（$PYR_{14}TFSI$）的混合物作为溶剂，LiTFSI 作为锂盐，发现相比于 TEGDME 体系电导率提高了 4 倍，电解液氧化电位接近 4.8V(vs. Li^+/Li)，电池充电电位降低了 0.5V。

此外，聚合物电解质和全固态电解质由于具有较高的安全性、能够保护锂电极并抑制锂枝晶，也被应用于锂空气电池中，虽然其电解质电导率、接触电阻等问题需要显著改善，但这也是重要的研究方向。

（4）电解液添加剂

添加剂的特点是用量少但是能显著改善电解液某一方面的性能。商品锂离子电池一般包含 10 种以上的添加剂，它们的作用一般为提高电解液的电导率、提高电池的循环效率、增大电池的可逆容量、改善电极的成膜性能等。与锂离子电池不同，有关锂空气电池电解液添加剂的研究目前还比较少，其主要作用是增大电池的比容量和降低电池的充放电过电位等，下文将简单介绍。

由于锂空气电池放电产物 Li_2O_2 在电解液中溶解度较差，随着放电过程的进行和产物的累积，正极孔被堵塞，致使放电过程无法继续进行。因此，Li_2O_2 的累积限制了电池的比容量。中国科学院物理研究所的谢斌博士首先发现硼基阴离子受体化合物三（五氟苯基）硼（TPFPB）能够促进 Li_2O_2 和 Li_2O 在有机溶剂中的溶解。Qu 等发现 TPFPB 能够与 O_2^{2-} 络合，提高 Li_2O_2 的溶解度，并能够降低 Li_2O_2 的氧化电位，提高其氧化动力学。但是 Xu 等[65]经过实验验证，发现随着 TPFPB 浓度的增加，电解液的黏度增加电导率降低。因此，要很好地控制 TPFPB 的加入量，避免带来不利的影响。

锂空气电池充放电电位差较大，导致其能量效率较低（60%～80%），为了提高能量效率，降低充放电过电位是非常必要的。Bruce 等发现四硫富瓦烯（TTF）可以作为一种氧化还原媒介，它的加入可以大幅降低锂空气电池充电过电位，并提高电池的倍率性能。在低的充电电位下 TTF 可以被氧化形成 TTF^+，然后 TTF^+ 氧化 Li_2O_2 并被还原为 TTF，TTF 的反复作用可以促进充电过程的进行。最近，Kang 等发现一种效果更好的氧化还原媒介 LiI，它的作用机制与 TTF 一样，具有更低的氧化电位，能够显著降低充电过电位。

关于该类针对充电过程中能够促进 Li_2O_2 分解的氧化还原媒介（redox mediator）的选择，Kang 等提出了 3 条标准：①氧化电势需要与 Li_2O_2 的电位相匹配，略高于 Li_2O_2 形成的平衡电位；②氧化还原介质被氧化后的产物可以有效地分解 Li_2O_2；③在电解液中稳定性高，不会带来其他的副反应。由于 LiI 及其氧化产物 I_2 的强烈腐蚀性，LiI 还不是理想的氧化还原介质。但之前积累的这些研究结果为理解锂空气电池过电位的主要起因、降低分解正极反应物过电位提供了重要的思路。

10.1.2.2 空气正极

锂空气电池空气正极具有 4 个方面的作用：提供氧气扩散通道，使氧气能够顺利到达电极电解质界面；为锂空气电池放电反应提供场所，容纳放电产物；在锂空气电池充放电过程中起催化作用，提供反应活性位；影响反应产物的形貌。

空气正极的种类和结构对锂空气电池的比容量、功率密度和寿命有着重要的影响，所以它的选择一定要考虑多方面的因素。锂空气电池正极材料需要满足如下一些基本性能：①具有良好的电子电导率和离子电导率；②具有合适的孔隙率，能够保证快速的氧气扩散，实验证明 2~50nm 的孔尺寸有利于锂空气电池性能的发挥；③具有较大的比表面积和孔隙率，能够容纳固态反应产物；④电化学和化学性能足够稳定，不与反应气或者电解液发生反应，也不参与电池反应；⑤较低的成本。

针对不同电解质类型，空气正极的设计有所不同，目前非水电解质锂空气电池常用的空气正极活性材料主要由多孔碳材料和催化剂两部分组成。

(1) 多孔碳材料

多孔碳材料是目前研究最为广泛的空气正极材料，主要是因为它具有如下一系列优点：①电导率高，能够提供快速的电荷转移；②比表面积大，密度较小，使锂空气电池具有高的质量比能量；③孔隙率合适，能够提供较多的位置容纳放电产物；④由于缺陷位的存在能够在锂空气电池放电过程中催化氧气还原。目前常用的多孔碳材料主要有 Super P 和 Ketjen Black 两种，它们单独使用作为锂空气电池正极活性材料就能获得良好的性能。

除了上述两种碳材料之外，目前多孔正极的研究主要是由实验室合成的碳材料，包括中空的碳纤维、直壁碳纳米管和石墨烯基材料等。其中中空碳纤维直接生长在多孔陶瓷衬底上，不含黏结剂，有利于观察正极表面产物的形貌。直壁碳纳米管具有高度有序的结构，不仅可以观察正极表面放电产物的形貌，还可以用来原位观察放电产物的生长和分解过程。石墨烯基材料比表面积特别大，具有特殊的孔结构，电导率高且具有一定的催化活性，所以用来作锂空气电池正极材料能够产生特别高的比容量。石墨烯基材料的研究主要包括功能石墨烯层状材料和石墨烯氧化物等。经过这些新型材料的研究发现，锂空气电池的容量、倍率性能、循环性能和放电产物 Li_2O_2 的形貌与正极活性材料的种类和结构密切相关。在中空碳纤维和直壁碳纳米管表面，放电产物主要是环形的盘状颗粒，而在石墨烯表面产物主要是纳米岛状生长。

目前很多文献报道多孔碳材料在锂空气电池中是不稳定的，在现有的电压范围内，会与放电产物 Li_2O_2 反应生成 Li_2CO_3，也会与电解液发生反应生成各种副产物，如 Li_2CO_3 和羧酸锂等。为了解决多孔碳的不稳定问题，很多种解决方法被报道：①控制锂空气电池的充电电压在 3.5V 以下，这样碳材料是相对稳定的；②选用其他的正极材料，如疏水碳材料、多孔金正极材料、TiC 正极材料、Pt/TiN 和 Ru/ITO 等，实验证明疏水碳材料相对于亲水碳材料更稳定一些；③在碳材料表面包覆一层活性材料（如 RuO_2），避免碳与电解液以及放电产物的接触；④使用氮掺杂的碳材料。这些方法虽然能够在一定程度上解决问题，但是由于锂空气电池目前面临的问题还很多，将充电电压降到 3.5V 以下并不容易，还需要进一步协同解决。

（2）催化剂

关于锂空气电池催化剂的研究是从 1996 年 Armand 等报道的酞菁钴开始的，后面陆续有新的催化剂被报道。这些催化剂大部分都是早期基于碳酸酯体系电解质，主要包括过渡金属氧化物、贵金属和非贵金属等几种，以下简单介绍这 3 类催化剂。

① 过渡金属氧化物催化剂　过渡金属氧化物催化剂由于价格较低，催化性能较好，在有机碳酸酯体系作为电解液的时候引起了广泛关注。Bruce 等先是发现了 Co_3O_4 具有较好的催化性能，能够促使锂空气电池获得较高的比容量和容量保持率，又能降低过电位。后来经过研究发现相比于其他过渡金属氧化物，$\alpha\text{-}MnO_2$ 由于具有较高的比表面积，在提高比容量和降低充电过电位方面具有最好的催化效果，如图 10-7 所示。

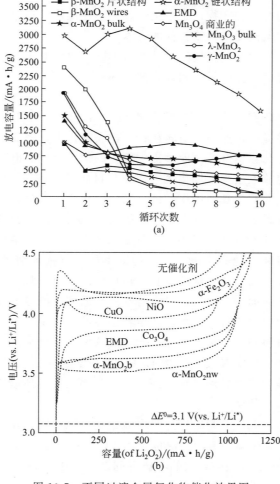

图 10-7　不同过渡金属氧化物催化效果图

Ogumi 等发现钙钛矿型金属氧化物材料也具有氧还原催化性能。此外，过渡金属硫化物、碳化物、氮化物、氧氮化物和碳氮化物也被发现在锂空气电池中具有较高的催化活性。

② 贵金属催化剂　目前研究的贵金属催化剂主要包括 Pt、Au、Ru 和 Pd 等，它们具有更优越的催化性能，如图 10-8 所示。Shao-Horn 等发现 Pt 的使用可以降低锂空气电池充电过电位，而 Au 能够降低放电过电位。使用 Pt-Au 纳米复合颗粒能够同时降低充电过电位和放电过电位，大大提高锂空气电池的能量效率。Zhou 等通过研究发现，直壁碳纳米管（MWCNT）负载 Ru 能够作为二元功能催化剂，明显降低锂空气电池的充放电过电位，提高其循环性能。Zhang 等研究了 Pd 修饰的中空碳球在锂空气电池中的催化性能，该电池使用比较稳定的 $LiCF_3SO_3$-TEGDME 电解液，发现 Pd 的修饰能够降低电解液的分解，提高倍率性能、电池比容量和寿命，其主要原因是 Pd 的修饰能够改善产物的形貌，使产物从圆盘状变为薄片状。

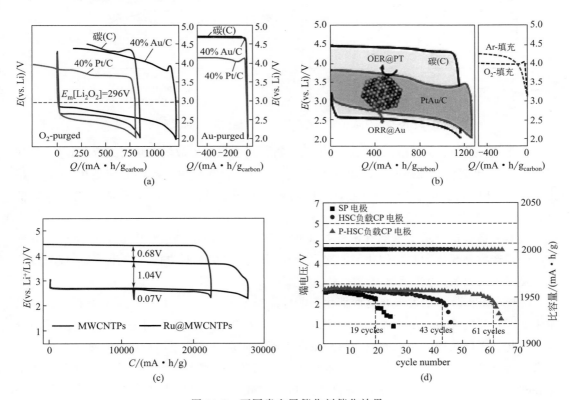

图 10-8　不同贵金属催化剂催化效果

③ 非贵金属催化剂　非贵金属催化剂目前研究较少，主要是由于其催化性能没有其他催化剂明显，但是该类催化剂价格较低，也引起了一些关注。目前非贵金属催化剂的研究主要是 Ren 等报道的 CuFe 合金催化剂，催化效果如图 10-9 所示。所用正极分别为 K-carbon 和 CuFe 催化的 K-carbon。该催化剂能够在一定程度上降低锂空气电池的放电过电位并提高其能量密度，在大的放电电流下效果更明显。

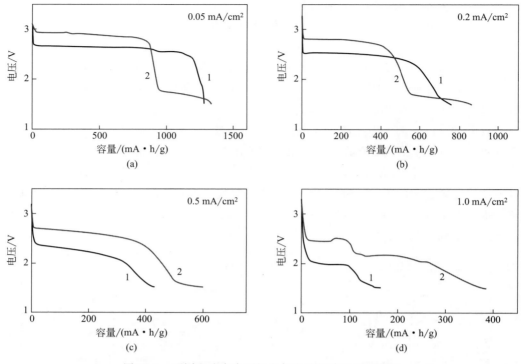

图 10-9 不同电流密度下锂空气电池的放电电压曲线

LiO_2 通常被认为是 Li_2O_2 形成的中间产物，无法稳定以放电产物形式存在，Lu 等使用 Ir 作为催化剂负载在还原石墨烯（rGO）上，控制放电产物为稳定的结晶相 LiO_2。相比 Li_2O_2，LiO_2 分解为单电子反应，充电时更容易分解成 Li 和 O_2，可将充电平台降低至 3.5V，提高锂空气电池的能量效率。

虽然大量报道展示了锂空气电池催化剂的研究，但是在非水有机电解质锂空气电池中，由于产物是固体颗粒，催化剂能起多少作用，具体在哪些方面起作用，如何起作用，仍需要深入细致的研究。

随着电解液体系的发展，这些催化剂的性能需要重新评估。McCloskey 等指出在碳酸酯体系锂空气电池中，由于放电产物不是 Li_2O_2，所以催化剂的作用是促进了电解液的分解。因此关于催化剂的研究需要在较稳定的电解液体系中重新开始。此外，如果今后不能解决高速透氧膜的问题，空气电极如果直接采用空气，需要解决其他产物的电化学或化学分解问题，如 LiOH、Li_2CO_3、Li_2O 等产物的分解，这或许需要结合电解质中的氧化还原介质及多功能催化剂技术。

（3）小结

目前，锂空气电池所使用的正极材料大部分是多孔碳材料，而对于是否需要催化剂目前还存在争议。首先是催化剂的催化机理目前尚不是很清楚；其次，随着放电反应的进行，催化剂有可能被产物完全覆盖，在放电反应后期可能难以继续起催化作用。但催化剂的存在也可能影响反应产物的形貌，这对于气体扩散和离子输运较为重要。空气电极催化剂的研究仍将是今后关注的重点。

10.1.2.3 负极

由于金属 Li 的质量非常小且具有最低的氧化还原电位,所以它非常适合应用于锂空气电池以取得较高的能量密度。一些锂合金材料如锂铝合金和锂硅合金等也取得了一些研究进展,但是其较大的体积形变限制了它们的应用。因此,目前的锂空气电池基本上是使用金属 Li 负极。

金属 Li 在锂空气电池中面临枝晶问题,该问题主要是由 Li 金属和电解质界面上不均匀的电流密度及浓度梯度造成的。金属 Li 在电解液中会形成一层 SEI 膜,SEI 膜的形成阻止了金属 Li 与电解液的进一步反应,但是 SEI 膜的各向异性及不均匀性容易引起 Li 表面的不均匀和 Li^+ 浓度的不同,从而造成 Li^+ 的不均匀沉积。随着锂枝晶的生长,会造成隔膜的刺穿和电池的短路,带来安全隐患。为了提高 Li 负极的安全性,一些解决方法被提出:①在 Li 金属表面镀上一层均匀的高锂离子导电的保护层,Seeo 公司发现聚合物电解质具有缓解锂枝晶的作用,他们使用聚苯乙烯作为骨架保持一定的机械稳定性,使用 PEO/锂盐混合物来提高离子电导率,这一思路在可充放电 Li/聚合物电解质/$LiFePO_4$ 电池中已经得到验证;②使用高锂离子电导率的玻璃或陶瓷材料作为固体电解质,如 Visco 等使用 LISICON 材料包裹金属 Li,从而防止锂枝晶的生长,但是该类材料容易与金属 Li 反应,需要在中间插入一层稳定的导电材料(如 Li_3N 或者 Li_3P),该材料具有良好的阻隔性能,但是易碎且增大电池内阻;③使用陶瓷和聚合物复合材料作为锂空气电池电解质,该材料可以满足上述两种材料的优势,既具有较好的柔韧性,又具有较好的阻隔性能。

如果金属 Li 的保护技术能够开发成功,空气电极对于透氧膜的需求可以减少甚至免除,那么,在复杂环境中锂空气电池金属 Li 的循环性、安全性也可以得到显著改善。

10.1.2.4 防水透气膜

目前的锂空气电池多数是在纯氧气中工作的,因为空气中的 H_2O、CO_2 和 N_2 等气体以及微细颗粒会引起副反应发生或电极钝化,对电池的性能产生影响。为了解决该问题,一种有效的方法便是使用氧气扩散膜,该薄膜能够阻止其他气体的透过,而不影响氧气的扩散。基于氧离子和电子混合传导的透氧膜一般在高温下工作,室温下高性能的透气膜开发目前还面临很大的困难。

氧气扩散膜的设计主要基于两种机制,第一种是基于尺寸控制机制。从表 10-1 空气中不同气体的物理参数可以看出来,相比于氧气分子,水分子的尺寸非常小。因此,通过尺寸控制机制(如诺森扩散和 Pouseille 流动)来促使氧气分子优先通过多孔膜的想法是难以实现的。

第二种设计是基于溶解/扩散机制,使气体通过无孔的聚合物膜,该机制满足方程:$P=DS$,其中,P 是气体在聚合物膜中的透过系数;D 是扩散系数;S 是溶解系数。气体的扩散系数跟气体的动力学直径(d_k)成反比,因此,水分子具有比氧气更高的扩散系数。在聚合物中,气体的溶解系数与气体的可压缩性存在 Arrhenius 关系,从表 10-1 可以看出,水分子的可压缩性远远高于氧气分子,因此,其溶解度要高于氧气。综上所述，水分子在聚合物膜中的透过率要远远高于氧气分子。

表 10-1　空气中不同气体的物理参数

气体	临界温度 T_c/K	临界体积 V_c/(cm³/mol)	动力学直径 d_k/Å
O_2	154.6	73.5	3.46
N_2	126.2	89.3	3.64
H_2O	647	55.9	2.65
CO	134.5	90.1	3.76
CO_2	304.2	91.9	3.30

注：1Å=0.1nm。

然而，可以使用高度亲水的透水膜先除去混合气体中的水分，剩余的气体再一一除去杂质气体即可作为锂空气电池的工作气体。

目前有一定效果的防水透气膜有张继光等研究的聚甲基硅氧烷、聚酯薄膜和高密度聚乙烯膜等。

10.1.2.5　Li_2O_2 基本性质

目前，Li_2O_2 电子结构、输运性质和磁学性质对锂空气电池电化学性能（尤其是充电过电位）以及电解液稳定性的影响是一个非常重要的研究方向，对 Li_2O_2 性质的理解有利于我们阐明锂空气电池充放电机理。

对于 Li_xO_y 整比化合物来说，热力学上结构最稳定的化合物为 Li_2O 和 Li_2O_2。通过理论预测，它们均为半导体或绝缘体，具有比较宽的带隙，如 Li_2O_2 体材料的带隙约为 4.9eV。如此高的带隙很难解释锂空气电池的可逆性和 Li_2O_2 可逆地生成和分解，因为一般来说，在电池系统中，要求参与氧化还原的各个物种具有好的电子电导。Hummelshoj 等提出了一种金属性的理论，他们认为在 Li_2O_2 中引入空位能够降低其带隙，随着空位浓度的增加，带隙能降低至约 3.0eV。在该带隙下，电子和电荷能够自由移动，这在一定程度上能够解释前面的问题。该理论被实验所证实，实验上观察到的 Li/O_2 电池放电产物 Li_2O_2 多为纳米颗粒，表面存在空位和缺陷。Lau 等通过 DFT 计算认为 Li_2O_2 表面存在类似超氧结构，该结构能够促进 Li_2O_2 的可逆分解，提高 Li_2O_2 表面的电子电导和引起电解液的表面反应，从而降低锂空气电池充电过电位、提高能量效率。Lu 等和 Yang 等通过实验结果进一步证实了超氧结构的存在。

产物 Li_2O_2 的形貌受多种因素的影响，首先与放电电流密度的大小有关。在电流密度较大的时候，Li_2O_2 产物的形貌趋向于在碳材料表面形成一层薄膜状；而在电流密度较小的时候，Li_2O_2 会由于表面导电的原因造成产物边缘与中间部分电子分布不同，最终产物形貌成环形的柿饼状。然而近期的文献报道发现柿饼状的产物出现不仅仅是电流密度大小的原因，还有电解液中含有水的原因。柿饼状形貌的产物出现的根本原因在于 Li_2O_2 与水发生反应，使得在形成 Li_2O_2 的过程中出现了分解的现象，尤其是当水含量较高的时候，可以发现产物是一层一层长成柿饼状的。再结合早期小电流密度的理论，就可以解释这种柿饼状 Li_2O_2 的产生原因。另外的一些工作报道了锂空气电池如果在含一定湿度的氧气中工作，在一定的水含量下，放电容量得到明显的增长，这也可以解释为 Li_2O_2 与水进一步发生反应形成 LiOH，使得放电继续进行，容量得到大大的提升。此外，Li_2O_2 的形貌还受催化剂的影响，Zhang 等发现在使用 Pd 修饰的情况下，产物更容易形成柿饼状。

10.1.3 锂空气电池中的科学问题

尽管锂空气电池具有非常高的理论能量密度，但是它所面临的诸多科学问题还没有搞清楚，以下简单小结这些问题。

10.1.3.1 负极

① 锂负极所面临的最主要的问题在前面第 10.1.2.3 节已经简单介绍，主要是枝晶生长与电极粉化的问题，该问题会导致循环效率的降低和电池安全性变差。

② 空气中水分和二氧化碳等杂质气体与金属 Li 的反应，导致金属 Li 表面生成 LiOH 和 Li_2CO_3 等副产物。这些副产物可能会导致锂枝晶的形成，从而影响锂空气电池中金属 Li 的循环性。

针对金属 Li 负极存在的问题，目前常用的解决方法是使用锂合金材料、表面采用聚合物与无机固体电解质（如 LATP 等）或者其复合物进行保护、使用多孔电极等。电池中如果使用聚合物电解质或无机陶瓷电解质，锂枝晶的问题将有可能被解决。

10.1.3.2 电解液

① 对于锂空气电池来说，很难寻找一种稳定的电解质体系。锂空气电池中间产物 O_2^- 是一种非常活泼的物质，它会与电解质体系中的有机溶剂或锂盐反应，造成电解质的不可逆分解。

② 目前的电解液主要使用有机溶剂，多数有机溶剂在工作温度范围内存在挥发问题，致使相应的锂空气电池无法处于开放状态。因此在目前的锂空气电池研究中分为密闭锂空气电池和开放锂空气电池。

目前新开发的 TEGDME、DMSO、PP13TFSI 和 LAGP 等体系虽然稳定性更高，但是也不能从根本上解决上述问题。

10.1.3.3 空气正极

① 从前面的论述可知，防水透气膜的开发是一个非常大的挑战。根据报道，目前能较好解决防水透气问题的正极材料当属 Zhou 等研发的交联网络凝胶复合正极，它是由单壁碳纳米管和离子液体复合而成，离子液体的疏水性致使水分子很难通过正极进入电池内部，从而使电池可以在空气中工作。

② 充放电机理，在不同的电解液、倍率、温度、催化剂、电极、气氛下，锂空气电池的放电产物和充电产物的种类、形貌、结构都可能存在差异，导致充放电反应路径有可能不对称。由于碳电极、催化剂也可能参与电极反应，导致具体体系的反应机理的确定存

在较大的困难,需要结合多种原位与非原位表征手段研究。此外,多电子转移过程目前还不是很清楚。上述问题正在广泛地开展。

③ 热力学与动力学特性。热力学包括反应过程中的开路电压、吸放热、熵变,反应路径中的始态、终态、亚稳态的中间产物的自由能等;动力学特性包括极化规律、极化起因、输运与反应动力学参数、暂态反应机理、电催化机理等。这些尚未获得系统的研究。

在多数锂空气电池中,充电过程和放电过程存在比较大的电位差,导致目前锂空气电池的能量损失大于 30%。该电位差可能是由动力学和热力学两种因素引起的,其中热力学电位差可以由反应路径不一致引起。如 10.1.2.2 节(2)中所述,催化剂的使用能够在一定程度上降低过电位,但是作用机理尚不清楚,且在非碳酸酯体系中作用不明显,有待进一步考证。目前 TTF 和 LiI 添加剂的使用能够在一定程度上缓解该问题。动力学引起的过电位可通过改进空气正极结构(如孔隙率)来改善。

目前锂空气电池倍率性能较差,很难使用大电流充放电,这主要是放电产物导电性差引起的。控制放电产物的形貌和空气正极的结构有利于提高锂空气电池的倍率性能。

④ 副产物。随着循环的进行,锂空气电池容量衰减比较严重。该问题是由放电产物的累积、电解液的分解和电池中各种副反应引起的,将会随着其他问题的解决而有所改善。在碳酸酯、醚类以及空气中有 CO_2 时,反应产物中包括 Li_2CO_3 等副产物。副产物对 Li/O_2 电池及锂空气电池的动力学特性、可逆性、反应机理的影响很大,这些也是目前研究的重点。

综上所述,锂空气电池的研究还处于初级阶段,在其应用之前还有很多问题需要解决(图 10-10),这需要该领域研究工作者进一步付出长期耐心的努力。

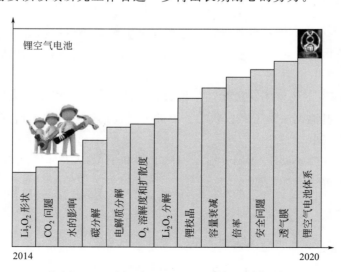

图 10-10 锂空气电池需要解决的科学问题

10.1.4 其他锂空气电池体系

空气中除了氧气以外还包含很多其他气体,这些气体中有一部分如 CO_2 和 H_2O 等也可以作为锂空气电池的正极反应气体，下面简要介绍 Li/CO_2 电池和 Li/H_2O 电池。

10.1.4.1 Li/CO_2 电池

从前面的论述可知，CO_2 的存在会对锂空气电池的性能产生影响，但是电池如果要在空气中工作，CO_2 的影响是很难避免的。所以人们希望能够通过一定的方法解决 CO_2 存在所带来的问题。据报道，如果反应气体中含有 CO_2，那么锂空气电池放电产物中将会有 Li_2CO_3 生成。相比于锂空气电池正常放电产物 Li_2O_2 来说，Li_2CO_3 的分解是非常困难的，在 DME 体系电解质中它的分解电位高达 4.8V，这就会进一步降低锂空气电池的能量效率。汪锐博士等在研究锂离子电池电极材料的过程中发现，Li_2MnO_3 和 NiO 的存在能够催化 Li_2CO_3 的可逆分解，并且能够大大降低 Li_2CO_3 的分解电位。

基于上述结果，刘亚利等研究了可充放电的 Li/CO_2：O_2（2：1）电池和 Li/CO_2 电池，并使用 XRD、FTIR、Raman、SEM 和 EELS 等手段表征了放电产物的种类和形貌。这些初步的研究结果说明，即使产物是 Li_2CO_3，也能够在合适的条件下分解。

10.1.4.2 Li/H_2O 电池

在 Li/H_2O 电池中，负极为锂片，正极为镍片或惰性金属，电解液为碱性水溶液。工作过程中，金属 Li 和水之间反应生成 LiOH 和 H_2。金属 Li 在水溶液中的理论比能量高达 8450W·h/kg，比非水电解质的锂一次电池高很多。特别是金属 Li 在碱性溶液中有很高的溶解速度，可以制造用于水下的高效率电池体系，用于如鱼雷等。锂水电池的研究主要集中在有效地控制金属 Li 的寄生腐蚀和解决电池的安全性问题。

可充放电的 Li/H_2O 电池目前研究还较少。Kim 等借助水系锂空气电池的思路和研究经验，设计了以金属 Li 作负极，$Li_{1+x+y}Ti_{2-x}Al_xP_{3-y}Si_yO_{12}$ 作固态隔膜，以溶解在溶液中的氧气和水作为正极材料，$Sr_{0.95}Ce_{0.05}CoO_{3-\delta}$-Cu（SCCO-Cu）的纳米复合物作催化剂的 Li/H_2O 电池，总反应 $Li + \frac{1}{2}H_2O + \frac{1}{4}O_2 \longrightarrow LiOH$。这种溶解部分氧气的 Li/$H_2O$ 电池放电容量可达到 1250mA·h/g，以 SCCO-Cu 作催化剂时充放电极化电压为 0.60V，循环 25 周略有衰减。如果以 Pt/C 作为催化剂，那么极化电压可以减小至 0.49V。

10.2 锂硫电池

10.2.1 锂硫电池基本工作原理

锂硫电池是另一种可充放电的金属锂电池，以 S 为正极、金属 Li 为负极。放电时负极 Li 失去电子变为 Li^+，Li^+ 迁移至正极与 S 单质及电子反应生成硫化物。锂硫电池的理论放电电压为 2.287V，如果按照所有单质 S 均完全反应生成 Li_2S 计算硫理论比容量为 1675mA·h/g，此外以锂理论比容量为 3860mA·h/g 计算所得锂硫电池体系的理论能量密

度为2600W•h/kg。锂硫电池充放电过程如下式所示

正极：$16Li - 16e^- \longrightarrow 16Li^+$ （10-3）

负极：$S_8 + 16e^- \longrightarrow 8S^{2-}$；$8S^{2-} + 16Li^+ \longrightarrow 8Li_2S$ （10-4）

硫电极的充电和放电反应过程比较复杂，存在一系列可逆的多硫离子中间产物，其电池结构示意图和典型的充放电曲线如图 10-11 所示。硫电极的放电过程主要包括两个步骤，分别对应两个放电平台。首先是 2.4~2.1V 的放电平台，对应着 S_8 的环状结构变为 S_n^{2-}（$4 \leqslant n \leqslant 8$）的链状结构，并与 Li^+ 结合生成 Li_2S_n；第二个是 2.1~1.8V 较长的放电平台，对应 S_n^{2-} 的链状结构变为 S^{2-} 与 Li^+ 结合生成 Li_2S，该平台是锂硫电池的主要放电区域。反之充电时硫电极中 Li_2S 按此逆过程逐步被氧化为 S_8 并释放出 Li^+，Li^+ 回到负极被还原成金属 Li。

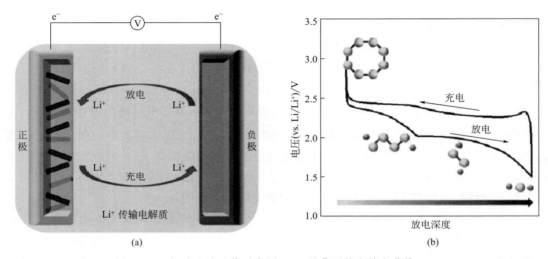

图 10-11　锂硫电池工作示意图（a）及典型的充放电曲线（b）

20 世纪 60 年代，Herbert 等首先申请专利提出了 S 作正极的想法，由于 S 自身是电子、离子的绝缘体，因此早期的研究为了提高 S 的反应活性采用了高温锂硫电池体系和有机溶剂溶解硫的方法，但都由于腐蚀性、高温下严重的自放电等原因而存在较为严重的问题，如今锂硫电池的研究重点集中在室温锂硫电池体系。

10.2.2　锂硫电池存在的基本问题

锂硫电池同样会存在着负极金属 Li 的稳定性问题，此外锂硫电池也存在其他的问题。一是由于充放电的产物 S 与 Li_2S 都是绝缘体，很难作为正极材料单独使用，因此借鉴锂离子电池的处理方法在电极制备过程中添加大量的导电添加剂炭黑或者复合另一种高电导率的材料；二是多硫离子的溶解问题与多硫离子穿梭效应，由于在充放电过程中存在着不同价态的硫离子，放电初期和充电末期产生的长链 Li_2S_n 易溶解于电解液，从而造成较大的可逆容量损失，电池的循环性受到影响。尤其是充电末期多硫离子 S_x^{2-}（$x \geqslant 4$）溶解于电解液后很容易扩散至金属 Li 负极一端并与金属 Li 发生还原反应而形成低价态的 S_y^{2-}

($4 \geqslant y \geqslant x$),产物随后又会扩散至正极表面并再度氧化成为 S_x^{2-}($x \geqslant 4$),如此反复地在正负极之间穿梭,形成了多硫离子的穿梭效应。穿梭效应使得锂硫电池充电平台延长,电池出现过充电现象,因此很多报道中锂硫电池的充电容量高于放电容量,库仑效率高于100%。

采用固态电解质可以解决金属 Li 的安全性以及多硫离子的溶解性问题,可以彻底避免因溶解带来的副反应导致的电池性能恶化。如 Scrosati 小组采用了纳米复合聚合物电解质 NCPE 作为锂硫电池的电解液,在90℃工作温度下使得可逆容量接近理论值。

此外,其他的一些方法也被用来抑制多硫离子穿梭效应,使用多孔结构的碳载体,让多硫离子尽可能限制在纳米结构的孔洞中,从而能起到一定效果的抑制作用;在电解液中加入 $LiNO_3$ 或 LiBOB 等添加剂,该添加剂能够在锂片表面形成一层致密且稳定的 SEI 膜,从而有效阻挡溶解于电解液中的多硫离子进一步与锂片反应,其中 $LiNO_3$ 作为电解液添加剂对提高锂硫电池库仑效率效果尤为明显,目前被锂硫电池研究人员广泛使用;将一些对多硫离子具有较强吸附作用的吸附剂添加在硫电极中,降低多硫离子在电解液中的溶解,常用的吸附剂包括纳米级粉体材料 $Mg_{0.6}Ni_{0.4}O$ 和 Al_2O_3 等金属氧化物,还有介孔结构的 SiO_2,但是这种方式能够改善的程度有限,尤其在低倍率下多硫离子的穿梭效应依然存在。Manthiram 等提出了在锂硫电池中间添加一个中间阻挡层,用于抑制多硫离子的扩散,避免与金属 Li 发生还原反应。中间阻挡层如果选择使用多硫离子难以通过的隔膜,势必会有效阻挡多硫离子的穿梭效应,实验中也取得了明显的效果。但是从能量密度的角度上说,添加一个中间层势必会降低锂硫电池的实际能量密度。

胡勇胜等提出了一种新型的双功能电解液 SIS(solvent in salt),在锂硫电池的使用中同时解决了多硫离子溶解性和金属 Li 负极稳定性两大问题。这种 SIS 电解液,通过大幅度提高锂盐浓度,采用传统醚类体系为溶剂。由于将大量自由溶剂分子与锂盐络合,从而可以有效地抑制多硫离子在电解液中的溶解,避免了充放电过程中多硫离子在正负极之间反复穿梭,防止了电池过度充电现象。而对于金属 Li 表面,由于 SIS 电解液具有高的阴阳离子浓度、高的 Li^+ 迁移数和高的黏度,高 Li^+ 浓度有利于金属 Li 负极的均匀物质交换,高的阴离子浓度可以防止金属 Li 表面由于阴离子耗尽所产生的空间电荷层,减小了非均匀沉积电场,此外高的黏度一定程度上阻碍了枝晶的生长,因此 SIS 电解液有效稳定了金属 Li 负极的表面。

锂硫电池需要解决的核心问题是金属 Li 保护、功能电解液、高负载硫正极技术。目前均取得了一定的进展。锂硫电池的发展目标是 500~600W·h/kg,循环 500~1000 次。实现这一目标,将有利于纯电动汽车、大规模储能等重要领域的技术发展,可显著降低电池成本,目前的研发水平离这一目标还有相当的距离。

10.3 本章结语

本章重点介绍了可充放电锂空气电池的工作原理、结构组成和所面临的挑战以及两种广义上的锂空气电池:Li/CO_2 电池和 Li/H_2O 电池。此外,本章的最后对锂硫电池进行了简要的论述。锂空气电池和锂硫电池是目前最为看好的高能量密度锂电池。高能量密度电池的开发将会导致储能成本的显著下降。

锂空气电池未来的发展需要重点解决以下问题：
① 锂空气电池反应路径的定量描述以及与电流密度的关系；
② 开发抗氧化性强的电解质体系和正极材料；
③ 理解催化剂的作用机理并开发低价有效的催化剂；
④ 设计新型的、有利于反应物输运的多孔空气正极；
⑤ 发展稳定的金属 Li 或者 Li 合金负极材料；
⑥ 研究具有高 O_2 透过率的疏水透气薄膜或者解决空气中其他气体的影响问题。

锂硫电池的发展需要重点解决金属 Li 保护、功能电解液、高负载硫正极技术。

在解决了关键的科学问题之后，还面临着工程制造问题，这些问题的解决任重而道远，但意义重大，值得更多的研究工作者付出长期不懈的努力。

本章需要进一步思考的基础科学问题：

1. 在空气中，液态醚类电解质体系，空气电极上可能有什么产物？
2. 随着空气正极不断形成固态产物覆盖在电极上，催化剂如何持续起作用？能否通过特殊的设计使其持续起作用？催化剂在气体还原反应过程中具体起什么作用？
3. 锂空气电池的动力学为何显著慢于燃料电池和锌空气电池的 ORR 反应？
4. 全固态锂空气电池中，什么样的固态电解质能在空气中工作？空气极上都可能有什么产物？产物是如何逐步演化并最终稳定的？产物形态是否显著区别于液态电解质体系？
5. 锂空气电池，如何显著提高金属锂负极的可逆性？
6. 提高锂硫电池中硫的利用率，允许硫的聚集体最大尺寸是多大？
7. 如何解决硫正极局部体积膨胀收缩引起的电子通道和离子通道不连续的问题？
8. 如何提高硫正极的反应动力学？
9. 采用液态电解质的锂硫电池全寿命周期安全性影响因素有哪些？能否全部解决？
10. 解决高面容量金属锂电极的安全性、倍率、循环性，液态电解质的锂硫电池，硫化物基或者聚合物基全固态电解质的锂硫电池，半液流的锂硫电池，哪一类更有优势？

参考文献

[1] Barber W A, Feldman A M, Fraioli A V. Composite paper electrode for a voltaic cell：US, 3551205A [P/OL]. 1970-12-29. http：//www.google.com/patents/US3551205.

[2] Moser J R, Schneider A A. Primary cells and iodine containing cathodes therefor：US, 3674562A [P/OL]. 1972-07-04. http：//www.google.com/patents/US3674562.

[3] Weast R. The Redox Potentials Against the Standard Hydrogen Electrode (SHE) for Various Reactions, Usually Called "Electrochemical Series" //Handbook of Chemistry and Physics [M]. 51st ed. Boca Raton：CRC Press Inc，1970.

[4] Rauh R D, Brummer S B. Effect of additives on lithium cycling in propylene carbonate [J]. Electrochimica Acta，1977，22 (1)：75-83.

[5] Newman G H, Francis R W, Gaines L H, et al. Hazard investigations of LiClO$_4$-dioxolane electrolyte [J]. Journal of the Electrochemical Society, 1980, 127 (9): 2025-2027.

[6] Nishi Y, Azuma H, Omaru A. Non aqueous electrolyte cell: US, 4959281A [P/OL]. 1990-09-25. http://www.google.com/patents/US4959281.

[7] Girishkumar G, Mccloskey B, Luntz A C, et al. Lithium-air battery: Promise and challenges [J]. Journal of Physical Chemistry Letters, 2010, 1 (14): 2193-2203.

[8] Abraham K M, Jiang Z. A polymer electrolyte-based rechargeable lithium/oxygen battery [J]. Journal of the Electrochemical Society, 1996, 143 (1): 1-5.

[9] Read J. Ether-based electrolytes for the lithium/oxygen organic electrolyte battery [J]. Journal of the Electrochemical Society, 2006, 153 (1): A96-A100.

[10] Ogasawara T, Debart A, Holzapfel M, et al. Rechargeable Li$_2$O$_2$ electrode for lithium batteries [J]. Journal of the American Chemical Society, 2006, 128 (4): 1390-1393.

[11] Visco S J, Katz B D, Nimon Y S, et al. Prevents the active metal from deleterious reaction with the environment on the other (cathode) side of the impervious layer, which may include aqueous or non-aqueous liquid electrolytes (catholytes) and/or a variety electrochemically active materials, including liquid, solid and gaseous oxidizers: US, 2005175894-A1 [P/OL]. 2005-08-11. http://www.google.com/patents/US20050175894.

[12] Kowalczk I, Read J, Salomon M. Li-air batteries: A classic example of limitations owing to solubilities [J]. Pure and Applied Chemistry, 2007, 79 (5): 851-860.

[13] Kumar B, Kumar J, Leese R, et al. A solid-state, rechargeable, long cycle life lithium-air battery [J]. Journal of the Electrochemical Society, 2010, 157 (1): A50-A54.

[14] Wang Y, Zhou H. A lithium-air battery with a potential to continuously reduce O$_2$ from air for delivering energy [J]. Journal of Power Sources, 2010, 195 (1): 358-361.

[15] Peng Z, Freunberger S A, Hardwick L J, et al. Oxygen reactions in a non-aqueous Li$^+$ electrolyte [J]. Angewandte Chemie-International Edition, 2011, 50 (28): 6351-6355.

[16] McCloskey B D, Scheffler R, Speidel A, et al. On the mechanism of nonaqueous LiO$_2$ electrochemistry on C and its kinetic overpotentials: Some implications for Li-air batteries [J]. Journal of Physical Chemistry C, 2012, 116 (45): 23897-23905.

[17] Laoire C O, Mukerjee S, Plichta E J, et al. Rechargeable lithium/TEGDME-LiPF$_6$/O$_2$ battery [J]. Journal of the Electrochemical Society, 2011, 158 (3): A302-A308.

[18] Lu Y C, Gallant B M, Kwabi D G, et al. Lithium-oxygen batteries: Bridging mechanistic understanding and battery performance [J]. Energy & Environmental Science, 2013, 6 (3): 750-768.

[19] Bruce P G, Freunberger S A, Hardwick L J, et al. LiO$_2$ and LiS batteries with high energy storage [J]. Nature Materials, 2012, 11 (1): 19-29.

[20] Zu C X, Li H. Thermodynamic analysis on energy densities of batteries [J]. Energy & Environmental Science, 2011, 4 (8): 2614-2624.

[21] Peng Jiayue, Zu Chenxi, Li Hong. Fundamental scientific aspects of lithium batteries (Ⅰ) —Thermodynamic calculations of theoretical energy densities of chemical energy storage systems [J]. Energy Storage Science and Technology, 2013, 2 (1): 55-62.

[22] Guo Xiangxin, Huang Shiting, Zhao Ning, et al. Rapid development and critical issues of secondary lithium-air batteries [J]. Journal of Inorganic Materials, 2014, 29 (2): 113-123.

[23] Lu J, Li L, Park J B, et al. Aprotic and aqueous LiO$_2$ batteries [J]. Chemical Reviews, 2014, 114 (11): 5611-5640.

[24] Xu K. Nonaqueous liquid electrolytes for lithium-based rechargeable batteries [J]. Chem Rev, 2004, 104 (10): 4303-4317.

[25] Aurbach D, Gofer Y, Langzam J. The correlation between surface-chemistry, surface-morphology, and cycling efficiency of lithium electrodes in a few polar aprotic systems [J]. Journal of the Electrochemical Society, 1989, 136 (11): 3198-3205.

[26] Aurbach D, Daroux M, Faguy P, et al. The electrochemistry of noble-metal electrodes in aprotic organic-solvents containing lithium-salts [J]. Journal of Electroanalytical Chemistry, 1991, 297 (1): 225-244.

[27] Mizuno F, Nakanishi S, Kotani Y, et al. Rechargeable Li-air batteries with carbonate-based liquid electrolytes [J]. Electrochemistry, 2010, 78 (5): 403-405.

[28] Zhang Z, Lu J, Assary R S, et al. Increased stability toward oxygen reduction products for lithium-air batteries with oligoether-functionalized silane electrolytes [J]. The Journal of Physical Chemistry C, 2011, 115 (51): 25535-25542.

[29] Xu W, Xu K, Viswanathan V V, et al. Reaction mechanisms for the limited reversibility of LiO_2 chemistry in organic carbonate electrolytes [J]. Journal of Power Sources, 2011, 196 (22): 9631-9639.

[30] Xu W, Viswanathan V V, Wang D, et al. Investigation on the charging process of Li_2O_2-based air electrodes in LiO_2 batteries with organic carbonate electrolytes [J]. Journal of Power Sources, 2011, 196 (8): 3894-3899.

[31] McCloskey B D, Bethune D S, Shelby R M, et al. Solvents' critical role in nonaqueous lithium-oxygen battery electrochemistry [J]. The Journal of Physical Chemistry Letters, 2011, 2 (10): 1161-1166.

[32] Freunberger S A, Chen Y, Peng Z, et al. Reactions in the rechargeable LiO_2 battery with alkyl carbonate electrolytes [J]. Journal of the American Chemical Society, 2011, 133 (20): 8040-8047.

[33] Bryantsev V S, Giordani V, Walker W, et al. Predicting solvent stability in aprotic electrolyte Li-air batteries: Nucleophilic substitution by the superoxide anion radical ($O_2^{\cdot-}$) [J]. Journal of Physical Chemistry A, 2011, 115 (44): 12399-12409.

[34] Bryantsev V S, Blanco M. Computational study of the mechanisms of superoxide-induced decomposition of organic carbonate-based electrolytes [J]. The Journal of Physical Chemistry Letters, 2011, 2 (5): 379-383.

[35] Peng Z, Freunberger S A, Hardwick L J, et al. Oxygen reactions in a non-aqueous Li^+ electrolyte [J]. Angewandte Chemie International Edition, 2011, 50 (28): 6351-6355.

[36] Xu D, Wang Z L, Xu J J, et al. Novel DMSO-based electrolyte for high performance rechargeable LiO_2 batteries [J]. Chemical Communications, 2012, 48 (55): 6948-6950.

[37] Lu Y C, Kwabi D G, Yao K P C, et al. The discharge rate capability of rechargeable LiO_2 batteries [J]. Energy & Environmental Science, 2011, 4 (8): 2999-3007.

[38] Xu W, Hu J, Engelhard M H, et al. The stability of organic solvents and carbon electrode in nonaqueous LiO_2 batteries [J]. Journal of Power Sources, 2012, 215: 240-247.

[39] Assary R S, Curtiss L A, Redfern P C, et al. Computational studies of polysiloxanes: Oxidation potentials and decomposition reactions [J]. The Journal of Physical Chemistry C, 2011, 115 (24): 12216-12223.

[40] Assary R S, Lau K C, Amine K, et al. Interactions of dimethoxy ethane with Li_2O_2 clusters and likely decomposition mechanisms for LiO_2 batteries [J]. The Journal of Physical Chemistry C, 2013, 117 (16): 8041-8049.

[41] Bryantsev V S, Faglioni F. Predicting autoxidation stability of ether-and amide-based electrolyte solvents for Li-air batteries [J]. The Journal of Physical Chemistry A, 2012, 116 (26): 7128-7138.

[42] Hsu C W, Chen P, Ting J M. Microwave-assisted hydrothermal synthesis of TiO_2 mesoporous beads having C and/or N doping for use in high efficiency all-plastic flexible dye-sensitized solar cells [J]. Journal of the Electrochemical Society, 2013, 160 (3): H160-H165.

[43] Mccloskey B D, Scheffler R, Speidel A, et al. On the efficacy of electrocatalysis in nonaqueous LiO_2 batteries [J]. Journal of the American Chemical Society, 2011, 133 (45): 18038-18041.

[44] McCloskey B D, Bethune D S, Shelby R M, et al. Limitations in rechargeability of LiO_2 batteries and possible origins [J]. Journal of Physical Chemistry Letters, 2012, 3 (20): 3043-3047.

[45] Xu W, Xiao J, Zhang J, et al. Optimization of nonaqueous electrolytes for primary lithium/air batteries operated in ambient environment [J]. Journal of the Electrochemical Society, 2009, 156 (10): A773-A779.

[46] Black R, Oh S H, Lee J H, et al. Screening for superoxide reactivity in LiO_2 batteries: Effect on $Li_2O_2/LiOH$ crystallization [J]. Journal of the American Chemical Society, 2012, 134 (6): 2902-2905.

[47] Jung H G, Hassoun J, Park J B, et al. An improved high-performance lithium-air battery [J]. Nature Chemistry, 2012, 4 (7): 579-585.

[48] Shui J L, Okasinski J S, Kenesei P, et al. Reversibility of anodic lithium in rechargeable lithium-oxygen batteries [J]. Nature Communications, 2013, 4: 2255.

[49] Zhang L, Zhang S, Zhang K, et al. Mesoporous $NiCo_2O_4$ nanoflakes as electrocatalysts for rechargeable LiO_2 batteries [J]. Chemical Communications, 2013, 49 (34): 3540-3542.

[50] Xu J J, Xu D, Wang Z L, et al. Synthesis of perovskite-based porous $La_{0.75}Sr_{0.25}MnO_3$ nanotubes as a highly efficient electrocatalyst for rechargeable lithium-oxygen batteries [J]. Angewandte Chemie International Edition, 2013, 52 (14): 3887-3890.

[51] Lei Y, Lu J, Luo X, et al. Synthesis of porous carbon supported palladium nanoparticle catalysts by atomic layer

deposition: Application for rechargeable LiO_2 battery [J]. Nano Letters, 2013, 13 (9): 4182-4189.

[52] Freunberger S A, Chen Y, Drewett N E, et al. The lithium-oxygen battery with ether-based electrolytes [J]. Angewandte Chemie International Edition, 2011, 50 (37): 8609-8613.

[53] Jung H G, Kim H S, Park J B, et al. A transmission electron microscopy study of the electrochemical process of lithium-oxygen cells [J]. Nano Letters, 2012, 12 (8): 4333-4335.

[54] Laoire C O, Mukerjee S, Abraham K M, et al. Influence of nonaqueous solvents on the electrochemistry of oxygen in the rechargeable lithium-air battery [J]. The Journal of Physical Chemistry C, 2010, 114 (19): 9178-9186.

[55] Lopez N, Graham D J, Mcguire R, et al. Reversible reduction of oxygen to peroxide facilitated by molecular recognition [J]. Science, 2012, 335 (6067): 450-453.

[56] Sun B, Zhang J, Munroe P, et al. Hierarchical NiCo$_2$O$_4$ nanorods as an efficient cathode catalyst for rechargeable non-aqueous LiO_2 batteries [J]. Electrochemistry Communications, 2013, 31: 88-91.

[57] Trahan M J, Mukerjee S, Plichta E J, et al. Studies of Li-air cells utilizing dimethyl sulfoxide-based electrolyte [J]. Journal of the Electrochemical Society, 2013, 160 (2): A259-A267.

[58] Chen Y, Freunberger S A, Peng Z, et al. LiO_2 battery with a dimethylformamide electrolyte [J]. Journal of the American Chemical Society, 2012, 134 (18): 7952-7957.

[59] Walker W, GiordaniV, Uddin J, et al. A rechargeable LiO_2 battery using a lithium nitrate/N,N-dimethylacetamide electrolyte [J]. Journal of the American Chemical Society, 2013, 135 (6): 2076-2079.

[60] Allen C J, Mukerjee S, Plichta E J, et al. Oxygen electrode rechargeability in an ionic liquid for the Li-air battery [J]. The Journal of Physical Chemistry Letters, 2011, 2 (19): 2420-2424.

[61] Peng Z, Freunberger S A, Chen Y, et al. A reversible and higher-rate LiO_2 battery [J]. Science, 2012, 337 (6094): 563-566.

[62] Cecchetto L, Salomon M, Scrosati B, et al. Study of a Li-air battery having an electrolyte solution formed by a mixture of an ether-based aprotic solvent and an ionic liquid [J]. Journal of Power Sources, 2012, 213: 233-238.

[63] Xie B, Lee H S, Li H, et al. New electrolytes using Li$_2$O or Li$_2$O$_2$ oxides and tris (pentafluorophenyl) borane as boron based anion receptor for lithium batteries [J]. Electrochemistry Communications, 2008, 10 (8): 1195-1197.

[64] Zheng D, Lee H S, Yang X Q, et al. Electrochemical oxidation of solid Li$_2$O$_2$ in non-aqueous electrolyte using peroxide complexing additives for lithium-air batteries [J]. Electrochemistry Communications, 2013, 28: 17-19.

[65] Xu W, Xiao J, Wang D, et al. Effects of nonaqueous electrolytes on the performance of lithium/air batteries [J]. Journal of the Electrochemical Society, 2010, 157 (2): A219-A224.

[66] Chen Y, Freunberger S A, Peng Z, et al. Charging a LiO_2 battery using a redox mediator [J]. Nature Chemistry, 2013, 5 (6): 489-494.

[67] Lim H D, Song H, Kim J, et al. Superior rechargeability and efficiency of lithium-oxygen batteries: Hierarchical air electrode architecture combined with a soluble catalyst [J]. Angewandte Chemie, 2014, 53 (15): 3926-3931.

[68] Xiao J, Mei D, Li X, et al. Hierarchically porous graphene as a lithium-air battery electrode [J]. Nano Letters, 2011, 11 (11): 5071-5078.

[69] Zhang S S, Foster D, Read J. Discharge characteristic of a non-aqueous electrolyte Li/O_2 battery [J]. Journal of Power Sources, 2010, 195 (4): 1235-1240.

[70] Xiao J, Wang D, Xu W, et al. Optimization of air electrode for Li/air batteries [J]. Journal of the Electrochemical Society, 2010, 157 (4): A487-A492.

[71] Yang X H, He P, Xia Y Y. Preparation of mesocellular carbon foam and its application for lithium/oxygen battery [J]. Electrochemistry Communications, 2009, 11 (6): 1127-1130.

[72] Kuboki T, Okuyama T, Ohsaki T, et al. Lithium-air batteries using hydrophobic room temperature ionic liquid electrolyte [J]. Journal of Power Sources, 2005, 146 (1-2): 766-769.

[73] Qin Y, Lu J, Du P, et al. In situ fabrication of porous-carbon-supported α-MnO$_2$ nanorods at room temperature: Application for rechargeable LiO_2 batteries [J]. Energy & Environmental Science, 2013, 6 (2): 519-531.

[74] Zhang J G, Wang D, Xu W, et al. Ambient operation of Li/air batteries [J]. Journal of Power Sources, 2010, 195 (13): 4332-4337.

[75] Mitchell R R, Gallant B M, Thompson CV, et al. All-carbon-nanofiber electrodes for high-energy rechargeable

LiO₂ batteries [J]. Energy & Environmental Science, 2011, 4 (8): 2952-2958.

[76] Mitchell R R, Gallant B M, Shao-Horn Y, et al. Mechanisms of morphological evolution of Li_2O_2 particles during electrochemical growth [J]. The Journal of Physical Chemistry Letters, 2013, 4 (7): 1060-1064.

[77] Gallant B M, Mitchell R R, Kwabi D G, et al. Chemical and morphological changes of Li/O₂ battery electrodes upon cycling [J]. The Journal of Physical Chemistry C, 2012, 116 (39): 20800-20805.

[78] Yoo E, Zhou H. Li-air rechargeable battery based on metal-free graphene nanosheet catalysts [J]. ACS Nano, 2011, 5 (4): 3020-3026.

[79] Zheng H, Xiao D, Li X, et al. New insight in understanding oxygen reduction and evolution in solid-state lithium-oxygen batteries using an in situ environmental scanning electron microscope [J]. Nano Letters, 2014, 14 (8): 4245-4249.

[80] Zhong L, Mitchell R R, Liu Y, et al. In situ transmission electron microscopy observations of electrochemical oxidation of Li_2O_2 [J]. Nano Letters, 2013, 13 (5): 2209-2214.

[81] Wang Z L, Xu D, Xu J J, et al. Graphene oxide gel-derived, free-standing, hierarchically porous carbon for high-capacity and high-rate rechargeable Li/O₂ batteries [J]. Advanced Functional Materials, 2012, 22 (17): 3699-3705.

[82] Gallant B M, Kwabi D G, Mitchell R R, et al. Influence of Li_2O_2 morphology on oxygen reduction and evolution kinetics in Li/O₂ batteries [J]. Energy & Environmental Science, 2013, 6 (8): 2518-2528.

[83] Ottakam T M, Freunberger S A, Peng Z, et al. The carbon electrode in nonaqueous Li/O₂ cells [J]. Journal of the American Chemical Society, 2013, 135 (1): 494-500.

[84] Ottakam T M, Freunberger S A, Peng Z, et al. A stable cathode for the aprotic LiO₂ battery [J]. Nat Mater, 2013, 12 (11): 1050-1056.

[85] Dong S, Chen X, Wang S, et al. 1D coaxial platinum/titanium nitride nanotube arrays with enhanced electrocatalytic activity for the oxygen reduction reaction: Towards Li-air batteries [J]. Chem Sus Chem, 2012, 5 (9): 1712-1715.

[86] Li F, Tang D M, Chen Y, et al. Ru/ITO: A carbon-free cathode for nonaqueous LiO₂ battery [J]. Nano Letters, 2013, 13 (10): 4702-4707.

[87] Jian Z, Liu P, Li F, et al. Core-shell-structured CNT@RuO₂ composite as a high-performance cathode catalyst for rechargeable LiO₂ batteries [J]. Angewandte Chemie: International Edition, 2014, 53 (2): 442-426.

[88] Kichambare P, Kumar J, Rodrigues S, et al. Electrochemical performance of highly mesoporous nitrogen doped carbon cathode in lithium-oxygen batteries [J]. Journal of Power Sources, 2011, 196 (6): 3310-3316.

[89] Kichambare P, Rodrigues S, Kumar J. Mesoporous nitrogen-doped carbon-glass ceramic cathodes for solid-state lithium-oxygen batteries [J]. ACS Applied Materials & Interfaces, 2012, 4 (1): 49-52.

[90] Cheng F, Liang J, Tao Z, et al. Functional materials for rechargeable batteries [J]. Advanced Materials, 2011, 23 (15): 1695-1715.

[91] Cheng F, Chen J. Metal-air batteries: From oxygen reduction electrochemistry to cathode catalysts [J]. Chemical Society Reviews, 2012, 41 (6): 2172-2192.

[92] Debart A, Bao J, Armstrong G, et al. An O₂ cathode for rechargeable lithium batteries: The effect of a catalyst [J]. Journal of Power Sources, 2007, 174 (2): 1177-1182.

[93] Debart A, Paterson A J, Bao J, et al. α-MnO₂ nanowires: A catalyst for the O₂ electrode in rechargeable lithium batteries [J]. Angewandte Chemie: International Edition, 2008, 47 (24): 4521-4524.

[94] Giordani V, Freunberger S A, Bruce P G, et al. H_2O_2 decomposition reaction as selecting tool for catalysts in LiO₂ cells [J]. Electrochemical and Solid State Letters, 2010, 13 (12): A180-A183.

[95] Miyazaki K, Kawakita K I, Abe T, et al. Single-step synthesis of nano-sized perovskite-type oxide/carbon nanotube composites and their electrocatalytic oxygen-reduction activities [J]. Journal of Materials Chemistry, 2011, 21 (6): 1913-1917.

[96] Yuasa M, Nishida M, Kida T, et al. Bi-functional oxygen electrodes using $LaMnO_3/LaNiO_3$ for rechargeable metal-air batteries [J]. Journal of the Electrochemical Society, 2011, 158 (5): A605-A610.

[97] Fu Z, Lin X, Huang T, et al. Nano-sized $La_{0.8}Sr_{0.2}MnO_3$ as oxygen reduction catalyst in nonaqueous Li/O₂ batteries [J]. J Solid State Electrochem, 2012, 16 (4): 1447-1452.

[98] Suntivich J, Gasteiger H A, Yabuuchi N, et al. Design principles for oxygen-reduction activity on perovskite oxide catalysts for fuel cells and metal-air batteries [J]. Nature Chemistry, 2011, 3 (7): 546-550.

[99] Chen Z, Higgins D, Yu A, et al. A review on non-precious metal electrocatalysts for PEM fuel cells [J]. Energy & Environmental Science, 2011, 4 (9): 3167-3192.

[100] Li F, Ohnishi R, Yamada Y, et al. Carbon supported TiN nanoparticles: An efficient bifunctional catalyst for non-aqueous Li/O_2 batteries [J]. Chemical Communications, 2013, 49 (12): 1175-1177.

[101] Kitaura H, Zhou H. Electrochemical performance of solid-state lithium-air batteries using carbon nanotube catalyst in the air electrode [J]. Advanced Energy Materials, 2012, 2 (7): 889-894.

[102] Shui J L, Karan N K, Balasubramanian M, et al. Fe/N/C composite in LiO_2 battery: Studies of catalytic structure and activity toward oxygen evolution reaction [J]. Journal of the American Chemical Society, 2012, 134 (40): 16654-16661.

[103] Lu Y C, Gasteiger H A, Parent M C, et al. The influence of catalysts on discharge and charge voltages of rechargeable Li-oxygen batteries [J]. Electrochemical and Solid State Letters, 2010, 13 (6): A69-A72.

[104] Lu Y C, Xu Z, Gasteiger H A, et al. Platinum-gold nanoparticles: A highly active bifunctional electrocatalyst for rechargeable lithium-air batteries [J]. Journal of the American Chemical Society, 2010, 132 (35): 12170-12171.

[105] Li F, Chen Y, Zhou H, et al. Performance-improved LiO_2 battery with Ru nanoparticles supported on binder-free multiwalled carbon nanotube paper as cathode [J]. Energy & Environmental Science, 2014, 7: 1648-1652.

[106] Xu J J, Wang Z L, Xu D, et al. Tailoring deposition and morphology of discharge products towards high-rate and long-life lithium-oxygen batteries [J]. Nature Communications, 2013, 4: 24-38.

[107] Ren X, Zhang S S, Tran D T, et al. Oxygen reduction reaction catalyst on lithium/air battery discharge performance [J]. Journal of Materials Chemistry, 2011, 21 (27): 10118-10125.

[108] Mccloskey B D, Scheffler R, Speidel A, et al. On the efficacy of electrocatalysis in nonaqueous Li/O_2 batteries [J]. Journal of the American Chemical Society, 2011, 133 (45): 18038-18041.

[109] Imanishi N, Hasegawa S, Zhang T, et al. Lithium anode for lithium-air secondary batteries [J]. Journal of Power Sources, 2008, 185 (2): 1392-1397.

[110] Hassoun J, Jung H G, Lee D J, et al. A metal-free, lithium-ion oxygen battery: A step forward to safety in lithium-air batteries [J]. Nano Letters, 2012, 12 (11): 5775-5779.

[111] Aurbach D. Review of selected electrode-solution interactions which determine the performance of Li and Li ion batteries [J]. Journal of Power Sources, 2000, 89 (2): 206-218.

[112] Cohen Y S, Cohen Y, Aurbach D. Micromorphological studies of lithium electrodes in alkyl carbonate solutions using in situ atomic force microscopy [J]. The Journal of Physical Chemistry B, 2000, 104 (51): 12282-12291.

[113] Choi N S, Lee Y M, Cho K Y, et al. Protective layer with oligo (ethylene glycol) borate anion receptor for lithium metal electrode stabilization [J]. Electrochemistry Communications, 2004, 6 (12): 1238-1242.

[114] Singh M, Gur I, Eitouni H B, et al. Solid Electrolyte material manufacturable by polymer processing methods: US, 20090075176A1 [P/OL]. 2009-03-19. http://www.google.com/patents/US20090075176.

[115] Visco S J, Katz B D, Nimon Y S, et al. Protected active metal electrode and battery cell structures with non-aqueous interlayer architecture: CN, 100568613C [P/OL]. 2009-12-09. http://www.google.com/patents/CN100568613C? cl=en.

[116] Zhang D, Li R, Huang T, et al. Novel composite polymer electrolyte for lithium air batteries [J]. Journal of Power Sources, 2010, 195 (4): 1202-1206.

[117] Baker R W. Future directions of membrane gas separation technology [J]. Industrial & Engineering Chemistry Research, 2002, 41 (6): 1393-1411.

[118] Zhang J, Xu W, Liu W. Oxygen-selective immobilized liquid membranes for operation of lithium-air batteries in ambient air [J]. Journal of Power Sources, 2010, 195 (21): 7438-7444.

[119] Zhang J, Xu W, Li X, et al. Air dehydration membranes for nonaqueous lithium-air batteries [J]. Journal of the Electrochemical Society, 2010, 157 (8): A940-A946.

[120] Crowther O, Keeny D, Moureau D M, et al. Electrolyte optimization for the primary lithium metal air battery using an oxygen selective membrane [J]. Journal of Power Sources, 2012, 202: 347-351.

[121] Fu Z, Wei Z, Lin X, et al. Polyaniline membranes as waterproof barriers for lithium air batteries [J]. Electrochimica Acta, 2012, 78: 195-199.

[122] Lau K C, Curtiss L A, Greeley J. Density functional investigation of the thermodynamic stability of lithium ox-

ide bulk crystalline structures as a function of oxygen pressure [J]. Journal of Physical Chemistry C, 2011, 115 (47): 23625-23633.

[123] Radin M D, Rodriguez J F, Tian F, et al. Lithium peroxide surfaces are metallic, while lithium oxide surfaces are not [J]. Journal of the American Chemical Society, 2011, 134 (2): 1093-1103.

[124] Gerbig O, Merkle R, Maier J. Electron and ion transport in Li_2O_2 [J]. Advanced Materials, 2013, 25 (22): 3129-3133.

[125] Garcia-Lastra J M, Bass J D, Thygesen K S. Communication: Strong excitonic and vibronic effects determine the optical properties of Li_2O_2 [J]. Journal of Chemical Physics, 2011, 135 (12): 121101.

[126] Chen J, Hummelsh J J S, Thygesen K S, et al. The role of transition metal interfaces on the electronic transport in lithium-air batteries [J]. Catalysis Today, 2011, 165 (1): 2-9.

[127] Ong S P, Mo Y, Ceder G. Low hole polaron migration barrier in lithium peroxide [J]. Physical Review B, 2012, 85 (8): 81105.

[128] Hummelsh J J S, Blomqvist J, Datta S, et al. Communications: Elementary oxygen electrode reactions in the aprotic Li-air battery [J]. The Journal of Chemical Physics, 2010, 132 (7): 71101.

[129] Lau K C, Assary R S, Redfern P, et al. Electronic structure of lithium peroxide clusters and relevance to lithium-air batteries [J]. The Journal of Physical Chemistry C, 2012, 116 (45): 23890-23896.

[130] Lu J, Jung H J, Lau K C, et al. Magnetism in lithium-oxygen discharge product [J]. Chem Sus Chem, 2013, 6 (7): 1196-1202.

[131] Yang J, Zhai D, Wang H H, et al. Evidence for lithium superoxide-like species in the discharge product of a Li/O_2 battery [J]. Physical Chemistry Chemical Physics, 2013, 15 (11): 3764-3771.

[132] Aetukurin B, McCloskey B D, Garc A J M, et al. On the origin and implications of Li_2O_2 toroid formation in nonaqueous LiO_2 batteries [EB/OL]. 2014-07-12. http://arxiv.org/abs/1406.3335.

[133] Zhang T, Zhou H. A reversible long-life lithium-air battery in ambient air [J]. Nature Communications, 2013, 4: 1817.

[134] Zhang T, Zhou H. From LiO_2 to Li-air batteries: Carbon nanotubes/ionic liquid gels with a tricontinuous passage of electrons, ions, and oxygen [J]. Angewandte Chemie-International Edition, 2012, 51 (44): 11062-11067.

[135] Takechi K, Shiga T, Asaoka T. A LiO_2/CO_2 battery [J]. Chemical Communications, 2011, 47 (12): 3463-3465.

[136] Gowda S R, Brunet A, Wallraff G M, et al. Implications of CO_2 contamination in rechargeable nonaqueous LiO_2 batteries [J]. Journal of Physical Chemistry Letters, 2013, 4 (2): 276-279.

[137] McCloskey B D, Speidel A, Scheffler R, et al. Twin problems of interfacial carbonate formation in nonaqueous LiO_2 batteries [J]. Journal of Physical Chemistry Letters, 2012, 3 (8): 997-1001.

[138] McCloskey B D, Bethune D S, Shelby R M, et al. Solvents' critical role in nonaqueous lithium-oxygen battery electrochemistry [J]. Journal of Physical Chemistry Letters, 2011, 2 (10): 1161-1166.

[139] Wang R, Yu X, Bai J, et al. Electrochemical decomposition of Li_2CO_3 in NiO-Li_2CO_3 nanocomposite thin film and powder electrodes [J]. Journal of Power Sources, 2012, 218: 113-118.

[140] Liu Y, Wang R, Lyu Y, et al. Rechargeable Li/CO_2-O_2 (2:1) battery and Li/CO_2 battery [J]. Energy & Environmental Science, 2014, 7 (2): 677-681.

[141] Pensado-Rodriguez O, Urquidi-Macdonald M, Macdonald D D. Electrochemical behavior of lithium in alkaline aqueous electrolytes - I. Thermodynamics [J]. Journal of the Electrochemical Society, 1999, 146 (4): 1318-1325.

[142] Urquidi-Macdonald M, Macdonald D D, Pensado O, et al. The electrochemical behavior of lithium in alkaline aqueous electrolytes [J]. Electrochimica Acta, 2001, 47 (5): 833-840.

[143] Kemp D D L E L, Momye W R, et al. Proceeding of the 11th IECECC [C]//Lake Tahoe, 1976.

[144] Ding Fei, Zhang Jing, Yang Kai, et al. Study on lithium electrode corrosion in KOH aqueous electrolytes (I) [J]. Chin J Power Sources, 2008, 32 (2): 91-94.

[145] Kim J K, Yang W, Salim J, et al. Li-water battery with oxygen dissolved in water as a cathode [J]. Journal of the Electrochemical Society, 2013, 161 (3): A285-A289.

[146] Danuta H, Juliusz U. Electric dry cells and storage batteries: US, 3043896A [P/OL]. 1962-07-10. http://www.google.com/patents/US3043896.

[147] Yao N P, HeredY L A, Saunders R C. Secondary lithium-sulfur battery [J]. Journal of the Electrochemical So-

ciety, 1970, 117 (8): 8.

[148] Rauh R D, Abraham K M, Pearson G F, et al. Lithium-dissolved sulfur battery with an organic electrolyte [J]. Journal of the Electrochemical Society, 1979, 126 (4): 523-527.

[149] Evers S, Nazar L F. New approaches for high energy density lithium-sulfur battery cathodes [J]. Accounts of Chemical Research, 2013, 46 (5): 1135-1143.

[150] Xiong S, Xie K, Diao Y, et al. Oxidation process of polysulfides in charge process for lithium-sulfur batteries [J]. Ionics, 2012, 18 (9): 867-872.

[151] KolosnitsynV S, Karaseva EV. Lithium-sulfur batteries: Problems and solutions [J]. Russ. J. Electrochem, 2008, 44 (5): 506-509.

[152] Mikhaylik YV, Akridge J R. Polysulfide shuttle study in the Li/S battery system [J]. Journal of the Electrochemical Society, 2004, 151 (11): A1969-A1976.

[153] Kumaresan K, Mikhaylik Y, White R E. A mathematical model for a lithium-sulfur cell [J]. Journal of the Electrochemical Society, 2008, 155 (8): A576-A582.

[154] Shin E S, Kim K, Oh S H, et al. Polysulfide dissolution control: The common ion effect [J]. Chemical Communications, 2013, 49 (20): 2004-2006.

[155] Zhang S S. Liquid electrolyte lithium/sulfur battery: Fundamental chemistry, problems, and solutions [J]. Journal of Power Sources, 2013, 231: 153-162.

[156] Aurbach D, Pollak E, Elazari R, et al. On the surface chemical aspects of very high energy density, rechargeable Li/S batteries [J]. Journal of the Electrochemical Society, 2009, 156 (8): A694-A702.

[157] Zhang S S. Role of $LiNO_3$ in rechargeable lithium/sulfur battery [J]. Electrochimica Acta, 2012, 70: 344-348.

[158] Xiong S, Xie K, Diao Y, et al. Properties of surface film on lithium anode with $LiNO_3$ as lithium salt in electrolyte solution for lithium-sulfur batteries [J]. Electrochimica Acta, 2012, 83: 78-86.

[159] Liang X, Wen Z, Liu Y, et al. Improved cycling performances of lithium sulfur batteries with $LiNO_3$-modified electrolyte [J]. Journal of Power Sources, 2011, 196 (22): 9839-9843.

[160] Xiong S, Kai X, Hong X, et al. Effect of LiBOB as additive on electrochemical properties of lithium-sulfur batteries [J]. Ionics, 2012, 18 (3): 249-254.

[161] Song M S, Han S C, Kim H S, et al. Effects of nanosized adsorbing material on electrochemical properties of sulfur cathodes for Li/S secondary batteries [J]. Journal of the Electrochemical Society, 2004, 151 (6): A791-A795.

[162] Zhang Y, Zhao Y, Yermukhambetova A, et al. Ternary sulfur/polyacrylonitrile/$Mg_{0.6}Ni_{0.4}O$ composite cathodes for high performance lithium/sulfur batteries [J]. Journal of Materials Chemistry A, 2013, 1 (2): 295-301.

[163] Choi Y J, Jung B S, Lee D J, et al. Electrochemical properties of sulfur electrode containing nano Al_2O_3 for lithium/sulfur cell [J]. Physica Scripta, 2007, 2007 (T129): 62-75.

[164] Ji X, Evers S, Black R, et al. Stabilizing lithium-sulphur cathodes using polysulphide reservoirs [J]. Nat Commun, 2011, 2: 325.

[165] Su Y S, Manthiram A. Lithium-sulphur batteries with a microporous carbon paper as a bifunctional interlayer [J]. Nat Commun, 2012, 3: 1166.

[166] Zu C, Su Y S, Fu Y, et al. Improved lithium-sulfur cells with a treated carbon paper interlayer [J]. Physical Chemistry Chemical Physics, 2013, 15 (7): 2291-2297.

[167] Su Y S, Manthiram A. A new approach to improve cycle performance of rechargeable lithium-sulfur batteries by inserting a free-standing MWCNT interlayer [J]. Chemical Communications, 2012, 48 (70): 8817-8819.

[168] Suo L, Hu Y S, Li H, et al. A new class of solvent-in-salt electrolyte for high-energy rechargeable metallic lithium batteries [J]. Nat Commun, 2013, 4: 1481.

第 11 章
表征方法

锂离子电池材料和器件的实验研究方法主要包括表征技术和电化学测量两部分。准确和全面地理解电池材料的构效关系需要综合运用多种实验技术。以正极材料的研究为例，图 11-1 小结了正极材料性质与性能之间可能存在的复杂多对多关系。电池材料关心的主要性质包括结构方面和动力学方面，均与材料的组成与微结构密切相关，对电池的综合性能有复杂的影响。每一项性能可能与材料的多种性质有关，每一类性质也可能影响多项性能，具体问题需要具体分析，没有特别统一的规律，这给电池的精细化研究带来了很大的挑战。

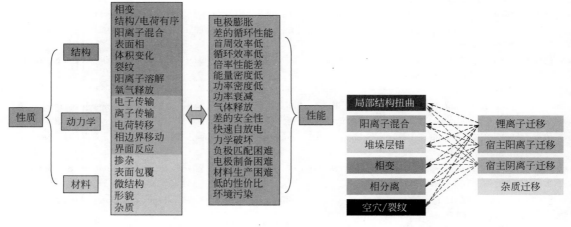

图 11-1 锂离子电池正极材料研究中性质与性能之间可能存在的复杂多对多关系

图 11-2 正极材料中由于离子迁移可能引起的结构问题

以结构演化为例，锂离子电池的正极材料通常是含有锂元素以及可变价过渡金属元素的材料，如六方层状结构 $LiCoO_2$、正交橄榄石结构 $LiFePO_4$、立方尖晶石结构 $LiMn_2O_4$ 等。在锂离子脱嵌过程中，可能带动其他骨架阳离子、阴离子、掺杂原子的迁移，进而对材料结构产生影响。如图 11-2 所示，离子的迁移可能会造成宿主材料局域晶格的扭曲、阳离子混占位、堆垛层错、相变、相分离、出现空穴与裂纹等情况。结构演化的过程还受电池充放电制度、倍率、温度、电解液、电极结构等因素影响。因此，锂离子电池正极材料结构演化的研究，需要十分仔细、系统地设计实验，否则容易得出片面的结论。

负极材料的研究同样面临复杂的情况。石墨负极已经广泛应用在锂离子电池中，但一些基本的问题，如膨胀、阶结构的形成、不对称动力学特性、溶剂共嵌入、表面SEI膜等问题，仍不是非常清楚。高容量硅负极材料需要考虑的问题更多。如图11-3所示，列出了硅负极材料研究一般需要考虑的性质与性能问题。硅负极主要存在体积形变很大，导致裂纹的产生，界面反应，黏结剂、电解液添加剂的选择等问题，这直接影响了硅负极的循环性能、首周效率、循环效率、倍率性能、功率、安全性等性能。电解液、隔膜的许多性质也对电池性能有重要影响，在此不再赘述。

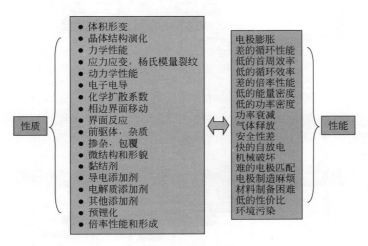

图11-3 硅负极材料研究的性质与性能的关系

液态锂离子电池中存在正极材料与电解液、负极材料与电解液的固液界面。电池中表、界面的性质与电池最后表现出来的性能息息相关，如图11-4所示。材料表面的晶体结构与体相不同，由于悬挂键、表面相的存在导致表面活性很高，与液体接触的界面可能发生很多副反应，影响电池的性能。另外，电池服役过程中，表面组成、结构发生演化，复杂溶剂、溶质在电极材料表面竞争吸附、发生电荷转移，在表、界面形成稳定或不稳定的新相，骨架原子溶出，都影响到材料的功率密度、循环寿命、服役寿命、自放电率、安全特性、温度特性等性能。事实上，锂离子电池电极材料中结构问题和表、界面问题是电池需要研究的核心问题。

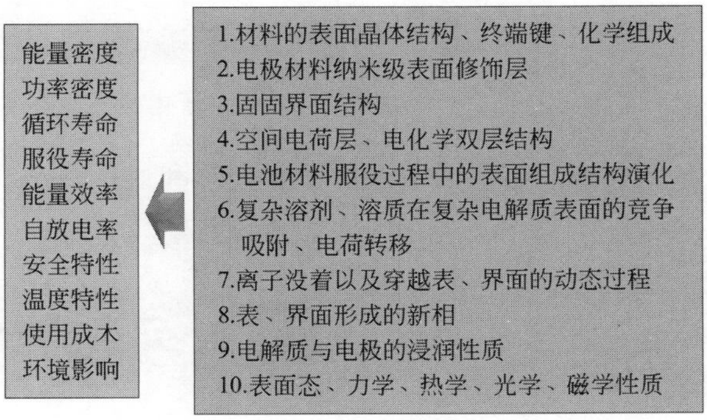

锂离子电池中涉及的动力学过程在空间尺度上大致如图11-5所示。有的过程在多个尺度体现，有的过程在较小的尺度发生，大体包括结构、输运、化学与电化学反应过程。这些现象和反应的时间尺度也是从皮秒到秒再到小时的跨度。

图11-4 电池中表、界面问题对电池性能的影响图示

既然Li^+的动力学过程涉及亚埃至毫米量级的空间跨度，对它进行研究时，也需要使

锂电池基础科学

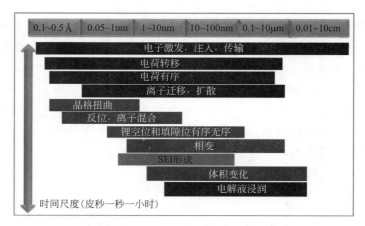

图 11-5　锂离子电池研究过程中空间尺度的跨度

用具有相应空间分辨、时间分辨、能量分辨能力的设备或表征技术。现有的锂离子电池实验技术的时间、空间、能量分辨能力大体如图 11-6 所示。研究者可以采用不同的表征技术，从不同尺度针对不同的问题研究锂离子电池中存在的现象及物理化学过程。

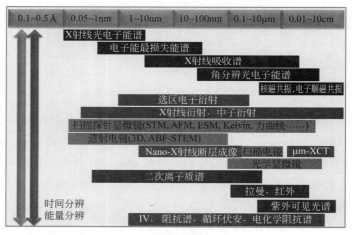

图 11-6　锂离子电池实验技术的空间分辨分布图

以下按照分析内容进行简要介绍和讨论。

11.1　元素成分及价态

材料成分的表征主要有能量散射 X 射线谱（EDX）、二次离子质谱（SIMS）、电感耦合等离子体（ICP）、X 射线荧光光谱仪（XRF），其中 SIMS 可以分析元素的深度分布且具有高灵敏度。元素价态的表征主要有 X 射线光电子谱（XPS）、电子能量损失谱（EELS）、扫描透射 X 射线显微术（STXM）、X 射线近边结构谱（XANES）等。由于价态变化导致材料的磁性变化,因此通过测量磁化率、顺磁共振(ESP)、核磁共振(NMR)也可以间接获得材料中元素价态变化的信息。若含 Fe、Sn 元素,还可以通过穆斯鲍尔(Mössbauer)谱来研究。杂质测量也有专门的分析技术。下文将介绍几类主要表征技术在锂电池领域的应用。

11.1.1 电感耦合等离子体

电感耦合等离子体(inductive coupled plasma,ICP)是一种用来分析物质的组成元素及各种元素含量的常用方法。能量分辨率较低,不具备空间分辨、时间分辨能力。根据检测方式的不同 ICP 分为 ICP-AES(atomic emission spectroscopy)和 ICP-MS(mass spectrometry)两种方法。ICP-AES 是根据每种原子或离子在热或电激发后,处于激发态的待测元素原子回到基态时发射出特征的电磁辐射而进行元素定性和定量分析的方法。ICP-MS 是根据运动的气态离子按质荷比(m/z)大小进行分离并记录其信息的分析方法。ICP-AES 可以很好地满足实验室主、次、痕量元素常规分析的需要;ICP-MS 相比 ICP-AES 是近些年新发展的技术,仪器价格更贵,检出限更低,主要用于痕量/超痕量分析。ICP 方法广泛用于锂离子电池正极材料[1-5]、负极材料[6-8]、电解液[9]、锂硫[10]及锂空电池[11,12]的材料化学元素组成分析。Aurbach 等[1]在研究正极材料与电解液的界面问题时,用 ICP 方法研究 $LiCoO_2$ 和 $LiFePO_4$ 在电解液中的溶解性。通过改变温度、电解液的锂盐种类、水含量、用碳包覆的正极材料等参数,用 ICP 测量改变参数时电解液中的 Co 和 Fe 含量的变化,从而找到减小正极材料在电解液中溶解的关键。精确测量过渡金属在电解液中随着充放电过程、储存过程、高温环境下的溶出,对正极材料、电解质材料、合成工艺的设计至关重要。

Park 等[2]在研究微波加热制备 $LiFePO_4$ 时,用 ICP 分析合成的 $LiFePO_4$ 中 Li/Fe 的比值随着制备条件改变时发生的变化,从而得到制备条件改变不会影响 $LiFePO_4$ 的成分变化。Meligrana 等[3]在研究表面活性剂对水热方法合成 $LiFePO_4$ 的影响时,用 ICP-AES 研究随着表面活性剂浓度的变化,合成的 $LiFePO_4$ 材料中 Li、Fe、P、C 各个成分百分比的变化。Wang 等[6]用 ICP 研究不同的加热方法制备 SnO_2 石墨复合物中 Sn 的含量变化。Fey 等[7]用 ICP 研究稻米壳热解制备碳的过程中,酸处理和热解温度等对得到的碳成分的影响。Yoshikawa 等[8]用喷雾干燥的方法制备富锂的 $Li_4Ti_5O_{12}$,并研究得到富锂 $Li_4Ti_5O_{12}$ 的电化学性质。为了确定富锂相中 Li 和 Ti 的原子比,作者用 ICP-MS 得到精确的 Li、Ti 比。在大量的正极材料研究中,偏离整比的材料设计经常可以见到,所以准确测定制备材料的摩尔比,特别是 Li 在材料中的摩尔比,对于检验目标材料是否达到设计要求以及了解性能优异材料的化学组成必不可少。

在新的锂电池研究中,Diao 等[10]在研究锂硫电池充放电过程中多硫离子在电解液中的溶解情况时,用 ICP-AES 确定电解液中 S 的含量,作者通过液相色谱-质谱联用(LC-MS)技术成功得到了电解液中溶解的多硫离子 Li_2S_4 和 Li_2S_6 的浓度。Kramer 等[9]研究将 1mol/L 的 LiTFSI EC/DEC 为 3:7 作电解液时,对正极集流体铝箔的腐蚀情况。通过 ICP-AES 和原位的电化学石英晶体微天平(electrochemical quartz crystal microbalance,EQCM)技术研究了电解液中 Al 的浓度随时间的变化,从而得到电解液对铝箔腐蚀的机理。Jang 等[11]在研究金属 Li 的量对锂空气电池限压测试时容量大小的影响时，用 ICP 测量第 1 周放电后金属 Li 在负极上的剩余量，来解释容量衰减与金属 Li 的量的关系。目前 ICP 广泛用于锂电池各类材料的研究中。

11.1.2　二次离子质谱

11.1.2.1　D-SIMS 及其在锂电池方面的应用

二次离子质谱（second ion mass spectroscopy，SIMS）分为 D-SIMS（dynamic secondary ion mass spectrometry，动态 SIMS）和 S-SIMS（static secondary ion mass spectrometry，静态 SIMS）。D-SIMS 是指通过发射热电子电离氩气或氧气等离子体轰击样品的表面，探测样品表面溢出的荷电离子或离子团来表征样品成分。由于离子束能量较高，轰击会对样品造成一定的破坏，另一方面也使其具有高空间分辨能力和较高的灵敏度，可以用于分析元素及其同位素在样品中的分布。

最早于 1980 年 Nelson 等[13]用 SIMS 表征技术探测 Li/Si 合金暴露在干燥空气中表面生成膜的成分，发现因为反应的活化能很高，Li/Si 合金与干燥空气中 N_2、O_2 在低温下几乎不发生反应，而主要是与干燥空气中的痕量水发生反应生成 Li_2O 和 LiOH。

SIMS 在探测样品成分的纵向分布上也具有广泛的应用，Castle 等[14]通过 SIMS 探测 V_2O_5 在嵌锂后电极表面到内部 Li^+ 的分布来研究 Li^+ 在 V_2O_5 中的扩散过程。发现 Li^+ 主要在界面聚集，并随着循环进行，Li^+ 聚集的界面向里层移动，并且含量越来越高[15]。随后，又有许多研究者对 Cr_2O_3 薄膜[16]、MnO_2 正极[17]、Sn-Co 合金负极[18]的脱嵌锂行为、非水烷基酯类溶剂[19]中集流体的电化学行为等作了详细的研究。

11.1.2.2　S-SIMS 及其在锂电池方面的应用

S-SIMS 是指采用初级离子源（可以是 Ga^+、Au^+、Bi^+、C_{60} 等），入射到样品表面激发出二次离子，主要有原子离子和分子离子等。S-SIMS 中应用最广的是飞行时间二次离子质谱仪（TOF-SIMS）。TOF-SIMS 的原理是指给所有激发出的离子相同的动能（3keV）加速，遵循能量守恒公式：不同质量的离子有不同速度，越重的离子飞行速度越慢，当飞行距离一定时，其飞行时间就越长，可以将时间换算成质量来区分不同的成分，所以只用测量每一个离子到达探测器的时间就可以换算出质量数，以表征材料表面的元素成分、分子结构、分子键接等信息。

由于多数的正负极材料均为复合结构材料，界面元素的分布是液态和固态锂电池重要的科学问题。SEI 是在锂离子电池第一周循环时形成的，起到了保护电极材料不继续与电解液反应的重要作用。而 SEI 在锂离子电池研究中是最重要也是最困难的研究对象，研究初期一直没有找到较为合适的研究手段。而 TOF-SIMS 既能分析样品的有机物成分信息又可以分析无机物信息，且 SIMS 具有高精度（10^{-6} 级）、同位素分辨、高空间分辨率的特性将使其在锂电池研究尤其是在 SEI 上有着广泛的应用。

Peled 等[20]采用飞行时间二次离子质谱（TOF-SIMS）研究了高定向裂解石墨（HOPG）负极表面形成固体电解质钝化膜（SEI 膜）的过程，首次得到了 SEI 在亚微米尺度的形成、组分等的侧向分布；HOPG 上的 SEI 生长不均匀，截面上的 SEI 主要成分是 Li 和 F，而基面处的 SEI 主要由有机物组成。TOF-SIMS 技术的应用使基面处的 SEI 成分含有聚合物的猜想首次得到直观的证实。Ota 等[21]研究了亚硫酸乙烯酯（ES）作为添加剂加到标准电解液中后，石墨负极和 $LiCoO_2$ 正极表面形成 SEI 膜的成分。发现负极表面膜主要是亚硫酸盐的复合物，而正极表面膜成分主要是磷酸盐、硫酸盐。

随后，Veryovkin 等[22]运用 TOF-SIMS 研究了正负极材料中形成的 SEI 在空气中暴露一段时间后的变化，发现短时间暴露于空气中的 SEI 元素具有更强的二次离子场信号，并且元素深度分布也被改变。Lu 等[23]、Nakai 等[24]、Lee 等[25]分别对 SEI 膜的 Li^+ 传输、添加剂影响以及全电池中 SEI 的分布进行了探讨。随着锂硫电池、锂空气电池的研究热潮，对金属 Li 的研究重新受到众多研究者的注意。李文俊等[26]运用 SIMS 观察了循环前后的 Li 表面的元素二维分布与深度分布（图 11-7），并对表面元素进行二次离子成像。如图 11-8 所示，验证了金属 Li 表面在循环后 F 含量显著升高，SEI 成分含有 Li、F、C、O 等元素，并提出金属 Li 表面孔洞的形成与 SEI 的分布有关。

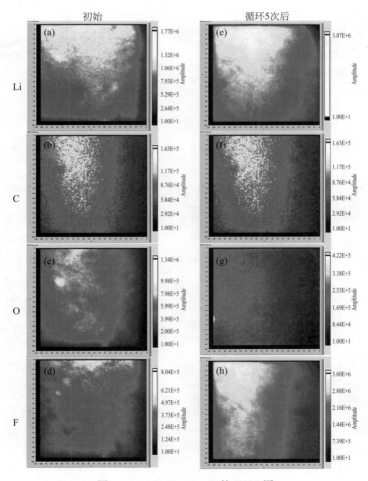

图 11-7　Li、F、C、O 的 ESM 图

(a)、(b)、(c)、(d) 为初始 Li 片；(e)、(f)、(g)、(h) 为循环 5 周后的 Li 电极[26]

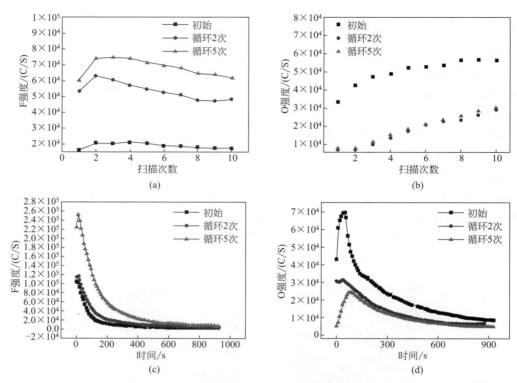

图 11-8 金属锂表面 F（a）、O（b）循环前后的深度分布及 F（c）、O（d）的分析[26]

由于 TOF-SIMS 具有超高的空间分辨率，使其在探测样品的 3D 成像上也得到了广泛的应用。Sven 等[27]为了进一步提高锂离子电池 $Li_{1+x}(Ni_{1-y-z}Co_yMn_z)_{1-x}O_2$（NCM）阴极活性材料的循环性能，提出了一种新的 NCM622（60%Ni）分子表面改性方法，通过使用有机磷三（4-硝基苯基）磷酸酯和三（三甲基硅烷）磷酸酯对 NCM622 进行改性，利用 TOF-SIMS 的元素 2D 分布以及 3D 成像等功能，可得出 NCM622 表面被一层磷酸盐包裹着，涂层几乎占据了 3D 模型的空隙空间（图 11-9）。Liu 等[28]通过 TOF-SIMS 的化学成像功能［即点或面扫描（2D 成像）得到成分分布图像］，表征了在 SEI 膜中，具有强 Li—F 键的 LiF 主要以插入 SEI 中的晶体颗粒的形式存在，通过离子成像直观显示了不同成分在表面的分布情况。此外，S-SIMS 与 D-

图 11-9 NCM622 改性物质 3D 重构图

SIMS 都有 3D 成像的能力，区别在于 S-SIMS 成像分辨率更高，通常可达到 70nm，而 D-SIMS 一般只能达微米级别。

11.1.2.3 D-SIMS 和 S-SIMS 的差异

D-SIMS 由于入射电流很高（$10\text{mA}/\text{cm}^2$），入射源离子浓度是大于 $10^{14}\text{atoms}/\text{cm}^2$，对表面主要是动态破坏作用（剥离速率 $100\mu\text{m/h}$）且产生的二次离子比率较高，因此探测灵敏度高，适于深度剖析。一方面由于其入射电流大，会将样品表面的大分子结构破坏，主要对样品产生剥离作用，生成原子离子，因此 D-SIMS 不适用于分析有机物，而更适用于分析无机物。另一方面，由于 D-SIMS 在剥离样品时，溅射和分析是同时进行，每次只能采集特定的几种元素，并且只适用于深度剖析。S-SIMS 由于入射电流很低（$1\text{nA}/\text{cm}^2$）（见图 11-10），入射源离子浓度在 $10^{12}\sim10^{13}\text{atoms}/\text{cm}^2$ 左右，只作用单分子层表面（剥离速率 0.1nm/h），几乎对表面没有破坏作用，使其既可以分析有机物也可以分析无机物；也因脉冲模式分析，且电流小，所以产生的二次离子比率相对少，灵敏度相比 D-SIMS 弱，但成像和表面分析能力强；深度剖析时溅射和分析交替进行，可以采集所有元素和化学成分信息，可进行表面分析和深度剖析。

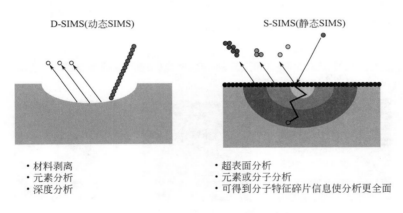

图 11-10 D-SIMS 和 S-SIMS 对比

TOF-SIMS 除了在探测样品成分的纵向分布应用外，通过采集离子质谱图用于表面元素（原子离子），同位素，分子化学结构（通过分子离子碎片）的表征，图 11-11 是利用 TOF-SIMS 对电极片进行质谱分析，可以准确地得到元素及分子结构之间的信息。

随着锂离子电池在清洁能源中的发展和广泛应用，借助相关技术手段对其展开更深入的机理研究成为了研究的热点。由于 SIMS 具有高空间分辨率、超高的表面灵敏度等优点，使其在表征锂离子电池的研究中有着独特的优势。TOF-SIMS、XPS、AES（俄歇电子能谱）是三种常见的表面成分分析技术，分别用这三种技术测试同样的锂电池材料样品（见图 11-12），可以比较出三种技术的分析能力；XPS 虽然可以得出化学态的信息（Li 的不同化学态：单质锂，氧化锂等），但是其空间分辨较弱（成像模糊）；AES 的空间分辨率较高，空间成像能力较好，可清楚地表征电极表面的颗粒的形貌，但是其检出限比 SIMS

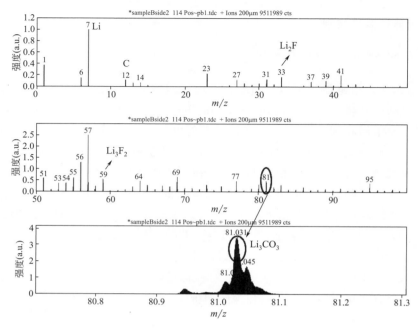

图 11-11 锂电池材料质谱图

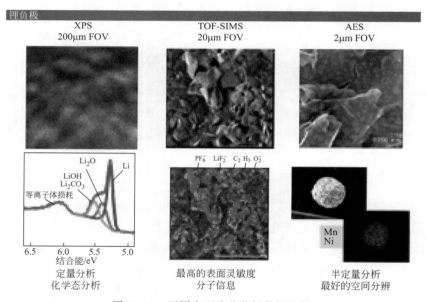

图 11-12 不同表面成分分析数据比较

高；TOF-SIMS 有着极高的表面灵敏度，可以清晰地检测分子信息（比如电解液的六氟磷酸 PF_6^-，氟化锂 LiF_2^- 以及 $C_2H_3O_2^-$ 等），还可通过高分辨图得到不同成分的分布。TOF-SIMS 的总离子质谱图和分布图是同时采集的，所以分析效率较高，通常获得较完整的图谱和成像信息只需要几秒到几分钟。

11.1.3　X射线光电子能谱

X射线光电子能谱（X-ray photoelectron spectroscopy，XPS）由瑞典Uppsala大学物理研究所Kai Siegbahn教授及其小组在20世纪50～60年代逐步发展完善。X射线光电子能谱不仅能测定表面的组成元素，而且还能给出各元素的化学状态信息，能量分辨率高，具有一定的空间分辨率（目前为微米尺度）、时间分辨率（分钟级）。因此XPS广泛用于锂电池的研究中[29,30]，如锂电池的负极材料[29,31-34]、正极材料[35-38]、高能量密度的锂硫电池[39,40]、锂空气电池[41,42]。1996年Aurbach等[29]在研究石墨和金属Li在各种电解液中性能时，用XPS研究金属Li和石墨表面生成的SEI膜的成分，发现SEI膜的主要成分有F、C、O、Li，至于As、P、Cl等元素则与电解液中锂盐的种类有关。胡勇胜等[43]研究用亚乙烯碳酸酯（VEC）作添加剂用在石墨电极中，在首周较高电压下VEC先在石墨表面还原生成一层SEI膜，防止丙烯碳酸酯（PC）的共嵌入。在该篇文章中，作者用XPS研究了在高电压下VEC在石墨表面生成的SEI的成分，主要还是以C、O、Li为主，联合FTIR发现其中主要成分为烷氧基锂盐。吴晓东等[44]用XPS研究了沉积了Ag的负极硅材料颗粒表面SEI的成分，发现SEI层中有1.9%的Ag，解释了为何Ag表面SEI膜较厚的原因。McCloskey等[41]研究非水锂氧气电池界面上碳酸盐形成的问题时，利用XPS、同位素标记和微分电化学质谱（DEMS），观察到大量的碳酸盐在锂氧气电池充放电过程中生成。Yang等[42]在研究将沉积在铜颗粒表面的钙钛矿结构的$Sr_{0.95}Ce_{0.05}CoO_{3-\delta}$作为锂空气电池的双功能添加剂时，用XPS表征该催化剂发现该颗粒表面同时存在Ce^{3+}、Co^{2+}、Co^{3+}、Co^{4+}等元素和相应的价态，说明在该催化剂的表面存在钙铁石相的$CeCoO_{2.5}$，Co^{4+}/Co^{3+}对是一个很好的OER催化剂。Agostinia等[40]开发了一种用包含Li_2S_8基于TEGDME的电解液的半液体锂电池体系，作者利用XPS发现Li_2S_8沉积在金属Li表面，因此通过减小电压和容量可能会提高电池的循环性能。锂电池的研究中，为了防止样品暴露在湿气/空气中，XPS技术发展了样品传送装置，实现了气敏性样品的测量，即可在真空或惰性气体环境下从其他设备或手套箱中转移样品，保护样品不受大气、氧气和水汽的影响，也可通过传送管上的法兰接口对其抽真空或连接惰性气体源，如图11-13所示。

图11-13　样品传送装置

用传送装置保护和暴露空气下的锂金属片样品表面成分对比图谱如图11-14所示，红线是通过转移装置保护下，没有暴露空气的样品图谱，蓝线为空气暴露状态的样品图谱，

C 1s、O 1s 和 Li 1s 的化学态随空气暴露而发生变化。对其分析，发现暴露在空气中的 Li 金属片样品，Li 和 Li_2O 的化学状态都消失了。

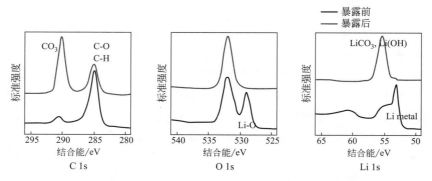

图 11-14　Li 金属片暴露空气前（转移装置保护）和后表面的 C1s、O1s 和 Li1s 光谱图

L. Castro 等[45]利用 XPS 表征方法分析研究了 $LiFePO_4$//石墨细胞的老化机理，在这项工作中，作者通过使用不同的探针/核峰，检测了 Fe^{2+}/Fe^{3+} 氧化还原反应的演变，以及循环时两个电极的界面层发生的变化。结果表明，Fe^{3+}/Fe^{2+} 氧化还原反应在循环和高温下缺乏可逆性。通过对两个电极上的界面层的深度剖析，可以进一步阐明活性锂的有害损失的来源是电解质中沉积或部分溶解的少量锂物质。该工作为了防止样品暴露在湿气/空气中，XPS 光谱仪通过一个转移腔直接连接到氩气干燥箱。样品从干燥箱内的包装中取出，放在样品盒上，没有任何污染。Seon-Hong Lee 等[46]使用 XPS 的氩离子刻蚀技术研究石墨负极表面 SEI 膜的厚度，SEI 测厚原理示意图如图 11-15 所示。图 11-15(a)所示相对负极表面氩离子枪的入射角是 45°，当入射光束照射到负极表面时，所有的石墨颗粒都暴露在蚀刻区域下，只有周围石墨颗粒的阴影区域没有暴露，如图 11-15(b)所示。经过适当的刻蚀时间后，暴露在刻蚀区域下的 SEI 层几乎被除去，然后暴露出石墨表面。在暴露和未暴露之间的步进区（台阶区）可以观察到 SEI 层的厚度。根据 XPS、XHR-SEM 和 TEM 结果，作者们提出了石墨电池上形成 SEI 层的示意图模型，如图 11-16 所示。SEI 的最外层（含粒子的第 2 层）主要由具有 $LiPF_6$ 还原产物的聚合物（来自 VC）组成，而 SEI 的内层（在石墨上的第 1 层）主要由具有 $LiPF_6$ 还原产物的聚合物（来自 EC）组成。外层厚度大致在 20~30nm 范围内，在石墨表面形成了厚度约在 25~30nm 范围内的复杂海绵型层（致密膜，第 1 层），并呈现出不均匀的层状结构。

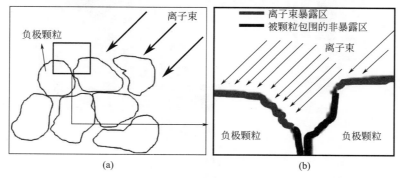

图 11-15　XPS 中离子束刻蚀观察 SEI 厚度原理示意图

最新发展的 XPS 技术已经可以通过环境腔的引入实现原位的测量，这为锂电池界面研究提供了更加有力的武器。Mao Baohua 等[47]采用常压 X 射线光电子能谱技术（AP-XPS），构建了一种全固态钠氧电池，并对其充放电循环进行了研究。固态钠氧电池的原理图如图 11-17(a)所示。它由三部分组成：正极、固体电解质和负极。如图 11-17(b)所示，Na-O_2 电池被放置在装有电触点的样品盒上，以便在 AP-XPS 分析室内进行电化学测量。本研究采用恒流 50nA 充放电来

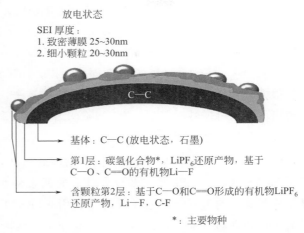

图 11-16　石墨电极上 SEI 模型

匹配 AP-XPS 数据采集时间。在此电流下，放电-充电循环时间约为 20h。由于在 1h 内可以获得一组 Na 1s、O 1s 和 C 1s 谱，因此每个循环可以获得 20 组数据点。因此，通过分析光谱可以研究电池不同放电-充电阶段的表面化学变化。目前来看，XPS 技术已经在锂电池领域应用十分广泛，可以说锂电池材料的改进、机理的研究已经离不开 XPS 技术的支持。

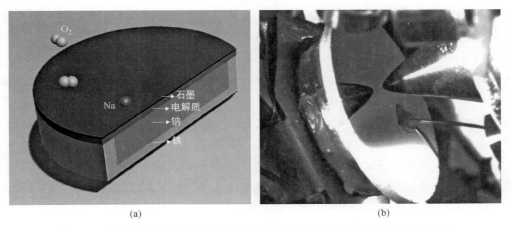

图 11-17　固态钠氧电池的原理图（a）及在 AP-XPS 分析室中测量照片（b）

11.1.4　电子能量损失谱

电子能损失谱（electron energy loss spectroscopy，EELS）也是材料学研究领域的重要手段。EELS 是利用入射电子引起材料表面电子激发、电离等非弹性散射损失的能量，通过分析能量损失的位置可以得到元素的成分。EELS 相比 EDX 对轻元素有更好的分辨效果，能量分辨率高出 1～2 个量级，空间分辨能力由于伴随着透射电镜技术，也可以达到 10^{-10}m 的量级，同时可以用于测试薄膜厚度，有一定时间分辨能力。通过对 EELS 进行

密度泛函（DFT）的拟合，可以进一步获得准确的元素价态甚至是电子态的信息。

Xu 等[48]在研究 Li（Ni$_x$Li$_{1/3-2x/3}$Mn$_{2/3-x/3}$）O$_2$（0＜x＜1/2）在高电压下表面结构的变化时，通过联合 EELS、XRD 及 STEM 技术和理论计算发现该材料在高电压充放电时会在表面生成一种类尖晶石的新固相结构。EELS 结果表明，充放电之后材料表面 O 的缺失很少＜5%，但 Li 会缺失 50%以上。王峰等[49]在研究二元金属氟化物 FeF$_2$ 和 CuF$_2$ 表现出不同的可逆性时，通过 TEM-EELS 联用并利用 EELS 非常高的分辨率（1nm）研究充放电前后各个元素在颗粒表面的分布，从而揭示 FeF$_2$ 与 CuF$_2$ 截然不同的可逆性原因。Rom 等[50]在研究纳米结构和微米结构的 Sn/SnSb 复合电极电化学性质巨大差异时，利用 EELS 对 Sn/SnSb 粉末及复合电极的元素分布进行了研究。Al-Sharab 等[51]在研究氟化铁和碳的纳米复合物电极材料时利用 STEM-EELS 联合技术研究了不同充放电状态时氟化铁和碳的纳米复合物的化学元素分布、结构分布及铁的价态分布。Schuster 等[52]通过联合 EELS/NEXAFS 技术研究了溶剂热处理合成的 LiFePO$_4$ 的充放电行为。其中，用 EELS 得到了不同充放电状态的颗粒表面和内部 Li 的分布，发现 Li 在高维的缺陷处不易嵌入和脱出。Naji 等[53]在 1999 年就利用 TEM 和 EELS 联用研究了用 MClO$_4$-EC（M=Li、Na）作电解液时在嵌锂或者嵌钠时碳表面生成的 SEI 的成分，发现碳表面存在 M$_2$CO$_3$、ROCO$_2$M 等化合物。Kohandehghan 等[54]发现用 ALD 方法在硅的纳米线表面包覆一层 TiN 之后能够大幅提高材料的循环性能。通过 EELS 确定在充放电过程中硅纳米线材料表面及截面的 Si、C、N、Ti、O 的分布，发现材料的循环性提高的原因是在包覆 TiN 之后能够明显地减少 SEI 的生成，从而提高库仑效率和循环性能。Gu 等[55]研究了硅负极在嵌锂过程中生成的无定形的 Li$_x$Si 如何向晶态的 Li$_{15}$Si$_4$ 转变的过程，文章中作者利用 EELS、原位的 TEM 和 DFT 理论模拟，发现硅在嵌锂过程中先生成 a-Li$_x$Si，当 Li 的量达到 x=3.75 时，a-Li$_{3.75}$Si 开始向 c-Li$_{3.75}$Si 转变。其中 EELS 清楚地显示了在嵌锂过程中 Li 和 Si 在 Li$_x$Si 中的分布情况及相应的浓度比。由此可见具有超高分辨率的 EELS 在锂电池中有大量应用，从 20 世纪末对碳负极表面 SEI 成分的探究，与原位的 TEM 结合研究充放电过程的机理都有应用。

随着球差校正电镜技术的发展，STEM-EELS 联用将有望对之前研究中无法涉及的电荷有序、局域互占位等问题给出更为精确的结果。环境 TEM 的发展则可能促进原位 EELS 技术在锂电池研究中的发展。

11.1.5　扫描透射 X 射线显微术

扫描透射 X 射线显微术（scanning transmission X-ray microscopy，STXM）是基于第三代同步辐射光源以及高功率实验室 X 光源、X 射线聚焦技术的新型谱学显微技术。采用透射 X 射线吸收成像的原理，STXM 能够实现具有几十纳米的高空间分辨的三维成像，同时能提供一定的化学信息，这方面软 X 射线更为适用。通过硬 X 射线衍射成像，易于聚焦，穿透性强，可以获得更高的空间分辨率，极限目标是 1nm，目前最新的实验技术达到了 5nm。STXM 能够实现无损伤三维成像，对于了解复杂电极材料、固体电解质材料、隔膜材料、电极以及电池可以提供关键的信息，而且这些技术可以实现原位测试的功能。具有较低空间分辨率（微米尺度）的仪器（micro-XCT）（Zeiss 公司，岛津公司等）已能提供，用于电池内部叠层电极结构的分析。纳米尺度分辨的实验室仪器，nano-XCT 技术也已商业化，可以对单颗粒实现 3D 成像，但不如同步辐射光源成像质量好。对于 3D 成像的结果，通过软件分析，还可以提供电极层、颗粒的孔隙率、孔的通道弯曲程度等信息。

Sun 等[56]研究碳包覆的 $Li_4Ti_5O_{12}$ 与未包覆之前相比，具有更好的倍率性能和循环性能。作者利用 STXM-XANES 和高分辨的 TEM 确定了无定形的碳层均一地包覆在 LTO 颗粒表面，包覆厚度约为 5nm。其中通过 STXM 作者获得了单个 LTO 颗粒的 C、Ti、O 分布情况，其中 C 包覆在颗粒表面。Zhou 等[57]在研究 $LiMn_{0.75}Fe_{0.25}PO_4$ 纳米棒与石墨烯复合后在充电状态的价态分布图及对应的内部变化时，作者用 STXM 观察到 Fe 的不同价态的分布图，从而观察到了 LMFP-C 在充电过程中的价态变化。Chueh 等[58]在研究 $LiFePO_4$ 在充电过程中许多纳米颗粒的脱锂情况时，利用 STXM 得到不同充电状态时颗粒上 Fe^{3+} 和 Fe^{2+} 的分布情况，从而发现对于纳米 $LiFePO_4$ 相转变的脱锂过程是非常快的，而触发一个 $LiFePO_4$ 颗粒脱锂却是决定该材料倍率性能的步骤。

由于渴望在三维尺度上了解材料和电池结构在充放电过程中的动态演化信息，STXM 实验技术在锂电池领域的研究具有十分重要的作用。只是目前仪器设备昂贵，光源的实验还不是很方便。相信未来该技术在锂电池领域的应用会越来越多。

11.1.6 X 射线近边结构谱

X 射线近边结构谱（XANES）是标定元素及其价态的技术，不同化合物中同一价态的同一元素对特定能量 X 射线有高的吸收，我们称为近边吸收谱。XANES 分为软 X 射线吸收谱（s-XAS）与硬 X 射线吸收谱（h-XAS），空间分辨尺度在数纳米到毫米范围。通过 h-XAS 近边谱的位置，比对参考化合物，可以得到元素价态的信息；通过对 h-XAS 实验同时获得的扩展 X 射线吸收精细结构（EXAFS）的数据分析，可以获得局域键长、结构有序度、配位环境变化的信息。由于硬 X 射线穿透能力较强，可以实现原位电池的观察。s-XAS 对电子结构更为敏感，通过 DFT 的辅助拟合分析，可以获得精确的价态，甚至是电子在特定电子轨道填充的信息，这对于理解电池充放电过程中的电荷转移非常重要。s-XAS 在高真空下完成，之前无法实现原位技术。目前通过引入环境腔和 Si_3N_4 窗口，已经可以实现原位技术，这对于电池基础研究是个重大利好消息。在 XAS 实验中，还分电子产率（TEY）、荧光产率（TFY）两种模式。TEY 更多获得表面信息，TFY 模式获得体相信息。

在锂电池领域中，XAS 主要用于电荷转移研究，如正极材料过渡金属变价问题。Uchimoto 等[59,60]研究了 $LiNiO_2$、$LiCoO_2$ 中 Li^+ 脱出后 Ni、Co、O 的变价情况。采用软 X 射线观察 Ni、Co 的 L 边吸收谱，O 的 K 边吸收谱。发现 Li^+ 脱嵌后的空穴补偿主要位于 O 的 2p 态，并不是 Ni、Co 价态的升高补偿。后来研究者们对掺杂的 $LiMO_2$ 型化合物正极材料进行了研究。Rosolen 等[61]研究了 $LiCo_{0.5}Ni_{0.5}O_2$ 化学脱锂过程中的变价情况，得到与电化学脱锂后一致的结构。文闻等[62]在研究 $LiM_{0.5-x}Mn_{1.5+x}O_4$ 正极材料过程中发现在 5V 的充放电范围内，Zn^{2+} 也没有变价。Myung 等[63]、Holzapfel 等[64]、Kobayashi 等[65,66]分别研究了 $Li_{1-\delta}Mn_xCr_{1-x}O_2$、$Li_xCo_{1-y}Fe_yO_2$ 以及 $LiNi_{0.80}Co_{0.15}Al_{0.05}O_2$ 正极材料。$Li_{1-\delta}Mn_xCr_{1-x}O_2$ 材料中 Mn、Cr 在充放电过程中都变价；$Li_xCo_{1-y}Fe_yO_2$ 只有 Fe 会变价，并且富铁的情况下变价会比富锂情况困难些[65]；$LiNi_{0.80}Co_{0.15}Al_{0.05}O_2$ 与 $LiCoO_2$ 不同，XANES 检测到颗粒表面含有 Li_2CO_3 和其他额外立方相杂质[66]。Yang 等[67]运用 XANES 谱研究了 $LiFePO_4$ 中不同元素形态的存在方式，指明了元素在不同化合物中的特征谱线可以作为 $LiFePO_4$ 基正极材料充放电循环中相转变定量分析的手段之一。Tanaka 等[68]将其用于 $Li_4Ti_5O_{12}$ 负极材料的研究，验证了第一性原理的计算结果。Akai 等[69]将 XANES 应用到固体电解质界面膜的研究中，验证了石墨负极 SEI 膜中主要含有 Li_2SO_3、$ROSO_2Li$、Li_2SO_4 和含有 C—S—C 或 C—S—S—C 键的有机硫化物，而正极形成的 SEI 膜中主要含有 Li_2SO_4 以及含有 C—S—C 或 C—S—S—C 键的有机硫化物。并且在 120℃没有观察到充电态的负极表面 SEI 膜的溶解，而正极表面 SEI 膜的有机硫化物溶解程度在放电态比充电态更低。王小建等[70]采用原位的 XANES 技术研究了 $LiFePO_4$ 正极在充放电过程中 Fe 的 K 边变化，发现在充电过程的一开始，Fe 的 K 边便开始移动，表明充电一开始 Fe 的价态便发生变化，由+2 价变到+3 价；另外，还采用了软 X 射线吸收谱的部分电子场检测和荧光场检测模式分别探测了 Li 含量在表面和体相的分布，表明 Li 含量在充电过程中，体相的变化慢于表面的变化，这主要是由于扩散长度的影响。Bak 等[71]运用准原位的 X 射线吸收谱研究了 $Li_xNi_{0.8}Co_{0.15}Al_{0.05}O_2$ 材料中 Ni、Co 的变价对 O_2 释放的影响，表明 O_2 释放与 Ni、Co 阳离子在材料中的迁移密切相关。这些迁移影响了从层状到无序的尖晶石相再到岩盐相的转变，并且 Ni 优先变成岩盐结构。Nam 等[72]研究了正极材料的热不稳定性，XANES 表明 Ni 是最不稳定的元素，Co 比 Ni 要稳定，还原过程也稍慢，Mn 是最稳定的过渡金属。禹习谦等研究了 $Li_4Ti_5O_{12}$ 负极材料储 Na 机制，对于全部钠化的 $Li_4Ti_5O_{12}$，计算得到的 $Li_7Ti_5O_{12}$ 和 $Na_6LiTi_5O_{12}$ 比值为 1∶1，而 XANES 得到的结果与计算的曲线刚好匹配。Yu 等[73]研究了 $NaCrO_2$ 中的相转机制，XANES 数据表明充电到 3.6V 后，Cr^{3+} 将转变为 Cr^{5+}，整个充电过程中，Cr 都稳定在氧八面体位，且没有 Cr^{6+} 形成。马君等[74]研究了 Mo 掺杂的 Li_2MnO_3 材料，XAS 谱表明在 Mo^{5+} 取代的氧化物中，Mo^{5+} 和 O^{2-} 都在脱锂过程中贡献了电子。另外，Mo^{5+} 的取代提高了 O^{2-}、Mn^{4+} 迁移的电荷容忍性。马君等[75]还研究了纯 Li_2MoO_3 材料，XANES 谱表明充电过程中，Mo^{4+} 转变成 Mo^{6+}，并且 MoO_6 八面体出现局部扭曲；而放电过程中，材料并没有回到初始状态，在 Li 嵌入过程中，Mo 减少了，但平均价态仍然是+4 价，并且边前峰并没有回到初始值，这意味着完全放电态处于初始态和部分充电态之间。Liu 等[76]也利用软 X 射线吸收谱研究了 $LiFePO_4$ 在充放电过程中 Fe^{2+} 的变价情况，并深入研究了其相转变和锂化对电子结构的影响。具体将在 11.3.2 节介绍。

11.1.7 杂质测量

除了通过上述分析手段外，X 射线荧光光谱分析和振动样品磁强计也被广泛应用于杂质分析。

X射线荧光光谱（XRF）分析利用初级X射线光子或其他微观离子激发待测物质中的原子，使之产生荧光（次级X射线）而进行物质成分分析和化学态研究的方法。按激发、色散和探测方法的不同，分为X射线光谱法（波长色散）和X射线能谱法（能量色散）。根据色散方式不同，X射线荧光分析仪相应分为X射线荧光光谱仪（波长色散）和X射线荧光能谱仪（能量色散）。XRF被工业界广泛应用于锂离子电池材料主成分及杂质元素分析。对某些元素检出限可以达到10^{-9}的量级。

振动样品磁强计（vibrating sample magnetometer，VSM）可完成磁滞回线、起始磁化曲线、退磁曲线及温度特性曲线、IRM和DCD曲线的测量，具有测量简单、快速和界面友好等特点。适用于各种磁性材料的测量，对样品的形状没有要求。作为含过渡金属正极材料中检测游离的磁性过渡金属杂质是一个较为方便的实验技术。

11.1.8 俄歇电子能谱仪

俄歇电子能谱（auger electronic spectroscopy，AES）是通过将电子源入射到样品表面，激发出二次电子（用于形貌观察）和俄歇电子（用于成分分析），通过配备离子溅射枪可进行材料纵向深度分析。AES主要用于分析固体材料表面纳米深度的元素（部分化学态）成分组成，可以对纳米级形貌进行观察和成分表征。既可以分析原材料（粉末颗粒，片材等）表面组成，晶粒观察，金相分布，晶间晶界偏析，又可以分析材料表面缺陷如纳米尺度的颗粒物、磨痕、污染、腐蚀、掺杂、吸附等，还具备深度剖析功能表征钝化层、包覆层、掺杂深度、纳米级多层膜层结构等。AES的分析深度为4～50Å，二次电子成像的空间分辨可达3nm，成分分布像可达8nm，分析材料表面元素（Li～U）组成，是真正的纳米级表面成分分析设备。可满足合金、催化剂、半导体、能源电池材料、电子器件等材料和产品的分析需求。AES提供全面的分析能力，包括SEI成像，SAM成像，采谱和深度剖析功能。可以为分析人员提供明确的固体材料表面化学成分及其分布的信息。筒镜式能量分析器（CMA）的独特设计可实现全面的成分表征和纳米级表面特征，薄膜结构及表面污染物的二次电子成像，高能量分辨率模式能够从俄歇谱和成像中提高AES的化学态识别能力。

图11-18是运用AES进行表面元素分析后得到的微分谱，从中可以看出该样品表面有Li、C、O、N、Si、F这几种元素，通过半定量计算可以得出Li元素的相对含量为60.4%。

图11-19（a）是对特征区域进行二次电子成像后得到的形貌图片，图11-19（b）是该区域进行mapping成像表征得到的元素分布图像，其中目标元素为Li，可以看出高亮区域为Li富集区，与SEM图片比对可以发现，颗粒边界处Li含量较多。

由于俄歇电子能谱在元素周期表中可以从Li开始识别，其应用特征已经引起了学术界的广泛关注，尤其是与锂电相关的行业，利用其二次电子成像能力与表面纳米层级元素分辨能力相结合的优势对锂电池一系列机理问题的深挖与阐释被业界寄予了厚望。

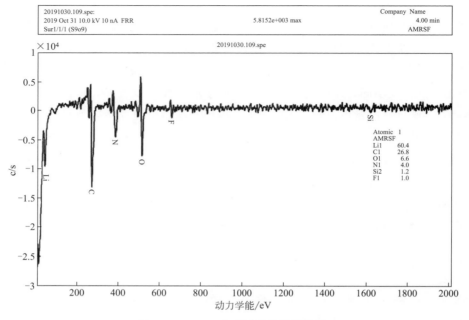

图 11-18 AES 表面元素分析微分谱

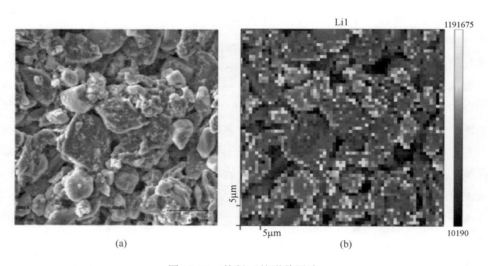

图 11-19 特征区的形貌图片

11.2 形貌表征

形貌一般采用扫描电子显微镜（SEM）、透射电子显微镜（TEM）、STXM、扫描探针显微镜（SPM）进行表征。SPM中的原子力显微镜（AFM）大量应用于薄膜材料、金属Li表面形貌的观察，主要用在纳米级平整表面的观察。这里主要介绍扫描电子显微镜

(SEM) 和透射电子显微镜 (TEM)。

扫描电子显微镜 (SEM) 和透射电子显微镜 (TEM) 作为传统的电子成像技术，在锂离子电池研究的诸多领域中都有广泛的应用。使用环境腔体后，二者都可以对实际体系进行观察，并具有一定的时间分辨能力，可以原位观察模型电池工作、材料受热、外电场影响下的形貌变化。SEM 收集的是样品表面的二次电子信息，它的衬度反映了样品的表面形貌和粗糙程度。SEM 的空间分辨率可以达到 10nm，实际分辨率受限于样品的导电性和电镜腔体的环境。虽然 SEM 的分辨率远小

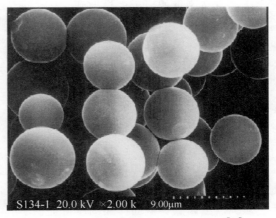

图 11-20　硬碳球的 SEM 二次电子像[74]

于 TEM，但它仍是表征电池材料的颗粒大小和表面形貌的最基本的工具（图 11-20）[77]。

扫描电镜对材料的分析，不仅要掌握材料表面的情况，有时候还需要对材料内部的结构和成分及性质也要分析，以了解材料的特性。比如锂离子电池正极材料 NCM 粉末颗粒，含有碳 (C)、氮 (N)、锂 (Li)、锰 (Mn)、钴 (Co)、镍 (Ni) 等元素，为镍钴锰三元电池正极材料。材料中含有活泼金属，在空气中容易氧化失活。在实际电池工作条件下，电池材料会发生团聚，形成枝晶、扩散等情况，导致性能下降，如图 11-21(a) 所示。这就要求研究者不仅要了解颗粒表面的信息，也要清楚截面的信息。对于截面样品可以使用离子束加工技术例如正极极片的 NCM 颗粒如图 11-21(b) 所示。

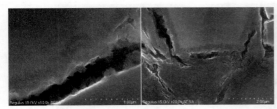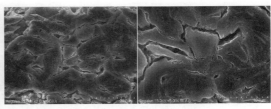

(a) 不同倍率下的负极极片的硅碳颗粒

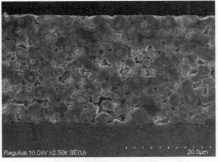

(b) 正极极片的NCM颗粒

图 11-21　扫描电镜图片示例

离子束加工技术是在真空环境下，Ar 气通过离子源高电压轰击成 Ar^+，经过加速聚焦形成具有高能量的离子束投射到工作区完成刻蚀、切割、抛光等工作，配合使用截面切割样品台、抛光样品台以及减薄样品台制备满足 SEM、TEM 观测的截面样品，去除表面覆盖物的抛光样品及具有极薄区的透射样品。与传统物理机械切割相比，离子束切割更适合制备 NCM 颗粒截面样品的原因有四点：其一，由于离子束轰击样品过程中载能离子进入基质，使基质内原子发生电离和原子碰撞，产生级联碰撞效应，在原子剥离过程遵循固体样品晶格排列顺序，从而使离子束在工作区域内产生的创伤性划痕更细微，SEM 图像衬度更好，对材料内部精细结构的观察极为有利。其二，作业过程中材料内部的孔隙或者沟壑区域不会被打磨下来的碎屑填埋，降低了不真实结构产生的可能性。其三，在加工过程中无水无氧的高真空环境有利于保护性质活泼的材料。其四，程序调节设置更为灵活，使工作区域的产生更具靶向性、高效性，产生的可观察区域范围更广、平整度更高、特征性更强。以上四点为截面样品的获得提供了有力支持。

同时，通过对循环后的电极材料进行 SEM 观察，可以直观地分析电极材料在循环过程中的形貌变化，找出电池失效的可能原因，有助于改进电极材料的设计。

笔者课题组利用密封转移盒在样品的转移过程中保护样品，避免样品在从手套箱转移到 SEM 电镜腔体的过程中接触空气。借助这个特点，在利用非原位 SEM 手段观察电极材料在循环过程中的形貌变化时做了一系列的工作。何宇等[78]利用这个装置，非原位地表征了硅柱子在循环过程中的形貌变化，直观地研究了硅的体积形变对电池性能带来的影响（图 11-22）。李文俊等[77]在利用密封转移盒转移样品的基础上，重新设计了针对金属锂电极的扫描电镜的样品托架，确保了在每次 SEM 观测中，探测的都是样品的同一个位置。利用这个特点，他研究了金属锂电极在 Li 的嵌入和脱出过程中表面孔洞和枝晶的形成过程。

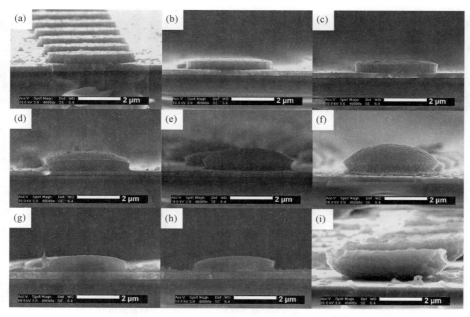

图 11-22　循环过程中硅柱子的形貌变化[75]

相对于 SEM，TEM 具有更高的分辨率，广泛被用来分析材料的表面和界面形貌与特性。在关于表面包覆以及阐述表面 SEI 的文献[79,80]中，都用到了 TEM 的表征技术。利用可操纵的原位样品杆和氮化硅窗口保护的液态样品池，科研者得以在 TEM 电镜腔体中组装原位电池，同时借助于 TEM 的高分辨特性，对电池材料在循环过程中的形貌和结构演化进行实时的测量和分析。黄建宇等[81]利用原位样品杆对 SnO_2 在离子液体中嵌脱锂过程中的形貌和结构演化进行了原位表征。随后，他们[82]对 TEM 原位电池实验的装置进行了改进，利用在金属 Li 上自然生产的氧化锂作为电解质，代替了原先使用的离子液体，提高了实验的稳定性，更好地保护了电镜腔体。Gu 等[83]利用氮化硅窗口保护的液态样品池，利用传统的电解液组装了原位电池，对硅纳米线的嵌脱锂行为进行了研究。但是这种技术对于样品池设计的要求极高，并且存在污染和损坏电镜腔体的潜在风险。Santhanugopalan 等[84]利用薄膜技术生长出了全固态电池，并利用 FIB 进行切割，将切片后的电池放入电镜腔体中，利用原位样品杆的针尖接触固态电池的两个电极，进行充放电循环。在这个工作中，他们研究了循环过程中锂硅界面的演化。这种技术稳定、安全，但同时也需要有制作全固态薄膜电池及切片的技术基础。同时，配合 SEM 和 TEM 仪器自带的能量耗散 X 射线谱（EDS），TEM 中的电子能量损失谱（EELS），还可以进一步分析元素组成。

聚焦离子束（focused ion beam，FIB）系统是利用静电透镜将离子束聚焦成尺寸非常小的微加工系统。通过高能离子轰击材料表面，实现材料的剥离、沉积、注入和改性。其中离子源是 FIB 的关键部件之一。目前商用系统的离子束多为液态金属离子源（liquid metal ion source，LMIS），其中镓元素具有低熔点、低蒸气压及良好的抗氧化能力而被广泛应用。FIB 中除了能够用 Ga^+ 离子源，还能通过使用 ExB 质量过滤技术的 Si、Cr、Fe、Co、Ni、Ge、In、Sn、Au、Mn、Pb 等专用金属离子源，以及利用电子回旋共振（ECR）的 Xe 或其他气体的等离子源。

FIB 具有两大功能，一是将液态金属离子源产生的离子束经加速、聚焦后用于轰击样品表面，使之产生二次电子信号以获得样品表面的电子像（类似于 SEM）；二是利用强电流离子束对表面原子进行剥离，以完成微、纳级别表面形貌加工。若在物理溅射的方式上搭配化学气体反应系统，则可实现有选择性的剥离金属、氧化硅层或沉积金属层。在 FIB 双束系统中，可以在离子束加工的同时，利用电子束实时监控加工的全过程。利用电子束成像分辨率高的特点（可达 0.7nm），原位观察样品截面和表面信息；利用电子束激发的特征 X 射线，对样品的截面和表面进行化学成分分析。利用电子束产生的背散射电子直接进行晶体取向分析。

在锂电池领域中，FIB 利用镓离子在很高的空间分辨率下切割去除材料。这样可以在样品特殊的位置制作剖面（断面）。样品既可以直接在 FIB 中研究，也可以转移到扫描电镜或者透射电镜中进行精细分析。当镓离子和一定气体作用，它也有可能沉积材料。因此 FIB 在很广阔的应用范围内能被用于多功能工具使用。在锂电池领域中，无论是正极材料还是负极材料以及隔膜材料，使用 FIB 技术都很频繁。如图 11-23 所示，循环过后镍钴锰酸锂的正极材料利用 FIB 进行截面切割，通过 SEM 对制备好的样品获得三维数据，如正极颗粒形貌。另外颗粒里面的晶粒以及气孔分布，还有黏结剂或者导电剂的分布，把相关的数据进行一个可视化处理以图片形式显示出来。

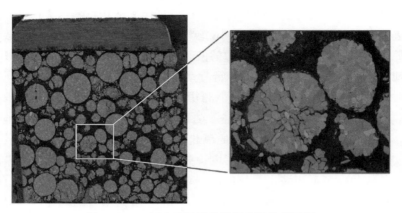

图 11-23　NMC 正极材料循环后极片截面图

其实在电池样品中,隔膜截面制样难度是最大的。因为隔膜很不耐高温及电子和离子的轰击而损伤。像陶瓷涂覆层隔膜中的涂覆层可能会影响离子传输,陶瓷跟隔膜之间有黏结剂,黏结剂气孔怎么分布的,我们需要通过对隔膜材料的截面进行切割后观看,通过了解其三维结构来做相关技术的分析。如图 11-24 所示,使用 FIB 对 Al_2O_3 陶瓷涂层 PE 隔膜截面进行切割得到的截面图,通过对这个三维图数据进行一个统计,得到我们想要的信息,这个在冷冻 FIB 中很容易实现。

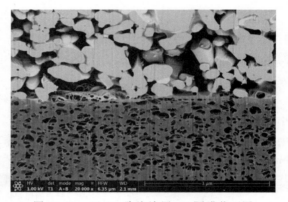

图 11-24　Al_2O_3 陶瓷涂层 PE 隔膜截面图

并且 FIB 切割隔膜要比 CP 切割隔膜的效果好,主要原因还是 CP 的切割束流是几百微安,而 FIB 束流仅为几百纳安,束流越小对隔膜产生的影响越小。

FIB 技术同样在 TEM 制样中应用广泛。对比普通的 TEM 制样方法,FIB 技术有速度快、精确度高、成品率高的优势特点。所以它能解决绝大部分材料中 TEM 制样的问题,FIB 技术被别认为是现今最有效的精确定位制样方法。像现今流行的在冷冻电镜中观看锂电池原位反应的过程研究就需要用到 FIB 对相关样品进行制备。由此可见,FIB 技术在锂离子电池中应用越来越广泛,但是由于其制样速度慢、价格昂贵所以锂电行业使用率不高。

冷冻电镜(cryo-electron microscopy,Cryo-EM)技术是在低温下使用透射电子显微镜观察样品的显微技术,即把样品夹成薄薄的水膜,并冻到液氮温度,冻的速度非常快,使水膜没有时间形成晶体,而是形成一种玻璃态的冰。样品保持低温放进显微镜里面,用高能的电子束作为光源,透过样品和附近的冰层,可以拍摄样品的形貌和它内部的结构信息[84]。利用探测器和透镜系统把信号成像记录下来,最后进行信号处理,经三维重构解析得到样品的结构。冷冻电镜技术作为一种重要的结构生物学研究方法,它与 X 射线晶体学、核磁共振一起构成了高分辨率结构生物学研究的基础。

金属锂由于其高理论比容量,是二次电池最理想的负极。然而,电化学沉积金属锂纳米结构的过程和原理仍需探索,对提升电化学性能和改善措施的思路也受限制。这是因为金属锂具有很高的化学反应活性。它不仅在空气中易氧化,也对强电子束敏感,这给表征过程带来了很多困难。在普通电镜下,金属锂会漂移、熔化并升华;无法得到其高分辨率图像。而在低温下,金属锂很稳定。最近,科学家利用冷冻电镜揭示金属 EDLi 的纳米结构表面和 SEI 的结构〔图 11-25(a)和(b)〕,发现金属形核状态为非晶态,后来慢慢长成结晶带[85]。随后,它的原子结构也被发现,与常规的面心立方锂〔图 11-25(c)〕很好地匹配[86]。另外,Kourkoutis 组结合了 Cryo-FIB 和 Cryo-TEM,发现 SEI 在含有结晶 LiF、Li_2O 和 Li_2CO_3 的 EDLi 表面上分布是不均匀的,存在 LiH 的 EDLi 和液体电解质之间维持完整的界面,并发现 LiH 也是另一种沉积物相〔图 11-25(d)~(f)〕[87]。

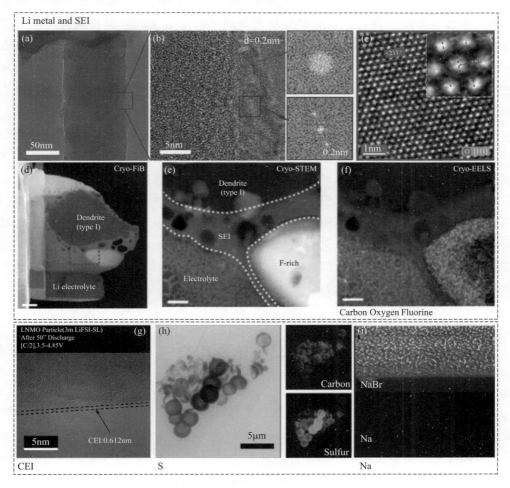

图 11-25 电池材料的 Cryo-EM 图像

(a)和(b)纳米图像,(a)局部放大图像,(b)体积和表面快速 EDLi 在 $0.5mA/cm^2$ 下 5min 的傅里叶变换结果;(c) EDLi 在 $2.0mA/cm^2$ 下 30min 的原子结构;(d) 电子透明的低温 FIB 提升薄片;(e) 低温 STEM 成像;(f) 电子电解质对 EDLi 的能量损失谱元素映射,比例尺为 300nm;(g) 50 次循环后 $LiMn_{1.5}Ni_{0.5}O_4$ 的 Cryo-TEM 图像;(h) 碳球/S 复合材料的 STEM 图像以及 C 和 S 的元素分布;(i) 涂有 NaBr 的 Na 金属的截面 SEM 图像,比例尺为 $5\mu m$

对于正极材料来讲，稳定的正极-电解液界面（CEI）对于抑制阴极材料与电解质之间的连续反应也很重要，尤其是在高电压下。Meng Shirley 组利用冷冻电镜发现循环 50 周后 $LiMn_{1.5}Ni_{0.5}O_4$ 表面形成了一层均匀的 CEI（0.612nm）［图 11-25（g）］。这层稳定的 CEI 保证了 $LiMn_{1.5}Ni_{0.5}O_4$/C 电池在 4.85V 的高压下可以循环 1000 多周[88]。

各种纳米结构的硫（S）正极可以实现高性能的 Li-S 电池。但是，这些正极中的 S 分布几乎是未知的，因为 S 在常规 TEM 中会升华。低温保护有助于稳定 S 并在微观尺度上绘制 S 分布图。通过冷冻扫描电子显微镜（STEM），我们可以得到硫在几百纳米尺度下的分布［图 11-25（h）][89,90]。如果需要进一步对 S 及其中间多硫化物进行纳米成像，以了解其未来的反应机理，例如作为多硫化物的形成和吸附原理。但是，这可能比 Li 金属更困难，因为 S 是绝缘体，并且更容易被局部热量蒸发。我们需要再次降低或者减小 TEM 加速电压到 80kV 来稳定样品硫。

除了锂离子/金属电池以外，研究人员将冷冻电镜扩展应用于其他金属电池，比如钠电池。Lynden Archer 的小组使用 Cryo-FIB 研究了 Na 电极的深入界面信息，发现 NaBr 被均匀地涂覆在 Na 阳极的表面上，厚度为 2mm［图 11-25（i）][90]。与 Li 金属比较，其他金属具有较重的原子质量，因此具有较高的 TEM 对比度，这也使得能够通过能量色散 X 射线光谱法探测其分布。注意到某些金属（例如 Na 和 K）比 Li 金属更具反应性，因此在转移和成像样品时应格外小心和谨慎，以减轻污染和辐射损伤。

冷冻电镜是表征对电子束敏感材料结构的强大工具。未来研究发现着重于那些由 Li、C、O、F、P 和 S 等非过渡金属元素组成的电池材料。也可以通过冷冻电镜捕捉电池反应中间产物，比如超氧化锂，多硫化物以及可溶的反应氧化还原电对。通过冷冻电镜，我们可以表征金属锂、纳米硫正极、固体电解质以及各种电极界面。这为研究下一代高比能量电池提供了必不可少的工具。

11.3 材料晶体结构表征

表征材料晶体结构的主要有 X 射线衍射技术（X-Ray diffraction，XRD）、扩展 X 射线吸收精细谱（extended X-Ray absorption fine spectroscopy，EXAFS）、中子衍射（neutron diffraction）、核磁共振（nuclear magnetic resonate，NMR）以及球差校正扫描透射电镜等。此外，Raman 散射也可以通过涉及晶格振动的特征峰及峰宽来判断晶体结构及其对称性。表 11-1 给出了这几类技术的一个简单的总结比较。

综上可知，同是表征材料结构的技术，得到的信息、针对的物质元素形态仍有不同，运用这些技术相互弥补，能够得到锂离子电池材料精细的结构及其变化情况。以下对其简单介绍、总结与探讨。

表 11-1　锂电池研究中所采用的材料晶体结构表征技术比较

技　术	探针粒子	探测粒子	得到的信息	特　点
X 射线衍射	光子	光子	材料内部平均结构信息	反映块体材料平均晶体结构性质,平均的晶胞结构参数变化,拟合后可以获得原子占位信息
扩展 X 射线吸收精细谱	光子	光子	材料内部局部的结构变化	同步辐射光源辐射的高强度 X 光源,探测局部区域的结构变化,获得配位环境、近程结构信息
中子衍射	中子	中子	材料的结构和内部粒子的运动学信息	与原子核作用,可以区分邻近原子及同位素,探测 Li 等轻原子,中子穿透强,可以做原位实验;变温后可以获得原子在晶格中的扩散轨迹
核磁共振	电磁波	电磁波	材料内部的局部结构和电子性能	对正极材料掺杂对电子结构的影响灵敏;可以探测离子扩散系数
球差校正扫描透射电镜	电子	电子	局部区域的原子结构、排布	原子级实空间成像,可清晰看到晶格与原子占位;对样品要求高;可以实现原位实验
Raman 光谱	光	光	特征晶体结构的识别、结晶度、晶格畸变等信息	可以无损、非原位检测,也可以 2D 成像、时间分辨;对样品要求不高,但要有 Raman 活性

11.3.1　X 射线衍射

　　X 射线衍射是通过入射的 X 射线光子与样品中原子的电子相互作用,出射的衍射可以反映材料的结构信息,是一个平均的统计结果。通过 XRD,可以获得材料的晶体结构、结晶度、应力、结晶取向、超结构等信息。微聚焦 XRD 具有一定的空间分辨能力,目前基于同步辐射光源技术可以实现 20nm 以上区域的衍射,今后发展的目标是空间分辨率达 1nm。目前 XRD 衍射探测器可以实现毫秒量级的记录,今后会朝着更快的方向发展。对于锂离子电池研究而言,现有的时间分辨水平已经能够完成大多数的原位 XRD 实验。此外,快速记录的能力还被用来发展高通量样品表征手段,有利于实现多批次样品的筛选表征。

　　李国宝等[92]利用 XRD 区分了人造石墨、1600℃ 处理的炭黑、1600℃ 处理的焦炭、1200℃ 处理的石油焦、1600℃ 处理的石油焦种碳族负极材料,并讨论了嵌锂容量与晶粒尺寸的关系。Peres 等[93]对 $Li_{0.98}Ni_{1.02}O_2$ 电化学脱锂后的 $Li_{0.63}Ni_{1.02}O_2$ 正极材料进行了 XRD 精修,发现与层状的 $NaNiO_2$ 相比,结构扭曲很微弱,John-Teller 效应并不是晶格扭曲的驱动力。Thomas 组[94,95]利用 XRD 直接观察了过渡金属氧化物嵌锂过程中 $Li_xV_6O_{13}$ 相($x=0$、0.5、1、2、3、6)的变化序列。为了研究连续嵌锂过程中晶体发生的一系列变化,分析电池循环中的电化学反应,Thurston 等[96]首次将原位的 XRD 技术应用到锂离子电池中。通过利用同步辐射光源的硬 X 射线探测原位电池装置中的体电极材料,直观地得到了晶格膨胀和收缩、相变、多相形成的结果。Eriksson 等[97]通过原位的 XRD 研究了 $LiMn_2O_4$ 正极材料的结构和动力学特性,发现在不同倍率下充放电峰位有所移动但峰形都没有宽化,解释了球磨的 $LiMn_2O_4$ 基正极材料有高的充放电倍率性能的原因。杨晓青等[98~99]基于同步辐射的原位 XRD 研究了 $LiMn_{0.5}Ni_{0.5}O_2$ 正极材料和碳包覆的 Si 负极材料,$LiMn_{0.5}Ni_{0.5}O_2$ 材料 2V 以上的充放电伴随着类似 $LiNiO_2$ 材料 H1 到 H2 的可逆相转变,而 $LiNiO_2$ 材料 4.3V 以上的 H3 相被抑制。Orikasa 等[100]首次利用时间分辨 XRD 对 LiFePO$_4$ 的相转变进行分析，发现在 XRD 图谱中 19.35°左右出现了一个新相，证明 LiFePO$_4$ 和 FePO$_4$ 两相之间存在亚稳态的中间相，而正是这一中间相使 LiFePO$_4$ 展现出优异的倍率性能（图 11-26）。

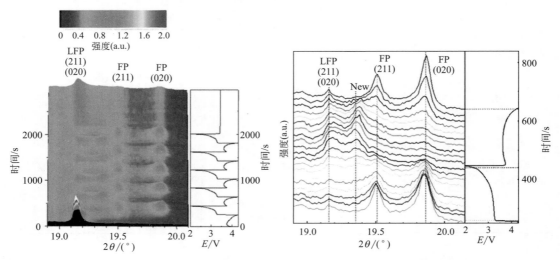

图 11-26　LiFePO$_4$ 在 10C 下的时间分辨 XRD 图谱（右侧是测量时的电压曲线）

负极材料中，研究发现碳包覆的 Si 负极材料的脱嵌锂反应首先发生在石墨中，然后在 Si 中，解释了该材料高比容量和循环性能的原因。Hatchard 等[101]利用原位 XRD 研究了非晶硅（a-Si）与 Li 合金化的反应，研究表明 2.5nm 是 a-Si 薄膜厚度的一个临界尺寸，只有大于这个厚度时，Li$_{15}$Si$_4$ 的晶相才会在放电到 30mV 时形成。Misra 等[102]通过原位同步辐射 X 射线衍射技术研究了硅纳米线负极中的锂离子嵌入/脱出机制，研究表明，亚稳态 Li$_{15}$Si$_4$ 在较低的电位下就可以形成，而硅纳米线的生长温度会影响该相的形成。此外，Yu 等[103]采用原位 XRD 技术研究了 Li$_4$Ti$_5$O$_{12}$ 在钠离子电池循环过程中的结构演变，并证明其中钠离子电池的容量与 Li$_4$Ti$_5$O$_{12}$ 的粒子尺寸有着紧密的联系。随反应的进行，粒径尺寸为 44nm 的 Li$_4$Ti$_5$O$_{12}$ 新产生的（400）和（440）峰不断加强，直至 2h 后才反应完全［图 11-27(a)］。而粒径尺寸增大至 120nm 以后，反应 24 h 后仍未完全进行。进一步增大尺寸至 440nm 后发现，在反应开始的 8h，Li$_4$Ti$_5$O$_{12}$ 没有表现出任何的相变［图 11-27(b)~(d)］。经对比发现，44nm 的 Li$_4$Ti$_5$O$_{12}$ 展现出最高的储钠容量［图 11-27(e)］，这有力地证明了 Li$_4$Ti$_5$O$_{12}$ 的储钠性能与其颗粒尺寸密切相关，这一发现也为以后 Li$_4$Ti$_5$O$_{12}$ 在钠离子电池中的进一步发展指明了方向，并奠定了良好的基础。随后几年中，原位 XRD 技术被广泛用来研究 Ag$_{52}$Sn$_{48}$ 纳米复合物、LiM$_{0.5-x}$Mn$_{1.5+x}$O$_4$、（1−x）LiNiO$_2$·xLi$_2$TiO$_3$、Li$_{1-x}$Ni$_{0.8}$Co$_{0.15}$Al$_{0.05}$O$_2$、LiCo$_{1/3}$Ni$_{1/3}$Mn$_{1/3}$O$_2$、MnO$_2$、Nb$_2$O$_5$、ZrO$_2$ 包覆的 LiMn$_{0.5}$Ni$_{0.5}$O$_2$、LiFePO$_4$、TiSnSb 等多种正负极材料[62,104~113]。

此外，有些电极材料在非常温条件下易发生相变反应，这会对材料的储能性能和安全性能造成严重影响。Yao 等[114]利用 in-situ XRD 技术研究了锂空气电池中 Li$_2$O$_2$ 和 Li$_2$O 的热稳定性。Li$_2$O$_2$ 的晶格参数在 280℃时显著减小，c/a 比升高。随着温度升高 Li$_2$O$_2$ 转化为 Li$_2$O$_{2-\delta}$，在 300℃以上出现 Li$_{2-\delta}$O 相。Sun 等[115]使用原位高能量的 XRD（HEXRD）

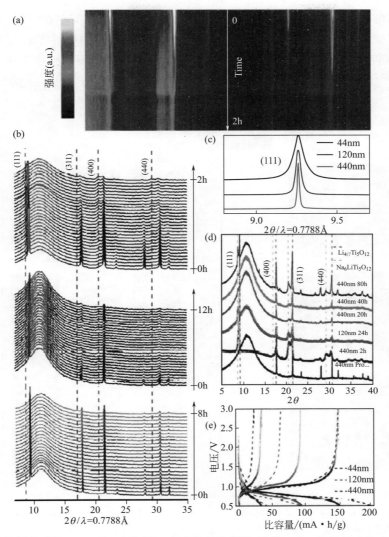

图 11-27 粒径尺寸为 44nm 的 $Li_4Ti_5O_{12}$ 反应 2 h 的 XRD 图谱（a）；粒径尺寸为 44nm（上），120nm（中），亚微型（下）的 $Li_4Ti_5O_{12}$ 在电化学反应中的原位 XRD 谱图（b）；不同晶粒尺寸的 $Li_4Ti_5O_{12}$ 的（111）晶面的 XRD 放大图（c）；不同尺寸的 $Li_4Ti_5O_{12}$ 在电化学反应不同时间的 XRD 谱图（d）；不同晶粒尺寸的 $Li_4Ti_5O_{12}$ 的充放电曲线（0.1C，0.5~3.0V）（e）

研究正极材料 $LiNi_{0.75}Co_{0.10}Mn_{0.15}O_2$（FCG）、脱锂后的正极材料（IC：$Li_{1-x}Ni_{0.86}Co_{0.10}Mn_{0.04}O_2$）与电解液在温度从室温加热到375℃时的热分解反应，证明FCG材料比IC材料具有更高的温度范围和更好的安全性。近来，Mai 等[116]在电池控温系统发现 $LiFePO_4$ 与 $FePO_4$ 两相在低温下的相转变存在中间相。研究表明，该中间产物可以有效地阻止随着温度降低离子扩散速率的衰减，而且由于 $FePO_4$ 到 $LiFePO_4$ 之间较高的活化能，放电过程比充电过程有更多的中间产物生成。293K 和 273K 温度下，电池在不同扫描速率（分别对应于1.4mV/s，2.8mV/s，4.2mV/s）下，41.7°~42.8°范围内持续强峰的出现[图11-28(a)~(c)]或峰形的变宽、峰位向高角度的偏移[图11-28(d)~

(f)]）均说明电池循环过程中 $LiFePO_4$ 与 $FePO_4$ 之间的相变确实存在中间产物。而273K下的离子扩散速率相比其他温度而言衰减较小，而这恰恰是273K温度下的中间产物使得扩散速率衰减较慢的原因［图11-28(g)］。此外，循环过程中不同温度下的晶格参数［b值，图11-28(h)］的变化也可以进一步说明二者之间的固液反应。这一方法对其他相关电化学测试的原位变温研究提供了宝贵的借鉴意义。

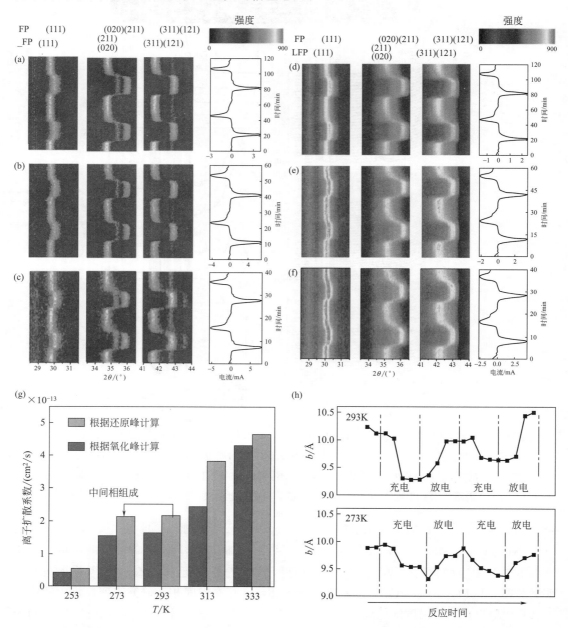

图11-28　293K时，在不同扫描速率下(111)，(211)，(020)，(311)，(121)的衍射图像［(a)1.4mV/s，(b) 2.8mV/s，(c) 4.2mV/s］；273K时，在不同扫描速率下(111)，(211)，(020)，(311)，(121)的衍射图像［(d) 1.4mV/s，(e) 2.8mV/s，(f) 4.2mV/s］；不同温度下 $LiFePO_4$ 的离子扩散速率（g）；273K和293K下精修得到的晶胞参数 b 随反应时间的变化（h）

11.3.2 扩展X射线吸收精细谱

扩展X射线吸收精细谱（extended X-Ray absorption fine spectroscopy，EXAFS）是通过X射线与样品的电子相互作用，吸收部分特定能量的入射光子，来反映材料局部结构差异与变化的技术，具有一定的能量和时间分辨能力，主要获得晶体结构中径向分布、键长、有序度、配位数等信息；通常需要同步辐射光源的强光源来实现EXAFS实验。

Rougier等[117]采用扩展X射线吸收精细谱（EXAFS）研究了准二维的$LiNiO_2$嵌锂后NiO_6八面体由于Jahn-Teller效应导致局部的形变，其中Jahn-Teller是由于Ni^{3+}处在低自旋态引起的。其与$NaNiO_2$不同，没有宏观形变，只有微观的形变，这对后期嵌锂有至关重要的作用。随后研究者们对其变体（如$Li_{0.65-z}Ni_{1+z}O_2$[86]、$LiCo_{1-y}Ni_yO_2$[61,64]、$LiCo_{1-y}Fe_yO_2$[64]等）分别作了类似的研究。Chadwick等[118]在对$LiMnO_2$的研究中没有发现Jahn-Teller效应引起的Mn离子周围配位原子的畸变。Paolone等[119]对$LiMn_2O_4$这类价格便宜、有一定应用前景的材料做了研究，表明存在两种Mn—O键长，当在高温测量时只存在一种Mn—O键长，正好与中子散射的数据一致。Chadwick等[118]则在$LiMn_2O_4$的基础上研究了掺杂Co、Ni、Ga、In样品的结构，发现Co、Ni替代了$LiMn_2O_4$中Mn的位置，而In、Ga则既占据了Li位，也占据了Mn位。除了正极材料，Jung等[120,121]通过用EXAFS分析研究了嵌SnO_x/CuO_x的碳纳米负极材料的电化学性质，表明嵌SnO_x/CuO_x的碳纳米纤维具有一个无序的结构，形成了SnO_x颗粒的特殊分布，由此导致电化学性能有所提升。由于硬X射线主要用于探测过渡金属K吸收边的电子，若要探测更轻元素（如C、N、P、O、F等）的变价需要软X射线吸收谱。另外，软X射线也能探测过渡金属L吸收边的电子。刘啸嵩等[76,122]通过软X射线给出了相变以及Li_xFePO_4纳米颗粒和单晶脱嵌锂对电子结构影响的研究，原位软X射线Fe的L边吸收谱形象地显示了脱嵌锂的过程，谱的两相拟合与两相转变反应一致，提供了Li在电极上成分和分布梯度的精确信息。

元素敏感的、时间分辨的Quick-EXAFS技术可以研究当多种可发生电荷转移的元素共存时，各自的响应动力学。目前似乎只有这种技术能实现这一功能。最近，禹习谦等[123]在研究富锂正极材料$Li_{1.2}Ni_{0.15}Co_{0.1}Mn_{0.55}O_2$ [$0.5Li(Ni_{0.375}Co_{0.25}Mn_{0.375})O_2 \cdot 0.5Li_2MnO_3$]动力学特性时，通过XAS谱发现除了Ni、Co还原反应贡献的容量外，一大部分可逆容量来源于富锂层状材料。Ni、Co的电荷转移动力学较快，与Mn相关的过程中脱锂的动力学相比于Ni、Co的都很差，可以判断，影响富锂材料动力学的组分是Li_2MnO_3。

11.3.3 中子衍射

从原理上讲，X射线衍射是光子与原子的核外电子相互作用，核外电子越多，作用强度越大。因此，当锂离子电池材料中有较大的原子存在时，X射线将难以对锂离子占位进行精确的探测。与X射线不同，中子衍射是中子与原子的原子核相互作用，其作用强度对于特定的原子是特定值。中子对锂离子电池材料中的锂较敏感。因此中子衍射在锂离子电池材料的研究中发挥着重要作用。如图 11-29 所示，给出了 1994～2013 年，中子衍射应用到锂离子电池中产生的文章数量变化趋势。由此可见，中子衍射在锂离子电池中应用越来越广泛。

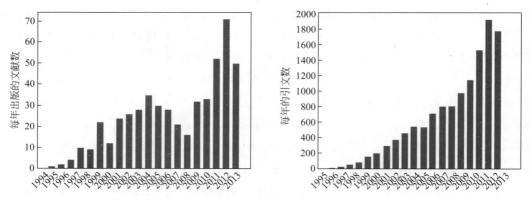

图 11-29　中子衍射应用到锂离子电池中产生的文章数量

图 11-30 为化学脱锂前后 Li_2MnO_3 的中子衍射谱[117]。中子衍射能准确给出锂离子电池材料中的 Li^+ 占位。2013 年，Arbi 等[124]通过中子衍射确定了锂离子电池固态电解质材料 LATP 中的 Li^+ 占位，他们使用中子衍射数据反演得到傅里叶图，从而给出了此材料中 Li^+ 的准确占位。由于中子与锂原子核相互作用时存在半波损，所以 Li^+ 的中子散射长度为负值（中子散射长度为 -0.214×10^{-12} cm），而其他大部分过渡金属原子的中子散射长度为正值，利用此特性可以很好地测定锂离子电池材料中的互占位问题，在提高材料的脱嵌锂速率研究中发挥作用。2014 年，Li 等[125]借助中子衍射与 X 射线衍射相互补充，确定了锂离子电池正极材料 $LiFePO_4$ 中存在 Li/Fe 互占位，利用 Rietveld 精修定量给出了互占位的比例，进一步证实了此材料中 Li/Fe 互占位。中子衍射在锂离子电池材料中的另一个重要应用是确定材料中的 Li^+ 运动通道。利用中子衍射结合最大熵模拟计算方法是目前唯一可以间接观察锂离子电池材料中 Li^+ 运动通道的实验方法。2008 年 Nishimura

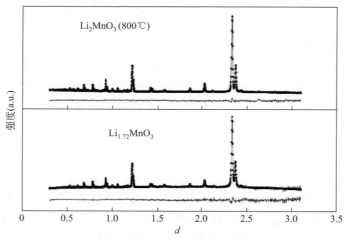

图 11-30　化学脱锂前后 Li_2MnO_3 的中子衍射谱[117]

等[126]利用常温和高温（620K）中子衍射数据结合最大熵方法给出了目前广泛应用的锂离子电池正极材料之一的 LiFePO$_4$ 中的一维弯曲 Li$^+$ 运动通道。2012 年 Han 等[127]利用一系列温度梯度的中子衍射数据结合最大熵方法给出了锂离子电池材料 Li$_7$La$_3$Zr$_2$O$_{12}$ 中的 Li$^+$ 运动通道。汪锐等[128]利用中子衍射技术研究了富锂相 Li$_2$MnO$_3$ 化学脱锂前后结构的变化。并通过计算验证了化学脱锂后 Li$_{1.75}$MnO$_3$ 中 1.75 个 Li 占据 2c 位和 4h 位的量分别为（0.7±0.3）和（0.9±0.1），初步发现 Mn、O 也存在在晶格中移动的可能。

做一次中子衍射实验需要较长时间，通常需要较大的样品量。最近发展的散裂中子源把中子的通量提高了几个数量级，这不仅可以提高数据的质量，还可以缩短实验时间。伴随着英国的 ISIS 升级完成，美国 SNS 和日本 J-PARC 的相继投入运行，中国散裂中子源（CSNS）也紧跟世界的步伐，2018 年已投入运行，我国目前为世界四大散裂中子源研究中心之一。此外，欧洲散裂中子源（ESS）也已经宣布动工开建，预计 2023 年建成运行。随着世界各大散裂中子源的相继修建和投入运行，必然有越来越多的中子散射研究人员投入到锂离子电池的研究之中，这对于锂电池的发展具有重大助益。

11.3.4 核磁共振

核磁共振（nuclear magnetic resonate，NMR）技术是观测原子的方法，额外施加电磁场来干涉低能态的核自旋转向高能态，再回到平衡态便会释放出射频，产生 NMR 信号。1946 年，费利克斯·布洛赫和爱德华·珀塞尔发现,将具有奇数个核子（包括质子和中子）的原子核置于磁场中，再施加以特定频率的射频场，就会发生原子核吸收射频场能量的现象，这是人们最初对核磁共振现象的认识。NMR 具有高的能量分辨、空间分辨能力，能够探测材料中的化学信息并成像。Grey 等[129]对 NMR 在锂离子电池正极材料中的研究开展了大量的研究工作，并进行了总结。表明从正极材料的 NMR 谱中可以得到丰富的化学信息及局部电荷有序无序等信息，并可以探测顺磁或金属态的材料，还可以探测掺杂带来的电子结构的微弱变化来反映元素化合态信息。另外结合同位素示踪还可以研究电池中的副反应等。NMR 也能用于负极材料的研究，Key 等[130]通过原位的 NMR 研究了 Si 的放电过程，发现放电过程中首先在 Li 阵列中形成独立的 Si 原子和小的 Si—Si 团簇，在放电末态 Si—Si 键断裂，全部形成独立的 Si 原子，同时锂硅化物与电解液自发反应，解释了自放电和容量损失的原因，该效应在加入 CMC 添加剂后显著改善。NMR 也具有时间分辨的能力，Bhattacharyya 等[131]开发了一种时间分辨的原位 NMR 技术，定量分析金属 Li 表面枝晶的形成和脱出。这一实验的前提假设是 Li 的总量是一定的，NMR 信号的强度变化完全源于 Li 表面枝晶的形成。Wiaderek 等[132]将 NMR 谱与 PDF 卡片数据深度分析相结合，给出了对 FeOF 负极材料充放电机理的深刻认识，认为在循环一周后，虽然表现出容量可逆的性能，但 NMR 谱及相关 PDF 数据的键价和分析表明充电末态得到的材料并不是初始材料的相，而是变成了富 F 的金红石相和富 O 的岩盐相，且 Fe 优先在富 F 的金红石相中发生反应。NMR 技术还可以用于探测材料的电导率。

目前采用 NMR 的成像技术可以对一个颗粒内部相转变反应进行研究，Ilott 等[133]利用时间分辨率足够高的原位核磁共振成像（*in-situ* MRI）技术成功获得了电化学双层电容器电池（EDLCs）中工作电极离子占位与化学位移信息的实时变化图谱，详细地说明了多孔碳电极吸附环境的化学位移机制，并通过化学位移图像可以得到充放电过程中占主导地位的电荷储存机制[133]。

11.3.5 球差校正扫描透射电镜

球差校正扫描透射电镜（spherical aberration-corrected scanning transmission electron microscopy，STEM）技术用球差校正器连续减小物镜球差系数至零值或负值。球差系数绝对值很小时，电镜的空间分辨本领大大提高，达到甚至超过电镜的信息极限。空间分辨本领高于信息极限时，像的最佳分辨率将取决于信息极限[134]。因此，球差校正的扫描透射电镜可以用来观察原子的排布情况。Yang 等[135]首次利用亚埃级透射电镜观察到 Li、Co、O 的原子级排布的晶格条纹，但这并不是实空间真实的原子排布。Oshima 等[136]利用环形明场成像的球差校正扫描透射显微镜（ABF-STEM）观察到了 Li_2VO_4 中 Li、V、O 在实空间的原子排布，其中明暗强度中菱形晶格中的 V、O、Li 的比例为 4∶2∶1。Lu 等[137]制备了 $TiNb_2O_7$ 负极材料，研究了它的电化学脱锂嵌锂的性质，循环伏安法检测到 Nb^{5+}/Nb^{4+}、Ti^{4+}/Ti^{3+} 的氧化还原电对，但是 Nb^{4+}/Nb^{3+} 的氧化还原电对没有被检测到。STEM 观察到的嵌锂位与第一性原理计算得到的结果一致。Huang 等[138]通过对 PLD 制备的 $LiCoO_2$ 外延膜退火前后原子在晶格的排布变化的研究，发现退火过程中，Li 层和 Co 层的原子在移动，而 O 层的原子保持不动。同时退火导致界面变粗糙，限制了 $LiCoO_2$ 和 Al_2O_3 基底的界面反应。Hayashi 等[139]对商业的电池做了研究。STEM 表明正极材料表面在循环后存在一层很厚的退化的表面层。笔者课题组与谷林合作应用此技术在锂电池材料方面开展了一些研究工作，从原子级别的角度观察了 $LiCoO_2$[140]、Li_2MnO_3[141]、Li_2MoO_3[75]等正极材料的脱嵌锂机制。如图 11-31 所示，从原子级的视角直观地观察到了 Li_2MnO_3 在充放电过程中 Li、Mn、O 原子的占位的变化过程。然后通过 O 平面、Li 平面以及 LiMn 平面的分析，得出 Li 可以从 LiMn 层脱出又重新嵌回局部区域。Mn 也会在充放电过程中迁移到 Li 层。

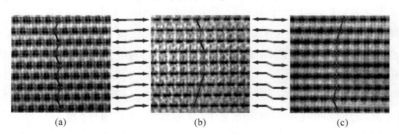

图 11-31　ABF-STEM 电镜图[117]

(a) 原始 Li_2MnO_3；(b) 充电到 4.8V 的 Li_2MnO_3；(c) 充电到 4.8V 后放电到 2V 的 Li_2MnO_3

锂离子电池的动力学特性、结构稳定性、表面反应均与材料中的原子占位有关，特别是 Li^+、O^{2-}。原子尺度的观察结果可以与宏观属性直接关联，具有十分重要的意义。此外，结合高空间分辨的 EELS，可以实现电荷有序的直接观察。Lu 等[141]利用球差校正的高分辨扫描透射电镜获得了与理论计算一致的 $Li_4Ti_5O_{12}$ 晶格照片，环形明场相的照片清楚展示了脱嵌锂过程中，Li 占位随嵌锂含量的增多从 $8a$ 位迁移到 $16c$ 位。结合 EELS 谱表明，钛离子根据 Li^+ 与 e^- 的相互作用强弱在局部区域显示出不同的化合价，表现出 Ti^{3+}/Ti^{4+} 电荷有序性，这与理论计算结果一致。

固体离子学的终极梦想之一是直接观察固体中 Li^+ 的运动，相信今后时间分辨（fs）的球差电镜技术如果能有幸研制成功，将能回答长期存在的科学问题。

11.3.6　Raman 光谱研究晶体结构

Raman 光谱可以用于研究晶体结构，早期的研究集中关注于 Li_xCoO_2、$Li_xMn_2O_4$、$Li_xV_2O_5$ 等材料。$LiCoO_2$ 是最早商业化应用的正极材料，具有 α-$NaFeO_2$ 的构型，最早的 $LiCoO_2$ 的拉曼光谱由 Inaba 等[142]报道，对用 Ni 取代了 Co 的位置后进行了表征。$LiCoO_2$ 中有两种拉曼活性模式，Co—O 伸缩振动 A_{1g} 的峰与 O—Co—O 的弯曲振动 E_g 的峰[143]。另一种具有层状结构的正极材料 $LiNiO_2$ 以及二元材料 $LiNi_{1-y}Co_yO_2$（$0 < y < 1$）也获得了广泛研究，使用 Ni 替代部分 Co 之后并不改变原来材料的空间群，但随着 Ni 含量的增加，$LiCoO_2$ 原有的 A_{1g} 和 E_g 拉曼振动活性强度变弱，因此 $LiNiO_2$ 的拉曼散射强度很弱，这也和菱形结构的坍塌与 $LiNiO_2$ 电子电导更高有关系[144]。三元正极材料 Li($Ni_{1/3}Co_{1/3}Mn_{1/3}$)O_2 自从 2001 年被 Ohzuku[145]报道后，也得到了广泛的研究。Li($Ni_{0.8}Co_{0.1}Mn_{0.1}$)O_2 的拉曼光谱证实了在 3～4.5V 的充放电过程中原有的层状结构不会发生改变，此斜六方的层状结构在 500cm^{-1} 和 560cm^{-1} 的位置上有标志峰[146]。$LiNi_{0.5}Mn_{1.5}O_4$ 有两种晶体结构，空间群为 Fd-$3m$ 和 $P43_32$。Fd-$3m$ 结构，Li 原子占 8a 位置，Ni 原子和 Mn 原子随机占 16d 位置，O 原子占 32e 位置。因为 Ni/Mn 原子随机占位，这种结构也被称为"无序"结构。这种结构常含有岩盐相杂质，如 Ni_xO[147]、$Li_xNi_{1-x}O_2$[147]或 $LiNiMn_xO_2$[148]，同时其尖晶石相内有氧缺陷[149]，所以晶胞参数随着氧缺陷不同可从 8.165Å（1Å=0.1nm）到 8.277Å（因为 $LiNi_{0.5}Mn_{1.5}O_{4-\delta}$ 中，更多的氧缺陷，会产生更多较大半径的 Mn^{3+}）。空间群为 $P43_32$ 的结构，通常为化学计量比的 $LiNi_{0.5}Mn_{1.5}O_4$，Li 原子位于 8c 位置，Ni 原子在 4a 位置，Mn 原子在 12d 位置，O 原子在 8c 和 24e 位置。因此，这种结构也叫"有序"的结构。不同文献报道的晶胞参数从 8.117Å 到 8.170Å[149]。这些晶胞参数的不同跟合成的路径以及表征技术有关。由于 Ni 和 Mn 的电子散射因子相似，普通的 XRD 难以区分，长时间扫描或使用同步加速器 X 射线衍射利用精修可以达到目的。中子衍射是一种区分不同原子核的元素的有效方法，因为中子是与核相互作用的，不同于与电子的相互作用。拉曼光谱也可以成功地区分这两种结构。如图 11-32 所示，Fd-$3m$ 相的 Raman 光谱 Ni—O 振动峰不尖锐，只是一个小鼓包[150]。而对于锂离子电池常用的负极材

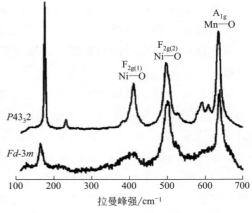

图 11-32　Fd-$3m$ 和 $P43_32$ 结构的 Raman 图谱[137]

料石墨，拉曼光谱也是一个极其有利的表征方式。对于碳材料，在拉曼光谱中有 D 峰和 G 峰这两个最显著的表征峰，D 峰代表的是碳材料内某些缺陷的存在或者形变，G 峰代表的是碳原子 sp^2 杂化的面内伸缩振动。Hardwick 等[151]利用原位拉曼光谱技术，对石墨表层嵌锂脱锂过程造成的石墨化的无序和畸变进行了研究，发现随着电位的降低 D 峰逐渐消失，意味着这种形变不均匀，即嵌锂是不均匀的，即使倍率降至很低（$C/37$ 以下），也不能使嵌锂均匀，这种在嵌锂过程造成的畸变是不可逆的。

11.4 物质官能团的表征

官能团又称官能基、功能团，是决定有机化合物化学性质的原子和原子团。常见官能团有烃基、含卤素取代基、含氧官能基、含氮官能基以及含磷、硫官能团 5 种。常用的表征官能团的技术有拉曼散射光谱（Raman scattering spectroscopy，RS）、傅里叶变换红外光谱（Fourier transform infrared spectroscopy，FTIR）、紫外光谱（ultra violet spectroscopy，UV）等。这些光谱技术都具有一定的能量分辨、时间分辨能力。紫外光谱主要用于溶液中特征官能团的分析，如溶液中多硫离子、多碘离子的分析。这里主要介绍拉曼散射光谱和傅里叶变换红外光谱。

11.4.1 拉曼散射光谱

拉曼光谱属于分子振动光谱，但它得到的振动能级信息不是来自样品对光的吸收，而是来自样品的散射。拉曼效应是由印度物理学家拉曼在单色光照射液体苯后散射出与入射光频率不同谱线的实验中发现的，从拉曼光谱可以得到分子振动和转动的信息。拉曼效应产生的原因在于入射光与物质发生相互作用，如果散射光的频率不发生变化，则发生瑞利散射；如果散射光的频率低于入射光，即一部分入射光把能量传递给物质，发生斯托克斯散射；如果散射光的频率高于入射光，即入射光从物质内部得到部分能量，发生的则是反斯托克斯散射。通常斯托克斯散射的强度要比反斯托克斯散射强得多，因此在拉曼分析中，通常使用斯托克斯散射光线。而斯托克斯光的频率与激发入射光源的频率之差称为拉曼位移。

拉曼光谱适用于对称结构、极性较小的分子，例如对于全对称振动模式的分子，在激发光子的作用下，会发生分子极化，产生拉曼活性，而且活性很强。通常准备测量拉曼光谱的样品无需特殊处理，在锂离子电池电极材料的研究中有着广泛的应用。Baddour-Hadjean 与 Pereira-Ramos 在 2010 年的一篇综述中较详细地讲述了拉曼光谱在锂电池中的应用[152]。在锂离子电池电极材料表征时，由于拆卸和转移过程难免人为或气氛原因对电极材料造成干扰，因此原位技术与拉曼光谱一起用在了电极材料的表征上。拉曼光谱对于材料结构对称性、配位与氧化态非常敏感，可用于测量过渡金属氧化物。

对于拉曼光谱的灵敏度不够的情况，可以使用一些 Au 和 Ag 等金属在样品表面进行处理，由于在这些特殊金属的导体表面或溶胶内靠近样品表面电磁场的增强导致吸附分子的拉曼光谱信号增强,称为表面增强拉曼散射(SERS)。SERS 可以很好地测量拉曼信号很弱的材料,李桂峰等[153]利用 SERS 对锂离子电池负极材料表面形成的 SEI 膜进行了研究。在研究中使用 Au 片和 Ag 片作为电极,发现银片低电位下形成的 SEI 膜主要为非晶态的 Li_2CO_3 和 $LiOH \cdot H_2O$,而在金片放电至 1.0V 的拉曼光谱中还观察到了 LiOH 和 $RCOCO_2Li$ 的存在。而在电解液微水状态下,$RCOCO_2Li$ 会和水发生反应生成 Li_2CO_3,因此 SEI 膜能稳定存在的成分为 Li_2CO_3、LiOH 和 $LiOH \cdot H_2O$ 以及拉曼光谱中难以观测到而在红外中已经证实的 LiF。在锂空气电池中,大家普遍认为放电过程会有中间产物 LiO_2 的产生,但由于 LiO_2 的不稳定性,现有的很多手段难以证实其存在。Peng 等[154]利用 SERS 的手段证实了锂空气电池充放电过程中确实存在着中间产物 LiO_2,而在充电过程中 LiO_2 并没有观测到,说明了锂空气电池的放电过程是一个两步反应过程,以 LiO_2 作为中间产物,而充电过程是不对称的一步反应,Li_2O_2 的直接分解,由于 Li_2O_2 导电性差分解困难,这也是导致充电极化大于放电极化的原因。

11.4.2 红外光谱

红外光谱使用的波段与拉曼类似,不少拉曼活性较弱的分子可以使用红外光谱进行表征,红外光谱也可作为拉曼光谱的补充,红外光谱也称作分子振动光谱,属于分子吸收光谱。当样品受到频率连续变化的红外光照射时,分子吸收其中的一些频率的辐射,分子振动或转动引起偶极矩的净变化,使振动或者转动的能级从基态跃迁至激发态,对应于这些区域的透射光光强减弱,记录下来的透过百分率对波数或者波长的曲线,也就是红外光谱。红外光谱主要用于化合物鉴定与分子结构表征,也可用于定量分析。与拉曼光谱相比,极性官能团的红外光谱更为强烈,如 C═O 伸缩振动的红外光谱比相应的拉曼光谱更为显著。依照红外光区波长的不同可以将红外光区分为三个区域:①近红外区,即泛频区,指的是波数在 $4000cm^{-1}$ 以上的区域,主要测量 O—H、C—H、N—H 键的倍频吸收;②中红外区,即基本振动区,波数范围在 $400\sim4000cm^{-1}$,也是研究和应用最多的区域,主要测量分子振动和伴随振动;③远红外区,即分子振动区,指的是波数在 $400cm^{-1}$ 以下的区域,测量的主要是分子的转动信息。但是由于水是极性很强的分子,它的红外吸收非常强烈,因此水溶液不能直接测量红外光谱,通常红外光谱的样品需要研磨制成 KBr 的压片。

通常红外光谱的数据需要进行傅里叶变换处理,因此红外光谱仪和傅里叶变化处理器联合使用,称为傅里叶红外光谱(FITR)。在锂离子电池电解液的研究中,使用红外光谱手段的工作较多。Mozhzhukhina 等[155]利用红外光谱对锂空气电池电解液常用的溶剂二甲基亚砜(DMSO)的稳定性进行了研究,发现 DMSO 在锂空气电池中无法稳定主要是由于超氧根离子(O_2^-)的进攻,而在红外光谱中观测到 SO_2 的信号存在,这个反应难以避免,即使在低至 3.5V 的电位下,DMSO 也无法稳定。与拉曼光谱相同,红外光谱也结合原位电池技术,实时地对电池充放电过程中电极材料的变化进行研究。Novák 等[156]利用原位红外技术,对石墨负极的表层进行了研究,随着电压的增大即放电深度或者说嵌锂的增加,可以看到原本可以表征石墨的特征峰逐渐消失,意味着嵌锂过程带来了石墨表层结构的畸变。

11.4.3 色谱技术

自 20 世纪 50 年代以来，随着气相色谱分离技术以及其他科学技术的发展，气相色谱 (GC) 开始广泛地应用到物质分析当中。GC 主要是利用物质的沸点、极性及吸附性质的差异来实现混合物的分离。配备检测器或联用其他设备，可较为方便地获得丰富的信息。随着电池技术的发展和深入研究，GC 在锂离子电池中的应用也越来越频繁，特别在电池胀气分析、电解液组成分析以及电解液老化机理的研究方面扮演着越来越重要的角色。

锂离子电池的产气主要源于 SEI 层形成过程中电解质溶剂的电化学分解，是限制其安全性能和寿命的关键问题之一。GC 配备检测器（如常见的热导检测器 TCD），可以对不同电解液电池的产气进行分析，如常见的气体 CO_2、CO、H_2、CH_4、C_2H_2、C_2H_4、C_2H_6、C_3H_6 和 C_3H_8 等。Teng 等使用 GC-TCD 原位研究软包电池不同碳酸盐的电解液的产气组成。研究发现复合电解质 GBL/EMC（γ-butyrolactone/ethylmethyl carbonate）中产气量明显减少，有较高的安全性和较长的寿命[157]。陈伟峰[158]使用气相色谱研究高温存储、过充和过放条件下的电池胀气组成及产气机理。研究发现，高温环境下存储时，胀气主要来源于正极钴酸锂和电解液之间的氧化反应，所占比例超过总体积的 80%，其主要成分是 CO_2。对于负极，由于 LiC_6 在不同温度下的活性变化很大，因而产生的气体所占比重随温度变化非常大。在过充条件下，正极的含锂量越也越低，正极材料的氧化性也越强，主要的气体成分是由电解液反应产生的 CO_2。产气量和温度在充电电压高过 4.9V 以后，同时剧烈上升。锂离子电池在过放条件下，在过放前期气体成分是来自于负极 SEI 膜氧化分解所产生的烷烃和烯烃类气体，产气量随电压的降低而增加。在电压达到 1.0V 以下急剧增加，其主要的成分为 CO_2。

GC 通过色谱分离技术分离物质。不同物质的保留时间不同，依次从色谱柱中流出后，由检测器确定物质含量，常见有热导检测器（TCD）和氢火焰离子化检测器（FID）等。为了进一步加强分析能力，可联用质谱（MS）检测器。气相色谱接常规 MS 检测器后可对电解液中的溶剂和添加剂进行定性及定量分析，图 11-33 为几种常见溶剂和添加剂在 GC-MS 中的出峰时间（乙腈溶液，GC 型号：安捷伦 7890B）。

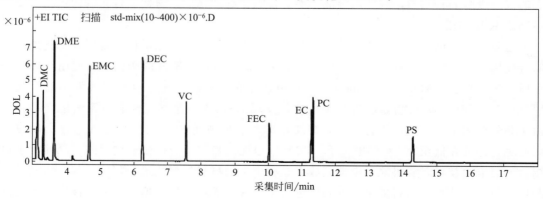

图 11-33 电解液常见溶剂和添加剂成分在 GC-MS 中的总离子流量图（TIES 测试分析中心）

电解液在循环过程中难免发生化学或电化学分解反应，导致电解液组分发生变化，为了进一步研究电解液老化过程的机理，需要对循环后的电解液组成进行分析。由于体系复杂，通常会综合使用多种分析技术，GC-MS 在其中同样也发挥着重要作用。

Ravdel 等[159]使用 GC-MS、^{19}F NMR 和 ^{31}P NMR 研究电解液 DMC-LiPF$_6$、DEC-LiPF$_6$ 和 EMC-LiPF$_6$ 热分解产物以研究不同电解液的热失效机理。Weber 等[160]使用 GC-MS 检测到了电解液的几种不同老化产物，其中一部分为氟磷酸盐，如二异丙基氟磷酸，另一部分为三烷基磷酸盐。Gachot 等[161]使用 GC-MS 结合 ESI-HRMS（电喷雾电离高分辨质谱）研究 EC-DMC/LiPF$_6$ 电解液的老化机理。ESI-HRMS 用于鉴定碳酸酯、烷氧基和环氧乙烷低聚物等中等挥发性化合物，提供电化学/化学驱动的老化机制。GC-MS 用于鉴定高挥发性有机化合物，两者结合建立完整的电解液老化机理。

相对于气相色谱，液相色谱法（LC）拥有不受样品挥发度和热稳定性限制的优点，在锂电池中也越来越广泛地用于电解液成分和杂质的分析。

Diao 等[162]在研究锂硫电池充放电过程中多硫离子在电解液中的溶解情况时，通过液相色谱-质谱联用（LC-MS）技术成功得到了电解液中溶解的多硫离子 Li$_2$S$_4$ 和 Li$_2$S$_6$ 的浓度。

Schultz 使用 GC-FID 和 HPLC-UV/vis（高效液相色谱-紫外/可见光分光光度计联用）两种方法分别研究从 EV 锂离子电池中提取的电解液的组成[163]。GC-FID 图中 DMC、EMC、DEC、EC 和电解液添加剂苯基环己烷的出峰时间分别为 3.56min、4.75min、6.13min、8.66min 和 11.02min（图 11-34）。DEC 的含量低于检测限。相较于 GC-FID，HPLC-UV/vis 对电解液添加剂和杂质物质的测量更加敏感。在 HPLC 谱图中（图 11-35），EC、DMC、EMC、DEC 和电解液添加剂苯基环己烷也均能检测出来，且出峰时间分别为 2.9min、3.6min、4.2min、5.0min 和 51.3min，DEC 的含量低于检测限。且 2.7min 和 3.2min 出现两个额外的峰，与标准谱图对比知 3.2min 的峰为 VC 的峰，使用 HPLC 与 ESI-MS 联用对 2.7min 的峰进行分析，得到其为 NMP 的峰。电解液中一些老化产物的峰也能在 HPLC 图中观察到，如 3.7min DMDOHC、5.1min DEDOHC。

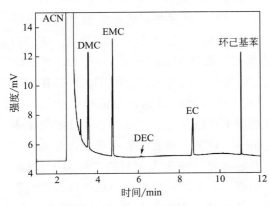

图 11-34　GC-FID 分析 EV 动力电池电解液[163]

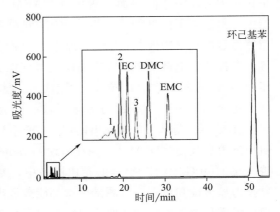

图 11-35　HPLC-UV/vis 分析 EV 动力电池电解液[163]

离子色谱法（IC）基于离子交换原理，可对共存的多种阴离子或阳离子进行分离、定性和定量。在锂电池中，IC 可用于原材料中元素含量的测定、电解液中阴阳离子的测定以及有机溶剂、黏结剂及隔膜等有机材料中离子的检测。

Liu[164]使用 IC 研究锂盐 $LiPF_6$ 溶液中 F^-、PO_4^{3-}、PF_6^- 的浓度，由测得的杂质离子 F^-，PO_4^{3-} 的存在与否判断 $LiPF_6$ 的纯度。Veluchamy[165]研究充电到不同电压的正极材料 Li_xCoO_2 表面的元素组成，使用 IC 测定得 F^-、Cl^- 和 CO_3^{2-} 的存在，由阴离子的种类推测出在 $LiCoO_2$ 电极材料的表面可能存在的物质是 LiF、Li_2CO_3 和 $LiCl$。朱彬和[166]为了研究五种常见锂盐：六氟磷酸锂（$LiPF_6$）、四氟硼酸锂（$LiBF_4$）、二草酸硼酸锂（LiBOB）、双（三氟甲基磺酰）亚胺锂（LiTFSI）、双氟磺酸亚胺锂（LiFSI）的含量，使用离子色谱法建立了五种锂盐阴离子（PF_6^-、BF_4^-、BOB^-、$TFSI^-$、FSI^-）的最佳分离条件。

11.5 材料离子输运的观察

表征离子输运特性的手段除了电化学方法之外，扫描隧道显微镜（scanning transmission microscopy，STM）、中子衍射（neutron diffraction，ND）、核磁共振（nuclear magnetic resonate，NMR）以及原子力显微镜（atomic force microscopy，AFM）系列技术也能提供相关信息。中子衍射结合最大熵模拟分析方法可以得到电极材料中的 Li^+ 扩散通道的信息[126]。Gobet 等[167]利用脉冲梯度场的 NMR 技术表征了 $\beta\text{-}Li_3PS_4$ 固体中 ^1H、^6Li、^7Li、^{31}P 核磁共振谱随热处理温度的变化，测得了 Li^+ 的自扩散系数，与之前报道的 Li^+ 电导率数量级一致。NMR 与中子衍射测量离子输运性质技术发展已经较为成熟，请参考文献 [168-179]。这里主要介绍原子力显微镜技术的应用。

原子力显微镜属于扫描探针显微镜一族，它是利用针尖原子与样品表面原子间的范德华作用力来反馈样品表面形貌信息。AFM 具备高的空间分辨率（约 0.1Å）和时间分辨能力，由于它不探测能量，并不具有能量分辨能力。AFM 以探测到的不同信息又分别有摩擦力显微镜（FFM）、电化学原子力显微镜（EC-AFM）、电化学应变显微镜（ESM）等，在锂离子电池材料研究中的应用得到极大的发展。最早 Hatzikraniotis 等[180]于 1996 年利用 AFM 观察 MoO_3 单晶嵌锂前后表面粗糙度的变化，认为嵌锂后的 MoO_3 的粗糙度增大主要源于锂嵌入导致局部体积域的形成。AFM 技术被应用到锂电池研究中。1997年，Hirasawa 等[181]首次提出使用 AFM 系列技术用于观察碳复合电极，多种技术测试表明动态力显微镜（DFM，即现在所说的 tapping mode AFM）最适于表征碳复合电极的形貌。在 HOPG/PVDF 混合的参比样品中，摩擦力显微镜、侧向力调制的摩擦力显微镜、黏弹力显微镜和仿真电流测试模型都能表征出黏结剂、石墨的明显界限。虽然没有看到石墨负极样品中与黏结剂的明显界限，但是黏弹力显微镜成的像确实显现出了衬度的变化。扫描探针显微镜在工作电极的形貌观察中具有无法估量的价值。由于制样及每次操作对样品的影响无法避免，原位观察的技术发展显得刻不容缓。Hirasawa 等[182]还首次采用实时的原位电化学原子力显微镜/摩擦力显微镜（in situ AFM/FFM）观察 HOPG 和 KS-44/PVDF 复合物电极在充放电过程中的形貌变化，第一次获得了在复合电极表面形貌和摩擦力实时变化的数据。在开路电压状态，HOPG 表面仅有一层极薄的、柔软的沉积物；当放电到 2V 时，在台阶附近一层很薄的 SEI 层被检测到，这层薄膜很不稳定，当再次充电时被分解消失不见了。当放电到 0.9V 以下时，形成了一层厚厚的 SEI 膜并观察到岛状生长物。Hirasawa 等的研究奠定了原位 AFM 的基础，Aurbach 等[183]将其应用到其他贵金属电极的研究中，开辟了原位 AFM 研究的热潮。Balke 等[184]在 2010 年发表了关于锂离子电池正极材料纳米尺度的离子扩散谱图的文章，利用射频溅射在 Al_2O_3 衬底上沉积了一层 500nm 厚的 $LiCoO_2$ 薄膜并在 800℃ 氧气气氛下退火 2 h。样品与衬底之间有一层薄薄的 Au 导电层，采用配有带励磁控制器、NI-6115 快速数据采集卡的 Veeco Dimension 的原子力显微镜进行表征，研究发现脉冲电压由于持续时间小于 Li^+ 扩散时间，不会造成 Li^+ 从 Li^+ 层脱出。文章另外还分析观察了高频激励和同时变化频率及振幅进行激励的两种情况，得到了包含表面形貌信息、表面 Li^+ 在晶界和晶粒传输信息的复杂图像。虽然没有提取出 Li^+ 传输的谱图，但这却是第一次在锂离子电池材料中采用原子力显微镜观察到 Li^+ 传输的图像。Chung 等[185]也使用电化学应力显微镜探测 $LiCoO_2$ 多晶薄膜的亚微米界面，研究了插层化合物的动力学特性。进一步的分析表明，多晶薄膜中 Li^+ 扩散路径是弯弯曲曲的，且 Li^+ 向电极的扩散程度依赖于针尖向样品施加的过电位频率。电化学滞后回线的形状和倾斜角随晶粒尺寸和过电位频率变化而变化。研究表明，表面晶粒晶向对回线的形状有着显著影响。另外，电化学应变显微镜（ESM）的谱图可以推断 Li^+ 在 $LiCoO_2$ 薄膜中面外的分布局域的扩散系数。为了更真实地得到电池内部的离子传输和最优扩散路径，Zhu 等[186]采用固态电解质通过磁控溅射的方法制备了一个全电池，再通过 in situ AFM 的手段检测 TiO_2 负极表面形貌随所加载的三角波形电压的变化。研究结果表明随着电压的反转，负极表面形貌也出现了可逆的变化。结合同步测得的幅值和相位的谱图，可以得到与负极薄膜中 Li^+ 分布相关的高空间分辨的纳米点阵图。可逆的形貌变化与其他电化学手段得到的结果有着类似的结果，这表明了简单的 in situ AFM 手段用于研究 Li^+ 在纳米尺度的嵌入/脱出、Li^+ 分布和电化学机制与电化学原子力显微镜、电化学应变显微镜一样有着重要的应用前景。Kostecki 等[187]观察到了尖晶石相的 $Li_2Mn_2O_9$ 不均匀的导电性，提出了在 Li^+ 的不可逆脱出形成了 MnO_2 相的系统中的一种降解机制。同样的方法用来研究尖晶石相的 $LiMn_2O_4$ 的导电性，进一步验证了这种机制[188]。Lipson 等和 Lee 等[189,190]通过扫描离子导电显微镜（SICM，也是扫描探针显微镜的一种）研究了形貌和电化学活性之间的关系，提出了 Li_2O_2 首先在 HOPG 台阶边缘处生长，形成垂直于台阶的纳米片，最后成膜的生长机制。为了排除 HOPG 上 C 的影响，作者在纳米多孔 Au 上观察了原位的 $Li-O_2$ 的电化学氧化还原反应。发现氧化还原电位低于 C 基电极，这是由于纳米多孔 Au 高的氧气吸附率允许致密的 Li_2O_2 薄膜的形成。

AFM 和 Raman 技术，特别是针尖增强（TERS）技术的联用正在开发，并已经有一些相关的研究，弥补了 AFM 无法确定成分的缺陷。AFM 技术的可拓展性在将来必然给锂电池研究带来新的突破。

11.6 材料微观力学性质

电池材料一般为多晶，颗粒内部存在应力。在充放电过程中锂的嵌入脱出会发生晶格膨胀收缩，导致局部应力发生变化，进一步会引起颗粒以及电极的体积变化、应力释放，出现晶格堆垛变化，颗粒、电极层产生裂纹。因此，锂离子电池中的力学问题开始引起广泛的关注，特别是高容量正负极材料的开发带来的力学问题更为突出，已经有超过百篇的论文发表。2014 年，与电池中力学特性有关的国际会议也首次召开。目前，基于 DFT 计算、反应分子动力学、有限元、相场、弹塑性模型及其耦合方法，可以计算材料的杨氏模量、模拟应力分布、预测裂纹演化等。锂离子电池的实验研究力学特性目前还不多，许多材料的力学特性参数，特别是充放电过程中力学参数的变化研究得较少。

用于探测微观力学特性的手段主要是 AFM 技术与纳米压印技术以及在 TEM 中与纳米探针、STM 探针联合的测试方法，采用固态电池时可以进行原位力学特性、应力的测量。

锂电池中研究较多的是采用 SPM 的探针来研究 SEI 膜的力学特性。在接触模式下，以恒力将探针扎入膜，便可得到该处扎入深度随力的响应曲线，进而可以得到杨氏模量等信息。锂离子电池材料的力学特性的研究主要集中在固体电解质界面（SEI 膜）方面。初始的研究还是利用 AFM 技术来观察形貌。如 Ogumi 等[191]在 2003 年利用 AFM 技术研究了石墨负极在 EC 基溶剂中不同共溶剂的影响，在不同 PC 基溶剂中成膜添加剂的影响。Jeong 等[192]采用 AFM 原位观察了 HOPG 基面在循环伏安过程中形成的表面膜的厚度，第二周充放电完成后从第一周循环后的 40nm 生长到 70nm。Inaba 等[193]采用原位的 AFM 技术研究了石墨负极的表面膜在高温下的变化，研究表明高温下 SEI 膜在放电态出现溶解和凝聚，SEI 膜性能恶化；充电态 SEI 生长严重。Becker 等[194]运用原位 AFM 观察了不同尺寸的硅柱子能经受的循环次数，发现尺寸最小的硅柱子形变最小。此外，还有许多文献通过原位的 AFM 技术成像讨论了 HOPG、石墨、硅负极表面的 SEI 的形成机制[195-198]。

Zhang 等[199]首次从力学性能和形貌结合的角度定量描述了 MN 负极上不均匀的 SEI 膜的形成，SEI 膜的杨氏模量。其定量的方法开启了直接观察在不同电极材料中 SEI 膜性能的途径。Zheng 等[200]在其方法的基础上，通过力曲线自动扫描技术，结合三维作图获得了硅负极表面的 SEI 膜的非均相三维覆盖情况以及不同杨氏模量 SEI 膜成分的空间分布情况，并通过杨氏模量的值在不同添加剂中以及充放电过程中的变化情况反映了成膜添加剂主要是有机的膜——高温时，杨氏模量小的有机物首先分解。这一部分在该系列讲座的锂电池基础科学问题（V）——电池界面[201]这一章节中有较为详细的介绍。目前，有关力曲线这一方法的应用正在不断地被探索与发展。

充放电过程中电极膨胀导致电池的体积膨胀，会产生较大的应力和直接的形变。对这种应力的应用曾经引起了一些研究者的注意，用来发展基于电池形变的巨电致伸缩器件。目前，从实用的角度看，循环次数还不能很好地满足要求。但是相对于压电陶瓷，锂电池作为电致伸缩产生的形变量甚至可以到 300%，具有诱人的应用前景。

11.7 材料表面功函数

Li^+ 和 e^- 嵌脱进入正负极材料后，会引起材料电子结构的变化，导致功函数出现变化。通过功函数的测量，可以间接反映表面电子结构的变化，进而了解表面新相的形成。关于锂离子电池材料功函数的研究较少，主要的表征手段是开尔文探针力显微镜（Kelvin probe force microscopy，KPFM）。KPFM 通过探测表面电势对探针的作用力，来得到样品表面的电势分布。Nagpure 等[202]利用开尔文探针显微镜技术（KPFM）测量了老化后的锂离子电池表面电势，老化后的电池具有更低的表面电势，这可以归因于颗粒尺寸、表面层的相变以及新沉积物的物理化学性质的影响。Zhu 等[203]也测量到了 TiO_2 负极材料的表面电势变低，表明充电维持能力下降[202]。极化后的表面出现势垒或势阱，这可以归因于晶界处不同的电学性能和电荷的累积。

另外还有电子全息（electron holography）、光发射电子显微镜（photoemission electronic microscopy，PEEM）的方法也可以得到表面电势的分布。Yamamoto 小组[204-206]通过电子全息的方法直接观测到了全固态锂离子电池充放电过程中电势的变化情况，成功地得到了不同体系下电势在界面的分布，验证了电势主要分布在正极/电解质界面的结论，且与 EELS 得到的结果一致。这些方法都具有较高的空间分辨率。今后全固态锂离子电池的兴起特别需要研究界面空间电荷层现象，对表面和界面势的测量将是研究此类问题的关键实验技术。

11.8 绝热加速量热仪在锂电领域中的应用

电池的安全性本质上是热安全，其程度可以用电池的热特性来进行评估，因此在电池安全性研究中，量热仪是最主要的手段。绝热加速量热法（ARC）是一种在近似绝热的情况下对样品热安全性进行测试分析的方法。绝热体系的获得：一是使量热腔与环境的温度完全相等，即不存在温差；二是使量热腔与环境间的热阻为无穷大，使两者之间没有热的交换，即分别从传导、对流和辐射三种方式上隔绝、阻止热量的传递。但在实际操作中，完全隔热是无法办到的，因而，绝热热量计均采用第一种方法，使量热腔与环境温度尽可能一致（如图 11-36 所示），以达到近乎绝热的目的。

ARC 基本操作模式逻辑如图 11-37 所示[207]。ARC 工作时采用加热（heat）-等待（wait）-搜索（seek）（H-W-S）的模式来探测样品的放热反应。在操作软件中设定起始温度、终止温度、温度梯度和灵敏度等主要参数，并且逐一设定其他安全参数。在 H-W-S 的操作模式下，加热炉的辐射加热器将炉体加热至起始温度，由于炉体与样品之间热平衡

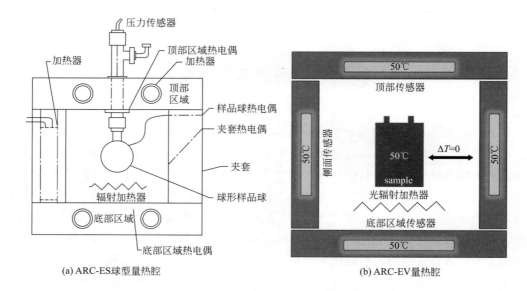

图 11-36 绝热加速量热仪设备原理

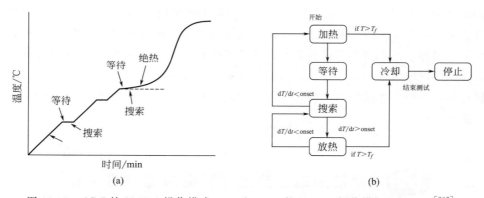

图 11-37 ARC 的 H-W-S 操作模式（a）和 ARC 的 H-W-S 操作模拟逻辑（b）[207]

需要一定时间，仪器进入等待模式，使得样品内部以及样品与炉体之间达到热平衡。等待时间结束后，仪器进入搜索模式，系统通过比较探测到的样品升温速率与预先设定的温度灵敏度（如 0.02℃/min）来判断样品是否存在放热现象。如果样品升温速率大于预先设定的灵敏度，则仪器进入放热状态。在放热阶段，ARC 控制器根据绝热炉各个区域温度与样品测试系统的温度差异调整绝热炉各个区域加热器的功率，从而维持绝热炉温度与样品测试系统温度的一致，保证绝热条件的实现，并实时记录系统的温度、升温速率、压力等数据。如果升温速率小于预先设定的灵敏度，则仪器进入加热状态，根据设定的温度梯度值和起始温度，仪器加热至下一温度梯度，开始下一轮的 H-W-S 模式，直到探测到放热或者达到预先设定的最终温度为止。

绝热加速量热仪测量灵敏度高，可达 0.005℃/min，能使用较大的样品数量，精确模拟样品内部热量不能及时散失时，样品放热反应的初始反应温度、放热速率等众多的热特性参数，尤其是能给出差示扫描量热法（differential scanning calarimetry，DSC）和差热分析法（differential thermal analysis，DTA）等无法给出的物质在热分解时压力随时间和温度的变化曲线。在进行电池热特性研究时，可以根据样品类型选择合适的量热腔，或在此基础上增加电压和内阻监测，使得ARC可同步提供电池热失控前后的电特性，为人们认识电池热失控过程提供了更丰富的信息，目前已成为世界上锂离子电池生产厂家和研究者进行安全性测试和热稳定性研究的主要手段。

Jiang等[208,209]使用ARC-ES研究了不同正极材料的热稳定性，满电态的钴酸锂、三元铁锂和磷酸铁锂的材料分别与EC、DEC溶剂充分混合，结果发现钴酸锂在150℃左右的时候开始自放热，三元材料自放热在220℃左右，而磷酸铁锂自放热温度在300℃以上，这个与其他测试经验也是相符的。同时对相同材料的不同粒径做热稳定性研究发现，对于钴酸锂这种正极材料，在EC、DEC体系中，随着粒径的增大，材料的热稳定性逐步提高。Ma等[210]研究了不同配比和电压状态的三元材料的热稳定性，结果发现随着Ni含量的增加，自放热起始温度逐渐降低，材料热稳定性下降；而对于同种配比的三元材料，通过调整试验时电池的电压状态发现，随着电压的升高，热稳定性也是逐步降低的。除了EC、DEC体系的电解液外，Jiang等[208]也研究了不同浓度的六氟磷酸锂电解液的热稳定性，六氟磷酸锂浓度的增加使温度、温升曲线右移，自放热起始温度升高，即提高了电解质体系的热稳定性。同时也可以用于研究不同添加剂对正极材料的影响。

Feng等[212]使用EV-ARC对于某款25A·h三元锂离子动力电池进行绝热热失控测试，如图11-38（a）所示。通过多个位置的温度监测发现，电池内部最高温度为853.5℃。同时将热失控过程划分为阶段Ⅰ～阶段Ⅵ共6个阶段。图11-38（b）总结了绝热热失控测试各个阶段所发生的主要化学反应：负极表面固体介面膜（SEI膜）的分解，隔膜基材的收缩、熔化，正极分解，电解质分解，黏结剂分解，隔膜崩溃，电池正负极之间发生短路，热失控发生等。同时通过多个热电偶同时监测电池不同部位的温度，发现电池的产热和散热并不均匀，最高温差可达几百摄氏度。各个阶段都分别有对应的温度和时间区间，有些反应可能是同时发生，需要对不同体系、荷电状态、健康状态的电池做系统研究。

王等[213]以5℃/台阶的温度，对某电池进行绝热热失控测试时，得到的温度时间曲线如图11-39所示。设定自放热的起始条件为温升速率≥0.02℃/min，并记录此时的温度为T_0，其对应的时间为t_1。T_0揭示了电池内部的热稳定性，T_0越高，说明电池的热稳定性越好。电池自放热开始后，内部会发生一系列复杂的化学、电化学反应并释放热量，使电池温度在不用外部加热的情况下继续升高。当温升速率达到1℃/min时，可以认为是热失控开始，此时的温度为T_c，标记时间为t_2。定性地说，T_c温度越高，说明电池安全性越好，热失控间隔时间$t=t_2-t_1$，时间越长，电池热安全性也越好；热失控时达到的最高温度T_{max}，标记此时的时间为t_3，T_{max}与电池的容量和能量密度有关，t_3与t_2之间的时间间隔越长，也可以说明电池的热安全性越好；除此之外，对于某些特殊设计的电池，如安全阀会在热失控触发，从而会引起一个短暂且小幅度的温度下降，定义此时温度为$T_阀$，对应时间为t_4，安全阀与t_4、t_2时间间隔越长也说明电池安全性越好。

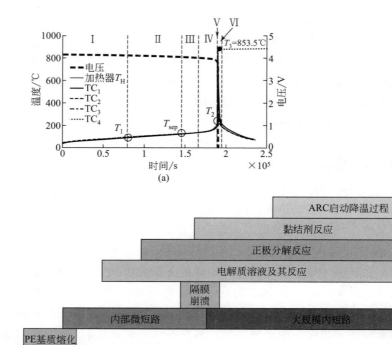

图 11-38 绝热热失控测试结果（a）和绝热热失控测试各阶段发生反应情况（b）[212]

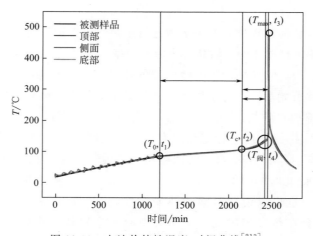

图 11-39 电池热特性温度-时间曲线[213]

在使用过程中，电池的电性能和安全性能都会出现不同程度的衰减，因此，研究锂离子电池全寿命周期的安全性对保障电池在使用过程中的安全性至关重要。邵晓挺等[214]对锂离子电池全生命周期安全性演变问题的国内外研究进展进行了综述，分析了国内外关于电池安全性能在循环老化和存储老化两种工况下的演变规律研究，总结发现锂离子电池全生命周期安全性演变规律与老化衰减途径密切相关，在常温/高温循环老化下，由于内阻的上升，电池在充放电下焦耳热增加，耐电滥用性能下降，电池热稳定性也会有一定程度的变化，变化规律与电池的材料体系和工艺水平相关；在常温/高温储存老化下，电池的耐电滥用性能也会降低，但由于负极的 SEI 膜在储存过程中稳定性提升，电池的热稳定性会得到提升；在低温循环工况下，电池的热稳定性急剧下降，主要原因是负极析锂，析出的锂金属非常活泼，在较低的温度下便可以与电解液发生反应，造成电池的自放热起始温度降低和自产热速率剧增，严重危害电池的安全性能。

李坤等[215]以 26A·h 软包型锂离子动力电池为研究对象，利用加速量热仪（ARC-EV+）研究不同循环周期下动力电池的电化学行为、电池的热稳定性和安全性及热失控行为。结果表明（表 11-2），随着循环次数的增加，电池自产热温度呈现总体下降，说明不断循环老化的电池 SEI 膜热稳定性逐渐变差。交流阻抗谱显示，SEI 膜与电解液的阻抗随着循环次数增加而增大，说明 SEI 膜与电解液结构和成分随着循环周期的变化是影响其热稳定性的关键因素。

表 11-2 不同循环周期下热失控各阶段温度[215]

循环次数	1	200	400	600	800	1000
$T_{自放热}$/℃	81.77	77.43	66.91	66.99	67.06	71.96
$T_{热失控}$/℃	263.37	199.72	230.16	228.94	228.72	252.23
T_{max}/℃	539.04	452.39	500.79	532.75	457.71	495.31

毛亚等[216]以钴酸锂/中间相碳微球为体系的锂离子电池为研究对象，采用绝热加速量热仪研究了不同工作电流、不同循环老化周期电池的产热特性和热失控行为，发现电池的发热量随着充放电倍率的增大而增大。通过比较不同循环老化周期的电池的产热速率，发现容量衰减速度与直流内阻、产热量之间存在很强的关联性，循环后电池的热失控过程中自放热和热失控起始温度稍有变化，但热失控起始时间大大缩短，见图 11-40。

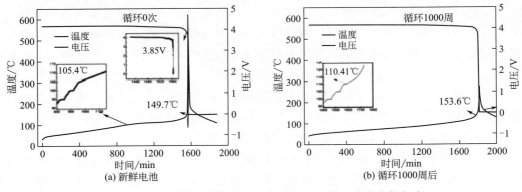

图 11-40 不同循环老化周期下 LCO/CMS 电池的热失控行为

赵学娟等[217]采用绝热加速量热仪与电池充放电循环系统联用的技术，对钴酸锂电池在不同倍率下循环的热行为进行研究，发现电池在放电末期的产热速率会明显增加，且各放电阶段的温升和温升速率随充放电倍率的增加总体呈增大的趋势；在绝热环境下，热量随电池充放电循环的进行而累积，最终会导致电池的热失控。而容量和环境温度对产热的影响研究结果则表明，钴酸锂电池在放电阶段的温升、温升速率和产热量都会随着电池容量的增加而增大，大容量电池发生热失控的时间也会相应缩短。钴酸锂电池在较高环境温度下的正常循环次数减少，且其在各放电阶段的温升、温升速率和产热量也相应地增大。这说明，循环倍率、电池容量以及环境温度的提高都会降低钴酸锂电池的热稳定性。

锂离子电池的比热容是电池热模拟领域的重要参数。庄等[218]介绍了 ARC 测试电池比热容的原理：通过已知功率的加热片在绝热条件下，给电池进行加热，加热片输出的热能完全被电池吸收并导致电池温度升高，根据热量衡算可计算出电池的比热容。

11.9 互联互通惰性气氛电池综合分析平台

锂电池及电极材料、电解质等对空气和水分敏感，若在空气中拆解电池，电极材料和电解质遇到空气或水后，发生化学反应，产生气体和热量，反应剧烈时可能会发生燃烧。为保证锂电池拆解过程的安全性，与锂电池测试分析结果的准确性，需要在惰性气氛下拆解电池，同时确保样品的测试表征在惰性气氛下完成。

中国科学院物理所和天目湖先进储能技术研究院（TIES）已建成互联互通惰性气氛电池材料与器件综合测试分析平台，该平台由互联互通手套箱系统与置于其中的仪器设备及通过惰性气氛转移装置可以实现不暴露空气和水分进行测试两部分组成。可以实现在惰性气氛下对电芯进行拆解、样品准备、转移以及部分测试（见后面介绍），可以实现锂电池及材料的综合测试分析，特别是适合于锂电池的失效分析。

平台的第一部分由 13 台手套箱组成，其内集成了 XRD、AFM、FTIR 显微镜、STA、DSC、3D 光学显微镜、金相显微镜、接触角测量仪、电池拆解仪、扣式电池组装等仪器设备，图 11-41(a) 为互联互通示意图，图 11-41(b) 为互联互通平台及内部部分设备。

平台内部包含圆柱电池拆解机、方型铝壳电池拆解机等，可实现惰性气氛下圆柱、方型、软包电池的拆解。且可以实现拆解过程实时录像与拍照，还可以实现方型电池拆解过程电压的监控，确保电池在拆解过程中不出现意外而使样品发生变化，导致测试分析数据不可靠。

平台配置了 3D 显微镜与金相显微镜，直接观察样品的形貌，可测量截面样品厚度和层数，可对极耳的毛刺和缺陷进行检测。特别是对气氛敏感的金属电极等进行观察。

平台还配置了桌面式 XRD，集仪器控制、数据采集系统于一体，可以实现在全惰性气氛下对空气及水分敏感的样品进行物相、结构分析。惰性气氛下的红外显微镜可对样品进行线扫、面扫，为采集样品表面微区的红外光谱提供了可能，并能够得到高质量的红外谱图。

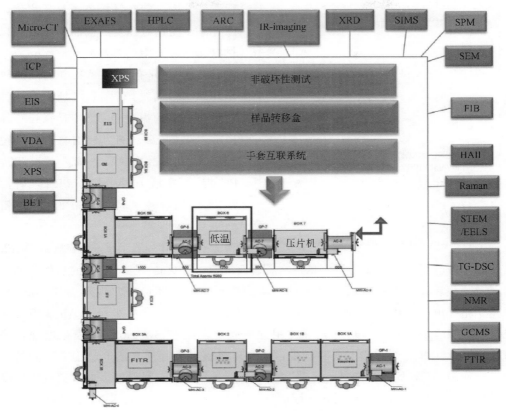

(a) TIES互联互通平台示意图

(b) 互联互通平台及内部设备实体照片

图 11-41　互联互通惰性气氛电池综合分析平台

平台中惰性气氛下的同步热分析仪（STA）和差分扫描量热仪（DSC），除了能够实现常规（空气下）的热分析功能。如对材料的热重或吸放热行为进行测试，分析不同材料的热行为，判断其热稳定性，还可以实现对空气及水分敏感样品如满电态负极的热行为进

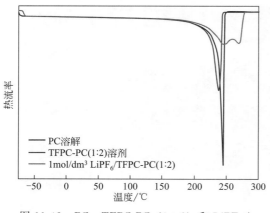

图 11-42 PC、TFPC-PC（1∶2）和 LiPF$_6$/TFPC-PC（1∶2）的 DSC 曲线[219]

行测试分析，DSC 还配备了高压坩埚，可以进一步测量可挥发的样品如电解液以及材料在电解液中的热稳定性。Yun 等[219]使用 DSC 研究有机溶剂及电解液的热稳定性，DSC 曲线如图 11-42 所示。三种溶液都在较宽的温度范围−55～240℃保持稳定。在 240℃（沸点）均有一个尖锐的与液-气转变相关的吸热峰。对于 TFPC-PC 混合溶液加入锂盐 LiPF$_6$ 后，在 265℃仍然存在一个与 LiPF$_6$ 热解相关的峰。

平台内还配置了原子力显微镜（AFM），它具有原子级分辨率，以原子尺寸观察物质表面结构，除具备一般 AFM 的功能，如对电极材料及隔膜表面形貌的观察，因测试环境为惰性气氛，还可以实现对 SEI 膜的研究，可以用于指导电解液的研究和筛选。1997 年，Hirasawa 等[220]首次利用 AFM 观察碳复合电极的形貌（图 11-43）。郑杰允等使用硅薄膜电极组装 Swagloke 型模拟电池，采用 LiPF6/（EC∶DMC＝1∶1）电解液体系，选取硅薄膜电极首周充放电过程中不同的电位截止点，使用 AFM 原位研究其 SEI 膜生成过程。并用 AFM 测得的力曲线研究 VC 添加剂对硅负极 SEI 膜的影响，得到 SEI 膜的三维杨氏模量图，试验表明 VC 的加入能使整个硅负极表面均被 SEI 膜所覆盖，为电解液的优化改性提供了实验依据。

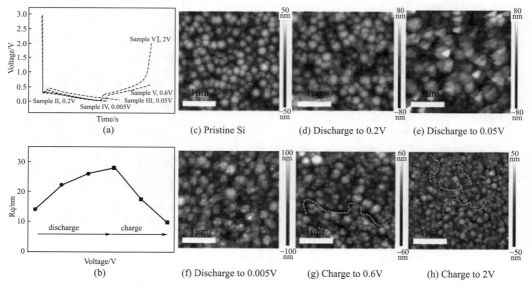

图 11-43 硅薄膜电极不同充放电状态表面 AFM 形貌图（接触模式成像）[220]

平台的第二部分为通过惰性气氛转移装置可以扩展到的仪器设备，即除了在平台内集成的仪器设备外，还可通过惰性气氛转移盒、样品转移袋等器件，实现与其他测试设备如

SEM、XRD、FIB、XPS、AES、TOF-SIMS、XRD、GC、GC-MS、NMR 等的连接，实现样品在转移和测试过程中不会接触到空气或水分，保证测试数据的真实可靠。图 11-44 为常用仪器 XPS、SEM 和 GC 的样品转移装置。

图 11-44　常用仪器的样品转移装置

借助于互联互通惰性气氛综合测试平台，使得锂电池拆解、样品的收集和准备与很多项目的测试都在惰性气氛下，或通过转移装置扩展到其他表征仪器，能够实现对电池材料的形貌、结构、化学组成、杂质信息、表面价态、热稳定性等进行分析，避免了暴露空气或水分导致材料变性、失效，确保数据真实可靠。再结合其他如 CT、ARC、IBC、Ramam、TEM 等仪器设备，可以实现对电池材料与器件的综合分析，特别是适用于锂电池的失效分析。

11.10　其他实验技术

除了前述的各类技术，还有一些先进的实验技术在锂电池中应用相对较少，但正在发展，并具有重要的、不可替代的价值。包括能直接测量材料能带结构的角分辨光电子能谱（ARPES），这一实验技术目前还需要大块单晶或外延薄膜样品。锂离子电池材料的电子结构主要是通过 DFT 计算获得，该信息对于理解多元素材料发生电荷转移时优先次序具有十分重要的意义。正电子淹没技术（positron annihilation technique，PAT）可以用于测量缺陷结构和电子结构，卢瑟福背散射（Rutherford backscattering spectrometry，RBS）可以测量薄膜组成，共振非弹性 X 射线散射（resonant incoherent X-ray scattering，RIXS）可以研究原子间磁性相互作用。

11.11 本章结语

锂电池的应用涵盖人类生产生活的方方面面，引起了全球范围的研究热潮。目前的研究已经积累了大量的数据，科技人员已经认识到电池材料与器件的电化学性能和材料的元素组分与分布、形貌结构、表面官能团与表面结构、材料微观力学性质、电势分布等多种物理化学性质密切关联。对这些关系的准确认识，是进一步发展电池技术、进行失效分析、比对技术先进性的关键。通过多种表征手段的结合，才能构筑起较为清楚的构效关系。更高空间分辨、时间分辨和能量分辨的实验技术以及各类实验技术的综合集成正在不断发展，以满足锂电池研发的广泛需求。工欲善其事，必先利其器。我国目前的锂电池研究与发达国家相比，在掌握先进仪器表征、开发新的表征测试技术方面还有显著的差距，需要所有研发人员的长期共同努力。面对种类繁多、价格不菲、不断进步的高精尖实验仪器，也需要研发人员仔细想清楚所需要研究的问题之后，再考虑应该通过什么样的最有效、最经济、最快速的实验方法去分析和解决问题。

本章需要进一步思考的基础科学问题：

1. 如何定量测量电极内的导电添加剂含量？如何判断导电添加剂分布是否均匀？是否连通？
2. 如何测量颗粒表面和电极表面 SEI 膜的组成、厚度、覆盖度、结构？如何测量 SEI 膜的热稳定性？
3. 如何设计实验，测量全电池充放电过程或任意工况下气体产生的量、种类？如何判断产生的气体是否被电池内其他物质吸收？
4. 如何测量颗粒和极片的孔隙率及其演化规律？
5. 如何判断电解液已经完全浸润电极？
6. 如何通过无损检测和拆解检测，定量判断极片膨胀率？并分解出膨胀的贡献率？例如颗粒直接膨胀、颗粒裂开、界面膜生长、极片开裂、极片溶胀等？
7. 通过哪些测量可以判断和比较电极材料及电解质的热稳定性以及热失控概率？
8. 如何测量电极活性层的粘接性？如何了解黏结剂的空间分布均匀性？
9. 如何定量检测材料中各类缺陷的密度和空间分布？
10. 如何测量材料表面结构演化规律？

参考文献

[1] Aurbach D, Markovsky B, Salitra G, et al. Review on electrode-electrolyte solution interactions, related to cathode materials for Li-ion batteries [J]. Journal of Power Sources, 2007, 165 (2): 491-499.
[2] Park K S, Son J T, Chung H T, et al. Synthesis of LiFePO$_4$ by co-precipitation and microwave heating [J]. Elec-

trochemistry Communications，2003，5（10）：839-842.

[3] Meligrana G，Gerbaldi C，Tuel A，et al. Hydrothermal synthesis of high surface LiFePO$_4$ powders as cathode for Li-ion cells [J]. Journal of Power Sources，2006，160（1）：516-522.

[4] Bakenov Z，Taniguchi I. Electrochemical performance of nanostructured LiM$_x$Mn$_{2-x}$O$_4$（M＝Co and Al） powders at high charge-discharge operations [J]. Solid State Ionics，2005，176（11-12）：1027-1034.

[5] Zheng Z H，Tang Z L，Zhang Z T，et al. Surface modification of Li$_{1.03}$Mn$_{1.97}$O$_4$ spinels for improved capacity retention [J]. Solid State Ionics，2002，148（3-4）：317-321.

[6] Wang Y，Lee J Y. Microwave-assisted synthesis of SnO$_2$-graphite nanocomposites for Li-ion battery applications [J]. Journal of Power Sources，2005，144（1）：220-225.

[7] Fey G T K，Chen C L. High-capacity carbons for lithium-ion batteries prepared from rice husk [J]. Journal of Power Sources，2001，97-98（1）：47-51.

[8] Yoshikawa D，Kadoma Y，Kim J M，et al. Spray-drying synthesized lithium-excess Li$_{4+x}$Ti$_{5-x}$O$_{12-\delta}$ and its electrochemical property as negative electrode material for Li-ion batteries [J]. Electrochimica Acta，2010，55（6）：1872-1879.

[9] Kramer E，Schedlbauer T，Hoffmann B，et al. Mechanism of anodic dissolution of the aluminum current collector in 1 M LiTFSI EC：DEC 3：7 in rechargeable lithium batteries [J]. Journal of the Electrochemical Society，2012，160（2）：A356-A360.

[10] Diao Y，Xie K，Xiong S，et al. Analysis of polysulfide dissolved in electrolyte in discharge-charge process of Li-S battery [J]. Journal of the Electrochemical Society，2012，159（4）：A421.

[11] Jang I C，Hidaka Y，Ishihara T. Li metal utilization in lithium air rechargeable batteries [J]. Journal of Power Sources，2013，244：606-609.

[12] Lee B G，Nam S C，Choi J. Anodic TiO$_2$ nanotubes as anode electrode in Li-air and Li-ion batteries [J]. Current Applied Physics，2012，12（6）：1580-1585.

[13] Nelson G C，Neiswander P A，Searcy J Q. Lithium compound identification in thermally activated batteries by iss and sims [J]. Journal of Vacuum Science & Technology，1981，18（3）：750-751.

[14] Castle J E，Decker F，Salvi A M，et al. XPS and SIMS study of the distribution of Li ions in thin films of vanadium pentoxide after electrochemical intercalation [J]. Surface and Interface Analysis，2008，40（3-4）：746-750.

[15] Swiatowska-Mrowiecka J，Martin F，Maurice V，et al. The distribution of lithium intercalated in V$_2$O$_5$ thin films studied by XPS and SIMS [J]. Electrochimica Acta，2008，53（12）：4257-4266.

[16] Li J T，Maurice V，Swiatowska-Mrowiecka J，et al. XPS，time-of-flight-SIMS and polarization modulation IR-RAS study of Cr$_2$O$_3$ thin film materials as anode for lithium ion battery [J]. Electrochimica Acta，2009，54（14）：3700-3707.

[17] Minakshi M，Thurgate S. Surface analysis on discharged MnO$_2$ cathode using XPS and SIMS techniques [J]. Surface and Interface Analysis，2009，41（1）：56-60.

[18] Li J T，Swiatowska J，Seyeux A，et al. XPS and ToF-SIMS study of Sn-Co alloy thin films as anode for lithium ion battery [J]. Journal of Power Sources，2010，195（24）：8251-8257.

[19] Myung S T，Sasaki Y，Sakurada S，et al. Electrochemical behavior of current collectors for lithium batteries in non-aqueous alkyl carbonate solution and surface analysis by ToF-SIMS [J]. Electrochimica Acta，2009，55（1）：288-297.

[20] Peled E，Tow D B，Merson A，et al. Composition，depth profiles and lateral distribution of materials in the SEI built on HOPG-TOF SIMS and XPS studies [J]. Journal of Power Sources，2001，97-98：52-57.

[21] Ota H，Akai T，Namita H，et al. XAFS and TOF-SIMS analysis of SEI layers on electrodes [J]. Journal of Power Sources，2003，119：567-571.

[22] Veryovkin I V，Tripa C E，Zinovev A V，et al. TOF SIMS characterization of SEI layer on battery electrodes [J]. Nuclear Instruments & Methods in Physics Research Section B-Beam Interactions with Materials and Atoms，2014，332：368-372.

[23] Lu P，Harris S J. Lithium transport within the solid electrolyte interphase [J]. Electrochemistry Communications，2011，13（10）：1035-1037.

[24] Nakai H，Kubota T，Kita A，et al. Investigation of the solid electrolyte interphase formed by fluoroethylene carbonate on Si electrodes [J]. Journal of the Electrochemical Society，2011，158（7）：A798-A801.

[25] Lee J T，Nitta N，Benson J，et al. Comparative study of the solid electrolyte interphase on graphite in full Li-ion

[26] Li W J, Zheng H, Chu G, et al. FD 176: Effect of electrochemical dissolution and deposition order on lithium dendrite formation: A top view investigation [J]. Faraday Discussions, 2014, doi: 10.1039/c4fd00124a.

[27] Sven N, Felix W, Thomas B, et al. Molecular surface modification of NCM622 cathode material using organophosphates for improved Li-ion battery full-cells [J]. Acs Applied Materials & Interfaces, 2018, acsami. 8b04405.

[28] Liu Z, L P, Zhang Q L, et al. A bottom-up formation mechanism of solid electrolyte interphase revealed by isotope-assisted time-of-flight secondary ion mass spectrometry [J]. The Journal of Physical Chemistry Letters, 2018, 9: 5508-5514.

[29] Aurbach D, Markovsky B, Shechter A, et al. A comparative study of synthetic graphite and Li electrodes in electrolyte solutions based on ethylene carbonate dimethyl carbonate mixtures [J]. Journal of the Electrochemical Society, 1996, 143 (12): 3809-3820.

[30] Aurbach D, Granot E. The study of electrolyte solutions based on solvents from the "glyme" family (linear polyethers) for secondary Li battery systems [J]. Electrochimica Acta, 1996, 42 (4): 697-718.

[31] Ruffo R, Hong S S, Chan C K, et al. Impedance analysis of silicon nanowire lithium ion battery anodes [J]. Journal of Physical Chemistry C, 2009, 113 (26): 11390-11398.

[32] Ahn H J, Choi H C, Park K W, et al. Investigation of the structural and electrochemical properties of size-controlled SnO_2 nanoparticles [J]. Journal of Physical Chemistry B, 2004, 108 (28): 9815-9820.

[33] Andersson A M, Herstedt M, Bishop A G, et al. The influence of lithium salt on the interfacial reactions controlling the thermal stability of graphite anodes [J]. Electrochimica Acta, 2002, 47 (12): 1885-1898.

[34] Dedryvère R, Laruelle S, Grugeon S, et al. Contribution of X-ray photoelectron spectroscopy to the study of the electrochemical reactivity of CoO toward lithium [J]. Chemistry of Materials, 2004, 16 (6): 1056-1061.

[35] Shaju K M, Rao G V S, Chowdari B V R. Performance of layered Li($Ni_{1/3}Co_{1/3}Mn_{1/3}$)O_2 as cathode for Li-ion batteries [J]. Electrochimica Acta, 2002, 48 (2): 145-151.

[36] Andersson A M, Abraham D P, Haasch R, et al. Surface characterization of electrodes from high power lithium-ion batteries [J]. Journal of the Electrochemical Society, 2002, 149 (10): A1358.

[37] Yabuuchi N, Yoshii K, Myung S T, et al. Detailed studies of a high-capacity electrode material for rechargeable batteries, Li_2MnO_3-$LiCo_{1/3}Ni_{1/3}Mn_{1/3}O_2$ [J]. Journal of the American Chemical Society, 2011, 133 (12): 4404-4419.

[38] Liu J, Manthiram A. Understanding the improved electrochemical performances of Fe-substituted 5 V spinel cathode $LiMn_{1.5}Ni_{0.5}O_4$ [J]. Journal of Physical Chemistry C, 2009, 113 (33): 15073-15079.

[39] Aurbach D, Pollak E, Elazari R, et al. On the surface chemical aspects of very high energy density, rechargeable Li-S batteries [J]. Journal of the Electrochemical Society, 2009, 156 (8): A694.

[40] Agostini M, Lee D J, Scrosati B, et al. Characteristics of Li_2S_8-tetraglyme catholyte in a semi-liquid lithium-sulfur battery [J]. Journal of Power Sources, 2014, 265: 14-19.

[41] McCloskey B D, Speidel A, Scheffler R, et al. Twin problems of interfacial carbonate formation in nonaqueous $Li-O_2$ batteries [J]. The Journal of Physical Chemistry Letters, 2012, 3 (8): 997-1001.

[42] Yang W, Salim J, Li S, et al. Perovskite $Sr_{0.95}Ce_{0.05}CoO_{3-\delta}$ loaded with copper nanoparticles as a bifunctional catalyst for lithium-air batteries [J]. Journal of Materials Chemistry, 2012, 22 (36): 18902.

[43] Hu Y, Kong W, Li H, et al. Experimental and theoretical studies on reduction mechanism of vinyl ethylene carbonate on graphite anode for lithium ion batteries [J]. Electrochemistry Communications, 2004, 6 (2): 126-131.

[44] Wu Xiaodong, Wang Zhaoxiang, Chen Liquan, et al. Ag-enhanced SEI formation on Si particles for lithium batteries [J]. Electrochemistry Communications, 2003, 5 (11): 935-939.

[45] Castro L, Dedryvere R, Ledeuil J-B, et al. Aging Mechanisms of LiFePO4//Graphite Cells Studied by XPS: Redox Reaction and Electrode/Electrolyte Interfaces [J]. Journal of The Electrochemical Society, 2012, 159 (4): A357-A363.

[46] Lee S, You H, Han K, et al. A new approach to surface properties of solid electrolyte interphase on a graphite negative electrode [J]. Journal of Power Sources, 2014, 247: 307-313.

[47] Mao B, Dai Y, Cai J, et al. Operando Ambient Pressure X-ray Photoelectron Spectroscopy Studies of Sodium-

Oxygen Redox Reactions [J]. Topics in Catalysis, 2018, 61: 2123-2128.

[48] Xu B, Fell C R, Chi M, et al. Identifying surface structural changes in layered Li-excess nickel manganese oxides in high voltage lithium ion batteries: A joint experimental and theoretical study [J]. Energy & Environmental Science, 2011, 4 (6): 2223-2233.

[49] Wang F, Robert R, Chernova N A, et al. Conversion reaction mechanisms in lithium ion batteries: Study of the binary metal fluoride electrodes [J]. Journal of the American Chemical Society, 2011, 133 (46): 18828-18836.

[50] Rom I, Wachtler M, Papst I, et al. Electron microscopical characterization of Sn/SnSb composite electrodes for lithium-ion batteries [J]. Solid State Ionics, 2001, 143 (3-4): 329-336.

[51] Al-Sharab J F, Bentley J, Badway F, et al. EELS compositional and valence mapping in iron fluoride-carbon nanocomposites [J]. Journal of Nanoparticle Research, 2013, 15 (4):

[52] Schuster M E, Teschner D, Popovic J, et al. Charging and discharging behavior of solvothermal $LiFePO_4$ cathode material investigated by combined EELS/NEXAFS study [J]. Chemistry of Materials, 2014, 26 (2): 1040-1047.

[53] Naji A, Thomas P, Ghanbaja J, et al. Identification by TEM and EELS of the products formed at the surface of a carbon electrode during its reduction in $MClO_4$-EC and MBF_4-EC electrolytes (M=Li, Na) [J]. Micron, 2000, 31 (4): 401-409.

[54] Kohandehghan A, Kalisvaart P, Cui K, et al. Silicon nanowire lithium-ion battery anodes with ALD deposited TiN coatings demonstrate a major improvement in cycling performance [J]. Journal of Materials Chemistry A, 2013, 1 (41): 12850.

[55] Gu Meng, Wang Zhiguo, Connell J G, et al. Electronic origin for the phase transition from amorphous Li_xSi to crystalline $Li_{15}Si_4$ [J]. ACS Nano, 2013, 7 (7): 6303-6309.

[56] Sun X C, Hegde M, Wang J, et al. Structural analysis and electrochemical studies of carbon coated $Li_4Ti_5O_{12}$ particles used as anode for lithium ion battery [J]. ECS Transactions, 2014, 58 (14): 79-88.

[57] Zhou J, Wang J, Hu Y, et al. Imaging state of charge and its correlation to interaction variation in an $LiMn_{0.75}Fe_{0.25}PO_4$ nanorods-graphene hybrid [J]. Chemical Communications, 2013, 49 (17): 1765-1767.

[58] Chueh W C, El Gabaly F, Sugar J D, et al. Intercalation pathway in many-particle $LiFePO_4$ electrode revealed by nanoscale state-of-charge mapping [J]. Nano Letters, 2013, 13 (3): 866-872.

[59] Uchimoto Y, Sawada H, Yao T. Changes in electronic structure by Li ion deintercalation in $LiNiO_2$ from nickel L-edge and OK-edge XANES [J]. Journal of Power Sources, 2001, 97-98: 326-327.

[60] Uchimoto Y, Sawada H, Yao T. Changes in electronic structure by Li ion deintercalation in $LiCoO_2$ from cobalt L-edge and oxygen K-edge XANES [J]. Journal of Synchrotron Radiation, 2001, 8: 872-873.

[61] Rosolen J M, Abbate M. XANES and EXAFS of chemically deintercalated $LiCo_{0.5}Ni_{0.5}O_2$ [J]. Solid State Ionics, 2001, 139 (1-2): 83-88.

[62] Wen W, Kumarasamy B, Mukerjee S, et al. Origin of 5V electrochemical activity observed in non-redox reactive divalent cation doped $LiM_{0.5-x}Mn_{1.5+x}O_4 (0 \leqslant x \leqslant 0.5)$ cathode materials—In situ XRD and XANES spectroscopy studies [J]. Journal of the Electrochemical Society, 2005, 152 (9): A1902-A1911.

[63] Seung-Taek M, Komaba S, Hirosaki N, et al. Structural investigation of layered $Li_{1-\delta}Mn_xCr_{1-x}O_2$ by XANES and in situ XRD measurements [J]. Journal of the Electrochemical Society, 2003, 150 (12): A1560-1568.

[64] Holzapfel M, Proux O, Strobel P, et al. Effect of iron on delithiation in $LixCo_{1-y}Fe_yO_2$. Part 2: In-situ XANES and EXAFS upon electrochemical cycling [J]. Journal of Materials Chemistry, 2004, 14 (1): 102-110.

[65] Kobayashi H, Emura S, Arachi Y, et al. Investigation of inorganic compounds on the surface of cathode materials using Li and OK-edge XANES [J]. Journal of Power Sources, 2007, 174 (2): 774-778.

[66] Kobayashi H, Shikano M, Koike S, et al. Investigation of positive electrodes after cycle testing of high-power Li-ion battery cells I. An approach to the power fading mechanism using XANES [J]. Journal of Power Sources, 2007, 174 (2): 380-386.

[67] Yang Songlan, Wang Dongniu, Liang Guoxian, et al. Soft X-ray XANES studies of various phases related to $LiFePO_4$ based cathode materials [J]. Energy & Environmental Science, 2012, 5 (5): 7007-7016.

[68] Tanaka S, Kitta M, Tamura T, et al. First-principles calculations of O-K ELNES/XANES of lithium titanate [J]. Journal of Physics D-Applied Physics, 2012, 45 (49): 494004.

[69] Akai T, Ota H, Namita H, et al. XANES study on solid electrolyte interface of Li ion battery [J]. Physica Scripta, 2005 (T115): 408-411.

[70] Wang Xiaojian, Jaye Cherno, Nam Kyungwan, et al. Investigation of the structural changes in $Li_{1-x}FePO_4$ upon charging by synchrotron radiation techniques [J]. Journal of Materials Chemistry, 2011, 21 (30): 11406-11411.

[71] Bak Seong Min, Nam Kyung Wan, Chang Wonyoung, et al. Correlating structural changes and gas evolution during the thermal decomposition of charged $Li_xNi_{0.8}Co_{0.15}Al_{0.05}O_2$ cathode materials [J]. Chemistry of Materials, 2013, 25 (3): 337-351.

[72] Nam Kyung Wan, Bak Seong Min, Hu Enyuan, et al. Combining in situ synchrotron X-ray diffraction and absorption techniques with transmission electron microscopy to study the origin of thermal instability in overcharged cathode materials for lithium-ion batteries [J]. Advanced Functional Materials, 2013, 23 (8): 1047-1063.

[73] Yu Xiqian, Pan Huilin, Wan Wang, et al. A size-dependent sodium storage mechanism in $Li_4Ti_5O_{12}$ investigated by a novel characterization technique combining in situ X-ray diffraction and chemical sodiation [J]. Nano Lett, 2013, 13 (10): 4721-4727.

[74] Ma Jun, Zhou Yong Ning, Gao Yurui, et al. Molybdenum substitution for improving the charge compensation and activity of Li_2MnO_3 [J]. Chemistry: A European Journal, 2014, 20 (28): 8723-8730.

[75] Ma Jun, Zhou Yongning, Gao Yurui, et al. Feasibility of using Li_2MoO_3 in constructing Li-rich high energy density cathode materials [J]. Chemistry of Materials, 2014, 26 (10): 3256-3262.

[76] Liu Xiaosong, Liu Jun, Qiao Ruimin, et al. Phase transformation and lithiation effect on electronic structure of Li_xFePO_4: An in-depth study by soft X-ray and simulations [J]. Journal of the American Chemical Society, 2012, 134 (33): 13708-13715.

[77] Wang Qing, Li Hong, Chen Liquan, et al. Monodispersed hard carbon spherules with uniform nanopores [J]. Carbon, 2001, 39 (14): 2211-2214.

[78] He Yu, Yu Xiqian, Li Geng, et al. Shape evolution of patterned amorphous and polycrystalline silicon microarray thin film electrodes caused by lithium insertion and extraction [J]. Journal of Power Sources, 2012, 216: 131-138.

[79] Li Hong, Huang Xuejie, Chen Liquan. Direct imaging of the passivating film and microstructure of nanometer-scale SnO anodes in lithium rechargeable batteries [J]. Electrochemical and Solid-State Letters, 1998, 1 (6): 241-243.

[80] Zhao Liang, Hu Yongsheng, Li Hong, et al. Porous $Li_4Ti_5O_{12}$ coated with N-doped carbon from ionic liquids for Li-ion batteries [J]. Advanced Materials, 2011, 23 (11): 1385-1388.

[81] Huang Jianyu, Zhong Li, Wang Chongmin, et al. In situ observation of the electrochemical lithiation of a single SnO_2 nanowire electrode [J]. Science, 2010, 330 (6010): 1515-1520.

[82] Liu Xiaohua, Huang Jianyu. In situ TEM electrochemistry of anode materials in lithium ion batteries [J]. Energy & Environmental Science, 2011, 4 (10): 3844.

[83] Gu M, Parent L R, Mehdi B L, et al. Demonstration of an electrochemical liquid cell for operando transmission electron microscopy observation of the lithiation/delithiation behavior of Si nanowire battery anodes [J]. Nano Letters, 2013, 13 (12): 6106-6112.

[84] Wang X, Zhang M, Alvarado J, et al. New insights on the structure of electrochemically deposited lithium metal and its solid electrolyte interphases via cryogenic TEM [J]. Nano Lett, 2017, 17: 7606-7612

[85] Li Y, Pei A, Yan K, et al. Atomic structure of sensitive battery materials and interfaces revealed by cryoelectron microscopy [J]. Science, 2017, 358: 506-510.

[86] Zachman M J, Tu Z, Choudhury S, et al. Cryo-STEM mapping of solid-liquid interfaces and dendrites in lithium-metal batteries [J]. Nature 2018, 560: 345-349

[87] Alvarado J, Schroeder M A, Zhang M, et al. A carbonate-free, sulfone-based electrolyte for high-voltage Li-ion batteries [J]. Mater Today, 2018, 21: 341-353.

[88] Levin B D, Zachman M J, Werner J G, et al. Characterization of sulfur and nanostructured sulfur battery cathodes in electron microscopy without sublimation artifacts [J]. Microsc Microanal, 2017, 23: 155-162.

[89] Choudhury S, Wei S, Ozhabes Y, et al. Designing solid-liquid interphases for sodium batteries [J]. Nat Commun, 2017, 8: 898.

[90] Wang X, Li Y, Meng Y S, Cryogenic Electron Microscopy for Characterizing and Diagnosing Batteries [J]. Joule, 2018, 2 (11): 2225-2234.

[91] Santhanagopalan D, Qian D, McGilvray T, et al. Interface limited lithium transport in solid-state batteries

[J]. The Journal of Physical Chemistry Letters, 2014, 5 (2): 298-303.

[92] Li G B, Lu Z G, Huang B Y, et al. An evaluation of lithium intercalation capacity into carbon by XRD parameters [J]. Solid State Ionics, 1995, 81 (1-2): 15-18.

[93] Peres J P, Demourgues A, Delmas C. Structural investigations on $Li_{0.65-z}Ni_{1+z}O_2$ cathode material: XRD and EXAFS studies [J]. Solid State Ionics, 1998, 111 (1-2): 135-144.

[94] Bergstrom O, Bjork H, Gustafsson T, et al. Direct XRD observation of oxidation-state changes on Li-ion insertion into transition-metal oxide hosts [J]. Journal of Power Sources, 1999, 81: 685-689.

[95] Bjork H, Gustafsson T, Thomas J O. Direct observation of XRD redox processes in TMO's [C] //Sweden: 1998 MRS fall meeting, 1999: 203-211.

[96] Thurston T R, Jisrawi N M, Mukerjee S, et al. Synchrotron X-ray diffraction studies of the structural properties of electrode materials in operating battery cells [J]. Applied Physics Letters, 1996, 69 (2): 194.

[97] Eriksson T, Hjelm A K, Lindbergh G, et al. Kinetic study of $LiMn_2O_4$ cathodes by in situ XRD with constant-current cycling and potential stepping [J]. Journal of the Electrochemical Society, 2002, 149 (9): A1164-A1170.

[98] Yang Xiaoqing, McBreen J, Yoon Won Sub, et al. Structural studies of the new carbon-coated silicon anode materials using synchrotron-based in situ XRD [J]. Electrochemistry Communications, 2002, 4 (11): 893-897.

[99] Yang Xiaoqing, McBreen J, Yoon Won Sub, et al. Crystal structure changes of $LiMn_{0.5}Ni_{0.5}O_2$ cathode materials during charge and discharge studied by synchrotron based in situ XRD [J]. Electrochemistry Communications, 2002, 4 (8): 649-654.

[100] Orikasa Y, Maeda T, Koyama Y, et al. Direct Observation of a Metastable Crystal Phase of Li_xFePO_4 under Electrochemical Phase Transition [J]. Journal of the American Chemical Society, 2013, 135 (15): 5497-5500.

[101] Hatchard T D, Dahn J R. In situ XRD and electrochemical study of the reaction of lithium with amorphous silicon [J]. Journal of the Electrochemical Society, 2004, 151 (6): A838-A842.

[102] Misra S, Liu N, Nelson J, et al. In Situ X-ray diffraction studies of (de) lithiation mechanism in silicon nanowire anodes [J]. Acs Nano, 2012, 6 (6): 5465-5473.

[103] Yu, XQ, Pan, HL, Wan, W, et al. A Size-Dependent Sodium Storage Mechanism in $Li_4Ti_5O_{12}$ Investigated by a Novel Characterization Technique Combining in Situ X-ray Diffraction and Chemical Sodiation [J]. Nano Letters, 2013, 13 (10): 4721-4727.

[104] Ronnebro E, Yin J T, Kitano A, et al. Structural analysis by synchrotron XRD of a $Ag_{52}Sn_{48}$ nanocomposite electrode for advanced Li-ion batteries [J]. Journal of the Electrochemical Society, 2004, 151 (10): A1738-A1744.

[105] Zhang L Q, Wang X Q, Noguchi H, et al. Electrochemical and ex situ XRD investigations on $(1-x)$ $LiNiO_2$ center dot $xLi_2TiO_3 (0.05 \leqslant x \leqslant 0.5)$ [J]. Electrochimica Acta, 2004, 49 (20): 3305-3311.

[106] Yoon Won Sub, Balasubramanian M, Yang Xiaoqing, et al. Time-resolved XRD study on the thermal decomposition of $Li_{1-x}Ni_{0.8}Co_{0.15}Al_{0.05}O_2$ cathode materials for Li-ion batteries [J]. Electrochemical and Solid-State Letters, 2005, 8 (2): A83-A86.

[107] Wu H M, Tu J P, Yuan Y F, et al. Electrochemical and ex situ XRD studies of a $LiMn_{1.5}Ni_{0.5}O_4$ high-voltage cathode material [J]. Electrochimica Acta, 2005, 50 (20): 4104-4108.

[108] Bowden W, Grey C P, Hackney S, et al. Lithiation of ramsdellite-pyrolusite MnO_2: NMR, XRD, TEM and electrochemical investigation of the discharge mechanism [J]. Journal of Power Sources, 2006, 153 (2): 265-273.

[109] Kodama R, Terada Y, Nakai I, et al. Electrochemical and in situ XAFS-XRD investigation of Nb_2O_5 for rechargeable lithium batteries [J]. Journal of the Electrochemical Society, 2006, 153 (3): A583-A588.

[110] Yang Xiaoqing, Yoon Won Sub, Kyung Yoonchung, et al. A comparative study on structural changes of $LiCo_{1/3}Ni_{1/3}Mn_{1/3}O_2$ and $LiNi_{0.8}Co_{0.15}Al_{0.05}O_2$ during first charge using in situ XRD [J]. Electrochemistry Communications, 2006, 8 (8): 1257-1262.

[111] Hwang Bing Joe, Hu Shao Kang, Chen Ching Hsiang, et al. In-situ XRD investigations on structure changes of ZrO_2-coated $LiMn_{0.5}Ni_{0.5}O_2$ cathode materials during charge [J]. Journal of Power Sources, 2007, 174 (2): 761-765.

[112] Zhang Dongyun, Zhang Peixin, Yi Juan, et al. XRD simulation study of doped $LiFePO_4$ [J]. Journal of Alloys and Compounds, 2011, 509 (4): 1206-1210.

[113] Sougrati M T, Fullenwarth J, Debenedetti A, et al. TiSnSb a new efficient negative electrode for Li-ion batteries: Mechanism investigations by operando-XRD and Mossbauer techniques [J]. Journal of Materials Chemistry, 2011, 21 (27): 10069-10076.

[114] Yao K P C, Kwabi D G, Quinlan R A, et al. Thermal stability of Li_2O_2 and Li_2O for Li-air batteries: In situ XRD and XPS studies [J]. Journal of the Electrochemical Society, 2013, 160 (6): A824-A831.

[115] Sun Y K, Chen Z, Noh H J, et al. Nanostructured high-energy cathode materials for advanced lithium batteries [J]. Nature Materials, 2012, 11 (11): 942-947.

[116] Yan M, Zhang G, Wei Q, et al. In operando observation of temperature-dependent phase evolution in lithium-incorporation olivine cathode [J]. Nano Energy, 2016, 22: 406-413.

[117] Rougier A, Delmas C, Chadwick AV. Noncooperative jahn-teller effect in $LiNiO_2$—An exafs study [J]. Solid State Communications, 1995, 94 (2): 123-127.

[118] Chadwick AV, Savin S L P, Packer R J, et al. EXAFS studies of lithium manganese oxides [J]. Physica Status Solidi C-Conferences and Critical Reviews, 2005, 2 (1): 657-660.

[119] Paolone A, Castellano C, Cantelli R, et al. Evidence of a splitting of the MnO distance and of a large lattice disorder in the charge-ordered phase of $LiMn_2O_4$ obtained by EXAFS [J]. Physical Review B, 2003, 68 (1): 14108.

[120] Jung Hongryun, Lee Wanjin. Electrochemical characterization of electrospun SnO x-embedded carbon nanofibers anode for lithium ion battery with EXAFS analysis [J]. Journal of Electroanalytical Chemistry, 2011, 662 (2): 334-342.

[121] Jung Hongryun, Cho Sungjune, Kim K N, et al. Electrochemical properties of electrospun Cu_xO ($x=1, 2$)-embedded carbon nanofiber with EXAFS analysis [J]. Electrochimica Acta, 2011, 56 (19): 6722-6731.

[122] Liu X, Wang D, Liu G, et al. Distinct charge dynamics in battery electrodes revealed by in situ and operando soft X-ray spectroscopy [J]. Nat Commun, 2013, 4: 2568.

[123] Yu Xiqian, Lyu Yingchun, Gu Lin, et al. Understanding the rate capability of high-energy-density Li-rich layered $Li_{1.2}Ni_{0.15}Co_{0.1}Mn_{0.55}O_2$ cathode materials [J]. Advanced Energy Materials, 2014, 4 (5): 1300950.

[124] Arbi K, Hoelzel M, Kuhn A, et al. Structural factors that enhance lithium mobility in fast-ion $Li_{1+x}Ti_{2-x}Al_x(PO_4)_3$ ($0 \leqslant x \leqslant 0.4$) conductors investigated by neutron diffraction in the temperature range $100 \sim 500$ K [J]. Inorganic Chemistry, 2013, 52 (16): 9290-9296.

[125] Li X Y, Zhang B, Zhang Z G, et al. Crystallographic structure of $LiFe_{1-x}Mn_xPO_4$ solid solutions studied by neutron powder diffraction [J]. Powder Diffraction, 2014, 29 (3): 248-253.

[126] Nishimura S, Kobayashi G, Ohoyama K, et al. Experimental visualization of lithium diffusion in Li_xFePO_4 [J]. Nat Mater, 2008, 7 (9): 707-711.

[127] Han J, Zhu J, Li Y, et al. Experimental visualization of lithium conduction pathways in garnet-type $Li_7La_3Zr_2O_{12}$ [J]. Chem Commun (Camb), 2012, 48 (79): 9840-9842.

[128] Wang Rui, He Xiaoqing, He Lunhua, et al. Atomic structure of Li_2MnO_3 after partial delithiation and Re-lithiation [J]. Advanced Energy Materials, 2013, 3 (10): 1358-1367.

[129] Grey C P, Dupre N. NMR studies of cathode materials for lithium-ion rechargeable batteries [J]. Chemical Reviews, 2004, 104 (10): 4493-4512.

[130] Key B, Bhattacharyya R, Morcrette M, et al. Real-time NMR investigations of structural changes in silicon electrodes for lithium-ion batteries [J]. Journal of the American Chemical Society, 2009, 131 (26): 9239-9249.

[131] Bhattacharyya R, Key B, Chen H, et al. In situ NMR observation of the formation of metallic lithium microstructures in lithium batteries [J]. Nature Materials, 2010, 9 (6): 504-510.

[132] Wiaderek K M, Borkiewicz O J, Castillo-Martinez E, et al. Comprehensive insights into the structural and chemical changes in mixed-anion FeOF electrodes by using operando PDF and NMR spectroscopy [J]. J Am Chem Soc, 2013, 135 (10): 4070-4078.

[133] Ilott A J, Trease N M, Grey C P, et al. Multinuclear in situ magnetic resonance imaging of electrochemical double-layer capacitors [J]. Nat Commun, 2014, 5: 4536.

[134] Lentzen M, Jahnen M, Jia B, et al. High-resolution imaging with an aberration-corrected transmission electron microscope [M]. Amsterdam, PAYS-BAS: Elsevier, 2002.

[135] Shao-Horn Y, Croguennec L, Delmas C, et al. Atomic resolution of lithium ions in $LiCoO_2$ [J]. Nat Mater,

2003, 2 (7): 464-467.

[136] Oshima Y, Sawada H, Hosokawa F, et al. Direct imaging of lithium atoms in LiV_2O_4 by spherical aberration-corrected electron microscopy [J]. Journal of Electron Microscopy, 2010, 59 (6): 457-461.

[137] Lu Xia, Jian Zelang, Fang Zheng, et al. Atomic-scale investigation on lithium storage mechanism in $TiNb_2O_7$ [J]. Energy & Environmental Science, 2011, 4 (8): 2638-2644.

[138] Huang R, Hitosugi T, Fisher C A J, et al. Phase transitions in $LiCoO_2$ thin films prepared by pulsed laser deposition [J]. Materials Chemistry and Physics, 2012, 133 (2-3): 1101-1107.

[139] Hayashi T, Okada J, Toda E, et al. Degradation mechanism of $LiNi_{0.82}Co_{0.15}Al_{0.03}O_2$ positive electrodes of a lithium-ion battery by a long-term cycling test [J]. Journal of the Electrochemical Society, 2014, 161 (6): A1007-A1011.

[140] Lu Xia, Sun Yang, Jian Zelang, et al. New insight into the atomic structure of electrochemically delithiated O_3-$Li_{1-x}CoO_2 (0 \leqslant x \leqslant 0.5)$ nanoparticles [J]. Nano Letters, 2012, 12 (12): 6192-6197.

[141] Lu X, Zhao L, He X, et al. Lithium storage in $Li_4Ti_5O_{12}$ spinel: The full static picture from electron microscopy [J]. Adv Mater, 2012, 24 (24): 3233-3238.

[142] Inaba M, Todzuka Y, Yoshida H, et al. Raman spectra of $LiCo_{1-y}Ni_yO_2$ [J]. Chemistry Letters, 1995, 24 (10): 889-890.

[143] Itoh T, Sato H, Nishina T, et al. In situ Raman spectroscopic study of Li_xCoO_2 electrodes in propylene carbonate solvent systems [J]. Journal of Power Sources, 1997, 68 (2): 333-337.

[144] Julien C. Local cationic environment in lithium nickel-cobalt oxides used as cathode materials for lithium batteries [J]. Solid State Ionics, 2000, 136: 887-896.

[145] Ohzuku T, Makimura Y. Layered lithium insertion material of $LiCo_{1/3}Ni_{1/3}Mn_{1/3}O_2$ for lithium-ion batteries [J]. Chemistry Letters, 2001, 7: 642-643.

[146] Saavedra-Arias J J, Karan N K, Pradhan D K, et al. Synthesis and electrochemical properties of Li ($Ni_{0.8}Co_{0.1}Mn_{0.1}$) O_2 cathode material: Ex situ structural analysis by Raman scattering and X-ray diffraction at various stages of charge-discharge process [J]. Journal of Power Sources, 2008, 183 (2): 761-765.

[147] Patoux S, Daniel L, Bourbon C, et al. High voltage spinel oxides for Li-ion batteries: From the material research to the application [J]. Journal of Power Sources, 2009, 189 (1): 344-352.

[148] Pasero D, Reeves N, Pralong V, et al. Oxygen nonstoichiometry and phase transitions in $LiMn_{1.5}Ni_{0.5}O_{4-\delta}$ [J]. Journal of the Electrochemical Society, 2008, 155 (4): A282-A291.

[149] Alcántara R, Jaraba M, Lavela P, et al. Optimizing preparation conditions for 5V electrode performance, and structural changes in $Li_{1-x}Ni_{0.5}Mn_{1.5}O_4$ spinel [J]. Electrochimica Acta, 2002, 47 (11): 1829-1835.

[150] Kunduraci M, Al-Sharab J F, Amatucci G G. High-power nanostructured $LiMn_{2-x}Ni_xO_4$ high-voltage lithium-ion battery electrode materials: Electrochemical impact of electronic conductivity and morphology [J]. Chemistry of Materials, 2006, 18 (15): 3585-3592.

[151] Hardwick L, Buqa H, Novak P. Graphite surface disorder detection using in situ Raman microscopy [J]. Solid State Ionics, 2006, 177 (26-32): 2801-2806.

[152] Baddour-Hadjean R, Pereira-Ramos J P. Raman microspectrometry applied to the study of electrode materials for lithium batteries [J]. Chem Rev, 2010, 110 (3): 1278-1319.

[153] Li Guifeng, Li Hong, Mo Yujun, et al. Study on solid electrolyte interphase film of anode materials for lithium batteries by SERS [J]. Chinese Journal of Light Scattering, 2003, 14 (4): 224-230.

[154] Peng Zhangquan, Freunberger S A, Chen Yuhui, et al. A reversible and higher-rate Li-O_2 battery [J]. Science, 2012, 337 (6094): 563-566.

[155] Mozzhukhina N, Méndez De Leo L P, Calvo E J. Infrared spectroscopy studies on stability of dimethyl sulfoxide for application in a Li-air battery [J]. The Journal of Physical Chemistry C, 2013, 117 (36): 18375-18380.

[156] Novák P, Goers D, Hardwick L, et al. Advanced in situ characterization methods applied to carbonaceous materials [J]. Journal of Power Sources, 2005, 146 (1-2): 15-20.

[157] Teng X, Zhan C, Bai Y, et al. In-situ analysis of gas generation in lithium ion batteries with different carbonate-based eletrolytes [J]. Applied Materials and Interfaces, 2015, 7 (41): 22751-22755.

[158] 陈伟峰. 软包装锂离子电池产气机理研究和预测 [D]. 北京: 清华大学, 2012.

[159] Ravdel B, Abrahan K M, Gitzendanner R, et al. Thermal stability of lithium-ion battery electrolytes [J]. Journal of Power Sources, 2003, 119-121: 805-810.

[160] Weber W, Kraft V, Grutzke M, et al. Identification of alkylated phosphates by gas chromatography-masss pee trometric investigations with different ionization principles of a thermally aged commercial lithium ion battery electrolyte [J]. Journal of Chromatography A, 2015, 1394: 128-136.

[161] Gachot G, Ribiere P, Mathiron D, et al. Gas Chromatography/Mass Spectrometry As a Suitable Tool for the Li-Ion Battery Electrolyte Degradation Mechanisms Study [J]. Analytical Chemistry, 2011, 83 (2): 478-485.

[162] Diao Y, Xie K, S Xiong, et al. Analysis of polysulfide dissolved in electrolyte in discharge-charge process of Li-S battery [J]. Journal of The Electrochemical Society, 2012, 159 (4), A421-A425.

[163] Schultz C, Kraft V, Pyschik M, et al. Separation and Quantification of Organic Electrolyte Components In Lithium-Ion Batteries via a Developed HPLC Method [J]. Journal of Electrochemical Society, 2015, 162 (4), A629-A634.

[164] Liu J, Li X, Wang Z, et al. Preparation and characterization of lithium hexafluorophosphate for lithium-ion battery electrolyte [J]. Transactions of Nonferrous Metals Society of China, 2010, 2 (20): 344-348.

[165] Veluchamy A, Doh C-H, Kim D-H, et al. Thermal analysis of Li_xCoO_2 cathode material of lithium ion battery [J]. Journal of Power Sources, 2009, 1 (189): 855-858.

[166] 朱彬和. 离子色谱在锂离子电池电解液组分中的方法研究和检测 [D]. 杭州: 浙江大学, 2016.

[167] Gobet M, Greenbaum S, Sahu G, et al. Structural evolution and Li dynamics in nanophase Li_3PS_4 by solid-state and pulsed-field gradient NMR [J]. Chemistry of Materials, 2014, 26 (11): 3558-3564.

[168] Jerliu B, Dorrer L, Huger E, et al. Neutron reflectometry studies on the lithiation of amorphous silicon electrodes in lithium-ion batteries [J]. Physical Chemistry Chemical Physics, 2013, 15 (20): 7777-7784.

[169] Manke I, Markotter H, Totzke C, et al. Investigation of energy-relevant materials with synchrotron X-rays and neutrons [J]. Advanced Engineering Materials, 2011, 13 (8): 712-729.

[170] Oudenhoven J F M, Labohm F, Mulder M, et al. In situ neutron depth profiling: A powerful method to probe lithium transport in micro-batteries [J]. Adv Mater, 2011, 23 (35): 4103.

[171] Sharma K, Bilheux H Z, Walker L M H, et al. Neutron imaging of ion transport in mesoporous carbon materials [J]. Physical Chemistry Chemical Physics, 2013, 15 (28): 11740-11747.

[172] Saito Y, Yamamoto H, Nakamura O, et al. Determination of ionic self-diffusion coefficients of lithium electrolytes using the pulsed field gradient NMR [J]. Journal of Power Sources, 1999, 81: 772-776.

[173] Reddy M J, Chu P P. Li-7 NMR spectroscopy and ion conduction mechanism in mesoporous silica (SBA-15) composite poly (ethylene oxide) electrolyte [J]. Journal of Power Sources, 2004, 135 (1-2): 1-8.

[174] Cahill L S, Chapman R P, Britten J F, et al. Li-7 NMR and two-dimensional exchange study of lithium dynamics in monoclinic $Li_3V_2(PO_4)_3$ [J]. Journal of Physical Chemistry B, 2006, 110 (14): 7171-7177.

[175] Wilkening M, Muhle C, Jansen M, et al. Microscopic access to long-range diffusion parameters of the fast lithium ion conductor Li_7BiO_6 by solid state Li-7 stimulated echo NMR [J]. Journal of Physical Chemistry B, 2007, 111 (30): 8691-8694.

[176] Hayamizu K, Tsuzuki S, Seki S. Molecular motions and ion diffusions of the room-temperature ionic liquid 1,2-dimethyl-3-propylimidazolium bis (trifluoromethylsulfonyl) amide (DMPImTFSA) studied by H-1, C-13, and F-19 NMR [J]. Journal of Physical Chemistry A, 2008, 112 (47): 12027-12036.

[177] Wilkening M, Gebauer D, Heitjans P. Diffusion parameters in single-crystalline Li_3N as probed by (6) Li and (7) Li spin-alignment echo NMR spectroscopy in comparison with results from (8) Li beta-radiation detected NMR [J]. Journal of Physics-Condensed Matter, 2008, 20 (2): 5.

[178] Wilkening M, Kuhn A, Heitjans P. Atomic-scale measurement of ultraslow Li motions in glassy $LiAlSi_2O_6$ by two-time (6) Li spin-alignment echo NMR correlation spectroscopy [J]. Physical Review B, 2008, 78 (5): 9.

[179] Epp V, Wilkening M. Fast Li diffusion in crystalline $LiBH_4$ due to reduced dimensionality: Frequency-dependent NMR spectroscopy [J]. Physical Review B, 2010, 82 (2): 4.

[180] Hatzikraniotis E, Samaras I, Paraskevopoulos K M, et al. Lithium Intercalation Studies on MoO_3 single crystals [J]. Ionics, 1996, 2 (1): 24-28.

[181] Hirasawa K A, Nishioka K, Sato T, et al. Investigation of graphite composite anode surfaces by atomic force microscopy and related techniques [J]. Journal of Power Sources, 1997, 69 (1-2): 97-102.

[182] Hirasawa K A, Sato T, Asahina H, et al. In situ electrochemical atomic force microscope study on graphite electrodes [J]. Journal of the Electrochemical Society, 1997, 144 (4): L81-L84.

[183] Aurbach D, Moshkovich M, Cohen Y, et al. The study of surface film formation on noble-metal electrodes in

alkyl carbonates/Li salt solutions, using simultaneous in situ AFM, EQCM, FTIR, and EIS [J]. Langmuir, 1999, 15 (8): 2947-2960.

[184] Balke N, Jesse S, Morozovska A N, et al. Nanoscale mapping of ion diffusion in a lithium-ion battery cathode [J]. Nature Nanotechnology, 2010, 5 (10): 749-754.

[185] Chung D W, Balke N, Kalinin S V, et al. Virtual electrochemical strain microscopy of polycrystalline $LiCoO_2$ films [J]. Journal of the Electrochemical Society, 2011, 158 (10): A1083.

[186] Zhu Jing, Feng Jinkui, Lu Li, et al. In situ study of topography, phase and volume changes of titanium dioxide anode in all-solid-state thin film lithium-ion battery by biased scanning probe microscopy [J]. Journal of Power Sources, 2012, 197: 224-230.

[187] Kostecki R, Kong F P, Matsuo Y, et al. Interfacial studies of a thin-film $Li_2Mn_4O_9$ electrode [J]. Electrochimica Acta, 1999, 45 (1-2): 225-233.

[188] Matsuo Y, Kostecki R, McLarnon F. Surface layer formation on thin-film $LiMn_2O_4$ electrodes at elevated temperatures [J]. Journal of the Electrochemical Society, 2001, 148 (7): A687.

[189] Lipson A L, Ginder R S, Hersam M C. Nanoscale in situ characterization of Li-ion battery electrochemistry via scanning ion conductance microscopy [J]. Advanced Materials, 2011, 23 (47): 5613-5617.

[190] Lee Wonyoung, Prinz F B, Chen Xi, et al. Nanoscale impedance and complex properties in energy-related systems [J]. MRS Bulletin, 2012, 37 (07): 659-667.

[191] Ogumi Z, Jeong S K. SPM analysis of surface film formation on graphite negative electrodes in lithium-ion batteries [J]. Electrochemistry, 2003, 71 (12): 1011-1017.

[192] Jeong S K, Inaba M, Abe T, et al. Surface film formation on graphite negative electrode in lithium-ion batteries: AFM study in an ethylene carbonate-based solution [J]. Journal of the Electrochemical Society, 2001, 148 (9): A989.

[193] Inaba M, Tomiyasu H, Tasaka A, et al. Atomic force microscopy study on the stability of a surface film formed on a graphite negative electrode at elevated temperatures [J]. Langmuir, 2004, 20 (4): 1348-1355.

[194] Becker C R, Strawhecker K E, McAllister Q P, et al. In situ atomic force microscopy of lithiation and delithiation of silicon nanostructures for lithium ion batteries [J]. ACS Nano, 2013, 7 (10): 9173-9182.

[195] McAllister Q P, Strawhecker K E, Becker C R, et al. In situ atomic force microscopy nanoindentation of lithiated silicon nanopillars for lithium ion batteries [J]. Journal of Power Sources, 2014, 257: 380-387.

[196] Tokranov A, Sheldon B W, Li C, et al. In situ atomic force microscopy study of initial solid electrolyte interphase formation on silicon electrodes for Li-ion batteries [J]. ACS Applied Materials & Interfaces, 2014, 6 (9): 6672-6686.

[197] Cresce A, Russell S M, Baker D R, et al. In situ and quantitative characterization of solid electrolyte interphases [J]. Nano Letters, 2014, 14 (3): 1405-1412.

[198] Wang Lixin, Deng Da, Lev L C, et al. In-situ investigation of solid-electrolyte interphase formation on the anode of Li-ion batteries with atomic force microscopy [J]. Journal of Power Sources, 2014, 265: 140-148.

[199] Zhang J, Wang R, Yang X, et al. Direct observation of inhomogeneous solid electrolyte interphase on MnO anode with atomic force microscopy and spectroscopy [J]. Nano Letters, 2012, 12 (4): 2153-2157.

[200] Zheng J, Zheng H, Wang R, et al. 3D visualization of inhomogeneous multi-layered structure and Young's modulus of the solid electrolyte interphase (SEI) on silicon anodes for lithium ion batteries [J]. Physical Chemistry Chemical Physics: PCCP, 2014, 16 (26): 13229-13238.

[201] Zheng Jieyun, Li Hong. Fundamental scientific aspects of lithium batteries (Ⅴ) —Interfaces [J]. Energy Storage Science and Technology, 2013, 2 (5): 503-513.

[202] Nagpure S C, Bhushan B, Babu S S. Surface potential measurement of aged Li-ion batteries using Kelvin probe microscopy [J]. Journal of Power Sources, 2011, 196 (3): 1508-1512.

[203] Zhu Jing, Zeng Kaiyang, Lu L. In-situ nanoscale mapping of surface potential in all-solid-state thin film Li-ion battery using Kelvin probe force microscopy [J]. Journal of Applied Physics, 2012, 111 (6): 063723.

[204] Yamamoto K, Iriyama Y, Asaka T, et al. Dynamic visualization of the electric potential in an all-solid-state rechargeable lithium battery [J]. Angew Chem-Int Edit, 2010, 49 (26): 4414-4417.

[205] Yamamoto K, Iriyama Y, Asaka T, et al. Direct observation of lithium-ion movement around an in-situ-formed-negative-electrode/solid-state-electrolyte interface during initial charge-discharge reaction [J]. Electrochemistry Communications, 2012, 20: 113-116.

[206] Yamamoto K, Hirayama T, Tanji T. Development of advanced electron holographic techniques and application to industrial materials and devices [J]. Microscopy, 2013, 62: S29-S41.

[207] 王浩, 李建军, 王莉, 等. 绝热加速量热仪在锂离子电池安全性研究方面的应用 [J]. 新材料产业, 2013 (01): 59-64.

[208] Jiang J, Dahn J R. ARC studies of the thermal stability of three different cathode materials: $LiCoO_2$; Li[$Ni_{0.1}Co_{0.8}Mn_{0.1}$]O_2; and $LiFePO_4$, in $LiPF_6$ and LiBoB EC/DEC electrolytes [J]. Electrochemistry Communications, 6 (1): 39-43.

[209] Jiang J, Dahn J R. Effects of particle size and electrolyte salt on the thermal stability of $Li_{0.5}CoO_2$ [J]. Electrochimica Acta, 2004, 49 (16): 2661-2666.

[210] Ma Lin, Nie Mengyun, Xia Jian. A systematic study on the reactivity of different grades of charged Li[$Ni_xMn_yCo_z$]O_2 with electrolyte at elevated temperatures using accelerating rate calorimetry [J]. Journal of Power Sources, 2016, 327: 145-150.

[211] Hilary McMillan, Alberto Montanari, Christophe Cudennec. Panta Rhei 2013-2015: Global perspectives on hydrology, society and change [J]. Hydrological Sciences Journal/journal Des Sciences Hydrologiques, 2016, 61 (7): 1174-1191.

[212] Feng Xuning, Fang Mou, He Xiangming, et al. Thermal runaway features of large format prismatic lithium ion battery using extended volume accelerating rate calorimetry [J]. Journal of Power Sources, 2014, 255: 294-301.

[213] 王莉, 冯旭宁, 薛钢, 等. 锂离子电池安全性评估的 ARC 测试方法和数据分析 [J]. 储能科学与技术, 2018, 7 (6): 313-322.

[214] 邵晓挺, 冯浩. 锂离子电池全生命周期安全性演变研究进展 [J]. 化工设计通讯, 2018, 44 (12): 201.

[215] 李坤, 王敬, 王芳, 等. 不同循环周期锂离子动力电池热失控特性分析 [J]. 电源技术, 2017 (4): 544-547.

[216] 毛亚, 白清友, 马尚德, 等. 循环老化对锂离子电池在绝热条件下的产热及热失控影响 [J]. 储能科学与技术, 2018, 7 (6): 172-179.

[217] 赵学娟. 锂离子电池在绝热条件下的循环产热研究 [D]. 合肥: 中国科学技术大学, 2014.

[218] 庄宗标, 徐秀娟, 姚卿敏, 等. 加速量热仪在锂离子电池热安全性能方面的研究 [J]. 电子质量, 2015 (4): 4-8.

[219] JiaojiaoYun, Li Zhang. A binary cyclic carbonates-based electrolyte containing propylene carbonate and trifluoropropylene carbonate for 5V lithium-ion batteries [J]. Electrochimica Acta, 2015 (10): 151-159.

[220] Karen Akemi Hirasawa, Keiko Nishioka. Investigation of graphite composite anode surfaces by atomic force microscopy and related techniques [J]. Journal of Power Sources, 1997 (1/2): 97-102.

第 12 章

电化学测量方法

锂离子电池电极过程一般经历复杂的多步骤电化学反应,并伴随化学反应,商品化的锂离子电池电极通常是非均相多孔粉末电极。为了获得可重现的、能反映材料与电池热力学及动力学特征的信息,需要对锂离子电池电极过程本身有清楚的认识。

电池中电极过程一般包括溶液相中离子的传输,电极中离子的传输,电极中电子的传导,电荷转移,双电层或空间电荷层充放电,溶剂、电解质中阴阳离子迁移,气相反应物或产物的吸附脱附,新相成核长大,与电化学反应耦合的化学反应,体积变化,吸放热等过程。这些过程有些同步进行,有些先后发生。

电极过程的驱动力包括电化学势、化学势、浓度梯度、电场梯度、温度梯度。影响电极过程热力学的因素包括理想电极材料的电化学势,受电极材料形貌、结晶度、结晶取向、表面官能团影响的缺陷能,温度等因素。影响电极过程动力学的因素包括电化学与化学反应活化能,极化电流与电势,电极与电解质相电位匹配性,电极材料离子、电子输运特性,参与电化学反应的活性位密度、真实面积,离子扩散距离,电极与电解质浸润程度与接触面积,界面结构与界面副反应,温度等。

为了理解复杂的电极过程,一般电化学测量要结合稳态和暂态方法,通常包括 3 个基本步骤,如图 12-1 所示。

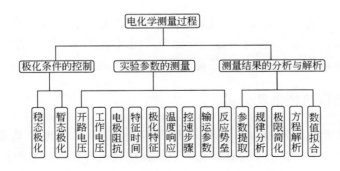

图 12-1 电化学测量的基本步骤

12.1 电化学测量概述[1]

12.1.1 测量的基本内容

电化学测量主要研究电池或电极的电流、电势在稳态和暂态的激励信号下随外界条件变化的规律,测量反映动力学特性的参数。

12.1.2 测量电池的分类及特点

电化学测量一般采用两电极电池或三电极电池,较少使用四电极电池。

12.1.2.1 两电极电池

如图12-2所示,虚线框所示是一个典型的两电极电池的测量示意图,其中W表示研究电极,亦称为工作电极(working electrode);C是辅助电极(auxiliary electrode),亦称为对电极(counter electrode)。锂电池的实际研究中多采用两电极电池,两电极电池测量的电压(voltage)是正极电势(potential)与负极电势之差,无法单独获得其中正极或负极的电势及其电极过程动力学信息。

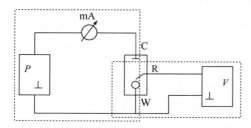

图 12-2 两电极与三电极测量体系示意图

12.1.2.2 三电极电池

图12-2是一个三电极电池示意图,W和C分别是工作电极和对电极(同12.1.2.1),R是参比电极(reference electrode)。W和C之间通过极化电流,实现电极的极化。W和R之间通过极小的电流,用于测量工作电极的电势。通过三电极电池,可以专门研究工作电极的电极过程动力学。

由于在锂离子电池中,正极和负极的电化学响应特性存在较大差异,通过测量两电极电池电压电流曲线,并对曲线进行dQ/dV处理,同时结合熵的原位测量,也能大致判断电池的电流或电压响应主要是与负极还是与正极反应有关。

12.1.3 参比电极的特性及分类

参比电极的性能直接影响电极电势的准确测量,通常参比电极具备以下基本特征:①参比电极应为可逆电极;②不易被极化,以保证电极电势比较标准和恒定;③具有较好的恢复特性,不发生严重的滞后现象;④具有较好的稳定性和重现性;⑤快速暂态测量时,要求参比电极具有较低的电阻,以减少干扰,提高测量系统的稳定性;⑥不同的溶液体系,采用相同的参比电极的,其测量结果可能存在差异,误差主要来源于溶液体系间的相互污染和液体接界电势的差异。

常用的水溶液体系参比电极有可逆氢电极、甘汞电极、汞-氧化汞电极、汞-硫酸亚汞电极等；常用的非水溶液体系参比电极有银-氯化银电极、Pt 电极、金属锂电极、金属钠电极等。此外，也可以用银丝、铂丝做准参比电极，或者采用电化学反应电位稳定的溶解于电解液的二茂铁氧化还原电对。关于准参比电极细节可参考 A. J. Bard 编著的 "Electrochemical Methods Fundamentals and Applications" 第 2.4 节。

12.1.4　研究电极的分类及特性

电化学测量中常用的研究电极主要有固体电极、超微电极和单晶电极。一般电化学研究所指的固体电极主要有 Pt 电极和碳电极。其中碳电极包括热解石墨、高定向热解石墨（HOPG）、多晶石墨、玻璃化碳、碳纤维等。固体电极在使用时需要对其表面进行特殊处理，以期达到较好的重复性。常规的处理步骤为：①浸泡有机溶剂，除去表面吸附有机物；②机械抛光，初步获取较高的表面光洁度；③电化学抛光，除去电极表面氧化层及残留吸附物质；④溶液净化，保证溶液的纯度，消除溶液中的杂质对测量结果的影响。

此外，超微电极和单晶电极以其独特的性质，也得到了较广泛的应用。前者可以快速获得动力学参数，且对待测材料的量要求很低，可以避免黏结剂、导电添加剂的干扰。后者可以精确获得溶剂吸脱附、表面结构、结晶取向等对电极过程动力学的影响。

在锂离子电池的研究中，固体电极包括含有活性物质的多孔粉末电极、多晶薄膜电极、外延膜薄膜电极、单颗粒微电极以及单晶电极等，多数测量时采用多孔粉末电极。

12.1.5　电极过程

电极过程一般情况下包括下列基本过程或步骤：①电化学反应过程：在电极/溶液界面上得到或失去电子生成反应产物的过程，即电荷转移过程；②传质过程：反应物向电极表面或内部传递，或反应产物自电极内部或表面向溶液中或向电极内部的传递过程，即迁移和扩散过程；③电极界面处靠近电解液一侧的双电层以及靠近电极内一侧的空间电荷层的充放电过程；④溶液中离子的电迁移或电子导体、电极内电子的导电过程。

此外，伴随电化学反应，还有溶剂、阴阳离子、电化学反应产物的吸附/脱附过程，新相生长过程以及其它化学反应等。

针对不同的电极材料及电极体系，锂离子电池电极过程可简化为锂离子电池中离子和电子的传输及存储过程。所涉及的电化学过程有电子、离子在材料的体相、两相界面和 SEI 的形成等过程。典型的电极过程及动力学参数有：①离子在电解质中的迁移电阻（R_{sol}）；②离子在电极表面的吸附电阻和电容（R_{ad}，C_{ad}）；③电化学双电层电容（C_{dl}）；④空间电荷层电容（C_{sc}）；⑤离子在电极电解质界面的传输电阻（$R_{incorporation}$）；⑥离子在表面膜中的输运电阻和电容（R_{film}，C_{film}）；⑦电荷转移电阻（R_{ct}）；⑧电解质中离子的扩散阻抗（$Z_{diffusion}$）；⑨电极中离子的扩散阻抗（$Z_{diffusion}$）——体相扩散电阻（R_b）和晶粒晶界中的扩散电阻（R_{gb}）；⑩宿主晶格中外来原子/离子的存储电容（C_{chem}），相转变反应电容（C_{chem}）；⑪电子的输运电阻（R_e）。

上述基本动力学参数涉及不同的电极基本过程,因而具有不同的时间常数。典型的电池中的电极过程及时间常数如图 12-3 所示,一般离子在电极、电解质材料内部的扩散以及固相反应是速率控制步骤(rate determining step,RDS)。

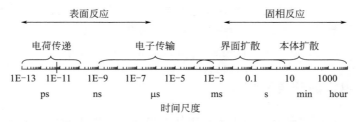

图 12-3　储能电池中的动力学过程及典型时间常数

12.1.6　极化的类型及影响因素

12.1.6.1　极化的类型及其特征

在施加了外来电场后,电池或电极逐渐偏离平衡电势的状态,称为极化。在不具有流动相的电池中,存在着 3 种类型的极化:①电化学极化——与电荷转移过程有关的极化,极化的驱动力是电场梯度;②浓差极化——与参与电化学反应的反应物和产物的扩散过程有关的极化,极化的驱动力为浓度梯度;③欧姆极化——与载流子在电池中各相输运有关的极化,驱动力是电场梯度。

此外,若还存在其他基本电极过程,如匀相或多相化学反应过程,则可能存在化学反应极化。

极化电势与平衡电势差值的大小被称为过电势。

12.1.6.2　极化的影响因素

各类极化的影响因素如下。

① 电化学极化的大小是由电化学反应速率决定的,电化学极化电阻(R_{ct})的大小与交换电流密度(i_0)直接相关。受多种因素影响,包括电极电位、电极电位与电解质电化学势差、反应物与产物的活度、参与电化学反应的电极的真实表面积、结晶取向、有序度、表面电导、反应温度、催化剂催化特性、电化学反应的可逆性等。

电化学极化的电流与过电势在一定的电流电压范围内一般符合 Tafel 关系,lg(i)与过电势成正比。

② 浓差极化与传质粒子的扩散系数有关。电池中的扩散过程可以发生在电极材料内部、多孔电极的孔隙中,以及电解质相中,参与扩散的可以是带电或中性粒子。涉及扩散的粒子流的流量一般符合 Fick 扩散定律,与扩散系数及浓度梯度有关(Fick 第一定律)。由于电池是非均相体系,扩散系数与浓度梯度是空间位置的函数,在电化学反应的过程中,会随时间变化。传质的快慢与传质距离的平方成正比(Fick 第二定律)。

浓差极化过电势 η_{con} 与电流 i、极限电流 i_1 的关系符合对数关系,$\eta_{con}=RT/(nF) \times \ln[(i_1-i)/i_1]$。在过电势较小时,$\eta_{con}=-RTi/(nFi_1)$。

③ 欧姆极化的大小是由电池内部涉及电迁移的各类电阻之和，即欧姆电阻决定的。欧姆极化过电势与极化电流密度成正比。

12.2 测量方法

12.2.1 稳态测量

12.2.1.1 稳态过程与稳态系统的特征

一个电化学系统，如果在某一时间段内，描述电化学系统的参量，如电极电势、电流密度、界面层中的粒子浓度及界面状态等不发生变化或者变化非常微小，则称这种状态为电化学稳态。

稳态不等同于平衡态，平衡态是稳态的一个特例。同时，绝对的稳态是不存在的，稳态和暂态也是相对的。稳态和暂态的分界线在于某一时间段内电化学系统中各参量的变化是否显著[2]。

12.2.1.2 稳态极化曲线的测量方法

稳态极化曲线的测量按照控制的自变量可分为控制电流法和控制电势法。

控制电流法亦称为恒电流法，施加恒定电流测量相应电势。控制电势法亦称为恒电势法，通过控制研究电极的电势来测量响应电流。

本质上恒电流法和恒电势法在极化曲线的测量方面具有相同的功能，如果电化学体系中存在电流极大值时选择恒电势法，则存在电势极大值时选择恒电流法。

12.2.1.3 稳态测量方法的应用

稳态极化曲线是研究电极过程动力学最基本的方法，在电化学基础研究方面有着广泛的应用。可根据极化曲线判断反应的机理和速率控制步骤；可以测量体系可能发生的电极反应的最大反应速率；可以测量电化学过程中的动力学参数，如交换电流密度、传递系数、标准速率常数和扩散系数等；可以测定 Tafel 斜率，推算反应级数，进而获取反应进程信息；此外，还可以利用极化曲线研究多步骤的复杂反应，研究吸附和表面覆盖等过程。

12.2.2 暂态测量

12.2.2.1 暂态过程与暂态系统的特征

暂态是相对稳态而言的，随着电极极化条件的改变，电极会从一个稳态向另一个稳态转变，在此期间所经历的不稳定的、电化学参量显著变化的过程称为暂态过程。

暂态过程具有如下基本特征：①存在暂态电流——该电流由双电层充电电流和电化学

反应电流组成，前者又称为非法拉第电流或电容电流，后者常常称为法拉第电流；②界面处存在反应物与产物粒子的浓度梯度——即电极/溶液界面处反应物与产物的粒子浓度，如前所述，不仅是空间位置的函数，同时也是时间的函数。

12.2.2.2 暂态过程中的等效电路分析及其简化

由于暂态过程中的各参量是随时间变化的，与稳态过程比较，更为复杂。为便于分析和讨论，将各电极过程以电路元件组成的等效电路的形式来描述，等效电路施加电流后的电压响应，应与电极过程的电流电压响应一致。典型的两电极测量体系等效电路如图12-4所示。

图12-4中，A和B分别代表研究电极和辅助电极（两电极体系），R_A和R_B分别表示研究电极和辅助电极的欧姆电阻；C_{AB}表示两电极之间的电容；R_u表示两电极之间的溶液电阻；C_d和C_d'分别表示研究电极和辅助电极的界面双电层电容；Z_r和Z_r'分别表示研究电极和辅助电极的法拉第阻抗。

若A、B均为金属电极，则R_A和R_B很小，可忽略；由于两电极之间的距离远大于界面双电层的厚度，故C_{AB}比双电层电容C_d和C_d'小得多，当溶液电阻R_u不是很大时，由C_{AB}带来的容抗远大于R_u，故C_{AB}支路相当于断路[$C \propto S/(4\pi kd)$，$R \propto 1/(\omega C)$]，可忽略；此外，若辅助电极面积远大于研究电极面积，则C_d'远大于C_d，此时，C_d'很小，相当于短路，故等效电路（图12-4）最终可简化为如图12-5所示。这相当于在电池中一个电极的电阻很小时的情况，如采用金属锂负极的两电极电池。

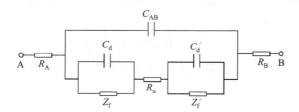

图12-4 两电极测量体系电解池的等效电路

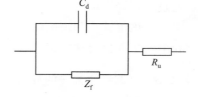

图12-5 两电极测量体系电解池的简化电路

由于电极过程的多步骤和复杂性，不同速率控制步骤下，电极体系的等效电路不尽相同，有时可以进一步简化，常见的有如下三种情形。

(1) 传荷过程控制下的等效电路

暂态过程中由于暂态电流的作用使得电极溶液界面处存在双电层充电电流，该双电层类似于平行板电容器，可用C_d表示，相应的充电电流的大小用i_c来表示。此外，界面处还存在着电荷的传递过程，电荷的传递过程可用法拉第电流来描述，由于电荷传递过程的迟缓性，导致法拉第电流引起了电化学极化过电势，该电流-电势的关系类似于纯电阻上的电流-电势关系，因而电荷传递过程可以等效为一个纯电阻响应，用R_{ct}表示。由于传荷电阻两端的电压是通过双电层荷电状态的改变而建立起来的，因而，一般认为R_{ct}与C_d在电路中应属于并联关系，传荷过程控制下的简化等效电路如图12-6所示。需要指出的是，这一简化模型基于传统电化学体系，在锂离子电池中，电极在多数状态下，大量电荷存储在电极内，造成电容效应，可以称为化学电容 C_{chem}，与 C_{dl} 应该是串联关系。因此，在实验上与 R_{ct} 并联显示在阻抗谱半圆上的到底应该是电双层电容还是化学电容还是两种电容之和取决于哪一个电容值更低。

（2）浓差极化不可忽略时的等效电路

暂态过程中，对于惰性电极，由于电极/溶液界面处存在暂态电流，因此开始有电化学反应的发生，界面处不断发生反应物消耗和产物积累，开始出现反应物产物浓度差。随着反应的进行，浓度差不断增大，扩散传质过程进入对流区，电极进入稳态扩散过程，建立起稳定的浓差极化过电势，由于浓差极化过电势滞后于电流，因此电流-电势之间的关系类似于一个电容响应。可以用一个纯电阻 R_w 串联电容 C_w 表示。该串联电路可用半无限扩散模型来模拟，如图 12-7 所示。这种情况在电池中经常出现。

图 12-6　传荷过程控制下的简化等效电路　　图 12-7　半无限扩散阻抗等效电路

上述 R_w 和 C_w 的串联结构可用一个复数阻抗 Z_w 来表示，Z_w 可理解为半无限扩散阻抗。由于扩散传质过程和电荷传递过程同时进行，因而两者具有相同的电化学速率，在电路中应属于串联关系。一般在阻抗谱上表现为 45° 的斜线。

在锂离子电池中，锂离子在电极材料内部的扩散或者在电极层颗粒之间的孔隙或者含孔颗粒内电解质相的扩散往往是速率控制步骤，具体何种过程是速率控制步骤和实际电极材料颗粒尺寸的大小及孔隙率的大小密切相关。由于存在边界条件约束，往往显示出有限边界条件下的扩散。在浓差极化不可忽略的情形下，可以如图 12-8 所示。有限边界条件下扩散的等效电路元件只是将 Z_w 换为相应的等效电路扩散元件。

（3）溶液电阻不可忽略时的界面等效电路

当溶液电阻不可忽略时，由于极化电流同时流经界面和溶液，因而溶液电阻与界面电阻应属于串联关系，典型的浓差极化不可忽略、溶液电阻不可忽略时的等效电路如图 12-9 所示。在锂离子电池中，由于电极通常是多孔粉末电极，电极的欧姆电阻一般也不可忽略，与电解质电阻是串联关系，往往合并在一项中。

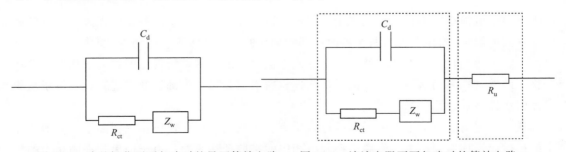

图 12-8　浓差极化不可忽略时的界面等效电路　　图 12-9　溶液电阻不可忽略时的等效电路

12.2.2.3 暂态测量方法的分类及其特点

暂态过程测量方法按照自变量的控制方式可分为控制电流法和控制电势法；按照自变量的给定方式可分为阶跃法、方波法、线性扫描法和交流阻抗法。

通常，用暂态测量能比稳态测量给出更多的电化学参量信息。一般来说，暂态测量法具有如下特点：①暂态法可以同时测量双电层电容 C_d 和溶液电阻 R_u；②暂态法能够测量电荷传递电阻 R_{ct}，因此，能够间接测量电化学过程中标准速率常数和交换电流的大小；③暂态法可研究快速电化学反应，通过缩短极化时间，如以旋转圆盘电极代替普通电极，并加快旋转速度，可以降低浓差极化的影响，当测量时间小于 10^{-5} s 时，暂态电流密度可高达 10 A/cm^2；④暂态法可用于研究表面快速变化的体系，而在稳态过程中，由于反应产物会不断积累，电极表面在反应时不断受到破坏，因而类似于电沉积和阳极溶解过程，很难用稳态法进行测量；⑤暂态法有利于研究电极表面的吸脱附结构和电极的界面结构，由于暂态测量的时间非常短，液相中的杂质粒子来不及扩散到电极表面，因而暂态法可用于研究电极反应的中间产物和复杂的电极过程。

上述 12.2.1 及 12.2.2 两小节介绍的内容主要适用于传统的电化学体系，氧化还原反应发生在电极表面，电极为惰性电极，电解质为稀浓度电解质情形，更详细准确的描述可参考相关的电化学教科书。

锂电池与传统电化学测量体系显著不同之处是氧化还原反应发生在电极内部而非电极表面，离子的扩散、电荷转移，相变可以发生在电极内部。锂电池的电极一般是非均相多孔粉末电极，孔隙之中存在着电解液，电解液中离子的浓度达到 1mol/L 甚至更高，这些不同导致获得可靠的锂离子电池电极过程动力学参数非常困难，需要仔细谨慎地设计测量电池体系，如锂空气电池的研究涉及多种中间产物的分析，圆盘电极和环盘电极等暂态测量被广泛应用。

12.3 典型的测量方法及其在锂电池中的应用

12.3.1 锂离子电池电极过程动力学及其测量方法

锂离子电池在电池充放电过程中一般经历以下几个步骤：①溶剂化的锂离子从电解液内迁移到电解液/固体电极的两相界面；②溶剂化的锂离子吸附在电解液/固体电极的两相界面；③去溶剂化；④电荷转移，电子注入电极材料的导带，吸附态的锂离子从电解液相迁移至活性材料表面晶格；⑤锂离子从活性材料表面晶格向内部扩散或迁移；⑥电子从集流体向活性材料迁移。

通常，锂离子电池中的电子输运过程比离子扩散迁移过程快很多，锂离子在电解液相中的扩散或迁移速度远大于锂离子在固体相中的扩散迁移速度；由于锂离子在固相中的扩散系数很小，一般在 $10^{-14} \sim 10^{-9}$ cm^2/s 数量级，而颗粒尺寸一般在微米量级，因此，锂离子在固体活性材料颗粒中的扩散过程往往成为二次锂电池充放电过程的速率控制步骤。由于电极过程动力学直接关系到电池的充放电倍率、功率密度、内阻、循环性和安全性等性质。定量掌握电池与电极过程动力学反应特性以及动力学参数随着充放电过程的演化过程，对于理解电池中的电化学反应、监控电池的状态、设计电源管理系统具有重要的意义。

锂离子电池的典型工作曲线与测试内容如图 12-10 所示，主要涉及热力学、动力学及稳定性三个维度，具体参数包括开路电压、倍率特性、温度响应、充放电深度、电压-容量特性、阻抗、能量密度、能量转换效率、功率密度、循环寿命、日历寿命、自放电、热行为等。

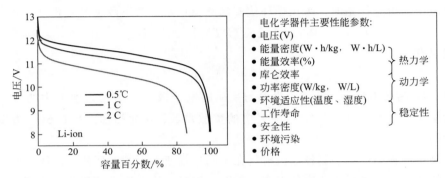

图 12-10　锂离子电池的典型工作曲线与测试内容（室温下）

目前，已有多种方法被开发并相继用于锂离子电池电极过程动力学信息的测量，如循环伏安法（cyclic voltammetry，CV），阻抗谱（impedance spectroscopy，IS），恒电流间歇滴定技术（galvanostatic intermittent titration technique，GITT），恒电位间歇滴定技术（potentiostatic intermittent titration technique，PITT），电流脉冲弛豫（CPR），电位阶跃计时电流（PSCA），电位弛豫技术（potential relax technique，PRT）等。

12.3.2　稳态测量技术——线性电势扫描伏安法

12.3.2.1　电势扫描伏安法及其特点

线性电势扫描法通过控制电极电势连续线性变化，同时测量通过电极的响应电流来分析电极过程。

线性电势扫描法可分为单程线性电势扫描法（LSV）和循环伏安法（CV），对应的电势扫描波形为单程线性电势波和连续三角波。

线性电势扫描法在电化学测量中有着广泛的应用，常用于：①判断电极体系中可能发生的电化学反应；②判断电极过程的可逆性；③判断电极反应的反应物来源；④研究电极活性物质的吸脱附过程。在锂电池的研究中，更多的是使用循环伏安法。

12.3.2.2　循环伏安法及其应用

循环伏安法（cyclic voltammetry，CV）是常见的电化学研究方法之一。在传统电化学中，常用于电极反应的可逆性、电极反应机理（如中间体、相界吸/脱附、新相生成、偶联化学反应的性质等）及电极反应动力学参数（如扩散系数、电极反应速率常数等）的探究。

典型的循环伏安过程为：电势向阴极方向扫描时，电活性物质在电极上还原，产生还原峰；向阳极方向扫描时，还原产物重新在电极上氧化，产生氧化峰。因而一次扫描，完成一个还原和氧化过程的循环，其电流-电压曲线称为循环伏安曲线。通过循环伏安曲线的氧化峰和还原峰的峰高、对称性，氧化峰与还原峰的距离，中点位置，可判断电活性物质在电极表面反应的可逆程度和极化程度。

由 12.1.5 节中的描述可知，电极过程是包括多个步骤的复杂过程。当扩散过程为控制步骤且电极为可逆体系时，采用循环伏安法测量化学扩散系数满足公式（12-1）[3]

$$I_P = 0.4463zFA[zF/(RT)]^{1/2}\Delta C_0 (D_{Li})^{1/2} v^{1/2} \tag{12-1}$$

常温时有

$$I_P = 2.69 \times 10^5 n^{1/2} A (D_{Li})^{1/2} v^{1/2} \Delta C_0 \tag{12-2}$$

式中，I_P 为峰电流的大小；n 为参与反应的电子数；A 为浸入溶液中的真实电极面积；F 为法拉第常数；D_{Li} 为 Li 在电极中的扩散系数；v 为扫描速率；ΔC_0 为反应前后待测浓度的变化。

基本测量过程如下：①测量电极材料在不同扫描速率下的循环伏安曲线；②将不同扫描速率下的峰值电流对扫描速率的平方根作图；③对峰值电流进行积分，测量样品中锂的浓度变化；④将相关参数代入式（12-2），即可求得扩散系数。采用循环伏安法获得的不同电极材料的锂离子化学扩散系数见表 12-1。

表 12-1 不同电极材料的锂离子化学扩散系数（CV 测量结果）

材料	$D_{Li}/(cm^2/s)$	参考文献	可靠性
$LiCoO_2$	10^{-13}	[4]	√
硅	$10^{-13} \sim 10^{-12}$	[5]	√
$Li_3V_2(PO_4)_3$	$10^{-10} \sim 10^{-9}$	[6]	√
$Li_3V_2(PO_4)_3$	$10^{-13} \sim 10^{-8}$	[6]	√
$LiFePO_4$	10^{-11}	[7]	×
$Li_3V_2(PO_4)_3$	10^{-13}	[7]	√
$Li_4Ti_5O_{12}$	10^{-11}	[8]	×

需要指出的是，上述介绍的循环伏安法用于测量化学扩散系数时，需要满足反应受扩散控制这一前提条件。传统电化学中多用于液相参与反应的物质的扩散。锂离子电池中多数氧化还原反应涉及固体电极内部的电荷转移，伴随着锂离子嵌入/脱出电极，而这是速率控制步骤，因此多数情况下，峰电流与扫速的平方根在较宽的扫速范围内满足线性关系，实际测量的是电极内部锂离子与电子的扩散，但同时也包含了液相锂离子的扩散。此外，对于嵌脱锂引起的连续固溶体反应，化学扩散系数应该是随嵌脱锂量变化的数值，而循环伏安方法计算化学扩散系数时取峰电流值，只能得到表观意义上在峰值电流对应的反

应电位下的平均化学扩散系数。对于两相反应，固体内部不存在连续的浓度梯度，测到的化学扩散系数也是与相转变反应耦合的扩散过程的表观化学扩散系数。可以说循环伏安法测到的化学扩散系数并非电极材料内部本征的离子扩散系数。

除了以上影响测量的本征因素，计算化学扩散系数需要知道电极面积。如果是多孔粉末电极，其真实反应面积远大于电极几何面积，且难以精确测量，这给循环伏安方法测量固态电极中化学扩散系数带来了很大的不确定因素，导致数据难以重复，因此根据循环伏安测试结果计算化学扩散系数的绝对值在不同文献之间比较时需持谨慎态度。

循环伏安法除了可以获得表观化学扩散系数之外，还可以通过一对氧化还原峰的峰值电位差判断充放电（电化学氧化还原反应）之间极化电阻的大小、反应是否可逆。通常在氧化与还原反应的过电位差别不大的条件下，可以把一对氧化峰与还原峰之间的中点值近似作为该反应的热力学平衡电位值。

此外，恒电流充放电的电压-容量曲线进行微分处理后，以 dQ/dV 作为纵轴，电压为横轴，可以获得与 CV 相似的曲线，由于 CV 实验中的时间是线性的，电流乘以时间即为电量。因此，在学术文献中，常常用 V-dQ/dV 来代替 CV 曲线，二者有本质区别。

循环伏安方法，通过扫描电势来观察在某一电势下的电流响应。在扫速非常慢的情况下，某一电势下发生的电荷转移反应可以充分进行，这一过程获得的 CV 谱相当于测量一个电极材料的态密度谱，可以表征不同电位下的电子注入能力。

12.3.3 准稳态测量技术——交流阻抗谱

12.3.3.1 交流阻抗谱概述

交流阻抗谱（alternating current impedance spectroscopy，IS）包括电化学阻抗谱、交流伏安法等，广泛用于研究介电材料及各类电子元器件，测量时要求待测体系测量端之间没有电压，通过对阻抗谱的分析，获得各类待测元件的阻抗参数。

电化学阻抗谱（electrochemical impedence spectroscopy，EIS）是在电化学电池处于平衡状态下（开路状态）或者在某一稳定的直流极化条件下，按照正弦规律施加小振幅交流激励信号，研究电化学的交流阻抗随频率的变化关系，称为频率域阻抗分析方法。也可以固定频率，测量电化学电池的交流阻抗随时间的变化，称为时间域阻抗分析方法。

交流伏安（alternating current voltammetry）法，是指在线扫伏安的基础上，叠加小幅交流信号，再通过傅里叶变换，研究交流信号的振幅和相位随电势或电流扫描的变化关系。

锂离子电池的基础研究中多用频率域阻抗分析方法。EIS 由于记录了电化学电池不同响应频率的阻抗，而一般测量覆盖了较宽的频率范围（$\mu Hz \sim MHz$），因此可以分析反应时间常数存在差异的不同的电极过程。为了使 EIS 数据能够进行有意义的分析，要求待测电化学系统需满足线性、因果性和稳定性3个基本条件，通常可以采用 Kramers-Kronig 变换来判断阻抗数据的有效性。

12.3.3.2 电化学阻抗谱的显示分类

电化学阻抗谱可以有多种展示方法，最常用的为复数阻抗图和阻抗波特图。复数阻抗

图是以阻抗的实部为横轴，负的虚部为纵轴绘制的曲线，亦称为 Nyquist 图或 Cole-cole 图。阻抗波特图则由两条曲线组成，其中的一条曲线描述阻抗模量 $|Z|$ 随频率的变化关系，称为 Bode 模量图；另一条曲线描述阻抗的相位角随频率的变换关系，称为 Bode 相位图。一般测量时同时给出模量图和相位图，统称为阻抗 Bode 图。除此之外，还包括介电系数谱（ε'，$-\varepsilon''$），介电模量谱（M'，$-M''$）。

12.3.3.3 电化学阻抗谱在锂电池基础研究中的应用

当扩散过程为控制步骤且电极为可逆体系时，理想情况下，阻抗低频部分存在扩散响应曲线。此时，可以利用扩散响应曲线测量电池或电极体系的化学扩散系数。典型的采用电化学交流阻抗法测量化学扩散系数的公式见式（12-3）～式（12-5）[12.9]。

$$\mathrm{Im}(Z\omega) = B\omega^{-1/2} \tag{12-3}$$

$$\mathrm{Re}(Z\omega) = B\omega^{-1/2} \tag{12-4}$$

$$D_{\mathrm{Li}} = [V_{\mathrm{m}}(\mathrm{d}E/\mathrm{d}x)/(FAB)]^{1/2}/2 \tag{12-5}$$

式中，ω 为角频率；B 为 Warburg 系数；D_{Li} 为 Li 在电极中的扩散系数；V_{m} 为活性物质的摩尔体积；F 为法拉第常数（$F=96487\mathrm{C/mol}$）；A 为浸入溶液中参与电化学反应的真实电极面积；$\mathrm{d}E/\mathrm{d}x$ 为相应电极库仑滴定曲线的斜率，即开路电位对电极中 Li 浓度曲线上某浓度处的斜率。

基本测量过程如下：①通过阻抗谱拟合获得低频扩散部分的 B 值；②测量库仑滴定曲线；③将相关参数代入方程式（12-5）即可求出 Li 的扩散系数。

采用电化学阻抗谱技术测量不同电极材料中的锂离子化学扩散系数汇总，见表 12-2。

事实上，由于 $\mathrm{d}E/\mathrm{d}x$ 很难精确获得，真实面积也存在差异，EIS 测到的化学扩散系数的绝对数值往往重现性低，可靠度较差。但是如果是同一个电极，在充放电过程中表面积没有出现大的变化，比较不同充放电状态下同一个电极的扩散系数的变化是合理的。该变化值可以与通过充放电曲线观察过电位的差值变化相对比，从而判断是否电极电阻是扩散控制。尽管如此，采用 EIS 技术测量化学扩散系数也存在一定问题，后续章节会进一步讨论。

表 12-2 不同电极材料的锂离子化学扩散系数（EIS 测量结果）

材料	$D_{\mathrm{Li}}/(\mathrm{cm}^2/\mathrm{s})$	参考文献	可靠性
$\mathrm{Li}_{1-x}\mathrm{NiO}_2$	$10^{-9} \sim 10^{-8}$	[10]	√
$\mathrm{Li}_{1-x}\mathrm{CoO}_2$	$10^{-9} \sim 10^{-8}$	[10]	√
$\mathrm{Li}_{1-x}\mathrm{C}$	$10^{-11} \sim 10^{-8}$	[11]	√
$\mathrm{Li}_{0.3}\mathrm{Cr}_{0.1}\mathrm{Mn}_{1.9}\mathrm{O}_2$	$10^{-8} \sim 10^{-7}$	[12]	√
$\mathrm{Li}_{0.3}\mathrm{Mn}_2\mathrm{O}_4$	$10^{-10} \sim 10^{-9}$	[12]	√
carbon	10^{-10}	[13]	√
$\mathrm{LiNi}_{0.5}\mathrm{Mn}_{1.5}\mathrm{O}_4$	$10^{-12} \sim 10^{-10}$	[14]	√
$\mathrm{LiNi}_{1/2}\mathrm{Mn}_{1/2}\mathrm{O}_2$	10^{-10}	[15]	√

续表

材料	$D_{Li}/(cm^2/s)$	参考文献	可靠性
$LiCoO_2$	$10^{-13} \sim 10^{-11}$	[16]	√
$LiCoO_2$	$10^{-12} \sim 10^{-11}$	[17]	√
$Li_{1-x}FePO_4$	$10^{-18} \sim 10^{-16}$	[18]	×
硅	$10^{-13} \sim 10^{-12}$	[5]	√
硅薄膜	$10^{-14} \sim 10^{-13}$	[19]	√
$Li_{1-x}FePO_4$（包覆 C）	$10^{-18} \sim 10^{-14}$	[20]	×
$Li_{1-x}FePO_4$（包覆 ZnO）	$10^{-18} \sim 10^{-14}$	[21]	×
$Li_3V_2(PO_4)_3$（单相）	$10^{-10} \sim 10^{-9}$	[6]	√
$Li_3V_2(PO_4)_3$（两相）	$10^{-13} \sim 10^{-8}$	[6]	×
$Li_4Ti_5O_{12}$	$10^{-16} \sim 10^{-13}$	[22]	×
$Li_4Ti_5O_{12}$	$10^{-12} \sim 10^{-10}$	[8]	×
$LiFePO_4$	$10^{-18} \sim 10^{-14}$	[23]	×

12.3.4 暂态测量方法（Ⅰ）——电流阶跃测量

12.3.4.1 电流阶跃暂态过程及其特点

电流阶跃暂态测量通过控制流过研究电极的电流，测量电极电势的变化，从而分析电极过程的机理，计算电极过程的有关参数。

对于具有图 12-5 所示特性的电化学系统，在某一时刻施加的恒定电流如图 12-11(a)所示，电化学系统的电势响应如图 12-11(b)所示。

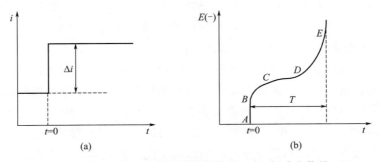

图 12-11 单电流阶跃极化下的控制信号（a）与响应信号（b）

该电极电势随时间响应曲线可做如下分析：AB 段为电池欧姆极化过电势，施加恒电流瞬间，界面处双电层电势来不及突变，浓差极化过电势尚未建立，因而表现为欧姆电阻的欧姆极化电势；BC 段为电荷传递迟缓性引起的电化学极化过电势，电池中极化电流变化后不久，电化学反应随之开始进行，初期由于电荷传递的滞后性，导致了 BC 段的过电势；CD 段为浓差极化过电势，随着电化学反应的进一步深入，反应物和反应产物开始出现浓度差，此时浓差极化过电势逐步建立；DE 段为双电层充电过电势。当反应持续一段时间后，电极表面反应物浓度逐渐趋于零，浓差极化过电势达到极大值，此时双电层开始快速充电，电极电势发生突变。

上述的定性描述实际是对电池中恒流充放电曲线的一般性描述。而多数电极的充电曲线也确实具有图 12-11(b) 的特征。如果 AB 段极化较大，一般超过 10mV，说明电极电阻较大，这可能与电极材料的电子电导低、导电添加剂电导率低、分散性差以及黏结剂较多、溶剂吸附导致的表面电阻变化有关。BC、CD、DE 段的长短和界限对于固溶体类反应很难区分，对于两相反应，BC 段往往与颗粒比表面积大小或者材料非计量比范围有关。CD 段一般电位恒定，DE 段一般斜率较大。

对于同一电极，比较不同电流密度下的充放电曲线，可以获得过电位与电流密度的关系曲线，对于两相反应，由于容易判断充放电平台电位，这一方法较为适用[24,25]。

12.3.4.2 恒电流间歇滴定技术

恒电流间歇滴定技术（galvanostatic intermittent titration technique，GITT）由德国科学家 W. Weppner 提出。基本原理是在某一特定环境下对测量体系施加一恒定电流并持续一段时间后切断该电流，观察施加电流段体系电位随时间的变化以及弛豫后达到平衡的电压，通过分析电位随时间的变化可以得出电极过程过电位的弛豫信息，进而推测和计算反应动力学信息。

当电极体系满足：①电极体系为等温绝热体系；②电极体系在施加电流时无体积变化与相变；③电极响应完全由离子在电极内部的扩散控制；④$\tau \ll L^2/D$，L 为离子扩散长度，D 为离子扩散系数；⑤电极材料的电子电导远大于离子电导等条件时，采用恒电流间歇滴定技术测量锂离子化学扩散系数的基本原理可用式（12-6）表达[26]

$$D_{Li} = 4[V_m/(nAF)]^2[I_0(dE/dx)/(dE/dt^{1/2})]^2/\pi \tag{12-6}$$

式中，D_{Li} 为 Li 在电极中的化学扩散系数；V_m 为活性物质的摩尔体积；A 为浸入溶液中的真实电极面积；F 为法拉第常数（96487C/mol）；n 为参与反应的电子数；I_0 为滴定电流值；dE/dx 为开路电位对电极中 Li 浓度曲线上某浓度处的斜率（即库仑滴定曲线）；$dE/dt^{1/2}$ 为极化电压对时间平方根曲线的斜率。

利用 GITT 方法测量电极材料中的锂化学扩散系数基本过程如下：①在电池充放电过程中的某一时刻，施加微小电流并恒定一段时间后切断；②记录电流切断后的电极电位随时间的变化；③做出极化电压对时间平方根曲线，即 $dE/dt^{1/2}$ 曲线；④测量库仑滴定曲线，即 dE/dx 曲线；⑤代入相关参数，利用式（12-6）求解扩散系数。

采用 GITT 技术获得的不同电极材料的锂离子化学扩散系数结果汇总见表 12-3。

表 12-3 不同电极材料的锂离子化学扩散系数（GITT 测量结果）

材料	$D_{Li}/(cm^2/s)$	参考文献	可靠性
V_2O_5	$10^{-14} \sim 10^{-12}$	[26]	√
$Li_{2/3}(Ni_{1/3}Mn_{2/3})O_2$	$10^{-11} \sim 10^{-10}$	[27]	√
$Li_{2/3}Co_{0.15}Mn_{0.85}O_2$	10^{-11}	[28]	√
$LiM_{1/6}Mn_{11/6}O_4$	$10^{-10} \sim 10^{-9}$	[29]	√

续表

材料	$D_{Li}/(cm^2/s)$	参考文献	可靠性
$Li_xCo_{0.5}Ni_{0.5}O_2$	$10^{-10}\sim10^{-9}$	[30]	√
$LiNi_{1/3}Co_{1/3}Mn_{1/3}O_2$	10^{-10}	[31]	√
$LiNi_{1/2}Mn_{1/2}O_2$	10^{-10}	[15]	√
$Li_{2-x}MMn_3O_8$	$10^{-13}\sim10^{-12}$	[32]	√
石墨	$10^{-13}\sim10^{-7}$	[33]	√
$LiNi_{0.5-x}Co_{2x}Mn_{1.5-x}O_4$	$10^{-13}\sim10^{-8}$	[34]	√
$LiFe_{1/4}Mn_{1/4}Co_{1/4}Ni_{1/4}PO_4$	10^{-15}	[35]	×
$LiMn_2O_4$	$10^{-14}\sim10^{-11}$	[36]	√
$LiNi_{0.5}Mn_{0.5}O_2$	$10^{-14}\sim10^{-13}$	[37]	√
$Li(Li_{0.23}Co_{0.3}Mn_{0.47})O_2$	$10^{-19}\sim10^{-14}$	[38]	√
$Li_3V_2(PO_4)_3$	$10^{-18}\sim10^{-13}$	[6]	√
$Li_{1-y}FePO_4$	10^{-12}	[39]	×
Li_yFeSO_4F	10^{-14}	[40]	×
$Li_3V_2(PO_4)_3$	$10^{-11}\sim10^{-10}$	[41]	√
$LiFePO_4$	$10^{-18}\sim10^{-14}$	[23]	×
$Li_4Ti_5O_{12}$	10^{-16}	[42]	×
$Li(Ni_{0.5}Mn_{0.3}Co_{0.2})O_2$	$10^{-11}\sim10^{-9}$	[43]	√
$Li_{1-x}VPO_4F$	$10^{-18}\sim10^{-12}$	[44]	√

固态离子学一般对离子在不同方向上的扩散系数不加区分。对于嵌入化合物而言，从结构的角度考虑，离子嵌入与脱出过程的扩散系数存在着区别。

以正极材料 $LiNiO_2$ 为例，在放电态时，锂离子已占满所有八面体空位 Li_xNiO_2 ($x=1$)，此时锂离子嵌入时受到 Li 层已占离子的库仑斥力，而脱出则较为容易。在充电态 $Li_{1-x}NiO_2$ ($x<1$)，Li 层中只有较少的 Li^+，这时 Li^+ 进一步脱出可能会破坏晶体结构，因此受到骨架离子较大的吸引力，而 Li^+ 的嵌入则有可能起到稳定结构的作用，此时嵌入阻力小于脱出阻力。

以负极材料石墨为例，Li^+ 的嵌入使石墨层发生膨胀，因此对于初始的石墨材料，Li^+ 的嵌入阻力要大于脱出阻力。当石墨层中已经占满 Li^+ 时（LiC_6），Li^+ 的嵌入要克服已占离子的库仑斥力，因此嵌入阻力仍应大于脱出阻力。

因此，根据上述分析，对于正极材料而言，离子嵌入或脱出扩散系数的大小与材料中 Li^+ 所占据的量有关；对于负极材料（石墨），锂离子嵌入的扩散系数应小于脱出的扩散系数。

对式（12-6）分析可以看出，假设嵌入与脱出过程的扩散系数不相等，我们可以通过对处于同一平衡态的电极施加正反向的电流来测得嵌入扩散系数 D_-（施加负向电流，即放电）与脱出扩散系数 D_+（施加正向电流，即充电）。

对于同一状态的同一电极，所有参数都一致，因此可以精确比较 D_+ 与 D_- 的区别，如式(12-7) 所示[45]

$$D_+/D_- = [(dE/dt^{1/2})_-]^2/[(dE/dt^{1/2})_+]^2 \quad (12-7)$$

不同 Li/Ni 名义组成合成的 Li_xNiO_2(x 为 0.5~1.4) 的 GITT 研究证明[46]，D_+ 与 D_- 的比值随 x 的增加确实出现了交叉变化的现象。在 x 较大时 D_- 较小，x 较小时 D_+ 较大。即在名义组成比 n(Li)/n(Ni)>1 时，锂离子的嵌入更困难；而 n(Li)/n(Ni)<1 时，锂离子的脱出更困难。与前述的设想一致。

图 12-12 为不同放电深度的石墨电极，开路状态时的 $LiMn_2O_4$ 的 GITT 曲线（这些状态的阻抗谱低频均为扩散控制）。每个电极在同一状态施加不同方向的电流，从图中大致可以看出，放电时电极电位的变化幅度要大于充电时电极电位的变化幅度。图 12-13 为 $LiMn_2O_4$ 施加电流后的 E-$t^{1/2}$ 曲线，在测量的范围之内基本为线性，满足 GITT 的要求。石墨电极的 E-$t^{1/2}$ 曲线也为线性。对图 12-12 中所有曲线的计算结果列于表 12-4 中。对于石墨电极，其脱出时的扩散系数最大可以超过嵌入时的 6 倍。$LiMn_2O_4$ 在开路状态离子脱出过程的扩散系数也大于嵌入过程的扩散系数。

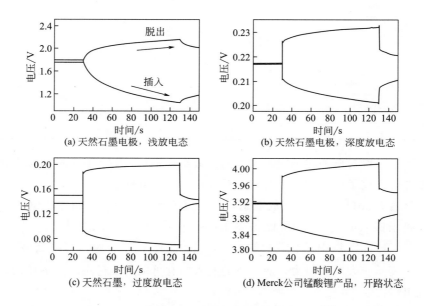

图 12-12　不同三电极电池体系计时电位曲线[45]

在实际锂离子电池中，发现锂离子电池可以快速放电但不能快速充电，充电的过程就是锂离子从正极脱出嵌入碳负极的过程，而锂离子电池的动力学主要受碳材料制约。这种现象与用 GITT 正反电流测试的结果一致，原因正是由于锂离子嵌入碳电极的速率要低于脱出的速率。因此，用这种方法。可以在较短的时间内测定电极材料嵌入脱出扩散系数的差异。结合库仑滴定，可以得到在不同嵌锂量时扩散系数的差异性，从而深入研究结构与动力学之间的关系。

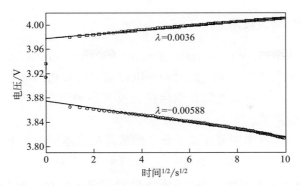

图 12-13　过电位与时间平方根曲线 ［图 12-13 结果，λ 为斜率］[45]

表 12-4　锂离子嵌入与脱出过程中扩散系数的比较（图 12-13 计算结果）[45]

电极	$dE/dt^{1/2}(+)$	$dE/dt^{1/2}(-)$	(−)/(+)	D_+/D_-
图(a)	0.0255	−0.0644	2.5	6.3
图(b)	0.0006311	−0.00106	1.7	2.9
图(c)	0.0011	−0.00218	2.0	4
图(d)	0.00336	−0.00588	1.8	3.3

12.3.5　暂态测量方法（Ⅱ）——电势阶跃测量

12.3.5.1　电势阶跃暂态过程及其特点

电势阶跃暂态测量是指控制流过研究电极的电势，按照一定的具有电势突变的规律变化，同时测量电流随时间的变化，从而分析电极过程的机理，计算电极过程的有关参数。

对于具有图 12-6 所示特性的电化学系统，在某一时刻施加的一恒定电势 η 如图 12-14(a)所示，其电流响应如图 12-14(b)所示。

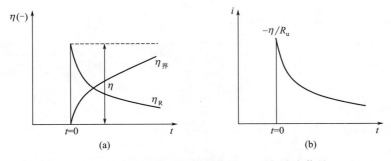

图 12-14　电势阶跃极化下的控制信号（a）与响应信号（b）

图 12-14 中的控制信号与响应信号之间的关系可依据 12.1.5 节中的 4 个基本过程进行分析：电势阶跃瞬间，由于双电层的作用，电路中的极化电流无法发生突变，极化电流只能慢慢变小，紧随其后的是系统欧姆极化过电势 η_R 逐步减小，界面双电层过电势 $\eta_界$ 逐步增加，两者之和为阶跃电势变化值。

12.3.5.2 恒电势间歇滴定技术

恒电势间歇滴定技术（potentiostatic intermittent titration technique，PITT）是通过瞬时改变电极电位并恒定该电位值，同时记录电流随时间变化的测量方法。通过分析电流随时间的变化可以得出电极过程电位弛豫信息以及其他动力学信息，类似于恒电位阶跃，只是 PITT 是多电位点测量。

使用恒电势间歇滴定技术测量锂离子化学扩散系数基本原理如下[8]

$$\ln(i) = \ln(2\Delta Q D_{Li}/d^2) - [\pi^2 D_{Li}/(4d^2)]t \tag{12-8}$$

式中，i 为电流值；t 为时间；ΔQ 为嵌入电极的电量；D_{Li} 为 Li 在电极中的扩散系数；d 为活性物质的厚度。

基本操作如下：①以恒定电位步长瞬间改变电极电位，记录电流随时间的变化；②利用方程式（12-8）做出 $\ln(i)$-t 曲线；③截取 $\ln(i)$-t 曲线线性部分的数据，求斜率即可求出锂离子化学扩散系数。

利用恒电位间歇滴定技术测量不同电极材料中的锂离子化学扩散系数汇总如下，见表 12-5。

表 12-5　不同电极材料中的锂离子化学扩散系数（PITT 测量结果）

材料	$D_{Li}/(cm^2/s)$	参考文献	可靠性
$LiMn_2O_4$	10^{-13}	[36]	√
$LiMn_2O_4$	10^{-9}	[47]	√
$LiFePO_4$	10^{-14}	[47]	√
$LiCoO_2$	$10^{-13} \sim 10^{-12}$	[17]	√
$Li[Li_{1/3-2x/3}Ni_xMn_{2/3-x/3}]O_2$	$10^{-14} \sim 10^{-13}$	[48]	√
$LiMPO_4$	$10^{-17} \sim 10^{-14}$	[49]	√
silicon film	10^{-13}	[19]	√
$LiMn_{0.5}Ni_{0.5}O_2$	$10^{-13} \sim 10^{-12}$	[50]	√
$LiFePO_4$(single phase)	$10^{-14} \sim 10^{-12}$	[39]	√
$LiFePO_4$(two-phase)	$10^{-18} \sim 10^{-15}$	[39]	√
Li_yFeSO_4F	10^{-14}	[40]	√
$LiFePO_4$(aqueous)	$10^{-14} \sim 10^{-12}$	[51]	√
$LiFePO_4$(non-aqueous)	$10^{-15} \sim 10^{-13}$	[51]	√
Li_2MnO_3-$LiMO_2$	10^{-12}	[52]	√
$Li_xMn_{1.95}Cr_{0.05}O_4$	$10^{-14} \sim 10^{-9}$	[53]	√
silicon	$10^{-14} \sim 10^{-13}$	[54]	√
MCMB	10^{-10}	[55]	√
$LiMnPO_4$	$10^{-13} \sim 10^{-12}$	[56]	√
$Li_4Ti_5O_{12}$	$10^{-12} \sim 10^{-11}$	[57]	√

续表

材料	$D_{Li}/(cm^2/s)$	参考文献	可靠性
$Li_3V_2(PO_4)_3/C$	$10^{-8} \sim 10^{-7}$	[58]	√
$Li_2MnO_3\text{-}LiMn_yNi_zCo_wO_2$	10^{-10}	[59]	√
$LiFe_{0.75}Zn_{0.25}PO_4$	$10^{-13} \sim 10^{-10}$	[60]	√
$LiFePO_4$	$10^{-15} \sim 10^{-12}$	[61]	√
$LiMn_2O_4$	$10^{-12} \sim 10^{-10}$	[62]	√
$Li_{1.08}Mn_{1.89}Al_{0.03}O_4$	$10^{-9} \sim 10^{-7}$	[63]	√
$LiFePO_4$	10^{-12}	[64]	√
$Li_xNi_yMn_yCo_{1-2y}O_2$	10^{-10}	[65]	√

对于在充放电过程中动力学明显存在差异的反应体系，电位阶跃能更加突出电流响应的差别，实际上这与循环伏安法的结果有类似之处。Chiang等[66]曾用PITT的方法展示磷酸铁锂电极嵌脱锂过程中电流变化跟随响应，从而说明其不同段反应的动力学。

12.3.6 暂态测量方法（Ⅲ）——电位弛豫技术

电位弛豫技术（potential relax technique，PRT）是在电池与外界无物质和能量交换的条件下研究电极电势随时间的变化关系，该方法属于电流阶跃测量方法中的断电流法，与GITT实验方法一致，不同的是分析弛豫过程中的电位变化。该方法最早由中国科学院物理研究所王庆等[67]运用于锂离子电池电极材料中的离子扩散动力学研究。

采用电位弛豫技术测量锂离子扩散系数的基本原理如式（12-9）所示[67]

$$\ln[\exp(\varphi_m-\varphi)\times F/(RT)-1]=-\ln N-(\pi^2 D_{Li}/d^2)t \quad (12-9)$$

式中，φ_m为平衡电极电位；φ为初始电位；R为气体常数[8.31 J/(mol·K)]；T为绝对温度；d为活性物质的厚度；D_{Li}为Li在电极中的扩散系数；t为电位达到平衡时的时间。

具体测量步骤如下：①对电池预充放电，使电池的库仑效率降至97%左右；②在电池充/放电到一定程度时，切断电流，采用CPT（chrono potentiometry technique）记录电压随时间的变化曲线；③运用式（12-9）对$\ln[\exp(\varphi_m-\varphi)\times F/(RT)-1]-t$作图，并对后半部分作线性拟合；④$\ln[\exp(\varphi_m-\varphi)\times F/(RT)-1]-t$曲线进行拟合，求解拟合曲线斜率，代入式（12-9）即可求得锂的化学扩散系数。

王庆等[67]运用该方法测量了MCMB样品中不同充放电状态下的锂离子化学扩散系数，嵌锂态化学扩散系数数量级在$10^{-8}\sim10^{-7}cm^2/s$，脱锂态数量级在$10^{-10}\sim10^{-9}cm^2/s$；与EIS、GITT、PITT、PSCA、CPR和CV方法不同的是，电位弛豫技术具有较高的精确度，同时可以测量电解质中的锂离子化学扩散系数，具体测量方法可参考文献[67]。吴川等[68]采用该方法测量了氟化物表面修饰的锰酸锂（$LiMn_2O_4$）样品中的锂离子化学扩散系数，其化学扩散系数测量值在$10^{-10}\sim10^{-9}cm^2/s$数量级，并发现了两相界面处的锂离子扩散系数要高于单相区域锂离子的扩散系数。随后，欧阳楚英等[69]

又对不同嵌锂量的 $Li_xMn_2O_4$（$x=0.1\sim0.9$）样品进行了锂离子化学扩散系数的测量，并结合了 Monte Carlo 方法模拟了锰酸锂样品中的锂离子电导率，证明了在两相界面处（$x=0.3\sim0.7$）锂的化学扩散系数要大于单相时的化学扩散系数。

运用电位弛豫技术测量电极过程动力学信息需要满足一定的前提条件。通常，锂离子电池在首周充放电过程中伴随着一些副反应，典型的副反应为 SEI 膜的形成，为避免副反应的发生对锂离子化学扩散系数测量所带来的干扰，通常电池需要进行几个充放电循环之后开始测量其化学扩散系数。此外，由于电位弛豫是一个非常缓慢的过程，一般在 8h 左右，在经过长时间的弛豫后，电位仍不能达到平衡状态，则有可能是仪器漏电所造成，需要特别注意。

12.3.7 不同电化学测量法的适用范围与精准性

不同的电化学测量方法具有不同的测量精度和适用范围，需要仔细讨论。

唐堃等[23]采用不同的电化学测量方法，系统地研究了 $LiFePO_4$ 薄膜电极中锂的化学扩散系数，并探讨了不同测量方法的适用性和可靠性。

12.3.7.1 循环伏安法测量 $LiFePO_4$ 薄膜电极中锂离子表观化学扩散系数

图 12-15(a) 为 $LiFePO_4$ 薄膜电极在不同电位扫描速率下的循环伏安曲线；图 12-15(b) 为峰值电流 I_P 对扫描速率平方根 $\nu^{1/2}$ 曲线。利用 12.3.2.2 节中式（12-2）的方法计算出的氧化还原过程中锂离子表观化学扩散系数分别为 2.1×10^{-14} cm^2/s、1.8×10^{-14} cm^2/s。该结果表明，脱锂与嵌锂时的电极过程动力学性质存在微弱差异，该扩散系数应该反映了嵌脱锂在峰值电流附近的平均化学扩散系数，该方法无法得到电极处于不同嵌锂量时的化学扩散系数。

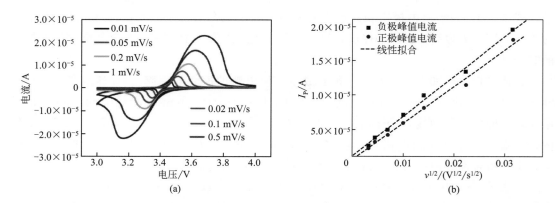

图 12-15 不同电位扫速下的循环伏安曲线（a）；氧化还原峰值电流与电位扫描速率平方根的关系（b）[23]

12.3.7.2 恒电流间歇滴定技术测量 $LiFePO_4$ 薄膜电极化学扩散系数

图 12-16(a) 所示为采用 GITT 法测量 $LiFePO_4$ 薄膜得到的开路电压曲线，图 12-16(b) 为利用开路电压曲线拟合得到的库仑滴定曲线。

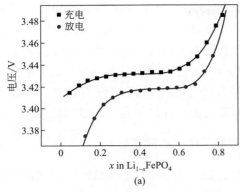
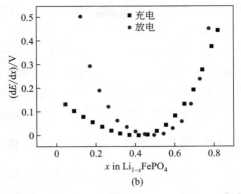

图 12-16　$LiFePO_4$ 充电和放电开路电压曲线（a）；通过 OCV 拟合计算所得 dE/dx 的值（b）[23]

图 12-17(a) 为电极电势对时间 t 的平方根曲线，图 12-17(b) 为不同锂含量下的 $dE/dt^{1/2}$ 曲线。

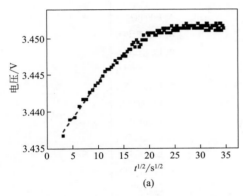
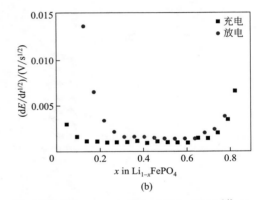

图 12-17　$Li_{0.44}FePO_4$ 薄膜电极的即时电压与 $t^{1/2}$ 的关系图（a）；充放电过程中 $dE/dt^{1/2}$ 与锂离子含量的对应关系图（b）[23]

利用 12.3.4.2 中式（12-6）计算出的锂离子化学扩散系数，如图 12-18 所示。

由图 12-18 可以看到锂离子扩散系数在充电过程中处在 $10^{-14} \sim 10^{-18} cm^2/s$，随着锂离子的含量变化，化学扩散系数呈 V 形分布，在 $x=0.4$ 左右有最小值；而对于放电过程锂的化学扩散系数的分布区间相对较大，在 $10^{-14} \sim 10^{-19} cm^2/s$，同样是 V 形分布，最小值出现在 $x=0.5$ 左右。这种 V 形分布的结果，尤其是充电过程最小值在 $x=0.4$ 处，放电过程最小值在 $x=0.5$ 处，与图 12-17(b) 中的 dE/dx

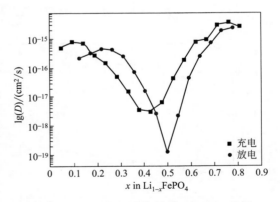

图 12-18　由 GITT 计算得到 $LiFePO_4$ 的锂化学扩散系数与锂含量的关系[23]

值对应。在这两个位置，反应的开路电压基本表现为一条水平线，得到的 dE/dx 无限接近于零。分析 12.3.4.2 节中式（12-6）可知，对于不同锂含量，材料的物质的量体积浓

度 V_m、法拉第常数 F、施加电流 I_0、电极表面积 A、电极厚度 d 均为定值，变量只有 dE/dx 和 $dE/dt^{1/2}$。$dE/dt^{1/2}$ 虽然是 U 形分布，但其随锂含量的变化仍在 0.001～0.01 之间，而 dE/dx 的数值却在 0.001～0.5 之间。这说明最终得到的锂离子化学扩散系数的变化规律主要是由于 dE/dx 的变化导致。而对于两相反应来说，dE/dx 在处于平衡态时应为零值，实验记录则不一定为零值，这导致平台区的 dE/dx 记录没有意义，所以最终得到的锂离子扩散系数随锂含量的变化规律并不能代表实际的锂离子扩散的快慢。

12.3.7.3　交流阻抗法测量 LiFePO$_4$ 薄膜中的锂离子化学扩散系数

图 12-19 为 LiFePO$_4$ 薄膜电极在不同充电状态的交流阻抗图，图 12-20 为阻抗的实部与虚部对 $\omega^{-1/2}$ 的曲线图。利用 12.3.3.3 节中式 (12-5) 得到的锂离子化学扩散系数如图 12-21 所示。

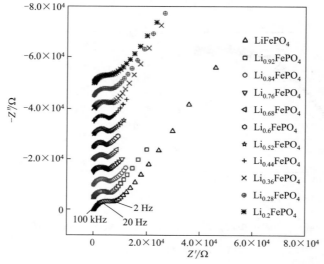

图 12-19　LiFePO$_4$ 薄膜电极在不同充电状态的交流阻抗图[23]

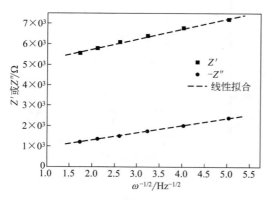

图 12-20　阻抗的实部（Z'）和虚部（Z''）对 $\omega^{-1/2}$ 关系图[23]

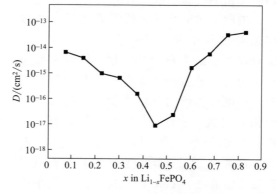

图 12-21　由 EIS 计算得到的锂离子化学扩散系数与锂含量的关系[23]

图 12-21 给出了由 EIS 计算得到的锂化学扩散系数与锂含量的关系图，可以看出，计算得到的锂离子扩散系数的变化规律与由 GITT 方法测得的变化趋势相同，即在充电的开始和末端有较大的扩散系数，而在中间平台段 $x=0.5$ 左右有最小的扩散系数。整个区间锂离子扩散系数在 $10^{-17}\sim10^{-13}$ cm^2/s 之间，数值相差达到 4 个数量级。分析 12.3.3.3 中式（12-5）可知，整个充电过程中的变量只有 Warburg 系数 B 和 dE/dx。D_{Li} 值与 B 的平方成反比，与 dE/dx 值的平方成正比，B 的值是随着 x 的变化呈 U 形分布，这一数值是可信的。如前所述，对于两相反应，dE/dx 会带来相当大的误差，从本质上对两相反应来讲，dE/dx 应该始终为零，但是由于测试中的极化问题，使得开路电压曲线只在中间部分表现的接近水平。这样就导致了 dE/dx 的值在两端和中间相差达到了 100 倍（0.004~0.445），在经过平方计算以后就相差 4 个数量级。因此，最终得到的锂化学扩散系数在充电的始末状态和中间状态数值相差 4 个数量级，应该主要归因于 dE/dx 值的变化，而 dE/dx 的值对于两相反应是不可信的。

12.3.7.4 恒电位间歇滴定技术测量 LiFePO$_4$ 薄膜中的锂离子化学扩散系数

图 12-22(a)所示为 Li$_{1-x}$FePO$_4$ 薄膜电极电位从 3.43V 变化到 3.435V，并记录电位恒定过程时的电流随时间变化的曲线，图 12-22(b)为 lg(i)-t 曲线，利用 12.3.5.2 节中式（12-8）计算得到的充放电过程中的锂离子化学扩散系数如图 12-23 所示。图 12-25 所示为不同锂含量情形下的 Li$_{1-x}$FePO$_4$ 薄膜电极中锂离子化学扩散系数。

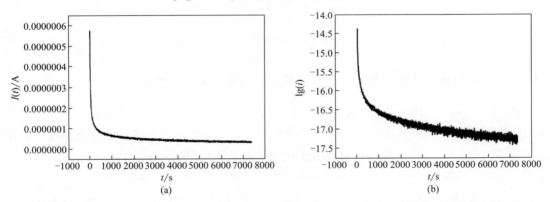

图 12-22 Li$_{1-x}$FePO$_4$ 薄膜电极电位从 3.43V 变化到 3.435V 并恒定过程电位记录的电流随时间变化的曲线（a）；Li$_{1-x}$FePO$_4$ 薄膜 lg(i)-t 曲线（b）[71]

图 12-23 表明充电过程和放电过程锂离子的化学扩散系数在 $10^{-15}\sim10^{-14}$ cm^2/s 内，变化规律也趋同，即在反应电位附近化学扩散系数有最小值。充电过程和放电过程锂离子的化学扩散系数随锂离子的含量没有明显的变化（图 14-24），对整个区间而言，放电过程的锂离子扩散系数要略高于充电过程的锂离子扩散系数。这一点与 12.3.7.1 节中的 CV 测量结果相悖，但考虑到测试精度，两者相差均在一个数量级范围以内，彼此的比较意义并不是很大。由于计算过程中没有像 GITT 和 EIS 方法一样使用到不确切的 dE/dx 数值，所以，整个锂化学扩散系数的区间处在同一个数量级之间变化。相对而言 PITT 方法对于 LiFePO$_4$ 这种两相反应的材料锂化学扩散系数的测量的准确性是优于 GITT 和 EIS 方法的。而 CV 方法也应是可靠的，虽然得到的是平均值。

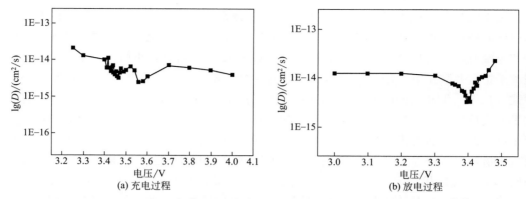

图 12-23　$Li_{1-x}FePO_4$ 薄膜电极中锂离子扩散系数随电压变化关系（PITT）[71]

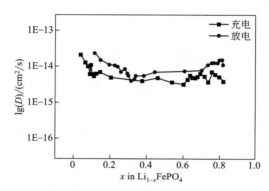

图 12-24　$Li_{1-x}FePO_4$ 薄膜电极锂离子扩散系数随锂含量的变化（PITT）[71]

12.3.7.5　电位弛豫技术测量 $LiFePO_4$ 薄膜中的锂离子化学扩散系数

图 12-25（a）为 $LiFePO_4$ 薄膜样品的电位随时间的弛豫曲线（嵌锂至 3.9V 时切断电流），图 12-25（b）为 $\ln[\exp(\varphi_m-\varphi)\times F/(RT)-1]-t$ 曲线，红色虚线为拟合曲线。利用 12.3.6 节中式（12-9）计算得到的锂离子化学扩散系数如图 12-26 所示。

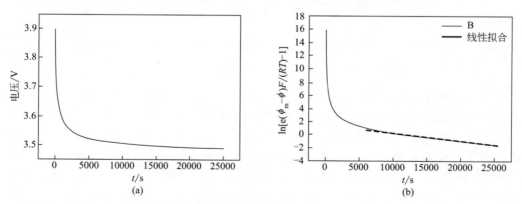

图 12-25　$LiFePO_4$ 薄膜样品电位随时间的弛豫曲线（a）；$\ln[\exp(\varphi_m-\varphi)\times F/(RT)-1]-t$ 曲线（b），红色虚线为拟合曲线[71]

利用电位弛豫技术所得锂化学扩散系数随锂离子含量的变化如图 12-26 所示。从图中可以看出，利用此方法得到的锂化学扩散系数随着充放电的进行并没有明显的变化，整个区域锂离子的化学扩散系数均在 10^{-15} cm^2/s 左右，整个区间的大小变化在一个数量级以内。充电过程和放电过程的锂离子扩散系数没有明显的区别，且与 PITT 测量基本一致。由于测量过程中，不涉及 dE/dx 项的运算，因此，电位弛豫技术对于两相反应类材料的离子化学扩散系数测量是适用的。

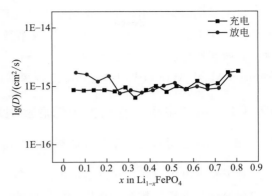

图 12-26　LiFePO$_4$ 薄膜中锂化学扩散系数（PRT）与锂含量的关系[71]

12.3.7.6　几种测量方法的比较与分析

循环伏安法计算得到 LiFePO$_4$ 薄膜中的表观化学扩散系数为 $1.8 \times 10^{-14} \sim 2.1 \times 10^{-14}$ cm^2/s。通过 GITT 和 EIS 方法得到的锂化学扩散系数分别在 $10^{-18} \sim 10^{-14}$ cm^2/s 和 $10^{-19} \sim 10^{-14}$ cm^2/s，这两种方法由于计算的过程中均用到了 dE/dx，而对于两相反应 dE/dx 的数值不可能准确测得，且在平台区域其数值近乎为零，所以所得结果均在反应的中段，即平台中部测得了与两端相差 4 个数量级的最小值，这主要是归因于使用了不准确的 dE/dx。计算所得的锂离子扩散系数在反应中段具有最小值这一现象并不能反映真实的物理过程。因此，GITT 和 EIS 方法本身只适用于固溶体体系。相应的，在回避使用 dE/dx 数值的 PITT 和 PRT 方法中，最终的计算结果并没有显示出锂离子扩散系数随锂含量的明显变化规律。PITT 方法测得的锂化学扩散系数基本处在 $10^{-14} \sim 10^{-13}$ cm^2/s 之间，而 PRT 方法测得的锂化学扩散系数在 10^{-13} cm^2/s 左右。这两种方法测得的扩散系数在整个区域范围内变化不大，尤其是由 PRT 计算所得结果基本在同一个数量级上，见图 12-27。

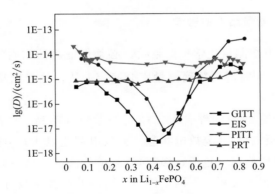

图 12-27　不同测试方法所得到的 Li$_{1-x}$FePO$_4$ 薄膜中锂离子的化学扩散系数[71]

以上举例可以看出，电化学测量方法在研究不同反应体系的电极过程动力学时需要慎重考虑。GITT、EIS 计算两相反应扩散系数是假设反应时扩散控制，服从 Fick 扩散定律。而两相反应材料的浓度梯度为一定值，非连续变化，因此数学处理上存在问题。采用 Johnson-Mehl-Avrami-Kolmogorov 方程来描述相转变动力学更为合适[70]。

12.3.8 影响电极过程动力学信息测量准确性的基本因素

12.3.8.1 电极材料对测量结果的影响

对于粉末电极材料，在制备电池的正负极时需要将电极活性物质粉末与黏结剂、导电添加剂混合后形成浆料涂覆在作为集流体的金属箔、网上或者多孔体中。对于这种多孔电极，由于制备时材料的粒度不同，导电添加剂、黏结剂的量及分散程度不同，同时也存在电势的不均匀分布，在测量过程中会引入较多的干扰和不确定性。

相对而言薄膜电极可以在很大程度上消除这些不确定因素。与粉体电极相比，薄膜电极由于直接沉积于金属集流体上，不需要有黏结剂和导电添加剂，所以更加的"干净"，薄膜的形貌可控，可以相对精确地得到其比表面积和离子传输路径（即薄膜厚度）；另外薄膜与衬底接触均匀，电势分布相对均匀。最理想的是外延膜电极或单晶电极，制备相对困难。

12.3.8.2 测量方法对测量结果的影响

对于固溶体即单相反应电极材料而言，在测量体系满足测量模型的前提下，采用循环伏安法、交流阻抗法、恒电流间歇滴定技术、恒电位间歇滴定技术和电位弛豫技术基本可以准确测量电极过程的动力学基本信息，锂电池中最重要的动力学参数为锂离子的化学扩散系数。

对于两相反应，如磷酸铁锂、钛酸锂等，采用 GITT、EIS 会由于 dE/dx 无法准确测量而产生较大的误差，循环伏安法只能测得表观化学扩散系数，不能测量扩散系数随锂含量的变化规律，有其局限性。因此，对于具体的电极材料和电化学体系，应合理选择测量方法，并构建满足测量方法的测试条件，才能得到真实的测量结果。

12.3.8.3 测量条件对测量结果的影响

使用 EIS、GITT 方法测量化学扩散系数，必须保证活性物质的摩尔体积 V_m 和与溶液的接触面积 A 恒定，同时需要精确测量库仑滴定曲线，测量结果的可靠性应结合其他测量手段重复定量测试。

此外，测量结果与样品中的锂含量、样品的形貌（单晶、多晶、薄膜）、电极类型（如一维传输，二维平板电极）、充放电状态（充放电截止电压）、电池循环次数、测试温度、相转变反应、样品的修饰（包覆、掺杂）、测试方法、计算参数的获取及测试体系中电解液的选取等都有着很大关系。不同条件下，测得的扩散系数可能存在 1～2 个数量级的变化。

12.4 本章结语

电化学表征技术在锂离子电池中有着非常广泛的应用，电化学表征方法也非常之丰富，除了文中介绍的几种方法外，还有诸如 PSCA、CPR、CITT、RPG 等。随着实际应用的需要，新的电化学表征方法，特别是与其他表征技术结合形成的各类原位电化学测量技术，正在迅速发展。

电极过程动力学研究的目的是获得能反映电极材料本征动力学特性的参数值，例如电荷转移电阻、扩散系数、交换电流密度、膜电阻等，并掌握该参数值随不同充放电深度（嵌脱锂量）以及温度的变化，从而能够理解、模拟、预测各类工况下及充电过程中电池极化电阻、电容的变化规律。而实验室在基础研究时往往采用粉末电极，导致难以可靠的比较不同材料之间动力学参数，除非材料的尺寸、粒度分布、表面官能团、导电添加剂、黏结剂、分散度、电极厚度、压实密度、体积容量得到了精确的控制和能实现高度的一致性。相对于手工制作的电极，自动化设备制作的电极往往具有较好的一致性，更适合用于来研究电极过程动力学。基础研究中最好采用薄膜电极、微电极或单晶电极。

对于批量生产的电池，通过比较充放电曲线，分析直流极化电阻、固定频率的交流阻抗、开路电压等，可以获得表观的动力学参数，采用这些参数通过电化学模拟软件，可以准确地预测电池各类工况下的荷电态、极化电阻、输出功率，这些将成为电源管理系统软件的核心内容[72,73]。

事实上，锂离子电池涉及的电化学为嵌入电极电化学，有别于传统的电极不发生结构演化，电化学反应主要发生在电极表面的溶液电化学。电化学双电层（electrical double layer，EDL）与空间电荷层（space charge layer，SCL）共存，在充放电过程中，离子将穿过 EDL 与 SCL，电荷转移往往发生在电极内部而非表面，电极为混合离子导体，电化学反应伴随着相变和内部传质，这与一般教科书上描述的电化学反应体系、研究方法、数学模型存在显著差异，需要发展新的理论与实验方法[74,78]。

本章需要进一步思考的基础科学问题：

1. 如何准确地测量电极在不同循环周次、不同 SOC 下的过电位？
2. 如何定量分解过电位的起因？并能用于不同倍率下的充放电曲线预测？
3. 能否基于全电池和半电池的充放电曲线，判断是否析锂？
4. 有些电极、半电池和全电池，随着循环，容量会有爬升的现象，都有哪些机制导致这一现象？对于全电池而言，半电池中测到的电极容量爬升现象对于全电池是否是有利的？
5. 如何判断和归属电极过程中的动力学速率控制步骤？
6. 对于材料在充放电过程中经历了固溶体或者相转变反应，如何获得材料和电化学反应的本征的动力学参数？例如电导率、电荷转移电阻、交换电流密度、化学扩散系数、迁移数、迁移率、载流子浓度、反应常数？
7. 如何根据过电位与电流密度的关系，判断电化学反应是欧姆极化、浓差极化和电化学极化？具体实验如何设计？
8. 对于同一个电池，采用 dQ/dV，CV，EIS，GITT，PITT 都有可能获得材料的表观化学扩散系数？请问每一种方法测到及推算出来的表观化学扩散系数的意义？针对不同类型的反应，哪种方法更精确？
9. 对于纳米结构的电极材料，可逆储锂有可能包括材料晶格、表面储锂，能否定量区分其贡献？
10. 液态电解质电池与全固态电池的电极过程动力学行为有何异同？

参考文献

[1] Jia Zheng, Dai Changsong, ChenLing. Electrochemical Measuring Method [M]. Beijing: Chemical Industry Press, 2006: 1-4.

[2] Jia Zheng, Dai Changsong, Chen Ling. Electrochemical Measuring Method [M]. Beijing: Chemical Industry Press, 2006: 46-188.

[3] Das S R, Majumder S B, Katiyar R S. Kinetic analysis of the Li ion intercalation behavior of solution derived nanocrystalline lithium manganate thin films [J]. Journal of Power Sources, 2005, 139: 261-268.

[4] Tang S B, Lai M O, Lu L. Li-ion diffusion in highly (003) oriented $LiCoO_2$ thin film cathode prepared by pulsed laser deposition [J]. Journal of Alloys and Compounds, 2008, 449 (1-2): 300-303.

[5] Ding N, Xu J, Yao Y X, et al. Determination of the diffusion coefficient of lithium ions in nano-Si [J]. Solid State Ionics, 2009, 180 (2-3): 222-225.

[6] Rui X H, Ding N, Liu J, et al. Analysis of the chemical diffusion coefficient of lithium ions in $Li_3V_2(PO_4)_3$ cathode material [J]. Electrochimica Acta, 2010, 55 (7): 2384-2390.

[7] Rui X H, Jin Y, Feng X Y, et al. A comparative study on the low-temperature performance of $LiFePO_4$/C and $Li_3V_2(PO_4)_3$/C cathodes for lithium-ion batteries [J]. Journal of Power Sources, 2011, 196 (4): 2109-2114.

[8] Rho Young Ho, Kanamura Kiyoshi. Li ion diffusion in $Li_4Ti_5O_{12}$ thin film electrode prepared by PVP sol-gel method [J]. Journal of Solid State Chemistry, 2004, 177 (6): 2094-2100.

[9] Wang C S, Appleby A J, Little F E. Electrochemical impedance study of initial lithium ion intercalation into graphite powders [J]. Electrochimica Acta, 2001, 46 (12): 1793-1813.

[10] Choi Y M, Pyun S I, Bae J S, et al. Effects of lithium content on the electrochemical lithium intercalation reaction into $LiNiO_2$ and $LiCoO_2$ electrodes [J]. Journal of Power Sources, 1995, 56 (1): 25-30.

[11] Levi M D, Aurbach D. Diffusion coefficients of lithium ions during intercalation into graphite derived from the simultaneous measurements and modeling of electrochemical impedance and potentiostatic intermittent titration characteristics of thin graphite electrodes [J]. Journal of Physical Chemistry B, 1997, 101 (23): 4641-4647.

[12] Tarascon J M, Armand M. Issues and challenges facing rechargeable lithium batteries [J]. Nature, 2001, 414 (6861): 359-367.

[13] Chang Y C, Sohn H J. Electrochemical impedance analysis for lithium ion intercalation into graphitized carbons [J]. Journal of the Electrochemical Society, 2000, 147 (1): 50-58.

[14] Mohamedi M, Makino A, Dokko K, et al. Electrochemical investigation of $LiNi_{0.5}Mn_{1.5}O_4$ thin film intercalation electrodes [J]. Electrochimica Acta, 2002, 48 (1): 79-84.

[15] Shaju K M, Rao G V S, Chowdari B V R. Li-ion kinetics and polarization effect on the electrochemical performance of $Li(Ni_{1/2}Mn_{1/2})O_2$ [J]. Electrochimica Acta, 2004, 49 (9-10): 1565-1576.

[16] Xia H, Lu L, Ceder G. Li diffusion in $LiCoO_2$ thin films prepared by pulsed laser deposition [J]. Journal of Power Sources, 2006, 159 (2): 1422-1427.

[17] Xie J, Imanishi N, Zhang T, et al. Li-ion transport in all-solid-state lithium batteries with $LiCoO_2$ using NASICON-type glass ceramic electrolytes [J]. Journal of Power Sources, 2009, 189 (1): 365-370.

[18] Gao Fei, Tang Zhiyuan. Kinetic behavior of LiFePO$_4$/C cathode material for lithium-ion batteries [J]. Electrochimica Acta, 2008, 53 (15): 5071-5075.

[19] Xie J, Imanishi N, Zhang T, et al. Li-ion diffusion in amorphous Si films prepared by RF magnetron sputtering: A comparison of using liquid and polymer electrolytes [J]. Materials Chemistry and Physics, 2010, 120 (2-3): 421-425.

[20] Joachin Humberto, Kaun Thomas D, Zaghib Karim, et al. Electrochemical and thermal studies of carbon-coated LiFePO$_4$ cathode [J]. Journal of the Electrochemical Society, 2009, 156 (6): A401-A406.

[21] Shenouda Atef Y, Liu Hua K. Studies on electrochemical behaviour of zinc-doped LiFePO$_4$ for lithium battery positive electrode [J]. Journal of Alloys and Compounds, 2009, 477 (1-2): 498-503.

[22] Yuan Tao, Yu Xing, Cai Rui, et al. Synthesis of pristine and carbon-coated Li$_4$Ti$_5$O$_{12}$ and their low-temperature electrochemical performance [J]. Journal of Power Sources, 2010, 195 (15): 4997-5004.

[23] Tang Kun, Yu Xiqian, Sun Jinpeng, et al. Kinetic analysis on LiFePO$_4$ thin films by CV, GITT, and EIS [J]. Electrochimica Acta, 2011, 56 (13): 4869-4875.

[24] Cui Z H, Guo X X, Li H. Equilibrium voltage and overpotential variation of nonaqueous Li-O$_2$ batteries using the galvanostatic intermittent titration technique [J]. Energy Environ Sci, 2015, 8: 182-187.

[25] Kaifu Zhong, Xin Xia, Bin Zhang, et al. MnO powder as anode active materials for lithium ion batteries [J]. Journal of Power Sources, 2010, 195 (10): 3300-3308.

[26] Bae J S, Pyun S I. Electrochemical lithium intercalation reaction of anodic vanadium-oxide film [J]. Journal of Alloys and Compounds, 1995, 217 (1): 52-58.

[27] Shaju K M, Rao GV S, Chowdari BV R. Electrochemical kinetic studies of Li-ion in O$_2$-structured Li$_{2/3}$(Ni$_{1/3}$Mn$_{2/3}$)O$_2$ and Li$_{(2/3)+x}$(Ni$_{1/3}$Mn$_{2/3}$)O$_2$ by EIS and GITT [J]. Journal of the Electrochemical Society, 2003, 150 (1): A1-A13.

[28] Shaju K M, Rao GV S, Chowdari BV R. EIS and GITT studies on oxide cathodes, O$_2$-Li$_{(2/3)+x}$(Co$_{0.15}$Mn$_{0.85}$)O$_2$ ($x=0$ and 1/3) [J]. Electrochimica Acta, 2003, 48 (18): 2691-2703.

[29] Shaju K M, Rao GV S, Chowdari BV R. Li ion kinetic studies on spinel cathodes, Li(M$_{1/6}$Mn$_{11/6}$)O$_4$ (M = Mn, Co, Co, Al) by GITT and EIS [J]. Journal of Materials Chemistry, 2003, 13 (1): 106-113.

[30] Montoro L A, Rosolen J M. The role of structural and electronic alterations on the lithium diffusion in Li$_x$Co$_{0.5}$Ni$_{0.5}$O$_2$ [J]. Electrochimica Acta, 2004, 49 (19): 3243-3249.

[31] Shaju K M, Rao GV S, Chowdari BV R. Influence of Li-ion kinetics in the cathodic performance of layered Li(Ni$_{1/3}$Co$_{1/3}$Mn$_{1/3}$)O$_2$ [J]. Journal of the Electrochemical Society, 2004, 151 (9): A1324-A1332.

[32] Schwenzel J, ThangaduraiV, Weppner W. Developments of high-voltage all-solid-state thin-film lithium ion batteries [J]. Journal of Power Sources, 2006, 154 (1): 232-238.

[33] Levi M D, Aurbach D. The application of electroanalytical methods to the analysis of phase transitions during intercalation of ions into electrodes [J]. Journal of Solid State Electrochemistry, 2007, 11 (8): 1031-1042.

[34] Ito Atsushi, Li Decheng, Lee Yunsung, et al. Influence of Co substitution for Ni and Mn on the structural and electrochemical characteristics of LiNi$_{0.5}$Mn$_{1.5}$O$_4$ [J]. Journal of Power Sources, 2008, 185 (2): 1429-1433.

[35] Wang X J, Yu X Q, Li H, et al. Li-storage in LiFe$_{1/4}$Mn$_{1/4}$Co$_{1/4}$Ni$_{1/4}$PO$_4$ solid solution [J]. Electrochemistry Communications, 2008, 10 (9): 1347-1350.

[36] Xie J, Kohno K, Matsumura T, et al. Li-ion diffusion kinetics in LiMn$_2$O$_4$ thin films prepared by pulsed laser deposition [J]. Electrochimica Acta, 2008, 54 (2): 376-381.

[37] Xia Hui, Lu Li, Lai M O. Li diffusion in LiNi$_{0.5}$Mn$_{0.5}$O$_2$ thin film electrodes prepared by pulsed laser deposition [J]. Electrochimica Acta, 2009, 54 (25): 5986-5991.

[38] Li Zhe, Du Fei, Bie Xiaofei, et al. Electrochemical kinetics of the Li Li$_{0.23}$Co$_{0.3}$Mn$_{0.47}$O$_2$ cathode material studied by GITT and EIS [J]. Journal of Physical Chemistry C, 2010, 114 (51): 22751-22757.

[39] Zhu Yujie, Wang Chunsheng. Galvanostatic intermittent titration technique for phase-transformation electrodes [J]. Journal of Physical Chemistry C, 2010, 114 (6): 2830-2841.

[40] Delacourt C, Ati M, Tarascon J M. Measurement of lithium diffusion coefficient in Li$_y$FeSO$_4$F [J]. Journal of the Electrochemical Society, 2011, 158 (6): A741-A749.

[41] Rui X H, Yesibolati N, Li S R, et al. Determination of the chemical diffusion coefficient of Li$^+$ in intercalation-type Li$_3$V$_2$(PO$_4$)$_3$ anode material [J]. Solid State Ionics, 2011, 187 (1): 58-63.

[42] Wunde F, Berkemeier F, Schmitz G. Lithium diffusion in sputter-deposited Li$_4$Ti$_5$O$_{12}$ thin films [J]. Journal of

Power Sources, 2012, 215: 109-115.

[43] Yang Shunyi, Wang Xianyou, Yang Xiukang, et al. Determination of the chemical diffusion coefficient of lithium ions in spherical Li Ni$_{0.5}$Mn$_{0.3}$Co$_{0.2}$O$_2$ [J]. Electrochimica Acta, 2012, 66: 88-93.

[44] Xiao P F, Lai M O, Lu L. Transport and electrochemical properties of high potential tavorite LiVPO$_4$F [J]. Solid State Ionics, 2013, 242: 10-19.

[45] 李泓. 锂离子电池负极材料及电极过程研究 [D]. 北京: 中国科学院物理研究所, 1999.

[46] 鲁中华. 锂离子电池富锂层状正极材料的制备和电化学性能研究 [D]. 北京: 中国科学院物理研究所, 1996.

[47] Dell'Era A, Pasquali M. Comparison between different ways to determine diffusion coefficient and by solving Fick's equation for spherical coordinates [J]. Journal of Solid State Electrochemistry, 2009, 13 (6): 849-859.

[48] Fell Christopher R, Carroll Kyler J, Chi Miaofang, et al. Synthesis-structure-property relations in layered, "Li-excess" oxides electrode materials Li Li$_{(1/3-2x/3)}$Ni$_x$Mn$_{(2/3-x/3)}$O$_2$ ($x=1/3$, 1/4, and 1/5) [J]. Journal of the Electrochemical Society, 2010, 157 (11): A1202-A1211.

[49] Meethong Nonglak, Kao Yu-Hua, Carter W Craig, et al. Comparative study of lithium transport kinetics in olivine cathodes for Li-ion batteries [J]. Chemistry of Materials, 2010, 22 (3): 1088-1097.

[50] Xie J, Imanishi N, Zhang T, et al. Electrochemical performance of all-solid-state Li batteries based LiMn$_{0.5}$Ni$_{0.5}$O$_2$ cathode and NASICON-type electrolyte [J]. Journal of Power Sources, 2010, 195 (24): 8341-8346.

[51] Manjunatha H, Venkatesha T V, Suresh G S. Kinetics of electrochemical insertion of lithium ion into LiFePO$_4$ from aqueous 2 M Li$_2$SO$_4$ solution studied by potentiostatic intermittent titration technique [J]. Electrochimica Acta, 2011, 58: 247-257.

[52] West W C, Soler J, Smart M C, et al. Electrochemical behavior of layered solid solution Li$_2$MnO$_3$-LiMO$_2$ (M= Ni, Mn, Co) Li-ion cathodes with and without alumina coatings [J]. Journal of the Electrochemical Society, 2011, 158 (8): A883-A889.

[53] Churikov Alexei V, Romanova Veronica O. An electrochemical study on the substituted spinel LiMn$_{1.95}$Cr$_{0.05}$O$_4$ [J]. Ionics, 2012, 18 (9): 837-844.

[54] Liu Li, Zhou Meng, Yi Lanhua, et al. Excellent cycle performance of Co-doped FeF$_3$/C nanocomposite cathode material for lithium-ion batteries [J]. Journal of Materials Chemistry, 2012, 22 (34): 17539-17550.

[55] Li Juchuan, Yang Fuqian, Xiao Xingcheng, et al. Potentiostatic intermittent titration technique (PITT) for spherical particles with finite interfacial kinetics [J]. Electrochimica Acta, 2012, 75: 56-61.

[56] Manjunatha H, Mahesh K C, Suresh G S, et al. Kinetics of lithium insertion into LiMnPO$_4$ from aqueous saturated LiOH: A study using galvanostatic and potentiostatic intermittent titration techniques [J]. Electrochimica Acta, 2012, 80: 269-281.

[57] Rho Y H, Kanamura K. Li ion diffusion in Li$_4$Ti$_5$O$_{12}$ thin film electrode prepared by PVP sol-gel method [J]. Journal of Solid State Chemistry, 2004, 177 (6): 2094-2100.

[58] Wu Feng, Wang Feng, Wu Chuan, et al. Rate performance of Li$_3$V$_2$(PO$_4$)$_3$/C cathode material and its Li ion intercalation behavior [J]. Journal of Alloys and Compounds, 2012, 513: 236-241.

[59] Amalraj Francis, Talianker Michael, Markovsky Boris, et al. Study of the lithium-rich integrated compound xLi$_2$MnO$_3$·(1−x) LiMO$_2$ (x≈0.5; M = Mn, Ni, Co; 2:2:1) and its electrochemical activity as positive electrode in lithium cells [J]. Journal of the Electrochemical Society, 2013, 160 (2): A324-A337.

[60] Baster Dominika, Zheng Kun, Zajac Wojciech, et al. Toward elucidation of delithiation mechanism of zinc-substituted LiFePO$_4$ [J]. Electrochimica Acta, 2013, 92: 79-86.

[61] Lee Jungbae, Kumar Purushottam, Lee Jinhyung, et al. ZnO incorporated LiFePO$_4$ for high rate electrochemical performance in lithium ion rechargeable batteries [J]. Journal of Alloys and Compounds, 2013, 550: 536-544.

[62] Xiao Liang, Guo Yonglin, Qu Deyu, et al. Influence of particle sizes and morphologies on the electrochemical performances of spinel LiMn$_2$O$_4$ cathode materials [J]. Journal of Power Sources, 2013, 225: 286-292.

[63] Dai Kehua, Mao Jing, Li Zitao, et al. Microsized single-crystal spinel LAMO for high-power lithium ion batteries synthesized via polyvinylpyrrolidone combustion method [J]. Journal of Power Sources, 2014, 248: 22-27.

[64] Lepage D, Sobha F, Kuss C, et al. Delithiation kinetics study of carbon coated and carbon free LiFePO$_4$ [J]. Journal of Power Sources, 2014, 256: 61-65.

[65] Li Zheng, Ban Chunmei, Chernova Natasha A, et al. Towards understanding the rate capability of layered transition metal oxides LiNi$_y$Mn$_y$Co$_{1−2y}$O$_2$ [J]. Journal of Power Sources, 2014, 268: 106-112.

[66] Meethong Nonglak, Huang Hsiao-Ying Shadow, Speakman Scott A, et al. Strain accommodation during phase

[67] Wang Qing, Li Hong, Huang Xuejie, et al. Determination of chemical diffusion coefficient of lithium ion in graphitized mesocarbon microbeads with potential relaxation technique [J]. Journal of the Electrochemical Society, 2001, 148 (7): A737-A741.

[68] Wu C, Wu F, Chen L Q, et al. Fabrications and electrochemical properties of fluorine-modified spinel $LiMn_2O_4$ for lithium ion batteries [J]. Solid State Ionics, 2002, 152: 327-334.

[69] Ouyang C Y, Shi S Q, Wang Z X, et al. Experimental and theoretical studies on dynamic properties of Li ions in $Li_xMn_2O_4$ [J]. Solid State Communications, 2004, 130 (7): 501-506.

[70] Yu Xiqian, Wang Qi, Zhou Yongning, et al. High rate delithiation behaviour of $LiFePO_4$ studied by quick X-ray absorption spectroscopy [J]. Chemical Communications, 2012, 48 (94) 11537-11539.

[71] 唐堃. 磷酸盐正极薄膜材料的制备和电化学性能研究 [D]. 北京: 中国科学院物理研究所, 2009.

[72] Santhanagopalan Shriram, Guo Qingzhi, White Ralph E. Parameter estimation and model discrimination for a lithium-ion cell [J]. Journal of the Electrochemical Society, 2007, 154 (3): A198-A206.

[73] Schmidt Alexander P, Bitzer Matthias, Imre Arpad W, et al. Experiment-driven electrochemical modeling and systematic parameterization for a lithium-ion battery cell [J]. Journal of Power Sources, 2010, 195 (15): 5071-5080.

[74] Bazant Martin Z. Theory of chemical kinetics and charge transfer based on nonequilibrium thermodynamics [J]. Accounts of Chemical Research, 2013, 46 (5): 1144-1160.

[75] Jamnik J, Maier J. Charge transport and chemical diffusion involving boundaries [J]. Solid State Ionics, 1997, 94 (1-4): 189-198.

[76] Jamnik J, Kalnin J R, Kotomin E A, et al. Generalised Maxwell-Garnett equation: Application to electrical and chemical transport [J]. Physical Chemistry Chemical Physics, 2006, 8 (11): 1310-1314.

[77] Bisquert J. Analysis of the kinetics of ion intercalation: Ion trapping approach to solid-state relaxation processes [J]. Electrochimica Acta, 2002, 47 (15): 2435-2449.

[78] Jamnik J, Maier J, Pejovnik S. A powerful electrical network model for the impedance of mixed conductors [J]. Electrochimica Acta, 1999, 44 (24): 4139-4145.

第 13 章
锂二次电池材料的计算研究

计算材料学是材料科学与计算机科学相互交叉而形成的一门新兴学科,主要用于材料中结构、物性的设计与计算模拟,涉及物理、化学、计算机、材料学、数学等多个学科领域[1]。随着基础理论的创新与计算机性能的大幅度提升,计算材料学也得到了快速的发展[2]。根据所研究问题的内容与所在的空间与时间尺度,计算材料学的模拟方法涵盖了从微观原子、分子水平,到介观微米级别,直至宏观尺度的各种理论。图 13-1 展示了不同尺度上的主要模拟方法[3]。

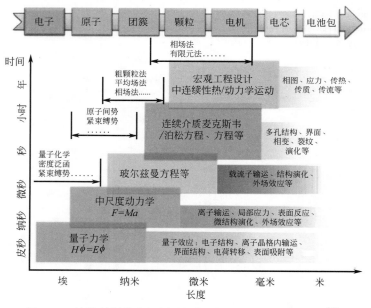

图 13-1 计算材料学在不同空间与时间尺度上的模拟方法[3]

计算材料学的迅速发展极大地促进了能源材料的研发进程[4]。锂二次电池作为绿色储能器件之一,虽已商业化,然而其中涉及的诸多基础科学问题仍不明朗,如 SEI 的生长机制、离子在电极材料中的扩散动力学特性、电极材料充放电过程中的结构演变、电位与结构的关系、空间电荷层分布等。实验探测手段如飞秒技术、高分辨透射电镜及 X 射线断层扫描（X-CT）等可以进行超快过程、原子尺度及三维立体原位成像等高分辨的时间与空间表征，但实验观测也常有难以观察、存在误差、实验困难及实验结果的片面性、不确定性等缺点，准确理解处于平衡态与非平衡态的锂电池内部的物理化学信息及演化规律，仍然需要结合理论上的解释，以获得更为本质的认识。理论计算模拟有助于人们更清楚地了解电池充放电过程中伴随着结构和性质变化所发生的电荷转移、质量转移、化学反应等过程；从而从微观上理解实验现象，寻找材料结构与性能之间的关系，同时也为电池材料的开发及电池行为的模拟提供理论依据。本章将从原子尺度、介观尺度和宏观尺度三个方面分别介绍理论模拟方法在锂二次电池研究中的应用。

13.1 原子尺度的模拟

材料微观尺度的计算模拟依赖于理论模型对电子、原子之间相互作用势的描述。根据势函数模型中涉及的参数数量及类型，可分为经验模型、半经验模型及量子力学模型。依赖的经验参数愈少，模型对体系的描述愈接近实际，理论可有效描述的范围也愈广。基于量子力学的理论方法，在一定的假设条件下，从薛定谔方程出发，其计算的可靠性已在各个材料领域的研究中得到证实[2]。伴随着计算精度的提高，计算量也迅速增加，近几十年来的理论创新与计算机技术的革新为基于量子力学的第一性原理计算提供了可能。

第一性原理计算的宗旨是不采用任何经验参数，只采用电子质量、光速、质子质量、元电荷、普朗克常数 5 个基本物理量计算研究微观粒子的行为。第一性原理计算方法有着半经验方法不可比拟的优势，只需知晓构成微观体系的元素及位置，即可应用量子力学来计算出该微观体系的总能量、电子结构等物理性质。一方面，第一性原理计算可作为真实实验的补充，通过计算可以更为深入地理解被模拟体系的特征和性质；另一方面，通过第一性原理计算可以在尚无实验的情况下对研究体系进行预测和设计[2]。

第一性原理计算，特别是基于密度泛函理论的第一性原理计算同分子动力学、蒙特卡罗方法相结合，在材料设计、合成和评价等诸多方面有许多突破性的进展，已经成为原子尺度上材料计算模拟的重要基础和核心技术。本部分将介绍第一性原理计算、分子动力学、蒙特卡罗等原子尺度的材料模拟方法在锂二次电池研究中的应用。

13.1.1 基于密度泛函理论的第一性原理计算

密度泛函理论的建立始于 20 世纪 60 年代，经过数十年的发展，基于密度泛函理论的第一性原理计算已成为当今计算材料科学的一个最为重要的工具。1985 年，Car 和 Parrinello 两人把分子动力学方法和第一性原理计算结合起来，使得第一性原理方法在处理问题时变得更为实际[5]。1998 年，Kohn 教授[6]因其对密度泛函理论（density functional theory，DFT）发展所做出的巨大贡献荣获了该年度的诺贝尔化学奖。自此，计算材料学特别是基于密度泛函理论的第一性原理计算在材料科学研究中迅速升温。本节将主要针对密度泛函理论的基本原理及其在锂电池中的应用展开论述。

13.1.1.1 密度泛函理论[7]

（1）绝热近似

根据量子力学理论，微观粒子的运动规律用薛定谔方程来描述

$$i\hbar \frac{\partial}{\partial t}\psi(r,t) = H\psi(r,t) \tag{13-1}$$

式中，H 为哈密顿量；ψ 为系统波函数；$\hbar = h/(2\pi)$ 为普朗克常数。直接求解薛定谔方程是困难的。然而，若体系哈密顿量不含时间 t 则可以通过分离变量，把式（13-1）简化为定态薛定谔方程

$$H\phi(r) = E\phi(r) \tag{13-2}$$

波函数随时间的演化为

$$\psi(r,t) = \phi(r)e^{-iEt/\hbar} \tag{13-3}$$

式中，E 为体系的总能量。上述定态薛定谔方程式（13-2）的求解仍然十分困难，特别是对于宏观体系原子数目高达 10^{23} 的数量级，精确求解方程式（13-2）几乎不可能。为此，必须做近似处理。由于原子核的质量比电子大得多，电子的运动速度要远高于原子核，可以近似认为原子核每移动一步，电子都能迅速弛豫到能量最低状态。这样就可以将原子核与电子的运动分开计算，这就是玻恩·奥本海默近似（Born Oppenheimer approximation），也称为绝热近似[8]。

玻恩·奥本海默近似的要点为：①将固体整体的平移、转动和核的振动运动分离出去；②考虑电子运动时，将坐标系原点设定在固体质心上，并令其随固体整体一起平移和转动，同时令各原子核固定在它们振动运动的某一瞬时位置上；③考虑核的运动时则不考虑电子在空间的具体分布。这样，通过分离变量即可写出电子分系统满足的薛定谔方程

$$H = T_e + V_{ee} + \sum_j v(\vec{r_j}) \tag{13-4}$$

式中，T_e 为电子动能项；V_{ee} 为电子间库仑相互作用；$\sum_j v(\vec{r_j})$ 为电子-原子核之间的相互作用。

尽管通过绝热近似将包含原子核与电子体系的定态薛定谔方程简化为只对电子求解，然而对于较大体系的精确计算仍然非常困难。密度泛函理论为简化这一问题提供了解决方案。

（2）Hohenberg-Kohn 定理

密度泛函理论的核心思想是体系的所有物理性质都由其电子密度的空间分布所决定，即所有性质都是电荷密度函数的泛函，该理论思想最早由 Thomas[9] 和 Fermi[10] 于 1927 年提出。对能量泛函变分可求得能量极小值对应的电荷密度分布，这时对应的能量也即基态能量。泛函极小问题也是对电子密度分布函数求解。这样的处理首先要从理论上证明的确存在总能对于电子密度分布的这样一个泛函。因此 Hohenberg 和 Kohn 基于他们的非均匀电子气理论，提出了如下两个定理[11]。

定理1：不计自旋的全同费米子系统的基态能量是粒子数密度函数 $\rho(r)$ 的唯一泛函。

定理 2：能量泛函 $E(\rho)$ 在粒子数不变的条件下，对正确的粒子数密度函数 $\rho(r)$ 取极小值，并等于基态能量。

这里所处理的基态是非简并的，多电子体系哈密顿量写为动能部分、多电子系统相互作用部分和多电子系统之外的外场部分

$$H = T + U + V \tag{13-5}$$

则 Hohenberg-Kohn 定理证明体系总能存在对基态电子密度分布函数的泛函形式

$$E[\rho] = [|\phi|T+V|\phi|] + \int dr v(r)\rho(r) \tag{13-6}$$

（3）Kohn-Sham 方程

尽管 Hohenberg-Kohn 定理证明了总能可通过求解最有利的基态电子密度分布函数得到，但总能对于电子密度分布函数的具体泛函形式以及如何才能利用上述泛函极值的性质求解总能的问题，Hohenberg-Kohn 定理并未给出回答。Kohn 和 Sham 随后提出的 Kohn-Sham 方程最终将密度泛函理论引入了实际应用的行列。

Kohn-Sham 方程的基本演绎过程如下[12]。

第一，将 $F[\rho]$ 这个泛函写成两部分泛函之和

$$F[\rho] = T(\rho) + V(\rho) \tag{13-7}$$

式中，$T(\rho)$ 和 $V(\rho)$ 分别是多体系统的未知形式的动能部分和势能部分。

第二，假设动能部分和势能部分可进一步显式的写成

$$F[\rho] = T(\rho) + \frac{1}{2}\iint dr dr' \frac{\rho(r)\rho(r')}{|r-r'|} \tag{13-8}$$

第三，引入一组单电子波函数的基底 $\varphi_i(r_i)$（$i=1，2，3\cdots$），电子密度分布函数和动能部分的泛函可以显式地表示成

$$\rho(r) = \sum_{i=1}^{N} |\varphi_i(r)|^2 \tag{13-9}$$

$$T(\rho) = \sum_{i=1}^{N} \int dr \varphi_i^*(r)[-\nabla^2]\varphi_i(r) \tag{13-10}$$

则整个能量泛函就可显式地表示为

$$E(\rho) = F(\rho) + \int dr v(r)\rho(r) \tag{13-11}$$

式（13-11）与真实的多体系统能量泛函相比是有差别的。

第四，加入未知形式的泛函项 $E_{xc}[\rho]$，修正泛函式与真实系统总能泛函之间的误差。最后的总能泛函表示为

$$E(\rho) = F(\rho) + \int dr v(r)\rho(r) + E_{xc}[\rho] \tag{13-12}$$

$E_{xc}[\rho]$ 的具体形式是未知的，它包含了多体系统的交换和关联效应，因而被称为交换关联势，它同时也是电子密度分布函数的泛函。对于耦合比较弱的系统（如稀薄电子气），可以预计交换关联势的数值较小。实际计算中通过拟合精确求解体系的能量和电荷密度分布来得到参数化的 $E_{xc}[\rho]$ 经验形式。

第五，利用泛函变分，寻求单电子态的 $\varphi_i(r_i)$ 最佳形式

$$\delta[E(\rho)]/\delta\rho = \delta\{F(\rho) + \int dr v(r)\rho(r) + E_{xc}[\rho]\}/\delta\rho = 0 \tag{13-13}$$

变分的结果得到单电子形式的方程组，称为 Kohn-Sham 方程

$$\{-\nabla^2 + V_{KS}[\rho(r)]\}\varphi_i(r) = E_i\varphi_i(r) \tag{13-14}$$

$$V_{KS}[\rho(r)] = v(r) + \int dr' \frac{\rho(r')}{|r-r'|} + \frac{\delta E_{xc}[\rho]}{\delta\rho(r)} \tag{13-15}$$

$$\rho(r) = \sum_{i=1}^{N} |\varphi_i(r)|^2 \tag{13-16}$$

式中，$V_{KS}[\rho(r)]$ 为 Kohn-Sham 势，至此完成了密度泛函理论计算多电子系统总能和电荷密度空间分布的演绎过程。

(4) 交换关联函数

密度泛函理论框架中包含未知形式的交换关联势 $E_{xc}[\rho]$，实际应用中通过拟合已被精确求解系统的结果，将交换关联势以参数化的形式表示出来。因此，很大程度上密度泛函计算结果的精度，取决于交换关联势的准确度。实际应用中，局域密度近似（local density approximation，LDA）是一种最简单有效的交换关联势的近似。它最早由 Slater 在 1951 年提出并付诸应用，该近似甚至早于密度泛函理论。这种近似假定空间某点的交换关联能只与该点的电荷密度有关，且等于同密度的均匀电子气的交换关联能[13]。目前具体计算中最常用的局域密度交换关联势是根据 Ceperley 和 Alder 用 Monte-Carlo 方法计算均匀电子气的结果[14,15]。LDA 在大多数材料计算中展示了巨大的成功。经验显示，LDA 计算原子游离能、分子解离能误差在 10%~20%，对分子键长、晶体结构可准确到 1% 左右[16]。但是对于与均匀电子气或者空间缓慢变化的电子气相差太远的系统，LDA 则不适用。

更精确的考虑需要计入某处附近的电荷密度对交换关联能的影响，如考虑到密度的一级梯度对交换关联能的贡献，称为广义梯度近似（generalized gradient approximation，GGA）[17]。这种近似是半局域化的，一般地，它比 LDA 更能给出精确的能量和结构，对开放的系统更为适用。目前常用的 GGA 方法有 Becke、Perdew-Wang 91 以及 BLYP 等形式[18-21]。

更进一步地，还可以考虑到密度高阶梯度的近似，这称为 Meta-GGA 或 Post-GGA[22]；甚至考虑到非局域的交换关联作用，如 Van der Waals[23,24]作用。这两方面虽有研究，但仍未找到一个足够精确又简单的形式。

概括来说，LDA 与 GGA 各有所长，一般规律是 LDA 倾向于将体系描述得过于局域化，而 GGA 则过于离域化。因此，LDA 通常低估晶胞参数，而 GGA 则会高估晶胞参数。此外，在密度泛函理论基础上引入 Hartree-Fock 方程中精确的电子交换作用，被称为混合密度泛函（Hybrid-GGA）[25]。

(5) 自洽计算

在实际计算中，将多体系统原胞划分为足够细的网格点，在每个网格点上初始化一组试探波函数 $\varphi_i(r)$（通常设为随机数），然后根据式（13-15）可以算出网格上的 Kohn-Sham 势，本征方程式（13-14）立即可解。解出来的本征函数 $\varphi_i(r)$ 的值与初始化的

$\varphi_i(r)$ 值一般不相同，将新解出来的波函数的一部分叠加到初始值上，重新计算 Kohn-Sham 势［式（13-15）］，利用修正过的势再次求解本征方程。所得到的本征函数又用于修正上一步循环输入的波函数式 $\varphi_i(r)$……，循环叠代的结果是最终 $\varphi_i(r)$ 不再变化，计算得以收敛。利用收敛后的这组单电子波函数，立即得到体系总能量和电荷密度分布，图 13-2 示意了整个自洽计算求解 Kohn-Sham 方程的迭代过程[7]。目前使用 DFT 进行计算的软件非常多，如 ABINIT，CASTEP，DMol3，Gaussian，Quantum ESPRESSO，SIESTA，VASP 及 WIEN2k 等。

13.1.1.2 密度泛函理论在锂电池研究中的应用

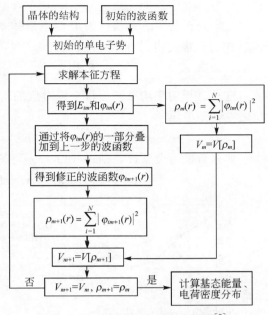

图 13-2 自洽计算迭代过程示意图[7]

密度泛函理论在锂二次电池基础研究中有着广泛的应用，常用于计算电极材料的结构稳定性、嵌锂电位、电子结构、能带、弛豫结构、缺陷生成能、迁移路径、活化能及锂离子传输动力学和脱嵌锂相变等性质[26]。

（1）嵌锂电位

电池的电压值是与电池反应的自由能变化相关联的参量。一个能自发进行的电化学反应，在等温可逆的条件下进行，当电池以无限小电流放电时，可对外做最大有用电功。由于电功＝电压×电量，电量可按电池反应计算，1 摩尔（mol）电子电量称为 1 法拉第（Faraday），以 F 表示。一个电子 e^- 的电量为 1.602×10^{-19} C，故有

$$F = N_A e^- = 96487.56 \text{C/mol} \tag{13-17}$$

设 n 为电池反应过程中转移的电荷数，则通过电池的电量为 nF，电池所做最大电功为

$$-W_r' = nFV_{AVE} \tag{13-18}$$

在等温等压条件下，体系的吉布斯自由能等于体系对外界所做的功

$$\Delta G = -W_r' \tag{13-19}$$

所以

$$\Delta G = nFV_{AVE} \tag{13-20}$$

又由于 Li^+ 带一个单位的电荷，因此

$$V_{AVE} = \Delta G / F \tag{13-21}$$

ΔG 是电池反应的吉布斯自由能，假设由电池嵌入反应引起的体积和熵的变化可忽略不计，则

$$\Delta G \approx \Delta E \tag{13-22}$$

故

$$V_{\text{AVE}} = \Delta E / F \tag{13-23}$$

对于正极为 LiA，负极为 B 的电池体系，假设充电后有 x 个锂从 LiA 中脱出并嵌入到 B 中（正极充电过程），其反应式为

$$\text{LiA} + \text{B} \longrightarrow \text{Li}_{1-x}\text{A} + \text{Li}_x\text{B} \tag{13-24}$$

这一过程的平均电压为每个锂对应的末态（Li_{1-x}A，Li_xB）与初态（LiA，B）的能量之差，则电压

$$V = [E(\text{Li}_{1-x}\text{A}) + E(\text{Li}_x\text{B}) - E(\text{LiA}) - E(\text{B})]/xe \tag{13-25}$$

如果负极为金属锂，则式（13-25）可简化为

$$V = [E(\text{Li}_{1-x}\text{A}) + E(\text{Li}_x) - E(\text{LiA})]/xe \tag{13-26}$$

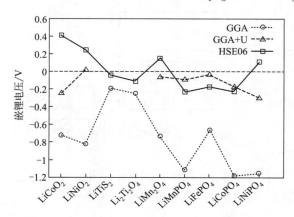

图 13-3 标准 GGA、GGA+U 和混合泛函 HSE06 计算得到各种锂二次电池正极材料的嵌锂电压[28]

研究早期人们发现，几乎对所有的正极材料，密度泛函理论计算得到的电压与实验值相比都偏低。Zhou 等[27]认为这是由于含有 3d 过渡金属离子的正极材料属于强关联电子体系，其 3d 轨道的电子是高度局域化的。而在单电子近似的密度泛函理论框架下，局域化的电子会引入自相互作用，因此导致嵌锂电位被低估。通过对标准的 GGA 进行强关联修正，也即采用 GGA+U 交换关联势，或采用杂化泛函 HSE06，计算得到了与实验值更为接近的电压（图 13-3）。在含有 3d 过渡金属离子的氧化物正极材料计算中，强关联效应已经公认为需要考虑的效应[28]。

（2）电极材料的稳定性

在锂二次电池材料中，目前所使用的正极材料仍是锂源的主要提供者。若材料的主体框架在锂离子脱嵌的过程中无法稳定存在，则会导致电池性能的迅速衰减，甚至带来安全隐患，故研究电极材料的稳定性十分必要。

电极材料稳定性的预测主要基于材料热力学性质的计算。大量实验证据已经表明，层状正极材料 LiCoO_2、LiNiO_2、LiMn_2O_4 中，当锂的脱出量达到一定程度，材料会发生分解，转变为热力学上更为稳定的尖晶石相或岩盐相[29]。Wang 等[29]计算了 Li_xCoO_2、Li_xNiO_2、$\text{Li}_x\text{Mn}_2\text{O}_4$ 材料在不同 Li 含量下的热力学相图，阐明了这 3 种材料的分解机制。以 LiNiO_2 为例，Wang 等计算了该材料在 0 K、220 K、730 K 的 Li-Ni-O$_2$ 三元相图（图 13-4）。通过分析相图中 Li_xNiO_2 稳定存在的区域及分解所经过的区域，认为 Li_xNiO_2 的分解是通过层状向尖晶石相的放热反应和尖晶石相向岩盐相的吸热反应两步来实现的。

除了脱嵌锂过程中的分解问题，电极材料在循环过程中的不稳定性还可能以氧析出的形式表现出来，这种现象在富锂相正极材料中较为常见[30]。富锂锰基 $x\text{Li}_2\text{MnO}_3 \cdot (1-x)\text{LiMO}_2$（M=Ni、Co、Mn 等）正极材料因其高容量而受到广泛的研究。然而，由于

氧气释放、低的电子电导等问题导致其具有高的不可逆容量和较差的倍率性能及循环性，阻碍了该类材料在实际中的应用。Xiao 等[31]通过第一性原理计算系统地研究了富锂锰基

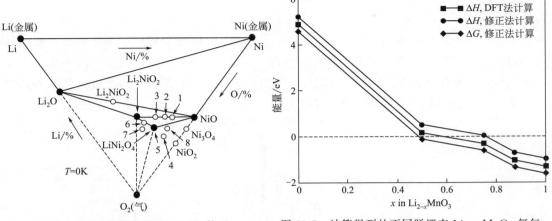

图 13-4　计算得到的 Li-Ni-O_2 体系在 0K 时的三元相图[29]

图 13-5　计算得到的不同脱锂态 $Li_{2-x}MnO_3$ 氧气释放过程中反应焓和吉布斯自由能[31]

正极材料的母相化合物 Li_2MnO_3 晶格中氧在不同锂含量时的热力学稳定性。图 13-5 为通过第一性原理计算得到的不同脱锂态 $Li_{2-x}MnO_3$ 氧气释放过程中的反应焓和吉布斯自由能。由图 13-5 可知，当脱锂量 $x \geqslant 0.5$ 时会有氧气产生，进而导致结构的变化。

为改善富锂锰基 $xLi_2MnO_3 \cdot (1-x)LiMO_2$（M＝Ni、Co、Mn 等）正极材料充放电过程中因氧析出导致的结构不稳定，包覆和掺杂通常成为必要的手段。理论计算方面，Gao 等[30]通过第一性原理计算研究了不同晶格位 Mo 掺杂 Li_2MnO_3 晶体的结构稳定性及 Li^+ 在晶格中不同跃迁方向的迁移能垒。图 13-6 是采用 PBE＋U 交换关联势计算得到的不同脱锂态的 Li_yMnO_3 和 $Li_yMn_{0.75}Mo_{0.25}O_3$ 的氧气释放过程中反应焓和吉布斯自由能[30]。分析图 13-6 可知，与未掺杂的 Li_2MnO_3 相比，掺 Mo 的 Li_2MnO_3 结构稳定性有了大幅度提升。

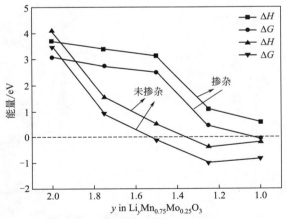

图 13-6　计算得到的不同脱锂态 Li_yMnO_3 和 $Li_yMn_{0.75}Mo_{0.25}O_3$ 的氧气释放过程中反应焓和吉布斯自由能[30]

(3) 电子结构

电池材料的电子结构也与材料的电化学性能有着密切的关系。电极材料中，脱嵌锂过程中电荷补偿的来源和氧的稳定性与过渡金属及氧的分波态密度的相对位置有关[32]，电极与电解质界面的化学稳定性则可从能态密度中做出定性预测[33]，此外电池的倍率性能也与电极的电子导电性有关[34]。电池中的固态电解质材料需要具有电子绝缘的特性，材

料的电化学窗口也与其能隙宽度相关[35]。

电极与电解质之间的界面问题一直是电池研究中的难点。Lepley 等[33]尝试通过理论计算理解 Li_3PS_4 和 Li_3PO_4 与金属 Li 负极之间的化学稳定性[33]。图 13-7 和图 13-8 分别为计算得到的 $\beta\text{-}Li_3PO_4$、$\gamma\text{-}Li_3PS_4$ 与金属 Li 界面模型的分波态密度图[33]。图中显示，当 $\beta\text{-}Li_3PO_4$ 与金属 Li 接触时，金属 Li 的电子态与 P、O 原子的电子态之间难以相互作用，因此 $\beta\text{-}Li_3PO_4/Li$ 界面的化学稳定性较高；而 $\gamma\text{-}Li_3PS_4/Li$ 的界面上，Li 的电子态与 S 的电子态之间有显著的重叠，二者之间容易发生电子转移，形成 Li—S 键，造成界面层形成新相，从而引起更为复杂的电化学过程。

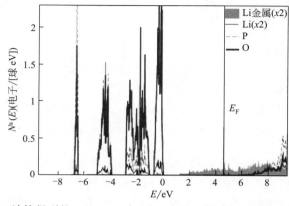

图 13-7　计算得到的 $\beta\text{-}Li_3PO_4$ 与金属 Li 界面模型的分波态密度图[33]

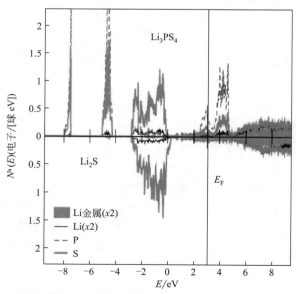

图 13-8　计算得到的 $\gamma\text{-}Li_3PS_4$ 与金属 Li 界面模型的分波态密度图[33]

（4）离子输运机制

锂离子的传输是锂二次电池中核心的输运过程，锂离子传输的路径、能量势垒等与电池的倍率性能、极化程度、离子互占位等现象有着直接的联系。锂离子在材料中的扩散性质一方面可以通过基于过渡态理论的弹性能带方法（NEB）获得，另外也可以采用基于第一性原理的分子动力学方法计算得到。

Ouyang 等[36]通过第一性原理计算研究了锂离子在橄榄石结构 $LiFePO_4$ 中的扩散机理。通过计算各种可能扩散路径的迁移势垒，表明对于 $LiFePO_4$、$FePO_4$ 及 $Li_{0.5}FePO_4$ 来说，扩散势垒沿着 c 轴方向分别为 0.6eV、1.2eV 和 1.5eV。而在沿着 a 轴、b 轴方向的迁移路径上，其扩散势垒非常大，以至于锂离子[37]难以扩散。这说明锂离子在 $LiFePO_4$ 中的扩散是一维的。这种一维的扩散行为通过第一性原理的分子动力学模拟得到了更进一步的验证，并且直观地观察到了锂离子的一维扩散行为[38]。

为了进一步解释 Cr 在 Li 位掺杂的 $LiFePO_4$ 材料电子电导率得到了大幅度提高而其电化学性能却没有得到明显改善的现象，Ouyang 等[36]计算研究了纯的和 Li 位掺 Cr 的 $LiFePO_4$ 中锂离子和铬离子沿着一维扩散通道输运的能量势垒（图 13-9 所示分别为 Li 与 Cr 沿着 c 轴方向的迁移势垒），结果发现锂离子可以很容易地沿着扩散通道扩散，但是铬离子很难离开本来的位置。这意味着铬离子堵塞了材料的一维扩散通道，如图

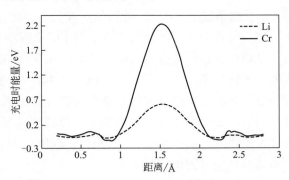

图 13-9　Li 与 Cr 在 $LiFePO_4$ 晶体中沿着 c 轴方向的迁移势垒[36]

13-10 所示。从不阻塞一维离子输运通道的角度出发，他们考虑了低价态 Na^+ 的 Li 位掺杂以及其他金属离子的 Fe 位掺杂对材料动力学性能的影响，并发现两种方法都能够不同程度地提高材料的电子电导率，且不会阻塞 Li^+ 的一维输运通道。

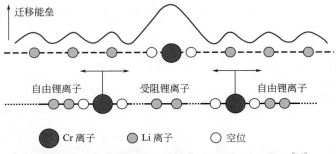

图 13-10　Cr 掺杂导致的锂离子输运阻塞效应模型[36]

(5) 缺陷生成能

材料中的缺陷和杂质会改变材料的许多性质，包括电子电导、离子电导等与锂电池性能紧密相关的性质。Hoang 等[39]通过第一性原理计算研究了橄榄石结构的 $LiFePO_4$ 中本征缺陷及掺杂对其电子传导和离子传导的影响。计算结果表明，在 $LiFePO_4$ 中本征的锂空位缺陷和空穴小极化子分别决定着材料的离子输运和电子输运过程。对一价掺杂元素（Na、K、Cu、Ag）、二价掺杂元素（Mg、Zn）、三价掺杂元素（Al）、四价掺杂元素（Zr、C、Si）和五价掺杂元素（V、Nb），Hoang 等[39]分别计算了各种元素掺入不同晶格位置的缺陷形成能，不同合成条件对掺杂浓度的调节以及掺杂后对体系输运性质的改

变，发现通过缺陷控制的合成手段可以有效地改善体系的电子导电性。

（6）晶体结构及演化

电极材料在脱嵌锂的过程中，有时会出现一些有趣的结构演化过程，如在石墨嵌锂过程中出现的"阶"结构[40]。随着透射电镜技术的发展，人们在 $LiFePO_4$ 的脱锂样品中观察到了三相界面，并且也发现了"阶"的现象[41]。借助原子尺度的计算模拟，该结构的演变过程及形成机制得到了进一步的阐释。

Sun 等[42]通过第一性原理计算研究了 $LiFePO_4$ 脱锂过程中出现的"阶"结构（图 13-11）。结果显示，"阶"结构是一种受 Li^+ 传输动力学控制的热力学亚稳态。其成因在于，当一层锂脱出后其最近邻锂层的 Li^+ 跃迁受到阻碍，因此充电过程中倾向于隔层脱锂而不是顺序脱锂。结合热力学条件，Sun 等[42]构建了一个用来描述 $LiFePO_4$ 充电过程的双界面脱锂模型：充电时，$LiFePO_4$ 颗粒中同时存在两个隔层脱锂过程，它们在空间上有一定的相位差；经历过两个隔层脱锂过程的区域为 $FePO_4$ 相，只经历过一个隔层脱锂过程的区域为"阶"结构，而两个隔层脱锂过程都没有经历过的区域为 $LiFePO_4$ 相。脱锂过程中"阶"结构位于 $LiFePO_4$ 与 $FePO_4$ 两相之间，该理论模型很好地解释了高分辨扫描透射电镜下观察到的现象[41]。

"阶"结构的形成表明，$LiFePO_4$ 中 Li^+ 的占据与否会影响其最近邻锂层的 Li^+ 传输。由于 $LiFePO_4$ 是一个强关联电子体系，Fe 3d 轨道电子高度局域，Li^+ 占据/缺失带来的电子/空穴会局域在最近邻的铁离子上。因此，Li^+ 可以通过改变其最近邻铁离子的价态来影响近邻锂层中的 Li^+ 传输行为。实际上这是一种以铁离子为媒介的层间 Li^+ 间接相互作用，而之前的计算研究主要局限于 Li^+ 之间的直接作用。孙洋等[26]的计算结果说明，对于电子高度局域的强关联电子体系，除了离子之间的直接相互作用，还需要考虑可能存在的间接相互作用。对于两/多相分离反应，在计算其离子传输性质时需要考虑相界面附近的离子相互作用。

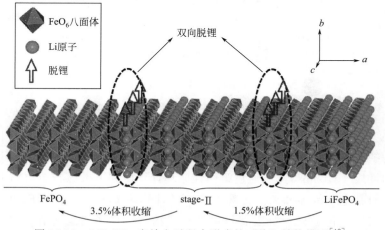

图 13-11 $LiFePO_4$ 充放电过程中形成的"阶"结构模型[42]

（7）材料力学性能

大多数电极材料在反复的脱嵌锂过程中会经历体积的变化，如层状材料在脱锂后会出

现沿 c 方向的膨胀，Si 负极在嵌锂之后则会出现多达 300% 的体积膨胀[43]。这就需要我们对材料的力学性能进行进一步研究，以理解材料在脱锂过程中形变产生的原因及对电池性能的影响。基于密度泛函理论的第一性原理计算能够通过求解力矩阵，获得材料的弹性常数，从而得到体模量、剪切模量、杨氏模量、泊松比等一系列力学性能的数值[44]。

Shenoy 等[45]计算了不同 Li 含量的 Li-Si 合金的力学性质，建立了 Si 负极嵌锂过程中形变与断裂过程的力学模型。通过计算晶态与非晶态 Li-Si 合金相的杨氏模量、剪切模量、体模量和泊松比（图 13-12），他们发现随着锂浓度的增加，上述模量值几乎呈线性降低，表明嵌锂引起了材料的弹性软化，同时也说明了材料的力学特性在电池的电化学过程中不是保持不变的，应力场的分布与形变及裂纹的扩展都与充放电状态相关。

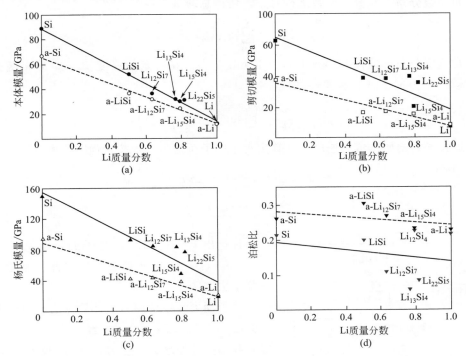

图 13-12　计算得到的晶态（蓝色点）与非晶态（红色点）Li-Si 合金 Li_xSi 在不同 Li 含量时的本体模量 (a)、剪切模量 (b)、杨氏模量 (c) 和泊松比 (d)[45]

13.1.2　分子动力学

基于密度泛函的第一性原理计算能够得到材料处于零温基态时的性质，这对理解材料的本征特性非常重要。在有限温度下材料中原子、离子的输运可通过分子动力学模拟进行研究。本节将介绍分子动力学方法及其在锂二次电池中的应用。

13.1.2.1　分子动力学方法介绍[46]

分子动力学方法是一种模拟经典系统平衡态和输运性质的一种技术，经典系统的粒子运动规律为牛顿力学

$$m\frac{\mathrm{d}^2\vec{r}}{\mathrm{d}t^2}=F \tag{13-27}$$

多粒子耦合

$$\begin{cases} m_i\dfrac{\mathrm{d}^2\vec{r}_i}{\mathrm{d}t^2}=F(\vec{r}_1,\cdots,\vec{r}_N)=-\dfrac{\partial E}{\partial \vec{r}_i} \\ \vec{v}_i(t)=\vec{r}_i(t) \end{cases} \tag{13-28}$$

求解式（13-28）方程组需设定粒子的初始位置和初始速度，通过实验，如 X 射线衍射以及核磁共振技术等，可以确定原子的初始位置。而初始速度通常按照一个温度关联的随机分布函数分布，对分布函数进行修正，使得总动量为零。

$$P=\sum_{i=1}^{N}m_i v_i=0 \tag{13-29}$$

通常速度分布可选择 Maxwell-Boltzmann 分布。在一定的温度 T 下，分布函数给出某个粒子在 x 方向上具有速度 v_x 的概率为

$$p=\left(\frac{m_i}{2\pi k_B T}\right)^{1/2}\exp\left(-\frac{1}{2}\frac{m_i v_{ix}^2}{k_B T}\right) \tag{13-30}$$

知道粒子的初始位置和速度，计算出每个原子所受的力，按牛顿定律离散形式

$$r(t+\Delta t)=2r(t)-r(t-\Delta t)+\frac{f(t)}{m}\Delta t^2$$

$$v(t+\Delta t)=\frac{r(t+\Delta t)-r(t-\Delta t)}{2\Delta t} \tag{13-31}$$

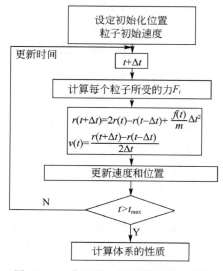

图 13-13　分子动力学模拟过程示意图

计算出 $t+\Delta t$ 时刻，粒子的位移 $r(t+\Delta t)$ 和速度 $v(t+\Delta t)$，更新粒子的位置和速度，如此重复下去（图 13-13）。

分子动力学中按照力的计算方法不同分为经验势方法（MD/MM 等）、半经验势方法（TBMD 等）和第一性原理分子动力学（FPMD）[47]。

$$F=-\nabla_{R_i}E_{\text{tot}}(R_j) \tag{13-32}$$

对于经验势方法（MD/MM 等），不直接考虑电子体系。使用经验势，可计算很大的接近宏观的体系。

$$E_{\text{tot}}^{\text{MD/MM}}=\sum_{i<j}\phi_2(r_i,r_j)+\sum_{i<j<k}\phi_3(r_i,r_j,r_k) \tag{13-33}$$

半经验势方法（TBMD 等），对电子采取紧束缚近似，可计算较大的体系。

$$E_{\text{tot}}^{\text{TB}}=E_{\text{BS}}^{\text{TB}}+\sum_{I,J}\phi(|R_I-R_J|) \tag{13-34}$$

第一性原理方法，如 QMD、CPMD、FPMD，电子和核都用比较准确的研究和计算

方法，计算量大，只能研究小体系。

$$E_{tot}^{DFT} = E_{BS}^{DFT} - \frac{1}{2}\iint dr'dr \frac{\rho(r')\rho(r)}{|r-r'|}$$
$$- \int dr V_{xc}(r)\rho(r) + E_{xc} + E_{nn} \tag{13-35}$$

计算出粒子所受力，计算 $t+\Delta t$ 时刻粒子的位置和速度也发展了许多种数值方法，最常用的有 Verlet 算法、Leap-frog 算法、速度 Verlet 算法以及 Beeman 算法等[46]。在第一性原理分子动力学过程中，应从以下几个方面合理地选择时间步长 Δt[37]：①和算法相关，但并不是绝对地依赖于算法；②每一步原子的运动应该保证原子在所处的势场变化不大，这首先要求原子运动的位移不能太大，而在力场的梯度变化比较大的情况下，这一点更要注意；③从经验上来说，通常每一步原子的运动距离应该小于该原子和其最近邻位原子距离的 1/20；④对于原子级系统来说，在非低温模拟情况下（$T>77K$），通常其步长最好应小于 4fs（$1fs=1\times10^{-15}s$）。

13.1.2.2 分子动力学模拟所获得信息

分子动力学模拟可以获得包括原子的位置和运动速度等原子尺度的信息。统计力学能把这些原子尺度的信息转换为宏观可观测量，如能量、比热容、压强等。这样的转换基于各态历经假设（ergodic hypothesis），即统计系统平均等效于对时间的平均。

$$\langle A \rangle_{ensemble} = \langle A \rangle_{time} \tag{13-36}$$

最基本的思想是：分子动力学的目的是通过模拟系统在足够长时间里系统随时间的演化过程，从而产生足够多的满足我们需要的相空间的各种构型，然后对这些构型进行平均，得到诸如结构、动力学、热力学等宏观性质。在锂二次电池中我们主要希望获得材料中离子的扩散系数和迁移路径[48]。

由 Fick 第二定律可知

$$\frac{\partial N(z,t)}{\partial t} = D \frac{\partial^2 N(z,t)}{\partial z^2} \tag{13-37}$$

上述方程的解为

$$N(z,t) = \frac{N_0}{A\sqrt{\pi Dt}} \exp\left(-\frac{z^2}{4Dt}\right)$$

扩散系数 D 的定义由爱因斯坦公式给出

$$2dD = \lim_{t\to\infty} \frac{\frac{1}{N}\left\{\sum_{i=1}^{N}[\vec{r}_i(t) - \vec{r}(0)]\right\}^2}{t} \tag{13-38}$$

d 为空间维度，忽略交叉项的平均得到

$$2dD = \lim_{t\to\infty} \frac{\frac{1}{N}\sum_{i=1}^{N}[|\vec{r}_i(t) - \vec{r}_i(0)|^2]}{t} \tag{13-39}$$

虽然分子动力学可以模拟系统粒子随时间的演化，查看粒子迁移的路径，计算粒子的扩散系数以及材料的稳定性，但是粒子运动的规律是牛顿力学，对质量较轻的粒子比如氢

和氦，由于低温存在量子效应因而计算效果不理想[49]。

13.1.2.3 分子动力学在锂电池中的应用

分子动力学不仅可以揭示材料中离子自扩散过程的运动路径、能量势垒，也为我们理解材料中离子的扩散机制提供了可能。Yang 等[50]通过计算 $LiFePO_4$ 的自扩散过程，发现该材料中 Li^+ 的迁移并不是连续发生的，而是如图 13-14 所示通过相邻晶格位置之间逐渐推进的跃迁发生的。Yang 等[50]的模拟不仅发现了 Li^+ 沿着一维方向的"之"字形路径前进的过程，还发现了 Li-Fe 离子的协同运动，这种协同运动导致了 Li/Fe 互占位且有利于 Li^+ 在通道之间的扩散。

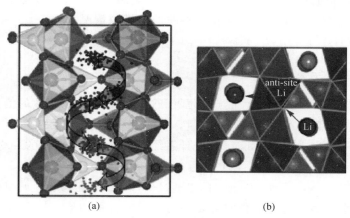

图 13-14　模拟得到 Li^+ 在 $LiFePO_4$ 中的运动轨迹（a）及 Li^+
通过 Li/Fe 互占位缺陷在通道之间的迁移（b）[50]

13.1.3　蒙特卡罗方法

13.1.3.1　蒙特卡罗方法的介绍

蒙特卡罗（Monte Carlo，MC）方法，又称随机抽样或统计实验方法，属于计算数学的一个分支，它是在 20 世纪 40 年代中期为了适应当时的原子能事业而发展起来的，学术界一般把 1949 年 Metropolis 和 Ulam 发表的论文作为 MC 方法诞生的标志[51,52]。

知道物理规律，就能用蒙特卡罗方法模拟出满足物理规律的大量随机事件，由这些随机事件统计出需要计算的物理量的值。有些实验实施非常昂贵或非常危险，还有一些物理过程目前还无法探测，这些都可以用蒙特卡罗方法对真实的过程进行模拟。从数学的观点而言，统计就是求平均，因此蒙特卡罗方法能通过抽样的方法求解定积分。求解积分的统计基础是独立同分布的中心极限定理（林德贝尔格-勒维定理），这个定理说明蒙特卡罗的积分标准方差和维度无关[53]。

如果已知事件的概率分布，需要采用抽样算法来获得满足这样概率分布的样本，以模拟真实的过程。常见的抽样方法有直接抽样法、舍选抽样法、复合抽样法、变换抽样法、近似抽样法、重要分布的随机抽样法和 Metropolis 抽样方法[54]。物理中常用的抽样方法是 Metropolis 抽样，是一种非归一化的抽样方法。详细介绍可见参考文献 [54]。Metropolis 抽样方法的考虑是：并不独立地选择相继各种状态点 x_l，而是构造一个马尔可夫（Markov）过程，每个状态 x_{l+1} 由前一个状态点 x_l 通过一个适当的跃迁概率 $W(x_l \rightarrow x_{l+1})$ 得到（$l = 1, 2, \cdots, m$）。当 $m \rightarrow \infty$，Markov 过程产生的状态的分布函数 $P(x_l)$ 趋于平衡分布。

平衡态的概率分布为

$$P_{eq}(x_i) = \frac{1}{Z} \exp[-H(x_i)/(kT)] \tag{13-40}$$

要达到上式的平衡分布，通常还必须加上细致平衡条件

$$P_{eq}(x_i) W(x_i \rightarrow x_j) = P_{eq}(x_j) W(x_j \rightarrow x_i)$$

$$\frac{W(x_i \rightarrow x_j)}{W(x_j \rightarrow x_i)} = \exp\left(-\frac{dH}{kT}\right) \tag{13-41}$$

即两个状态正向与反向的跃迁概率之比只依赖于两者的能量差 $dH = H(x_j) - H(x_i)$。式（13-41）不能唯一地确定跃迁概率 $W(x_l \rightarrow x_{l+1})$，通常可以选择

$$W(x_i \rightarrow x_j) = \begin{cases} \frac{1}{t_s} \exp[-dH/(kT)] & dH > 0 \\ \frac{1}{t_s} & \text{others} \end{cases} \tag{13-42}$$

t_s 为蒙特卡罗时间。也有其他方式确定选择跃迁概率。

对于正则系综，MC 模拟的基本步骤为：①随机选择一个初始位形；②利用随机数，产生一个新的位形；③计算能量变化 $dH = H(x_j) - H(x_i)$；④若 $dH < 0$，接受新位形并回到第 2 步，否则继续进行下一步；⑤计算 $e^{-dH/(k_B T)}$，并产生一个 [0，1] 之间的随机数 R；⑥如果 $R < e^{-dH/(k_B T)}$，同样接受新位形，并回到第 2 步；⑦否则，保持原位形作为新位形并回到第 2 步。

以上步骤表明，对于一个能量低于原位形的新位形直接接受，而对于能量升高的位形，就按 Boltzmann 概率接受。

13.1.3.2 蒙特卡罗方法在锂电池中的应用

SEI 膜对电池的安全性、循环性、首周库仑效率等有着重要的意义[55]。Methekar 等[56]通过动力学蒙特卡罗模拟充电放电过程中的副反应以及离子扩散等过程，观测到石墨表面 SEI 膜随着循环的次数增加而逐渐形成的过程。图 13-15(a)表示首周循环后负极表面形貌，图 13-15(b)为 100 周循环后的负极表面形貌，红色表示 SEI 膜，品红表示处于原始状态的石墨表面，绿色表示吸附了金属锂的位点[56]。

Ouyang 等[37,57]采用蒙特卡罗的方法模拟了 $Li_x Mn_2 O_4$ 在不同锂含量下材料的平衡态结构，计算了放电过程的电压，并模拟了升高温度后体系放电电压平台的变化（图 13-16），采用蒙特卡罗的方法模拟了外电流和电压的关系，结果符合欧姆定律。

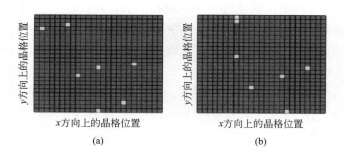

图 13-15 模拟得到的负极表面在首周循环后（a）和 100 周循环后（b）的形貌[56]

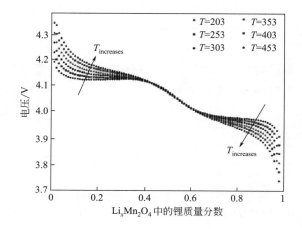

图 13-16 模拟得到的 $Li_xMn_2O_4$ 不同温度电压平台的变化[37]

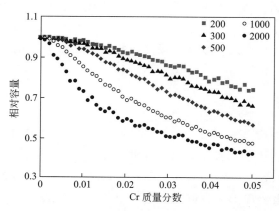

图 13-17 不同 Cr 含量体系的可逆容量（200、300、500、1000、2000 表示晶胞模型的大小）[58]

Ouyang 等[58] 还通过蒙特卡罗方法模拟了 $LiFePO_4$ 中掺杂 Cr 的情况，迁移出截面的 Li^+ 量随 Cr 含量变化，Cr 含量越多，迁出的 Li^+ 越少，电池容量越低，如图 13-17 所示。

基于密度泛函理论的第一性原理计算能从电子结构的层面揭示锂二次电池材料的结构、能带、缺陷、电导率、力学性质等本征特性，分子动力学和蒙特卡罗模拟等方法能在原子尺度研究材料的动力学性能，上述方法都有助于理解材料在电化学过程中表现出来的种种现象。这些计算方法所关注的是材料在空间尺度为 $10^{-10} \sim 10^{-9}$ m，时间尺度为皮秒量级的过程。如果需要考虑尺度在 $10^{-9} \sim 10^{-3}$ m，时间尺度在秒或者毫秒量级的问题时，则需要采用下一节介绍的介观尺度的模拟方法。

13.2 介观尺度的模拟

介观尺度指的是介于原子层次和连续体之间的区域，在这个空间尺度里主要关注的是微观组织随时间发生的演化，如晶体的生长凝固过程等。常用的模拟方法有相场模型方法[59]、分子力场方法[60]等。

13.2.1 相场模型

相场模型可用来处理纳米和微米尺度的结构演化，在该尺度范围内，材料的尺寸、形状、空间分布对材料的物理性质和化学性质起着重要的作用（图13-18）。微结构演化实际是自由能减少的过程，材料的自由能是体相化学能、界面能、弹性应变能、磁性能、静电能以及外加电场作用能的共同贡献。相场模型是建立在界面扩散的基础之上，微结构随时间的演化是连续性方程，即 Cahn-Hilliard 非线性方程和时间依赖的 Ginzburg-Landau 方程[61]。相场方法模拟的介观尺度的现象包括晶粒生长和粗糙化、材料成型、裂纹演化、位错和溶质的相互作用、电迁移和多组分互扩散等过程[62,63]，这些过程在锂二次电池中都可能发生，并直接影响电池的循环性、倍率性、安全性等。

13.2.1.1 相场模型的基本原理

相场方法中，微观结构用一个连续变量序参数 ϕ 来表示。如 $\phi=0$、$\phi=1$ 以及 $0<\phi<1$ 分别代表沉积相、基体、界面。在相场模型中，将总自由能表示为式（13-43）形式[49]

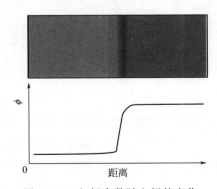

图 13-18　相场参数随空间的变化

$$F = \int \mathrm{d}r^3 \left[f(c_1, c_2, \cdots, c_n, \eta_1, \eta_2, \cdots, \eta_p) + \sum_{i=1}^{n} \alpha_i (\nabla c_i)^2 \right.$$
$$\left. + \sum_{i=1}^{3}\sum_{j=1}^{3}\sum_{k=1}^{p} \beta_{ij} \nabla_i \eta_k \nabla_j \eta_k \right] + \iint G(r-r') \mathrm{d}^3 r \mathrm{d}^3 r' \tag{13-43}$$

式中，f 表示局部自由能密度，是场变量 c_i 和 η_i 的泛函，代表短程的化学相互作用；α_i 和 β_{ij} 表示能量梯度的系数，源于界面能，仅在界面梯度非零时梯度项非零，具有各向异性，梯度积分项代表非局域相互作用，如静电相互作用、偶极相互作用等。

在相场模型中演化方程[59]

$$\frac{\partial c_i(\vec{r},t)}{\partial t}=\nabla M_{ij}\nabla\frac{\delta F}{\delta c_j(\vec{r},t)} \tag{13-44}$$

$$\frac{\partial \eta_p(\vec{r},t)}{\partial t}=-L_{pq}\nabla\frac{\delta F}{\delta \eta_q(\vec{r},t)} \tag{13-45}$$

式中，M_{ij} 和 L_{pq} 是原子或者界面的扩散速度。建立模型后就能根据式（13-44）、式（13-45）导出动力学方程，可将上述微分方程转化为代数方程，也可在实空间离散化[59]。通过相场模型虽然能够模拟晶体生长、固态相变、裂纹演化、薄膜上的相变、离子在界面处的迁移等，但是模拟结果缺少和时间观测量的定量比较，计算时界面厚度的设定通常大于实际情况，导致细节的缺失。

13.2.1.2 相场方法在锂电池研究中的应用

Yamakawa 等[65]通过相场方法模拟层状正极材料 $LiCoO_2$ 不同晶粒大小、不同晶粒取向对 Li^+ 扩散的影响。图 13-19 为模拟得到的在恒流放电时 $LiCoO_2$ 中 Li 浓度的空间分布，显示出 Li^+ 的扩散对晶界、晶体方向的空间分布及晶粒的尺寸都十分敏感。

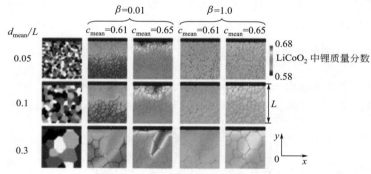

图 13-19　模拟得到在恒流放电条件下 $LiCoO_2$ 中 Li 浓度在不同颗粒尺寸（d_{mean}/L）及不同晶面的扩散系数比值（β）下的 Li^+ 浓度分布[65]

锂二次电池中还有很多现象可以通过相场模拟获得进一步的认识。如在纳米硅负极材料的充放电过程中伴随着材料的体积膨胀，会导致黏结剂的脱落、SEI 重新生成和容量衰减[66]。Zuo 等[64]通过相场模型研究硅薄膜在嵌锂过程中力场分布以及裂纹的演化过程。

13.2.2　分子力学

分子力学，也叫力场方法（force field method），常用于计算化学和生物大分子等的结构和能量，也可以用于研究锂二次电池中的电化学过程。

13.2.2.1　分子力学方法的介绍[60]

分子力学方法中，在考虑几何结构或力学特性等静态性质时，将原子与原子之间的作用视为主要的相互作用。因此在分子力场模型中，把组成分子的原子看成是由弹簧连接起来的球,然后用简单的数学函数来描述球与球之间的相互作用。

分子力场函数一般可以表达为

$$U = U_{bond} + U_{angle} + U_{torsion} + U_{electrostatic} + U_{Van-der-Waals} \tag{13-46}$$

式中,前3项分别代表键伸缩能、键角弯曲能和二面角扭曲能,是描述分子内成键的作用项。后两项为静电作用和范德华作用,是描述分子间的非键相互作用。

构成力场函数体系需要有一套联系分子能量和构型的函数,还需要给出各种不同原子在不同成键状况下的物理参数,如正常的键长、键角、二面角等,都是通过拟合特定分子的数据而生成的,这些被拟合的分子称为训练基(training set)。这些力场参数多来自实验或者量子化学计算。由于力场参数是拟合训练基分子得到的,并且所拟合的实验数据是常温常压下测量的,所以这些参数用于计算其他分子或者高温高压情况时,分子力学的准确性会降低。

13.2.2.2 分子力场方法在锂电池中的应用

Garofalini 等[67]利用分子力场方法研究了锂二次电池里,当金属锂嵌入纳米材料 FeF_2 发生转化反应的作用机理。通过利用合理的经验势函数模型以及相关参数模拟 $FeF_2 + 2Li \Longrightarrow 2LiF + Fe$ 反应过程中的动力学过程,图 13-20 显示了在 FeF_2(001)表面上发生的反应过程[67]。图中 Fe 原子、F 原子和 Li 原子分别用绿色、黄色和红色表示,整个 FeF_2 表面上有 4320 个原子,这 4 幅图分别表示有 72 个、288 个、360 个、864 个 Li 进入 FeF_2 表面进行反应的过程。可以明显地看到随着越来越多的 Li 进入,金属 Fe 团簇和晶体状的 LiF 都在各自形成。

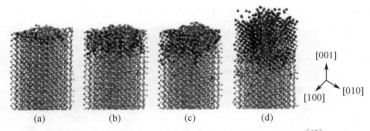

图 13-20 模拟得到 Li 在 FeF_2(001)表面的反应[67]

相场方法和分子力场方法可用于研究微观结构及其演化过程对锂二次电池性能的影响,对于更大空间尺度上($10^{-6} \sim 10^0$ m)的问题,体系的随机性不明显,可以采用下文介绍的基于连续体模型的宏观尺度模拟方法。

13.3 宏观尺度的模拟

在宏观尺度上,体系的运动遵循牛顿力学定律,热运动满足宏观的扩散方程,可以采用的数值模拟方法包括有限元方法[68]、有限差分方法[69]等,其中有限元方法常用来研究锂二次电池中的热流分布[70]、应力分布[71]等现象，本节将主要介绍有限元方法及其在电池材料研究中的应用。

13.3.1 有限元方法介绍

有限元方法（finite element method）是一种求解偏微分方程边值问题近似解的数值技术，如求解电池内部的热、力、电流、电压的分布等都需要求解在特定边界条件下的输运方程。而这些方程的精确求解几乎是不可能的。人们通常采用两种方法去逼近真实的解：全域的展开逼近，典型代表为傅里叶级数展开；基于子域的分段函数展开，如有限元[68]。后者将求解域看成是由许多简单而又相互作用的元素的互连子域组成（单元），对每一单元假定一个合适的（较简单的）近似解，然后推导求解这个总域的满足条件（如结构的平衡条件），从而得到问题的解。这两种方式各有特点，全域的方法采用高次连续函数，通过几个基底函数得到很好的效果。第二种方法要通过较大的工作量才能得到较好的效果。但是能将复杂方程的解化繁为简，使得描述和求解成为可能。

有限元方法模拟的流程可以概括为[68]：首先对求解问题区域离散化，然后确定这个区域的状态变量和变量的控制方法即变量时间空间的演化方程，然后对单个单元进行推导，之后对由单元组装起来的总体进行求解，最后联立方程组求解并获得结果。

13.3.2 有限元方法在锂电池研究中的应用

对于高功率的动力电池而言，电池内部温度分布对其安全性和寿命有重要影响。西安交通大学曹秉刚等[70]在考虑内阻、对流、外部耗散等条件下利用有限元方法模拟了电池工作时内部温度的空间分布，以及电池在155℃热炉中的温度分布（图13-21），模拟结果和VLP50/62/100S-Fe（3.2V/55A·h）LiFePO$_4$/石墨电池的实验观测相符合。

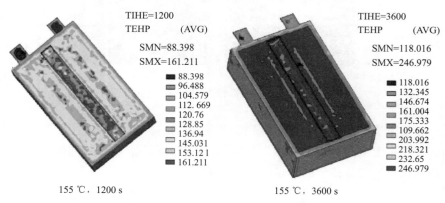

图13-21　有限元模拟在155℃，1200 s和3600 s时电池中的温度分布[70]

Bower等[71]利用有限元的方法研究了嵌锂后硅颗粒内的应力分布。图13-22显示了采用有限元方法模拟的薄膜非晶硅电极在充放电循环过程中的应力分布及形变和断裂过程。

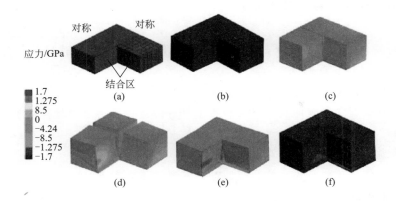

图 13-22　非晶薄膜 Si 电极在充放电循环过程中的断裂、形变和应力分布[71]

宏观尺度的材料模拟着眼于工程领域的问题，可以有助于理解锂二次电池中出现的应力、传热、流动以及多场耦合等宏观现象。

13.4　本章结语

随着各种理论的完善和计算机技术的发展，计算模拟已经成为锂二次电池研究中的重要方法之一。不同时间与空间尺度的模拟方法有助于我们从原子分子层次、介观层次和宏观层次理解材料结构与性能之间的关系。而这些不同尺度上的计算方法可以通过参数传递进行联系，从而实现材料的多尺度模拟[2]，为材料设计提供一种新的途径。通过实验与理论计算相结合的方法，未来锂二次电池研究中有望更深入地理解以下关键科学问题：①设计出新的电极与电解质材料，找到一些目前尚未发现的锂二次电池材料，并且具有更好的性能与更高的安全性，目前正在发展的材料基因组方法是锂电池计算研究方面的重要发展方向；②理解体相电极材料反应的热力学问题，通过精确计算材料的反应生成焓，预测材料在各种条件，如不同温度、压力、组分下的稳定性，甚至建议材料的合成方案；③理解电位与材料组成、微观结构的关系，通过计算与实验得到大量已知结构及新结构所对应的电化学电位，建立电位与结构关系的数据库，采用统计方法，寻找决定电位的因素，理解它们之间的物理关联；④理解电极材料充放电过程中组成与结构的演变，通过计算得到的生成焓数据，建立与电极变化过程有关的相图，分析电极材料结构演变对稳定性的影响，理解影响电极循环性能的因素以及电极材料的失效机制；⑤与电池有关的复杂过程的理解，如输运性质、尺寸效应、界面问题等复杂过程；⑥材料与器件制备过程、服役过程的数字化模拟与仿真。

对锂二次电池中这些基础科学与应用技术问题的理解，以及对全产业链过程的多尺度计算和模拟，最终有望加速锂二次电池以及下一代锂电池材料的开发过程，帮助我们更快找到高性能、高安全性的正极、负极与电解质材料，促进锂电池产业的发展。

本章需要进一步思考的基础科学问题:

1. 电子离域材料体系的费米能级,导带低和价带顶能级位置如何精确计算?如何与电极的热力学平衡电位关联起来?
2. 如何确定阴离子电荷转移的起始电位?
3. 如何计算材料嵌锂和脱锂的反应路径和宿主材料的结构稳定性?材料颗粒的临界破裂条件如何计算?
4. 阶结构能否通过热力学计算预测出来?
5. 如何预测嵌脱锂导致的组成和结构相图?如何预测相转变点?
6. 非晶态材料的嵌脱锂电位如何计算?
7. 能否计算材料的热失控起始反应温度?
8. 能否通过分子动力学重构 SEI 膜的形成过程?
9. 多尺度计算是否通过串联计算,参数传递就可以实现?
10. 如何通过高通量计算、人工智能及大数据分析方法获得电极和电解质材料的构效关系?例如针对离子电导率、电化学窗口、析氧电位等?

参考文献

[1] 坚增运,刘翠霞,吕志刚.计算材料学 [M].北京:化学工业出版社,2012.
[2] Curtarolo S, Hart G L W, Nardelli M B, et al. The high-throughput highway to computational materials design [J]. Nature Mater, 2013, 12: 191-201.
[3] Shi S Q, Gao J, Liu Y, et al. Multi-scale computation methods: Their applications in lithium-ion battery research and development [J]. Chin Phys B, 2016, 25 (1): 018212.
[4] Meng Y S, Dompablo M E. Frist principles computational materials design for energy storage materials in lithium ion batteries [J]. Energ Environ Sci, 2009, 2 (6): 589-609.
[5] Car R, Parrinello M. Unified approach for molecular dynamics and density-functional theory [J]. Phys Rev Lett, 1985, 55 (22): 2471-2474.
[6] Kohn W. Nobel lecture: Electronic structure of matter—Wave functions and density functionals [J]. Rev Mod Phys, 1999, 71 (5): 1253-1266.
[7] 施思齐.锂二次电池正极材料的第一性原理研究 [D].北京:中国科学院物理研究所,2004.
[8] Born M, Huang K. Dynamical Theory of Crystal Lattices [M]. Oxford: Oxford Universities Press, 1954.
[9] Thomas L H. The calculation of atomic fields [J]. Proc Cambridge Phil Soc, 1927, 23 (5): 542-548.
[10] Fermi E. Un metodo statistico per la Determinazione di alcune proprietà dell'Atomo [J]. Rend Accad Naz Lincei, 1927, 6: 602-607.
[11] Hohenberg P, Kohn W. Inhomogeneous electron gas [J]. Phys Rev, 1964, 136 (3B): 864-871.
[12] Kohn W, Sham L J. Self-consistent equations including exchange and correlation effects [J]. Phys Rev, 1965, 140 (4A): 1133-1138.
[13] Slater J C. A simplification of the Hartree-Fock method [J]. Phys Rev, 1951, 81 (3): 385-390.
[14] Ceperley D M, Alder B J. Ground-state of the electron-gas by a stochastic method [J]. Phys Rev Lett, 1980, 45 (7): 566-569.
[15] Perdew J P, Zunger A. Self-interaction corrction to density-functional appoximations for many-electron systems [J]. Phys Rev B, 1981, 23 (10): 5048-5079.
[16] Jones R O, Gunnarsson O. The density functional formalism, its application and prospects [J]. Rev Mod Phys, 1989, 61 (3): 689-746.
[17] Langreth D C, Perdew J P. Theory of nonuniform electronic systems. 1. Analysis of the gradient approximation and a generalization that works [J]. Phys Rev B, 1980, 21 (12): 5469-5493.

[18] Becke A D. Density-functional exchange-energy approximation with correct asymptotic-behavior [J]. Phys Rev A, 1988, 38 (6): 3098-3100.

[19] Perdew J P, Chevary J A, Vosko S H, et al. Atoms, molecules, solids, and surfaces—Applications of the generalized gradient approximation for exchange and correlation [J]. Phys Rev B, 1992, 46 (11): 6671-6687.

[20] Perdew J P, Wang Y. Accurate and simple analytic representation of the electron-gas correlation-energy [J]. Phys Rev B, 1992, 45 (23): 13244-13249.

[21] Perdew J P, Burke K, Ernzerhof M. Generalized gradient approximation made simple [J]. Phys Rev Lett, 1996, 77 (18): 3865-3868.

[22] Tao J, Perdew J P. Climbing the density functional ladder: Nonempirical meta-generalized gradient approximation designed for molecules and solids [J]. Phys Rev Lett, 2003, 91 (14): 146401.

[23] Andersson Y, Langreth D C, Lundqvist B I. Van der Waals interactions in density-functional theory [J]. Phys Rev Lett, 1996, 76 (1): 102-105.

[24] Kohn W, Meir Y, Makarov D E. Van der Waals energies in density functional theory [J]. Phys Rev Lett, 1998, 80 (19): 4153-4156.

[25] Becke A D. A new mixing of Hartree-Fock and local density-functional theories [J]. J Chem Phys, 1993, 98 (2): 1372-1377.

[26] 孙洋. 锂二次电池电极材料中离子传输与相变反应机理的第一性原理研究 [D]. 北京: 中国科学院物理研究所, 2014.

[27] Zhou F, Cococcioni M, Marianetti C A, et al. First-principles prediction of redox potentials in transition-metal compounds with LDA+U [J]. Phys Rev B, 2004, 70 (23): 235121.

[28] Chevrier V L, Ong S P, Armiento R, et al. Hybrid density functional calculations of redox potentials and formation energies of transition metal compounds [J]. Phys Rev B, 2010, 82 (7): 75122.

[29] Wang L, Maxisch T, Ceder G. A first-principles approach to studying the thermal stability of oxide cathode materials [J]. Chem Mater, 2007, 19 (3): 543-552.

[30] Gao Y R, Ma J, Wang X F, et al. Improved electron/Li-ion transport and oxygen stability of Mo-doped Li_2MnO_3 [J]. Journal of Materials Chemistry A, 2014, 2 (13): 4811-4818.

[31] Xiao R J, Li H, Chen L Q. Density functional investigation on Li_2MnO_3 [J]. Chemistry of Materials, 2012, 24 (21): 4242-4251.

[32] Goodenough J B, Kim Y. Challenges for rechargeable Li batteries [J]. Chemistry of Materials, 2010, 22: 587-603.

[33] Lepley N D, Holzwarth N A W, Du Y A. Structure, Li^+ mobilities, and interfacial properties of solid electrolytes Li_3PS_4 and Li_3PO_4 from first principles [J]. Phys Rev B, 2013, 88: 104103.

[34] Shi S Q, Liu L J, Ouyang C Y, et al. Enhancement of electronic conductivity of $LiFePO_4$ by Cr doping and its identification by first-principles calculations [J]. Phys Rev B, 2003, 68 (19): 195108.

[35] Quartarone E, Mustarelli P. Electrolytes for solid-state lithium rechargeable batteries: Recent advances and perspectives [J]. Chem Soc Rev, 2011, 40 (5): 2525-2540.

[36] Ouyang C Y, Shi S Q, Wang Z X, et al. The effect of Cr doping on Li ion diffusion in $LiFePO_4$ from first principles investigations and Monte Carlo simulations [J]. Journal of Physics: Condensed Matter, 2004, 16 (13): 2265-2272.

[37] 欧阳楚英. 锂二次电池正极材料离子动力学性能研究 [D]. 北京: 中国科学院物理研究所, 2005.

[38] Ouyang C Y, Shi S Q, Wang Z X, et al. First-principles study of Li ion diffusion in $LiFePO_4$ [J]. Phys Rev B, 2004, 69 (10): 104303.

[39] Hoang K, Johannes M D. First-principles studies of the effects of impurities on the ionic and electronic conduction in $LiFePO_4$ [J]. J Power Sources, 2012, 206: 274-281.

[40] Ohzuku T, Iwakoshi Y, Sawai K. Formation of lithium-graphite intercalation compounds in nonaqueous electrolytes and their application as a negative electrode for a lithium ion (shuttlecock) cell [J]. J Electrochem Soc, 1993, 140 (9): 2490-2498.

[41] Gu L, Zhu C B, Li H, et al. Direct observation of lithium staging in partially delithiated $LiFePO_4$ at atomic resolution [J]. J Am Chem Soc, 2011, 133 (13): 4661-4663.

[42] Sun Y, Lu X, Xiao R J, et al. Kinetically controlled lithium-staging in delithiated $LiFePO_4$ driven by the Fe center mediated interlayer Li-Li interactions [J]. Chem Mater, 2012, 24 (24): 4693-4703.

[43] Li H, Huang X J, Chen L Q, et al. A high capacity nano-Si composite anode material for lithium rechargeable batteries [J]. Electrochem Solid-State Lett, 1999, 2 (11): 547-579.

[44] Stadler R, Wolf W, Podloucky R, et al. Ab initio calculations of the cohesive, elastic, and dynamical properties of CoSi$_2$ by pseudopotential and all-electron techniques [J]. Phys Rev B, 1996, 54 (3): 1729-1734.

[45] ShenoyV B, Johari P, Qi Y. Elastic softening of amorphous and crystalline Li-Si Phases with increasing Li concentration: A first-principles study [J]. J Power Sources, 2010, 195 (19): 6825-6830.

[46] Frenkel D, Smit B. Understanding Molecular Simulation [M]. Newyork: Academic Press, 2001.

[47] Richard M M. Electronic Structure: Basic Theory and Practical Methods [M]. Cambridge: Cambridge University Press, 2004.

[48] Van derVen A, Ceder G, Asta M, et al. First-principles theory of ionic diffusion with nondilute carriers [J]. Phys Rev B, 2001, 64 (18): 184307.

[49] Chen J, Li X Z, Zhang Q F, et al. Quantum simulation of low-temperature metallic liquid hydrogen [J]. Nature Commun, 2013 (4): 1-5.

[50] Yang J J, Tse J S. Li ion diffusion mechanisms in LiFePO$_4$: An ab initio molecular dynamics study [J]. J Phys Chem A, 2011, 115 (45): 13045-13049.

[51] Metropolis N, Ulam S M. The Monte Carlo method [J]. J Am Statis Asoc, 1949, 44 (247): 335-341.

[52] Niederreiter H. Quasi-Monte Carlo methods and pseudo-random numbers [J]. Bull Amer Math Soc, 1978, 84 (6): 957-1041.

[53] Hoffmann K H, Meyer A. Parallel Algorithms and Cluster Computing: Implementations, Algorithms and Applations [M]. Berlin: Springer, 2006.

[54] Binder K. Applications of Monte Carlo methods to statistical physics [J]. Rep Prog Phys, 1997, 60 (5): 487-559.

[55] Zheng J Y, Zheng H, Wang R, et al. 3D visualization of inhomogeneous multi-layered structure and Young's modulus of the solid electrolyte interphase (SEI) on silicon anodes for lithium ion batteries [J]. Phys Chem Chem Phys, 2014, 16 (26): 13229-13238.

[56] Methekar R N, Northrop P W C, Chen K, et al. Kinetic Monte Carlo simulation of surface heterogeneity in graphite anodes for lithium-ion batteries: Passive layer formation [J]. J Electrochem Soc, 2011, 158 (4): A363.

[57] Ouyang C Y, Shi S Q, Wang Z X, et al. Temperature-dependent dynamic properties of Li$_x$Mn$_2$O$_4$ in Monte Carlo simulations [J]. Chin Phys Lett, 2005, 22 (2): 489-492.

[58] Ouyang C Y, Shi S Q, Wang Z X, et al. The effect of Cr doping on Li ion diffusion in LiFePO$_4$ from first principles investigations and Monte Carlo simulations [J]. J Phys: Condensed Matter, 2004, 16 (13): 2265-2272.

[59] Chen L Q. Phase-field models for microstructure evolution [J]. Ann Rev Mater Res, 2002, 32 (1): 113-140.

[60] Cornell W D, Cieplak P, Bayly C I, et al. A second generation force field for the simulation of proteins, nucleic acids, and organic molecules [J]. J Am Chem Soc, 1995, 117 (19): 5178-5197.

[61] Kobayashi R. Modeling and numerical simulations of dendritic crystal growth [J]. Phys D: Nonlinear Phenom, 1993, 63 (3-4): 410-423.

[62] 贾伟建. 凝固微观组织相场法模拟 [D]. 兰州: 兰州理工大学, 2005.

[63] Long Wenyuan. Phase-field simulations of dendritic growth in aluminum alloy solidification [D]. Wuhan: Huazhong University Science and Technology, 2004.

[64] Zuo P, Zhao Y P. A phase field model coupling lithium diffusion and stress evolution with crack propagation and application in lithium ion batteries [J]. Phys Chem Chem Phys, 2014, 17 (1): 287-297.

[65] Yamakawa S, Yamasaki H, Koyama T, et al. Numerical study of Li diffusion in polycrystalline LiCoO$_2$ [J]. J Power Sources, 2013, 223: 199-205.

[66] Beaulieu L Y, Hatchard T D, Bonakdarpour A, et al. Reaction of Li with alloy thin films studied by in-situ AFM [J]. J Electrochem Soc, 2003, 150 (11): A1457.

[67] Ma Y, Garofalini S H. Atomistic insights into the conversion reaction in iron fluoride: A dynamically adaptive force field approach [J]. J Am Chem Soc, 2012, 134 (19): 8205-8211.

[68] 王勖成. 有限单元法 [M]. 北京: 清华大学出版社, 2008.

[69] 张文生. 科学计算中的偏微分方程有限差分方法 [M]. 北京: 高等教育出版社, 2006.

[70] Guo G F, Bo L, Bo C, et al. Three-dimensional thermal finite element modeling of lithium-ion battery in thermal abuse application [J]. J Power Sources, 2010, 195 (8): 2393-2398.

[71] Bower A F, Guduru P R. A simple finite element model of diffusion, finite deformation, plasticity and fracture in lithium ion insertion electrode materials [J]. Modell Simu Mater Sci Eng, 2012, 20 (4): 45004.

第 14 章
总结和展望

自从摇椅式可充放电锂电池概念由 Armand M 等人在 1972 年提出，锂离子电池的基础研究历经 48 年，在材料体系、电化学反应机理、热力学、动力学、结构演化、表界面反应、安全性、力学行为等方面不断取得更为深入广泛的认识，并最终推动锂离子电池技术发展和成功实现了商业化。锂离子电池面临着电池性能需要全面提升、应用领域需进一步拓宽的强劲需求，因此要求基础研究能够提供创新的、更好的技术解决方案，对锂离子电池材料复杂的构效关系能精确认识，对于电池在制造和服役过程中的失效机制有全面的理解，对各种控制策略的效果能提供可靠的科学依据。同时，锂离子电池的发展也在促进着固态电化学、固态离子学、能源材料、能源物理、纳米科学等交叉基础学科的发展。作为本书的最后一章，本章将对锂离子电池基础研究的科学问题，存在的难点、发展趋势进行总结。

锂离子电池的基本概念，始于 1972 年 Armand 等提出的摇椅式电池（rocking chair battery）[1]，正负极材料采用嵌入化合物（intercalation compounds），在充放电过程中，Li^+ 在正负极之间来回穿梭。正负极材料的研发，是锂离子电池发展的关键，5 位杰出的

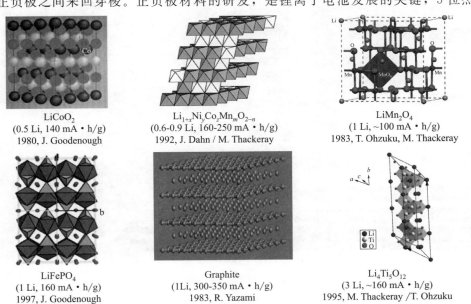

图 14-1　商业锂离子电池正负极材料的示意图、主要发明人、发明时间

科学家在此方面做出了重要的开创性贡献,如图 14-1 所示。Sony 公司的研究人员如西美绪等在 1989 年寻找到了合适的正负极材料、电解质材料的组合,最终推动了以 $LiCoO_2$ 作 Li 源正极、石油焦作负极、$LiPF_6$ 溶于丙烯碳酸酯(PC)和乙烯碳酸酯(EC)作电解液的可充放电二次锂电池,并在 1991 年开始商业化生产,标志着锂离子电池时代的到来[2]。

按照正负极材料的应用和发展,锂离子电池的研发大体可以分为三代,见表 14-1。目前第三代电池在锂离子电池的全部市场中占有比例还较低,全部使用液态有机溶剂电解质。是否还存在第四代锂离子电池,目前尚不清楚。随着第三代锂离子电池的发展,电池充电电压的上限逐渐从 4.25V 开始提升。针对不同的正极材料,充电电压从 4.35V 一直提高到 4.9V。针对 4.9~5V 电压工作的正负极材料、电解质、隔膜、黏结剂、导电添加剂、集流体都需要进一步的研发。

表 14-1 以正负极材料为区分标准的锂离子电池代际划分

代际	正极	负极	开始时间
第一代	$LiCoO_2$	针状焦	1991 年
第二代	$LiMn_2O_4$, $LiNi_{1/3}Co_{1/3}Mn_{1/3}O_2$ $LiFePO_4$	天然石墨 人造石墨 钛酸锂	1994 年
第三代	高电压 $LiCoO_2$ $LiNi_{x\geqslant 0.5}Co_yMn_zO_2$ $LiNi_{0.8}Co_{0.15}Al_{0.5}O_2$ $LiFe_{1-x}Mn_xPO_4$ $xLi_2MnO_3\text{-}Li(NiCoMn)O_2$ $LiNi_{0.5}Mn_{1.5}O_4$	软碳 硬碳 SnCoC SiO_x Nano-Si/C Si-M 合金	2005 年

图 14-2 展示了一个材料体系逐步扩展的发展趋势。

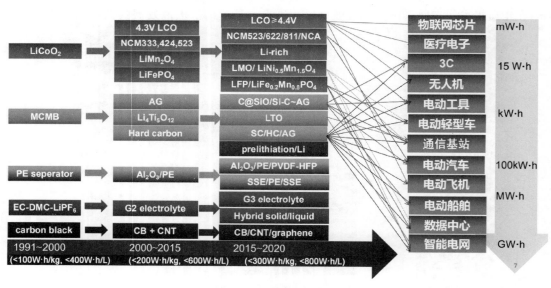

图 14-2 锂离子电池的材料发展过程示意图

14.1 锂离子电池中涉及的学科领域

作为一个电化学储能器件，锂离子电池是固态电化学与非水有机电化学的研究对象。锂离子电池中涉及离子在固体电极、界面中的储存与输运，这是固体离子学的重要内容。锂离子电池中采用了金属、无机非金属、有机物、聚合物等多种材料，涉及材料化学、固体化学、化学工程等领域。锂离子电池采用固体电极，也有用到固体电解质，因此也是固体物理研究的对象。图14-3小结了电化学储能（电池）研究的内容和涉及学科。可以看出，锂电池研究涉及24门一级或二级学科，从文章数量看，排在前5位的是化学、材料、电化学、能源、物理。

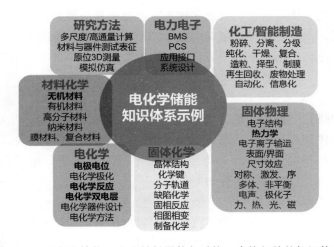

图 14-3 电化学储能（电池材料器件与系统）直接相关的知识体系

14.2 锂离子电池中基础科学问题讨论

锂离子电池中的基础科学问题主要包括材料体系、电化学反应机理、储锂过程热力学、动力学、结构演化、表界面反应、安全性、力学性质等，在本书中已分别阐述。本节强调锂离子电池基础研究相对于其他研究领域，特别是相对于经典溶液电化学的特色。

14.2.1 固态电化学

与经典的电化学研究体系采用惰性电极、电化学反应主要发生在界面和双电层不同，锂离子电池中的电极材料参与电化学反应时具有以下特点。

(1) 结构演化

在充放电过程中，表面、局部、体相晶体结构会发生演化，可能出现局部结构扭曲、新相、相分离、有序无序转化、断裂等。在不同的充放电速度下，可能会出现不稳定的亚稳态相结构。锂离子会从内部脱出或从外部进入，有时其他阳离子、阴离子、溶剂化阳离子也会进入负极或正极材料。

(2) 电荷转移

电极材料内部的某个或多个活性元素，通常是可变价过渡金属，有时也包括阴离子，参与电荷转移，实现电荷补偿。电荷转移不仅仅发生在电极与电解质界面，有时也会发生在电极内部；电荷转移在空间上未必是均匀分布，响应时间上有可能比典型液相电化学体系慢。

(3) 空间电荷层

锂离子电池电极材料包括半导体或绝缘体，在电极表面一般会覆盖固体电解质界面层，有时电池中采用固体电解质。因此在锂离子电池电极与电解质材料的固相区域有可能存在着空间电荷层，同时在液相区域有电化学双电层（浓电解质下的空间电荷层）。在充放电过程中，由于电场作用下锂离子要穿越界面嵌入脱出，因此会显著影响平衡态下的空间电荷层与双电层。

(4) 混合离子导体

固体电化学中研究电极或者是金属导体，或者是半导体。锂离子电池的电极材料要求电极同时是电子导体和离子导体。混合离子导体中，电子与离子的输运、储存过程伴随着复杂的相互作用，包括电子与电子、电子与晶格、电子与离子、电子与缺陷、离子与离子、离子与晶格、离子与缺陷等多种相互作用。

(5) 复杂反应产物

一般的溶液电化学产物为可溶物或沉积物。锂离子电池中，反应产物可以是与反应物同结构的连续固溶体，或是与反应物结构类似的另一相材料（如 $LiFePO_4$ 转变为 $FePO_4$，$Li_4Ti_5O_{12}$ 转变为 $Li_7Ti_5O_{12}$），也可以是分解产物（如 MnO 嵌锂后转变为 Li_2O 与 Mn 的两相共存）。与此同时，有机溶剂与锂盐还容易分解组成与结构复杂的固体电解质层，覆盖在电极材料表面。

(6) 物理化学性质的变化

锂离子的嵌入会引起材料晶体结构的变化，电子的注入会引起材料电子结构的变化。大量研究表明，在充放电过程中，电极材料的能带结构、电子输运和离子输运特性、介电性、磁性、应力、传热、光学等特性会发生明显的变化。通过对物性变化的测量，可以获得材料电子结构、晶体结构及其演化的信息，也可以根据物性随嵌脱锂变化的特点开发新的器件。

14.2.2 复杂的构效关系

实际应用的器件需要同时满足多项技术指标的要求，最终走向商业化应用的材料和器件，每一项指标应该能满足应用的最低要求，所有的设计都是为了避免出现各类问题。具体应用时突出若干主要的性能，寻求综合性能的平衡。

图 14-4 展示了采用液体电解质的锂离子电池电极材料需要考虑的构效关系。除了电极材料，在电池中，还需要考虑电极活性层内活性物质、导电添加剂、黏结剂的组成、比例、分布均匀性，极片的压实密度、电极孔隙率、颗粒之间的孔径大小、孔结构的连通性及其分布均匀性等。有时电极设计成梯度分布，从靠近电解液侧到集流体侧，可能有某种梯度分布或分层设计，这些结构特点也会影响到电极的电化学特性。

锂离子电芯中，除了正负极材料，还包括电解质材料、隔膜、集流体、封装材料、PTC 等。每一种材料均存在各自的构效关系，不再一一赘述。

在锂离子电池的基础研究中，往往主要关注其中的少数构效关系。在应用时，则须尽可能全面地了解电池中各类材料及材料之间的构效关系，而电池中的构效关系往往呈现出复杂的多对多关系，且相互影响，有时呈现出非线性特点。这经常会导致基础研究看起来性能突出的新材料体系在实际应用中往往并不占优势，甚至无法使用，主要的原因是未对材料的性质与性能进行全面考察，有一种或多种性能无法满足最低要求。经过多年的基础研究，仅有图 14-1 所示的 6 种正负极材料获得应用，可见锂离子电池研究开发的难度，主要的原因是满足各项性能要求、具有综合性能优势的材料难以寻找。而且材料发明之后需要仔细调整材料的多种物理化学性质，以期最后能够应用。除了在理解复杂的多对多构效关系方面存在巨大的挑战，锂离子电池的基础科学研究方面，至少还有如下文提到的难点。

体相晶体结构	可逆比容量
体相晶体度	不可逆比容量
体相平均组成	质量能量密度
组成空间分布	体积能量密度
表面结构	质量功率密度
表面组成	体积功率密度
理论密度	容量衰减速率
振实密度	能量衰减速率
球形度	电压衰减速率
颗粒流动性	功率衰减速率
杂质含量	初始库仑效率
电子电导	每周库仑效率
体相化学扩散系数	倍率特性
输运电阻	脉冲功率
电荷转移电阻	响应速率
相变速率	产气速率
SEI膜组成结构覆盖度	自放电率
应力分布	安全性
表面浸润性	环境适应性
热导率	全寿命周期成本

图 14-4 液态锂离子电池电极材料需要考虑的性质与性能之间的构效关系

14.3 锂离子电池共性基础科学问题研究难点

Goodenough 先生撰写了 "Challenges for Rechargeable Li Batteries" 的文章，讨论了锂电池中 HOMO/LUMO、电位控制、SEI 等问题[3]。本节从个人的理解角度，探讨了基础研究的难点。

14.3.1 SEI 膜

在本书中，讨论了锂离子电池中的 SEI 膜问题。SEI 膜一般为无定形结构，通常包含 Li_2CO_3、LiF、LiOH、ROLi、$ROCO_2Li$、低聚物、高聚物、正极溶出物、杂质等组成。在电极中，覆盖在活性物质颗粒、导电添加剂、黏结剂的表面。SEI 膜厚度可以从 2nm 到几微米，从靠近电解质侧到集流体侧往往分布不均，有时还沉积在隔膜上。SEI 膜的组成与微观结构受电池中电极材料、电解质、电解质中添加剂与杂质、充放电制度、深度、温度、寿命的影响，不断发生演化。两种以上活性材料共存时，一种材料的SEI膜可能会受到另一种材料的SEI膜的影响。上述因素导致对SEI膜的定性与定量的表征非常困难。SEI膜的性质对多种电化学性能有直接的影响，控制SEI膜的生长，使其致密、完全覆盖、厚度薄、充放电与高温储存中稳定、能适应正负极材料在充放电过程中的体积变化而不发生断裂、离子电导率高、电子绝缘且有一定的自我修复特点，是应用研究追求的目标。这方面的优化是建立在大量经验的基础上，现有的关于SEI膜研究的数据存在片面性、可靠性低的问题。SEI膜生长过程的理论模拟可以通过第一性原理计算判断反应路径和产物，通过Force Field及分子动力学模拟方法，获得溶剂分解、沉积的结果及动态演化。

14.3.2 结构演化

锂离子电池正负极活性材料的结构包括体相和表面的晶体结构、结晶度、缺陷、形貌。在嵌脱锂过程中，晶格中锂与锂空位的比例不断变化，其在空间排布的无序或有序导致出现固溶体或新相。有时骨架阳离子与阴离子也会随着锂的嵌入脱出发生位移，导致晶格扭曲或新相产生。结构的演化，来自于锂的脱出或嵌入，首先发生在材料的表面。如果锂离子从颗粒表面向内部晶格扩散速率较慢，就会出现过渡态、表面与内部相不一致、局部相分离、局部结构扭曲等现象。结构演化是动态过程，平衡态与非平衡态时结构可能不一致。结构演化涉及相边界的产生和移动。相边界结构大多在原子到几个纳米尺度，因此研究起来较为困难。此外，类似石墨嵌锂产生的系列"阶"结构（staging）及不同阶之间的结构演化，其机制和动态过程在原子尺度还不清楚。有些电极材料的嵌锂与脱锂的反应过程还会出现不对称反应路径的现象，这既与热力学因素有关，也与不对称动力学因素有关。

目前的X射线衍射与中子衍射，主要获得平均结构演化的信息，新相的检出需要较高的结晶度和一定的体积分数（>1%～5%）。选区电子衍射可以获得纳米及亚微米区域的晶体结构信息，高分辨球差电镜可以获得表面与近表面区域的原子尺度的结构信息。NMR与EXAFS可以获得平均局域结构、局域配位的信息。在高功率锂离子电池研究中，了解非平衡态材料表面与体相结构的演化，对全面掌握电极材料性能衰减和性能改进具有重要的意义，也有相当的难度。此外，结构演化的理论计算目前主要还是在热力学方面，通过寻找优化的处于基态的弛豫结构，来判断不同嵌锂量时的平衡态结构。从一个相到另一个相，在原子尺度上理解离子和电子是如何通过扩散、重排、电荷转移实现相变过程的动态模拟，尚未能实现。

14.3.3 多尺度复杂体系输运

锂离子电池的电极是多孔粉末非均相电极，颗粒存在一定的尺寸分布、结晶取向，活性材料往往经过表面包覆处理，活性材料颗粒之间分布着导电添加剂，黏结剂在分子水平上粘接着活性颗粒、导电添加剂与集流体。颗粒之间吸附着电解液。电解液还原时会产生气体，吸附在颗粒上或滞留于颗粒孔隙间。电池的充放电过程，涉及电子从集流体向电极层颗粒表面及内部的传输，离子从电解液相朝着电极层、活性颗粒内部传输。部分骨架离子也因为结构不稳定在骨架内迁移。离子的输运是原子尺度的迁移，电池的充放电是宏观尺度，因此，锂离子电池充放电过程中涉及多尺度非均相介质中的混合电子离子输运，而且输运过程中还伴随着表面的电化学与化学的副反应及离子与电子的储存过程。目前的电化学模拟已经基本能通过整合不同尺度的计算方法实现多尺度复杂输运过程的模拟，从而为电源管理软件的可靠性提供准确的科学基础。但是，实验上对于这一复杂体系，分别测到电子、离子在各相中的输运特性，并且分离在空间中每相的动力学参数，仍然具有较大的挑战。在本征输运参数与宏观动力学方面建立起可靠的依赖关系并非易事。在基础研究中，经常看到根据材料的一两个动力学参数来预测该类材料甚至电池的充放电速率，缺乏严谨的科学依据。但需要指出的是，由于锂离子电池速率控制步骤一般是离子在固相中的扩散，因此根据该材料的化学扩散系数来判断该材料是否能够大倍率充放电，能够达到或接近理论容量，在仅受扩散控制时具有一定的可信度。

14.3.4 材料表面反应

SEI 膜的形成是由于电极材料表面发生的电化学反应引起的。电池在高温搁置时，正负极材料表面的 SEI 膜、氧化物包覆层以及过渡金属，可能会部分溶出到电解质中。过充时，正极表面可以发生电解质氧化分解反应，产生气体，形成正极表面的 SEI 膜（电镜照片有时不能直接观察到 SEI 膜，可能是正极表面的 SEI 膜不稳定，或者电化学氧化分解产物不易沉积在正极颗粒表面）。负极表面可能析出金属锂。正负极材料在充放电过程和储存过程中，表面往往会出现与体相不一致的新相。此外，由于目前锂离子电解质中 $LiPF_6$ 遇到痕量水会产生 HF，HF 会进攻正负极材料的表面，导致表面氧化物会逐渐转化为氟化物。除了正负极活性材料表面发生的反应，导电添加剂、集流体的表面反应也很重要。对上述表面反应及反应产物的定量检测识别与准确认识具有相当的难度。

14.3.5 高倍率问题

锂离子电池分为能量型与功率型，能量型电池往往无法高倍率充放电，而大量的应用十分渴望电池能够在 10min 以内完成 100% 或至少 80% 的充放电。电池的高倍率问题涉及电极过程动力学以及高倍率下材料结构稳定性、电池放热、安全性等。电池的倍率特性与电极材料本征的电子电导、电接触电阻、表观化学扩散系数、电极活性材料平均粒径大小、电极厚度、电解液电导率有关。调控上述特征，有望提高电池的充放电倍率。在锂离子电池中，电解质相的离子输运电阻往往远小于电极材料表面和内部的离子输运电阻。高倍率下，在电极活性材料表面形成很高的电场梯度和浓度梯度，导致活性材料表面结构易于被破坏，使得电池循环性降低。能否高倍率充放电有时不完全取决于动力学因素，而是决定于是否会造成显著的结构破坏。

锂离子电池石墨类负极材料，脱锂过程的化学扩散系数高于嵌锂扩散系数，锂离子电池表现出可以快放不能快充的非对称动力学特点。在高倍率下，不同电极材料充放电过程中电极反应动力学参数是否对称需要系统研究。

14.3.6 正负极材料的电压调控

提高电池能量密度的有效途径之一是提高电池的平均工作电压。因此希望正极材料嵌脱锂电压高，负极材料嵌脱锂电压尽可能接近或略高于金属锂沉积电位，这与材料的电子结构有关，而电子结构取决于晶体结构。

对于相转变反应和电子局域的体系，材料与电池的电压可以通过查找热力学手册或第一性原理计算获得反应物与产物吉布斯生成自由能，然后通过能斯特方程较为精确地估计。此类材料的嵌脱锂电位无法通过掺杂进行显著的调控，形成固溶体后对应的氧化还原电位会有微弱变化，如 $Li(Fe_{1/4}Mn_{1/4}Co_{1/4}Ni_{1/4})PO_4$，$Li(Fe_{1/2}Mn_{1/2})PO_4$，可以理解为局域的电子相互作用引起。此类材料嵌脱锂的过程不会引起电极电位的变化，充放电曲线开路电压一般显示平台特征（如果没有明显的尺寸效应的话），典型的体系如相转变反应 MnO 负极、$Li_4Ti_5O_{12}$ 负极、$LiFePO_4$ 正极等。

对于电子离域的体系，如层状嵌入氧化物，脱锂时电极电位连续变化，可以采用点阵气体模型、平均场和密度泛函（为了精确计算其内能，在选择势函数时需要考虑交换关联能，一般采用 LDA+U 或 GGA+U 来计算其内能变化）来模拟。此类材料在嵌脱锂时可以引起电位的大范围变化，典型的正极材料如富锂层状结构正极材料 $[xLi_2MnO_3-yLi(NiCoMn)O_2]$，其嵌脱锂电位在 4.8~2.0V。实际应用时希望电极电位连续但小范围变化，这与能带结构中价带顶的电子态密度分布（DOS）有关系，而 DOS 与材料中晶体场导致过渡金属 d 轨道能级分裂的范围以及电子之间的相互作用有关。目前似乎还未形成系统的理论，可以准确预测未知的此类材料的 DOS 特征，并根据这一特征寻找合适的正极材料，这还需要后续进一步研究，较好的办法是借助于高通量计算，但前提是能够确定计算方法可以准确描述离域材料的电子结构。为了提高正极材料能量密度，考虑到目前电解质允许的充电电位范围低于 4.3V，较为理想的情况是正极材料在接近或低于 4.3V 时脱嵌锂，放电电压保持不变，这样可以获得高的能量密度。如前所述，相转变反应的材料可以不断脱出锂而不发生 E_F 价带顶降低的情况，符合这一要求的是 $LiMnPO_4$，但该材料的可逆容量为 170mA·h/g，能量密度不高。

掺杂可以显著影响电极电位，但目前还没有特别明确的结论。一般认为电负性更强的氟离子部分取代氧会提高电位。离域电子结构的材料中，阳离子掺杂应该可以调控材料的电极电位。尽管已有大量的实验和理论的努力，由元素掺杂引起的电子结构的变化还缺乏规律性的认识，需要进一步总结和研究。此外，同一种材料，不同的晶体结构，电压也会有显著的差距，最典型的例子是 $LiFeSO_4F$。

提高电压可以使层状化合物脱除更多的锂，显著提高正极材料的容量。但一般电压高于 4.3V 时，碳酸酯类电解液容易在正极材料表面发生氧化分解，产生气体；晶体结构中的氧可能会脱出。因此提高材料的充电电压，需要兼顾材料结构稳定性和电解质电化学窗口，导致问题更为复杂。

14.3.7 电荷有序

锂离子电池正负极材料在充放电的过程中经历了电子注入或逸出，引起骨架中具有可转移电子的原子得失电子。如果一种原子在材料中存在不同的氧化态，则有可能会出现有序分布的状态。嵌脱锂后电荷的有序或无序分布在电极材料中观察和研究的还较少。$Li_4Ti_5O_{12}$ 材料中的所有 Ti 均为四价。当三个电子和锂离子迁入后，形成 $Li_7Ti_5O_{12}$。高分辨球差电镜下的电子能量损失谱研究发现，其中 2 个 Ti 处于四价（Ti_a 原子柱），3 个 Ti 处于三价（Ti_b 原子柱），并且在空间中呈现出交替分布[4]。电荷有序会引起局部结构扭曲，典型的例子是 $LiMn_2O_4$ 中的 Mn^{3+} 与 Mn^{4+} 的电荷有序。由于协同 Jahn-Teller 扭曲效应，在 283.5 K 出现尖晶石（Fd-$3m$）（无序态）向正交结构（$I_{4/1}amd$）（电荷有序态）转变的现象，其原因是低温下出现电荷有序，引起局部结构周期性变化[5]。电荷有序出现的原因被认为与静电相互作用有关。电荷有序态或无序态对充放电过程的影响还不太清楚。

14.3.8 离子在固体输运中的驱动力

载流子输运的驱动力一般包括浓度梯度、电场梯度、温度梯度和对流，后两者较少出现在电池中。锂离子在电池中的传输包括在体相、表面、界面的传输。锂离子电池液体电解质一般为高浓度的电解质（约 1mol/L），在电解质相中没有浓度梯度，离子在电解质相中的驱动力是电场梯度。电场梯度作用在溶剂化的离子上，离子的输运服从欧姆定律。在固体电解质中，受结构因素的制约，在晶格内不会出现明显的浓度梯度；可以猜测，外电场作用在较大尺寸的固体电解质中时，晶格内的离子感受到的外电场较弱，很可能是电场作用在固体电解质的表面或界面，离子受电场梯度的作用，通过 kick-off 机制，从一侧接力传递到另一侧。但有意思的是，一般通过可逆电极实验测到的固体电解质的离子电导率的 I-V 曲线服从欧姆定律，并没有出现复杂的情况。在电极材料中，有的电极是导体，内部是等势体，因此离子传输的驱动力主要也应发生在界面或表面，上述考虑还缺乏实验验证和理论推导。Joachim Maier 一直致力于用空间电荷层理论解释异质结固体中离子的输运，在实际体系中的应用还相对较少。离子在电极或电解质材料中输运的微观机制、离子驱动力的微观作用机制，还存在一些未知的问题，相关的实验验证也比较缺乏。

14.3.9 寿命预测与失效分析

锂离子电池开始应用在电动汽车、规模储能、分布式储能、智能建筑上，这些应用需要较为准确地预测电池的寿命。相对简单的办法是精确地测定电池的容量、电压随着循环次数的衰减，然后外推制定循环次数后的容量；或者通过在高温、高倍率下测量电池的容量衰减速率，对比室温下的衰减速率，进行加速老化测试；或者把两者结合起来。寿命预测，还显著取决于电池的品质。高质量、一致性好、可靠性高的电池容易进行寿命预测。

锂离子电池的寿命衰减与多个因素有关，包括电极材料表面、内部晶体结构逐渐发生不可逆变化，有效化学组成钝化，颗粒出现裂纹，电子接触变差，电解液量减少，SEI 膜不断生长，极片从电极脱落，内部短路等。可以通过对电池的动力学参数（电压、电阻、容量、倍率等）测量获得电池动力学的信息，间接理解电池的失效机制，再通过多种测试手段，对材料、极片及电芯的物理、化学性质进行系统的分析，以便理解电池的失效。但是锂离子电池中的构效关系是多对多关系，物性测量的实验结果如何能唯一地解释某一种失效原因，需要大量的基础研究。随着动力电池、储能电池的发展，对于电池失效分析的研究将更加深入广泛。

14.3.10　材料的可控制备

锂离子电池中涉及金属、半导体、绝缘体、无机非金属、有机溶剂、锂盐、聚合物等多种形式的材料，为了兼顾热力学、动力学、稳定性、电极制造方面的性质要求，材料往往设计成复杂的核壳结构、多层结构、梯度结构等，并且需要由表及里控制元素、组分的组成，控制材料的结晶度、结晶取向，控制形貌、尺寸、孔结构。锂离子电池对杂质的控制尤其严格。为了降低成本，应该采用工序简单、可规模制备、易于工业放大的合成路线，选择最廉价合适的前驱体，制备过程尽量绿色环保低排放或无排放，实现原子经济。这就要求基础科学研究能够针对每类材料，发展出可控制备的方法，这经常是具有挑战性的问题，材料的质量品质不好实际上与对合成制备过程中的反应、变化过程没有充分的了解和精确的控制有关。

此外，目前第一性原理计算还无法准确预测含多种元素非整比化合物的物理化学性质，因此采用组合化学、高通量制备有助于加快研发速度。锂离子电池的材料研发已经引入了高通量制备，较为知名的例子包括 Jeff Dahn 课题组采用 64 通道组合化学的方法研制锡合金与硅合金负极以及 Wildcat 公司采用高通量方法优化三元材料。

此外，绿色化学、过程强化、制造全流程的数值模拟也逐渐开始引入到锂离子电池材料制备的研发和生产中。

14.4　锂离子电池基础研究发展趋势讨论

14.4.1　创新驱动

上述基础科学问题的深入研究的确引人入胜，有利于理解电池的行为，增加了大量新的知识。电池研究开发的目的最终还是获得应用，并且开发出更高、更快、更可靠、更安全的电池。改善电池电化学性能的关键是开发新的材料、新的材料体系组合、新的可逆储锂机制、新的材料改性办法。这方面的研究构成了锂离子电池基础研究的最活跃的部分。表 14-1 罗列出了第三代锂离子电池的正负极材料，再加上电池中其他各类电池材料，研究清楚每一种电池材料的物理化学性质、电化学性质、复杂的构效关系、材料之间的匹配行为，仍然需要大量的基础研究工作。

受体积膨胀因素的制约，硅负极材料的可逆容量在实际电池中远远无法达到理论容量，一般复合材料的比容量为 $450\sim600\mathrm{mA\cdot h/g}$。

为了提高电池的能量密度，有些学者提出了"Beyond Li-ion batteries""Post-Li-ion

batteries"的说法或口号，发展各类金属锂电池，如图14-5所示。其他电池的研究，如Mg、Al、Zn、Na电池的研究也逐渐活跃。从能量密度考虑，对于同一类电化学反应，锂体系的质量能量密度高于其他金属体系，Mg、Al的质量能量密度、体积能量密度也非常高，参见第1章。金属锂电池的发展，需要研究锂枝晶、孔洞的形成及界面与电解液的副反应，与之匹配的固体电解质、或固体电解质修饰层。

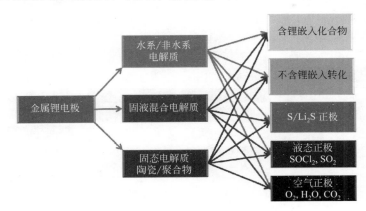

图14-5　可充放金属锂电池的种类

从能量密度发展角度考虑，未来电池发展趋势是第三代锂离子电池、可充放电金属锂电池、可充放电锂硫电池及可充放电锂空气电池，见图14-6。图14-6中的数据是根据热力学数据计算获得。从图14-6中可以看出，以金属锂为负极，嵌入化合物为正极的可充放金属锂电池体积能量密度最高。不考虑非活性物质，Li/H_2O体系的质量能量密度最高。

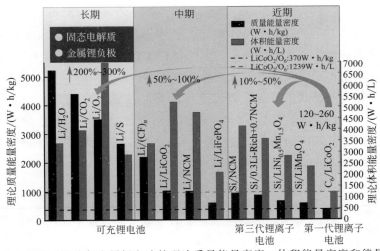

图14-6　可充放电锂离子电池与金属锂电池的理论质量能量密度、体积能量密度和能量密度提升路线

由于电池中非储能物质的存在，电池实际能量密度与理论能量密度还有显著差距，比例从10%～60%不等，不同的电池体系该比例不同[6]。电池中活性物质占的质量分数、体积分数越大，实际能量密度就越高。

采用金属锂电池，由于电解质、正极的不同，可以有多种设计。以锂空气电池为例，选择电解质、空气电极、金属锂负极、气体种类、电池形式、密封形式、温度中的一种要素组合，将会产生不同的锂空气电池，见图14-7。目前的基础研究主要对非水体系锂空气电池开展了大量研究。锂空气电池还有大量未知的问题和技术需要研究。

1.电解质	2.空气电极	4.气体	5.电池结构
水系/酸碱	多孔碳基载体	O_2	圆柱插芯
非水有机/添加剂	多孔金属载体	H_2O	圆柱卷绕
离子液体	多孔导电氧化物	CO_2	管式/微管式
非对称电解液	多孔导电氮化物	空气	纽扣式
聚合物固体电解质	多孔复合载体	其他气体	方型叠层
无机固体电解质	3.金属锂负极	6.密封形式	半液流
凝胶电解质	纯金属锂膜	密闭体系	双液流
聚合物无机电解质	锂合金薄膜	开口/纯氧罐	7.工作温度
固液复合电解质	双面保护锂电极	开口/透氧膜	<60℃
多层电解质	锂单面修饰	完全开口	>60℃

图14-7 锂空气电池的设计要素考虑

固态可充放电金属锂电池、锂硫电池也有各自的设计体系要素，在此不再重复。每类金属锂电池基本相似的地方，主要是理解和控制金属锂与电解质的界面，电解质与正极的界面。图14-7并未列举所有的可能的可充放金属锂电池。正极有很多选择，最后哪些电池能获得商业应用，需要综合考虑技术指标体系是否能满足实际需求。

14.4.2 指标驱动

电池技术的发展，某种程度上是不断提升技术指标的过程。表14-2列举了6种电化学储能器件的技术指标，每种储能技术，由于材料体系与设计不同，技术指标有一定的范围，而且每种技术还在不断发展。因此实际考虑每类储能技术的指标，需要仔细地了解具体的电池体系及其发展现状，不能一概而论。从表14-2来看，锂离子电池的能量密度最高。从14.4.1节的讨论可以看出，锂离子电池与可充放电金属锂电池的能量密度还有很大的提升空间。而能量密度的提高，意味着单位能量或功率的价格的下降，甚至锂离子电池价格几乎以每年30%的速度在下降。

表14-2 6种电化学储能器件的技术指标体系对比

储能技术	比能量/(W·h/kg)	比功率/(W/kg)	循环寿命/次数	单体电压/V	服役寿命/a	能量效率/%	自放电率/(%/月)	库仑效率/%	安全性	成本/[元/(W·h)]	工作温度/℃
液态锂离子电池	90~260	100~20000	1000~2×10⁴	3~4.5	5~15	90~95	<2	约95	中	1.5~10	-20~55
铅酸电池	35~55	75~300	500~5000	2.1	3~10	50~75	4~50	80	好	0.5~1	-40~60
镍氢电池	50~85	150~100	1000~3000	1.2	5~10	50~75	1~10	70	良	2~4	-20~60

续表

储能技术	比能量/(W·h/kg)	比功率/(W/kg)	循环寿命/次数	单体电压/V	服役寿命/a	能量效率/%	自放电率/(%/月)	库仑效率/%	安全性	成本/[元/(W·h)]	工作温度/℃
超级电容器	5～15	1000～10^4	5000～10^5	1～3	5～15	95～99	>10	99%	好	40～120	-40～70
钒液流电池	20～40	50～140	5000～10^4	1.4	5～10	35～82	3～9	80%	好	6～20	10～40
钠硫电池	130～152	90～230	4000～5000	2.1	10～15	75～90	0	约90	良	1～3	300～350

每种电池有其各自的技术特点，例如，镍氢电池目前循环寿命达到了5000次以上，且能在低温高倍率充放电。每种电池应该能找到最适合的目标应用。不同的储能技术针对相同的应用市场，则需要综合评价技术经济性指标。从技术发展的角度考虑，每种电池体系也应不断提升技术指标，拓宽应用范围。

单就锂电池而言，已经有很多的种类。在单一指标方面，各自显示了不同的特点，参考图14-8。如何不断提高电池每一项的技术指标，如何从其他电池实现最高技术指标的技术途径中获得启发，是基础和应用技术开发研究的重要工作。

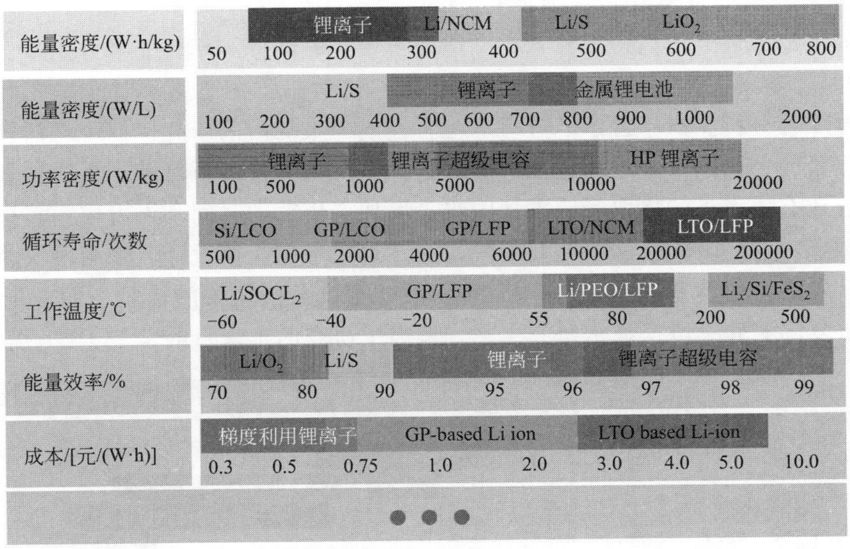

图14-8 不同锂电池的指标水平

HP Li ion—高功率锂离子电池，目前水平是20 kW/kg；LTO—钛酸锂负极；GP—石墨；NCM指各类三元正极材料；LFP—磷酸铁锂；LCO—锂钴氧。

这些指标水平可能会进一步提高

电池技术水平的提高，是综合指标的提升。每一种需求，提出了不同的指标体系，如图14-9所示。应用需求与电池技术的发展互动，不断提出更高的要求。掌握最先进技术同时能控制好成本的企业，在市场上产品最有竞争力，拥有知识产权，拥有产品定价权。指标驱动可以说是对于锂离子电池与金属锂电池的研发人员技术开发最大的驱动力。为了提高电池性能，需要全面优化关键材料、改进电池设计、改进电池制造装备水平。

14.4.3 方法驱动

由于锂离子电池在人类社会、工业生产、国家安全方面的重要性与影响日益提升，而且存在着非常多的技术解决方案，使得传统上作为电化学研究的一个子领域吸引了凝聚态物理、化学化工、电力电子、材料科学与技术、先进制造、尖端仪器、交通运输、先进能源等多个学科领域的科学家和工程人员的参与。可以明显看到，锂离子电池基础科学研究仪器水平不断提升，几乎各类先进科学仪器都在锂离子电池的研究中出现；针对锂离子电池的研究、制造也开发了许多专门的仪器设备。图14-10展示了不同空间分辨率的表征方法，发展原位

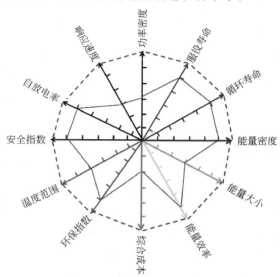

图14-9 电池指标体系12臂蜘蛛图及在能源领域的应用
（图中连线没有实际意义）

与非原位的表征方法，提高检测技术的空间分辨率、时间分辨率、能量分辨率（化学信息和电子结构），获得三维空间的组成、结构、物理化学性质在锂离子电池充放电过程中的演化，通过综合分析测试平台系统分析电池特性成为最前沿的研究。但可想而知，随之到来的是数据量的急剧增长，这一过程中必然会产生对研究帮助不是特别大、特别重要的大量数据，导致有效信息的提取凝练更加困难，增加了研究的时间和人力成本。岳飞曾经说过："运用之妙，存乎一心"。研究过程中，应该首先考虑简单设计，采用最快速、最低成本的实验，发挥人的智慧和创造性，获得最有价值最直接相关的信息，而非通过难以企及、不易重复的实验手段来显示研究水平，偏离了研究的本来目的。

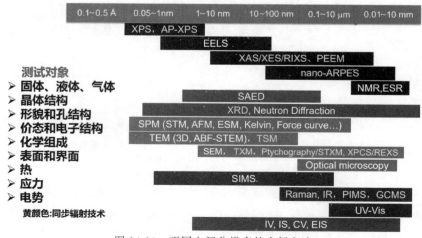

图14-10 不同空间分辨率的表征方法

在实验方法蓬勃发展的同时，理论计算也在不断前进。由于锂离子电池涉及多尺度、多场、非均相、多相共存、动态演化的特点，因次模拟锂离子电池各个尺度的平衡态与非平衡态行为非常有挑战性，不同空间、时间尺度的计算方法示意见图 14-11。理论上材料的基本性质可以在获得处于基态的弛豫结构后得以预测，在此基础上，逐渐往上，预测宏观体系的物理化学性质，模拟充放电过程中的行为。一方面，针对最基础的固体离子输运的问题，通过局域量子化、多体势、重整化群方法，解决非绝热近似下的涉及离子电子晶格复杂相互作用的问题；另一方面，引入相场、力场、有限元、连续介质力学等方法，模拟电极纳米尺度、微米尺度的反应、相变、输运、应力分布等问题。最近 ORNL 的科学家通过虚拟电池项目（VIBE），根据不同的模拟场景对锂电池的行为进行模拟预测，该模拟针对多孔三维结构，涉及正负极、电解质、其他组成，模拟充放电、热交换、电化学反应、机械应力等物理响应。可以预见，结合实际测量的数据，对大批量生产、一致性较好的电池进行全寿命周期、各类工况、滥用条件下电池行为的模拟，更易于获得可靠的数据、模型、数据库、算法，从而为电池管理系统提供可靠的数据和方法。

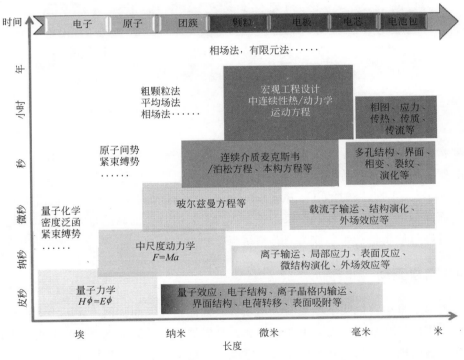

图 14-11　不同空间、时间尺度的计算方法示意图
（注：1 埃 = 0.1nm）

现有的多尺度理论和数值模拟可以根据确定材料的数据对电池进行数值模拟，但对于新材料体系，经常很难通过理论模拟来准确判断该材料是否具有应用前景。而且如何寻找新材料也没有特别明确的规律。

为了加快开发新材料的速度,美国的学者提出采用高通量计算的办法来筛选材料,最早将其引入锂离子电池研究的是 MIT 的 Gerbrand Ceder,他称之为材料基因组(material genome)方法。通过高通量计算,采取替代、掺杂等办法,在已有结构的基础上,创造新的材料。目前材料基因组的方法不局限于高通量计算,已经提出在全产业链中引入高通量方法,来加快电池材料与电芯开发的速度,提高制造过程中的控制精度、生产效率,见图 14-12。

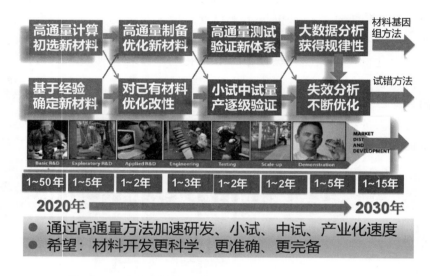

图 14-12 全产业链高通量开发材料与电池技术的示意图

需要指出的是,高通量方法,或者说材料基因组方法,并不完全意味着理性设计。如果计算和制备时没有明确的思想,也有可能成为漫无目标的撒网捕鱼(Fishing without target,Tarascon J M,2012,Gordon Conference)。有效的高通量方法,需要对电池的要求有全面深刻的了解,设计好约束条件,同时需要对具体的高通量方法的局限性有清楚的了解,否则可能事倍功半,成为"花拳绣腿"。

14.4.4 需求驱动

在本书第 1 章中,说明了电池的应用领域、方向正在不断扩展。实际上,电池在电子产品中的应用已经几乎到了无所不在的状态。高温器件、低温器件、可穿戴电子产品、自供电器件、透明电子产品、微小机器及航空航天、机器人等技术领域对电池的技术参数提出了不同于消费电子、电动汽车、规模储能的要求,需要进一步开发新的材料和电芯技术。即使采用相似的正负极材料,由于技术指标不尽相同,也需要在材料体系的匹配、优化与电池设计方面进行创新研究。因此,针对各类需求开发性能不同的电池一直推动着电池技术的不断前进(图 14-13)。

新科学：系统热力学计算，精准动力学模拟，创新材料体系，全体系稳定性，界面反应与调控，原子尺度到宏观尺度器件动态与长时间静态的能量储存、转换过程认识

新技术：高精度柔性化全产业链智能制造技术、规模储能系统集成与智能控制技术、全寿命周期储能系统数字模拟仿真、预测和失效分析技术；新型器件

新需求：智能汽车、电动船舶、航空航天、大规模储能、5G智能手机、机器人、国家安全、深空、深海、深地探索……

图 14-13　新科学，新技术，新需求推动着锂离子电池持续进步

14.5　本章结语

可充放电锂电池的研究从金属锂电池开始，兴盛于锂离子电池，目前又似乎回到了可充放电金属锂电池，早期的电解质是 PEO 聚合物电解质，后来是碳酸酯类液态电解质，现在又回到聚合物固态电解质，看起来像是个轮回。但这一次，却不是以 Moli 公司在 1989 年的不幸爆炸结束，而是即将迎来新的电池革命。这一次有了新的工具：材料基因组、多尺度多场模拟方法，为选择新材料、深入准确理解和模拟电极过程提供了理论依据；球差电镜、nano-CT、各类原位实验技术可以实现原子尺度、三维结构的实时观察；数字化工厂为精确控制复杂材料和电池的制造、对电池产品全寿命周期跟踪监控提供了技术保证。有理由相信，持续广泛的锂电池的基础研究，将会不断带来更加令人兴奋、造福社会的新的电池技术。

总结目前锂离子电池的基础研究活动，可以用以下顺口溜总结：

创新原理，妙想奇思；
深入理解，拨云见日；
创新方法，纵横两极；
纲举目张，归纳梳理；
创新体系，构造范式；
寻根究底，格物致理；
创造纪录，追求极致；
系统集成，服务社稷。

本章涉及的锂离子电池及锂电池基础研究的讨论，只是个人看法，受限于水平和视野，认识存在偏颇，没有能够覆盖锂离子电池基础研究的所有重要方面。欢迎专家、学者、学生以及对锂电池感兴趣的朋友来函交流，批评指正。在此特别感谢陈立泉、黄学杰两位老师的教诲和 E01 组、中国科学院物理研究所各位老师、同学十分有益的讨论及长

期合作带来的知识和启发。

参考文献

[1] Armand M, Murphy D, Broadhead J, et al. Materials for Advanced Batteries [M]. New York: Plenum Press, 1980: 145.
[2] Nishi Y. The development of lithium ion secondary batteries [J]. Chemical Record, 2001, 1 (1): 406-413.
[3] Goodenough J, Kim Y. Challenges for rechargeable Li batteries [J]. Chem Mater, 2010, 22: 587-603.
[4] Lu Xia, Zhao Liang, He Xiaoqing, et al. Lithium storage in $Li_4Ti_5O_{12}$ spinel: The full static picture from electron microscopy [J]. Adv Mater, 2012, 24: 3233-3238.
[5] Atsuo Yamada, Masahlro Tanaka. Jahn-Teller structural phase transition around 280K of $LiMn_2O_4$ [J]. Material Research Bulletin, 1995, 30 (6): 715-721.
[6] Zu Chenxi, Li Hong. Thermodynamic analysis on energy densities of batteries [J]. Energy Environ Sci, 2011, 4: 2614-2624.